Drug Delivery Systems: Advanced Technologies Potentially Applicable in Personalised Treatment

Advances in Predictive, Preventive and Personalised Medicine

Volume 4

Series Editor:

Olga Golubnitschaja

Managing Editor:

Kristina Yeghiazaryan

For further volumes:
http://www.springer.com/series/10051

Jorge Coelho
Editor

Drug Delivery Systems: Advanced Technologies Potentially Applicable in Personalised Treatment

Springer

Editor
Jorge Coelho
Chemical Engineering
University of Coimbra
Coimbra
Portugal

ISSN 2211-3495 ISSN 2211-3509 (electronic)
ISBN 978-94-007-6009-7 ISBN 978-94-007-6010-3 (eBook)
DOI 10.1007/978-94-007-6010-3
Springer Dordrecht Heidelberg New York London

Library of Congress Control Number: 2013931263

Printed on acid-free paper

Springer is part of Springer Science+Business Media (www.springer.com)

The quote below is dedicated to my friends
Maria Helena Gil, Margarida Figueiredo,
Pedro Simões, Pedro Gonçalves,
António Portugal and João Gabriel Silva,
who supported me from the very
beginning. . .
Thanks !

*"We must find time to stop and thank
the people who make a difference in
our lives."*
—John F. Kennedy

to Ana Clotilde,
for making me feel special every single
day. . .

What this Book Series is About . . .

Current Healthcare: What Is Behind the Issue?
For many acute and chronic disorders, the current healthcare outcomes are considered as being inadequate: global figures cry for preventive measures and personalised treatments. In fact, severe chronic pathologies such as cardiovascular disorders, diabetes and cancer are treated after onset of the disease, frequently at near end-stages. Pessimistic prognosis considers pandemic scenario for type 2 diabetes mellitus, neurodegenerative disorders and some types of cancer over the next 10–20 years followed by the economic disaster of healthcare systems in a global scale.

Advanced Healthcare Tailored to the Person: What Is Beyond the Issue?
Advanced healthcare promotes the paradigm change from delayed interventional to predictive medicine tailored to the person, from reactive to preventive medicine and from disease to wellness. The innovative Predictive, Preventive and Personalised Medicine (PPPM) is emerging as the focal point of efforts in healthcare aimed at curbing the prevalence of both communicable and non-communicable diseases such as diabetes, cardiovascular diseases, chronic respiratory diseases, cancer and dental pathologies. The cost-effective management of diseases and the critical role of PPPM in modernisation of healthcare have been acknowledged as priorities by global and regional organisations and health-related institutions such as the Organisation of United Nations, the European Union and the National Institutes of Health.

Why Integrative Medical Approach by PPPM as the Medicine of the Future?
PPPM is the new integrative concept in healthcare sector that enables to predict individual predisposition before onset of the disease, to provide targeted preventive measures and create personalised treatment algorithms tailored to the person. The expected outcomes are conducive to more effective population screening, prevention early in childhood, identification of persons at-risk, stratification of patients for the optimal therapy planning, prediction and reduction of adverse drug-drug or drug-disease interactions relying on emerging technologies, such as pharmacogenetics, pathology-specific molecular patters, sub/cellular imaging, disease modelling,

individual patient profiles, etc. Integrative approach by PPPM is considered as the medicine of the future. Being at the forefront of the global efforts, the European Association for Predictive, Preventive and Personalised Medicine (EPMA, http://www.epmanet.eu/) promotes the integrative concept of PPPM among healthcare stakeholders, governmental institutions, educators, funding bodies, patient organisations and in the public domain.

Current Book Series, published by Springer in collaboration with EPMA, overview multidisciplinary aspects of advanced bio/medical approaches and innovative technologies. Integration of individual professional groups into the overall concept of PPPM is a particular advantage of this book series. Expert recommendations focus on the cost-effective management tailored to the person in health and disease. Innovative strategies are considered for standardisation of healthcare services. New guidelines are proposed for medical ethics, treatment of rare diseases, innovative approaches to early and predictive diagnostics, patient stratification and targeted prevention in healthy individuals, persons at-risk, individual patient groups, sub/populations, institutions, healthcare economy and marketing.

Prof. Dr. Olga Golubnitschaja

Book Series Editor

Dr. Golubnitschaja, Department of Radiology, Medical Faculty of the University in Bonn, Germany, has studied journalism, biotechnology and medicine and has been awarded fellowships for biomedical research in Paediatrics and Neurosciences (Medical Centres in Austria, Russia, UK, Germany, the Netherlands, and Switzerland). She is well-cited in the research fields of "gene hunting" and "subtractive hybridisation" applied to predictive prenatal and postnatal diagnostics published as *O. Labudova* in years 1990–2000. Dr. Golubnitschaja is an expert in molecular diagnostics actively publishing in the fields of perinatal diagnostics, Down syndrome, diabetes mellitus, hyperhomocysteinemia, cardiovascular disease, neurodegenerative pathologies and cancer. She is the *cofounder* of the theory of multi-pathway organ-related blood fingerprinting with specific molecular patterns at epi/genomic, transcriptional and post/translational levels and author of fundamental works in *integrative medicine*. Dr. Golubnitschaja holds appointments, at the rank of Professor, at several European Universities and in International Programmes for Personalised Medicine and is author of more than 300 international publications in the field. Awards: National and International Fellowship of the Alexander von Humboldt-Foundation; Highest Prize in Medicine and Eiselsberg-Prize in Austria; She is *Secretary-General* of the "European Association for Predictive, Preventive and Personalised Medicine" (EPMA in Brussels, www.epmanet.eu), Editor-in-Chief of *The EPMA-Journal* (BMC in London); Book Editor of *Predictive Diagnostics and Personalised Treatment: Dream or Reality*, Nova Science Publishers, New York 2009; Book Co-editor *Personalisierte Medizin*, Health Academy, Dresden 2010; Book Series Editor *Advances in Predictive, Preventive and Personalised Medicine*, Springer 2012; *European Representative* in the EDR-Network at the NIH/NCI, http://edrn.nci.nih.gov/; and *Advisor and Evaluator* of projects dedicated to personalised medicine at the EU-Commission in Brussels, NIH/NCI, Washington, DC, USA, and at Foundations and National Ministries of Health in several countries worldwide.

Preface

The social and economic development attained over the last decades turned possible a complete change in terms of life style and life expectancy in developed countries. The scientific knowledge brought unprecedented capacities of diagnosis and health treatment. However, the current healthcare systems are essentially based on the identification of signs and symptoms of pathologies, which in most cases (e.g., cancer, diabetes, cardiovascular diseases) results in impossibility of finding effective solutions for the patients because it is already too late, and the progress/complications of the disease are already established. The change of mentalities and philosophies in the healthcare sector driven by the desire to address the mentioned problems that certainly will improve the life-quality and decrease the mortality is crucial. On this matter, since 2009, the European Association for Predictive, Preventive and Personalised (EPMA) (www.epmanet.eu) has done a pioneering work regarding the change of the actual paradigm and aiming to establish the platform for the medicine of future. This challenging task can only be accomplished with a strong and clear coordination of multidisciplinary networking areas involving: patients, healthcare institutions, physicians, researchers, pharmaceutics and governmental institutions. The management of the massive information available and its rapid dissemination is of outmost importance to create the basis for sustainable development and correct implementation of the preventive, predictive and personalized medicine (PPPM). The EPMA has strong commitment on the wide spread of updated information regarding the most relevant emerging innovations and developments on PPPM. On this matter, *The EPMA Journal* constitutes the platform of excellence to consolidate different professional groups in the concept of PPPM. The vision to excel in consolidation of the PPPM area is now witnessed by the publication of a series of books on advances in PPPM. This book belongs to that series and it is devoted to the "Drug delivery systems: advanced technologies potentially applicable in personalised treatments". This book intends to elucidate about essential topics involving the development and preparation of effective drug delivery systems, such as, polymers available, self-assembly, nanotechnology, pharmaceutical formulations, three dimensional structures, molecular modeling, tailor-made solutions, and technological tendencies. The scientific understanding of these areas constitutes a paramount route to establish personalized and effective solutions for specific diseases and individuals.

I would like to express my deepest thanks to all contributors that turned the preparation of this book into an enjoyable and exciting task, and accepted to share the excellence of their research involving in most cases, completely different areas that are essential to join in order to achieve the main goal of any part involved: the improvement of the patient's quality of life.

At the end of the day, the patients are and will be always the most important part in this complex network.

Prof. Dr. Jorge F. J. Coelho
Editor

Dr. Jorge F. J. Coelho was born in Figueira da Foz (Portugal) in 1978. He graduated in Chemical Engineering at the Faculty of Science and Technology of the University of Coimbra (FCTUC) in 2002. He received his Ph.D. in 2006 from the same University, in a project related to the development and industrial implementation of a new living radical polymerization method, which was awarded with CUF Prize 2007 (to distinguish the best Ph.D. project defended by a Portuguese citizen in the areas of Chemical Engineering, Environmental Engineering, Biological Engineering and Agriculture Engineering).

Since 2009, Dr. Coelho is Assistant Professor at the Chemical Engineering Department of FCTUC. He is involved in teaching different classes in the areas of polymer science and chemical engineering at undergraduate, postgraduate and doctoral programs in Chemical, Mechanical and Biomedical Engineering. His research interests involve the synthesis of tailor-made polymers (controlled/"living" radical polymerization), synthesis of biodegradable polymers (step-growth polymerization), self-assembly, nanotechnology, biomaterials and polymer characterization. His research activities have been supported by the Portuguese Science Foundation, National Strategic Reference Framework (NSRF), FP7 European program and private funds. He is co-author of 50 papers published in international peer-reviewed journals and several book chapters. He belongs to the editorial board of *The EPMA Journal* and he is an *ad hoc* reviewer for several scientific journals. He is the National Representative of EPMA in Portugal.

Contents

Contributors

Henrique A. Almeida Centre for Rapid and Sustainable Product Development, Polytechnic Institute of Leiria, Leiria, Portugal

Filipe E. Antunes Department of Chemistry, University of Coimbra, Coimbra, Portugal

Paulo J. Bártolo Centre for Rapid and Sustainable Product Development, Polytechnic Institute of Leiria, Leiria, Portugal

Ana Luísa Cardoso Center for Neuroscience and Cell Biology, University of Coimbra, Coimbra, Portugal

Rongjun Chen Department of Chemical Engineering, South Kensington Campus, Imperial College London, London, United Kingdom

Jorge Coelho Department of Chemical Engineering, University of Coimbra, Coimbra, Portugal

Rosemeyre Cordeiro Department of Chemical Engineering, University of Coimbra, Coimbra, Portugal

Joana Crisóstomo Laboratory of Physiology, Institute of Biomedical Research in Light and Image, Faculty of Medicine of University of Coimbra, Coimbra, Portugal

Research Centre for Chemical Processes Engineering and Forest Products, Chemical Engineering Department, Faculty of Sciences and Technology of University of Coimbra, Coimbra, Portugal

Maria C. Pedroso de Lima Center for Neuroscience and Cell Biology, University of Coimbra, Coimbra, Portugal

Department of Life Sciences, University of Coimbra, Coimbra, Portugal

Salomé dos Santos Department of Chemistry, University of Coimbra, Coimbra, Portugal

Tiago dos Santos Centre for BioNano Interactions, School of Chemistry and Chemical Biology, University College Dublin, Belfield, Dublin 4, Ireland

Sónia Duarte Center for Neuroscience and Cell Biology, University of Coimbra, Coimbra, Portugal

Henrique Faneca Center for Neuroscience and Cell Biology, University of Coimbra, Coimbra, Portugal

André Filipe Ferreira Department of Chemical Engineering, University of Coimbra, Coimbra, Portugal

Paula Ferreira Department of Chemical Engineering, University of Coimbra, Coimbra, Portugal

Margarida Figueiredo Department of Chemical Engineering, University of Coimbra, Coimbra, Portugal

Ana Fonseca Department of Chemical Engineering, University of Coimbra, Coimbra, Portugal

Joana R. Góis Department of Chemical Engineering, University of Coimbra, Coimbra, Portugal

Tamaz Guliashvili GE Power and Water, Water and Process Technologies, Trevose, PA, USA

Rodrigo José Lopes Department of Chemical Engineering, University of Coimbra, Coimbra, Portugal

Krzysztof Matyjaszewski Department of Chemistry, Carnegie Mellon University, Pittsburgh, PA, USA

Bruno Medronho Institute for Biotechnology and Bioengineering, Centre of Genomics and Biotechnology (IBB/CGB), Faculty of Sciences and Technology, University of Algarve, Faro, Portugal

Leonid G. Menchikov N.D. Zelinsky Institute of Organic Chemistry, Russian Academy of Sciences, Moscow, Russian Federation

Patrícia Mendonça Department of Chemical Engineering, University of Coimbra, Coimbra, Portugal

Pedro Miguel Moreno INEB – Instituto de Engenharia Biomédica, Universidade do Porto, Porto, Portugal

Hugo Oliveira CNRS, Laboratoire de Chimie des Polymères Organiques (UMR5629), Université de Bordeaux/IPB, ENSCBP, Pessac Cedex, France

Ana Paula Pêgo INEB – Instituto de Engenharia Biomédica, Universidade do Porto, Porto, Portugal

Rúben F. Pereira Centre for Rapid and Sustainable Product Development, Polytechnic Institute of Leiria, Leiria, Portugal

Anatoliy V. Popov Department of Radiology, University of Pennsylvania, Philadelphia, PA, USA

Antonio J. Ribeiro Centre for Pharmaceutical Studies, Faculty of Pharmacy, University of Coimbra, Coimbra, Portugal

Nuno Rocha Department of Chemical Engineering, University of Coimbra, Coimbra, Portugal

Raquel Seiça Laboratory of Physiology, Institute of Biomedical Research in Light and Image, Faculty of Medicine of University of Coimbra, Coimbra, Portugal

Arménio Serra Chemistry Department, University of Coimbra, Coimbra, Portugal

Pedro Nuno Simões Department of Chemical Engineering, University of Coimbra, Coimbra, Portugal

Sara Trabulo Center for Neuroscience and Cell Biology, University of Coimbra, Coimbra, Portugal

Francisco Veiga Centre for Pharmaceutical Studies, Faculty of Pharmacy, University of Coimbra, Coimbra, Portugal

Polymers in Drug Delivery: Concepts, Developments and Potential

Rongjun Chen

Abstract Targeted delivery of therapeutic agents, including small-molecule drugs and biopharmaceuticals, to the specific site of action can improve their safety and pharmaceutical efficacy. Controlled-release polymer technologies have played an important role in the advancement of targeted drug delivery research. Modern advances in polymeric drug delivery will depend on rational design of polymers tailored for carrying specific payload and exerting desirable biological functions as a result of a detailed and thorough understanding of biological processes. This chapter provides a unique coverage of the field of polymers in drug delivery. It addresses the foundations of drug delivery in a conceptual context, highlights the extracellular and intracellular barriers to effective polymeric drug delivery, critically reviews the recent developments in different classes of polymeric delivery agents (including polymer-drug conjugates, polymeric micelles, multi-component polyplexes, polymer-modified liposomes, polymersomes, and membrane-disruptive polymers), and discusses their applications for the treatment of various diseases including cancer.

Keywords Polymer therapeutics • Drug delivery • Intracellular delivery • Controlled release • Stimuli-responsive • Responsive polymers • Predictive, preventive and personalised medicine

1 Introduction

The pharmaceutical and biotechnology industries have increasing financial and regulatory pressures caused by continuing patent expirations of a number of block-buster drugs, dwindling product pipelines and increasingly cautious government

R. Chen (✉)
Department of Chemical Engineering, South Kensington Campus, Imperial College London, Prince Consort Road, London SW7 2AZ, United Kingdom
e-mail: rchen27@gmail.com

J. Coelho (ed.), *Drug Delivery Systems: Advanced Technologies Potentially Applicable in Personalised Treatment*, Advances in Predictive, Preventive and Personalised Medicine 4, DOI 10.1007/978-94-007-6010-3_1,

regulators [1]. Drug delivery technologies are widely expected to change the landscape of those industries in the foreseeable future [2, 3]. Targeted delivery of therapeutic agents to the specific site of action can improve their safety and pharmaceutical efficacy. The development of drug-delivery products may therefore create a new armamentarium of therapeutics.

1.1 Challenges and Opportunities of Polymers in Drug Development

Low water solubility, unfavourable stability, short circulation time in plasma, poor bioavailability, non-specific toxicity against normal tissue and cells, and multi-drug resistance of a large number of small-molecule drugs can seriously hinder their therapeutic efficacy. This prevents the approval of promising new agents for clinical use. Water-soluble, targeted drug delivery systems are therefore intrinsically attractive for conventional chemotherapy [4].

The advances in genomics and proteomics research and the improved understanding of genetic bases of diseases have enabled the development of a growing number of macrodrugs, including nucleic acids (e.g. DNA, RNA and oligonucleotides), peptides and proteins (e.g. antibodies, oligopeptides, growth factors, hormones and therapeutic enzyme). These biopharmaceuticals represent a powerful new class of medicine with potential for treatment of a wide variety of previously intractable human diseases. The therapeutic promise of the bioactive macromolecules as cell-targeted therapeutics is, however, tempered due to their problems including low stability, rapid clearance from the human body, low cellular uptake, susceptibility to enzyme degradation, and inefficient transport to the cell nucleus [5, 6]. The emergence of drug-delivery platforms can enable development and commercialisation of entirely new classes of biological drugs, which need precise delivery to their intracellular sites of biological action [7–9].

Current research into methods of drug administration can be categorised into three general delivery classes: physical, biological and chemical. Physical methods of administration include electroporation [10], particle bombardment [11], naked injection [12] and ultrasonic nebulisation [13]. These approaches could have good transfection efficiency. However, they are restricted by in vivo challenges and difficulties to target specific diseased cells. Biological methods have focused on utilising recombinant viruses that can efficiently infect diseased cells. Viral vectors are, to date, the most effective vectors for gene transfection. For clinical purpose, however, enthusiasm for viral delivery systems is dampened by concerns regarding serious safety issues, limited drug-carrying capacity, and difficulties in large-scale production and targeting [14]. Increasing interest is being focused on the research into chemical methods of administration using non-viral vectors such as peptides, liposomes, polymers and nanoparticles. Chemical delivery is considered the safest, simplest and most versatile of the three methods. Despite the current lack of excellent efficiency, chemical delivery agents can be rationally designed to mimic the biological activity of viruses to enable improved and optimised drug profiles. Among the non-viral chemical delivery technologies under development, "controlled-release polymer technologies and liposomes will likely continue to have the greatest clinical impact for the foreseeable future" [9].

This chapter focuses on polymers in drug delivery for treatment of various diseases including cancer. There are some basic requirements for design of polymers for targeted drug delivery and controlled release. The polymeric carriers should be inherently safe (non-toxic, non-immunogenic, and preferably biodegradable), and preferably cost-effective and amenable for industrial-scale manufacture [15, 16]. They are required to deliver therapeutic agents into specific target cells by traversing across various extracellular and intracellular barriers, and release them at a controlled rate at sites of action [7]. Increasingly sophisticated applications in the field of controlled release require that the polymers can respond in some way to physical stimuli (i.e. temperature, ultrasound, light, and magnetic and electrical fields) or chemical stimuli (i.e. pH, redox potential, ionic strength, and chemical agents) [8]. This chapter has a particular focus on pH-responsive polymers in view of acidic environment in cellular compartments such as endosomes and lysosomes [7].

1.2 Historic Overview of Polymers in Drug Delivery

Since Paul Ehrlich first described the concept of selective delivery of drugs by carriers to the site of action in 1906 with the phrase of "magic bullet" [17], researchers have focused on methods to direct delivery and improve the efficacy of therapeutic agents. The realisation by de Duve et al. in 1974 that cellular endocytosis is useful for lysosomotropic drug delivery [18], and the concept of a polymer-drug conjugate model proposed by Ringsdorf in 1975 [19], have stimulated great interest in polymeric drug delivery.

Initial work with polymeric carriers puts an emphasis on delivery of small-molecule drugs, such as doxorubicin, camptothecin and paclitaxel, to improve their clinical value. Although these agents are already in clinical use, their suboptimal therapeutic performance was impaired by a variety of factors, in particular, rapid clearance of drugs from the blood stream, low levels of drug accumulation in targeted tumour tissues and undesirable side effects [15]. In an effort to address these problems, Kopeček and Duncan first published breakthrough research on N-(2-hydroxypropyl)methacrylamide (HPMA) copolymer-anticancer drug conjugates throughout the 1970s and 1980s [20–23]. Systematic studies have been carried out on rational design, preclinical evaluation, manufacture and formulation validation, and clinical development of these conjugates to improve biodistribution and efficacy of drugs. It has been reported that there have been at least ten conjugates in early clinical development [15, 24].

Occurring in tandem with the research and developments in polymer-mediated delivery of small-molecule drugs, biomedical researchers were continuing to elucidate the genetic basis of disease. The emergence of bioactive macromolecules (such as nucleic acids and proteins) as drugs has sparked renewed interest in polymeric carriers as a means to deliver these macrodrugs to their sites of biological action. Early on, attempts were made to deliver protein-based drugs through conjugation to polymers for treatment therapies. Abuchowski et al. first described a method of covalently attaching polyethylene glycol (PEG) to bovine serum albumin in 1977 [25]. Since then, this PEGylation method has been widely employed to improve on the clinical usefulness of numerous proteins and other potentially therapeutic compounds by reducing their toxicity and immunogenicity as well as increasing their water solubility, stability and plasma half-life [26–28]. In 1990s, transfer of the first polymer-protein conjugates, such as Adagen (PEG-adenosine deaminase) and Oncaspar® (PEG-L-asparaginase), to market heralded the birth and great potential of a new class of anticancer agents [15, 29].

In addition to the advent of polymer conjugation to therapeutic agents, work in the late twentieth and early twenty-first centuries also led to developments of a wide variety of polymeric drug delivery technologies. They can be used to smuggle a large number of different types of therapeutic agents (including small-molecule drugs and biopharmaceuticals) to the site of action precisely and enable controlled release of payload over a prolonged period of time. The polymers can be designed and formulated in different forms, such as polymer-drug conjugates [15], polymeric micelles [30–32], nanoparticles [33, 34], polymer-modified liposomes [35, 36], multi-component polyplexes (complexes of polymers and genes or proteins) [7], and polymersomes [4], where drug payloads can be covalently bound, complexed or encapsulated. In 2004, a novel anticancer agent SP1049C containing doxorubicin and micelle-forming pluronic copolymers first entered to preclinical studies [37]. In 2005, Abraxane, the first therapeutic nanoparticle of albumin entrapped with paclitaxel, was approved as a treatment for metastatic breast cancer [15].

2 Barriers for Polymeric Drug Delivery

In order to escort drug payload to right target cells and to right intracellular sites of action, polymeric delivery agents need to get over a variety of barriers on both systemic and intracellular levels. This section highlights a set of extracellular and intracellular barriers to effective drug delivery, which may give us vital clues for rational design of polymeric carriers.

2.1 Extracellular Barriers

2.1.1 Serum Stability

Serum stability is a crucial factor for successful drug delivery. Serum components can inactivate many drugs. They may also interact with drug delivery agents and cause them to be rapidly eliminated by the reticuloendothelial system (RES) [38–40]. The RES, or the mononuclear phagocytic system (MPS), is a group of cells, including the macrophages of liver, spleen, and lymphatic system, responsible for removal of old cells, miscellaneous cellular debris, and foreign macromolecules and particles from the blood stream. Hydrophobic or charged surface may induce binding of opsonins, such as complement component C3b, on the surface of carriers, leading to receptor-mediated uptake by the macrophages of the RES [41]. Cationic delivery agents can bind to serum albumin and other negatively charged proteins, which as a result, causes aggregation and can lead to rapid clearance by the RES [7]. Extensive investigations have been carried out to reduce the RES uptake and prolong circulation time in the blood stream. One strategy is surface modification of colloidal carriers with inert, hydrophilic, non-ionic polymers such as PEG [42–46]. The protective steric effects from the hydrophilic corona can reduce the adsorption of serum components onto the surface of carriers, and therefore reduce the recognition by macrophages of the liver and spleen [7].

2.1.2 Renal Clearance

Renal glomerular filtration can result in elimination of polymeric delivery agents from the blood stream [47]. The rate of renal clearance depends on the molecular weight and structure of polymeric carriers. Polymers with higher molecular weight have a slower renal clearance. Accordingly, polymer molecular weight should be high enough for prolonged circulation time [15, 28]. However, non-biodegradable polymeric carriers may persist within the body as indigestible residues following drug delivery. The molecular weight of non-biodegradable polymers, therefore, should be lower than the renal excretion threshold to ensure their elimination out of the body by renal clearance [15, 48]. It has been reported that the non-biodegradable N-(2-hydroxypropyl) methacrylamide (HPMA) copolymer has a weight average molecular weight (Mw) threshold of ~45 kDa, and its Mw should be lower than this to allow the renal clearance [49]. In addition, polymers with different structures may differ in their renal excretion thresholds. Charged polymers with rigid and extended conformation can be filtered through glomerular pores more difficultly than uncharged polymers behaving as flexible coils [48, 50].

2.1.3 Tumour Targeting and Accumulation

Polymeric delivery of anticancer agents through passive tumour targeting can improve their therapeutic potential. Polymeric delivery agents are largely prevented from exposure to normal tissues due to their retention in the blood stream. They penetrate out of the blood compartment at the tumour site and accumulate in tumour at much higher concentrations than in normal tissues due to the enhanced permeation and retention (EPR) effect. The EPR effect, typical of solid tumours and missing in normal tissue, is attributed to increased vascular density resulting

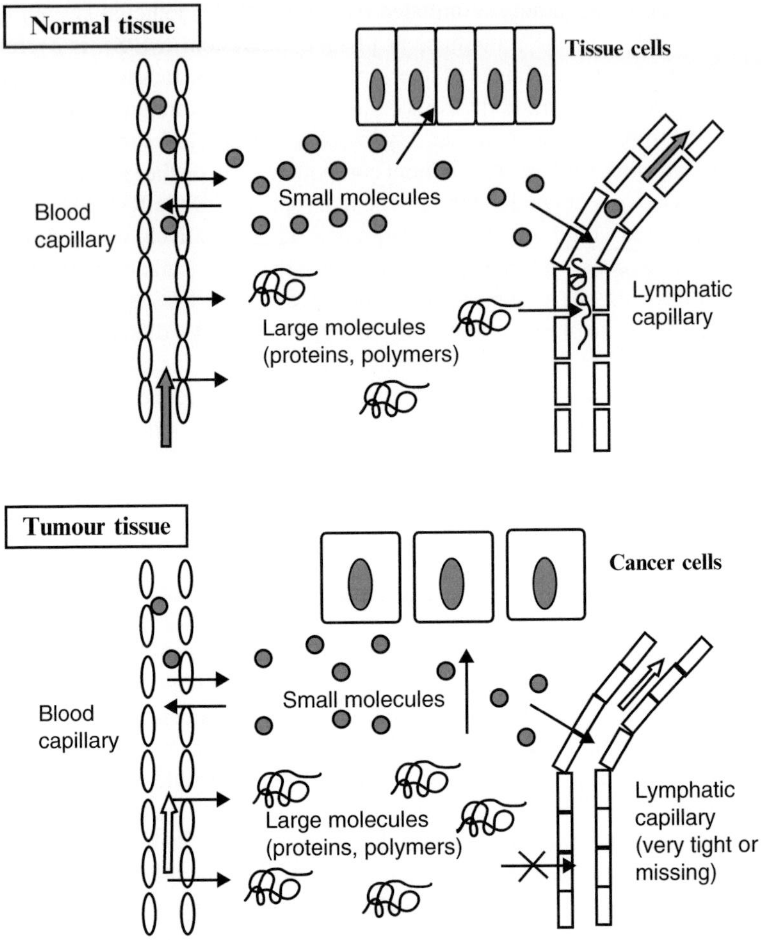

Fig. 1 Schematic illustration of the EPR effect principle (reproduced from Ulbrich and Šubr [52])

from rapid tumour angiogenesis and enhanced permeability of tumour vasculature, coupled with impaired or missing lymphatic clearance of macromolecular delivery systems in the tumour interstitium [51, 52] (see Fig. 1). Factors influencing EPR-mediated passive tumour targeting include polymer molecular weight or particle size [15, 53], polymer architecture [54] and plasma residence time [55]. The delivery systems can release drug payload in the extracellular fluid of the tumour tissue [39] or at an intracellular site [56].

Tumour targeting can be further enhanced by receptor-mediated delivery [21]. This active tumour targeting is realised through the attachment of targeting ligands, such as antibodies, peptides, lectins, saccharides, and hormones, onto the surface of polymeric delivery systems, and subsequent binding to corresponding receptors over-expressed on the surface of specific cells of interest [33, 57]. Receptors on

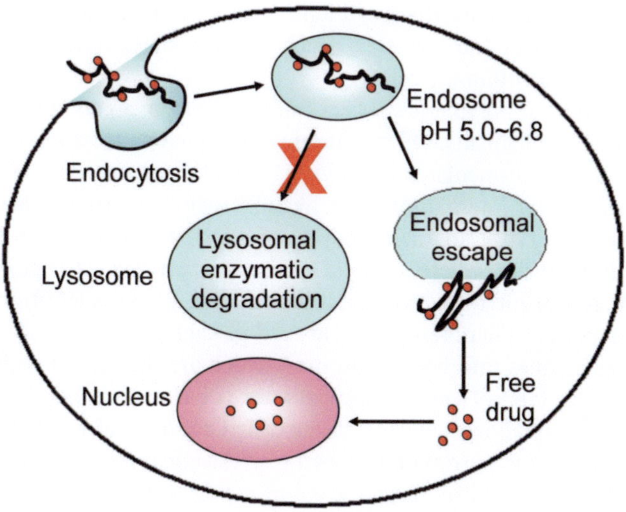

Fig. 2 Barriers to intracellular trafficking of polymeric delivery agents

the cell membrane can vary depending on the type of cells and are frequently recycled after receptor-mediated internalisation [39]. Active tumour targeting using ligand-attached polymeric delivery systems results in a higher degree of cell specificity, enhanced tumour accumulation and cell uptake, avoidance of multidrug resistance from the p-glycoprotein, and thus increased therapeutic efficacy drugs [7, 33, 58].

2.2 *Intracellular Barriers*

For effective drug delivery, polymeric delivery agents also need to cross over key intracellular barriers (see Fig. 2), including endocytosis of polymeric systems into an endosome, endosomal escape and intracellular targeting at the site of interest, such as cytoplasm, mitochondria or nucleus [59].

2.2.1 Endocytosis

One of the major challenges facing non-viral delivery agents, including polymers, is their low delivery efficacy as compared to viral vectors. Low cell uptake of polymeric drug delivery systems across the plasma membrane is a significant obstacle to effective drug delivery with such systems [60]. Although small molecules can diffuse through the plasma membrane, macromolecules are often too large and too polar to enter cells by passive diffusion. Those macromolecular systems enter cells usually through endocytosis, which consists of membrane envelopment of extracellular molecules into vesicles, pinch-off of vesicles from the cell surface, and their trafficking within the cells.

Rational design of polymeric carriers can benefit from a good understanding of the endocytotic mechanisms. Of the currently accepted endocytotic pathways, the highly studied clathrin-mediated endocytosis is widely conserved across cell types and is responsible for internalisation of ligands, such as transferrin and epidermal growth factor (EGF) [61]. In this endocytotic route, the scaffolding protein clathrin are recruited to the site, where a ligand binds the appropriate receptor on the cell surface. This, as a result, forms coated pits that mature into vesicles prior to pinching off from the plasma membrane. Vesicles are gradually acidified as they mature from early endosomes into late endosomes, which are then fused with lysosomes [62, 63]. Another type of endocytotic pathway is caveolin-mediated endocytosis, which involves formation of invaginations called caveolae at localised regions of the plasma membrane surrounding materials. Caveolae are typically 50–80 nm in diameter, characterised by the presence of caveolin protein family member, as well as high levels of sphingolipids, cholesterol, signalling proteins and glycosyl phosphatidylinositol-anchored proteins [64]. This type of endocytosis is a two-step transport pathway from caveolae, through a larger organelle (termed the caveosome) with a non-acidic pH, to a smooth endoplasmic reticulum compartment, therefore bypassing endosomes and the Golgi complex [65, 66]. There have been reported some clathrin- and caveolin-independent endocytotic routes, where cells can internalise materials without the involvement of either clathrin or caveolin [66–68]. Other accepted endocytoic pathways include phagocytosis and macropinocytosis, both involving formation of generally $>1\ \mu m$ diameter vesicles for uptake of large volumes [69, 70]. Phagocytosis involves the uptake of large particles, such as microorganisms and apoptotic cells, whilst macropinocytosis involves fluid-phase sampling of the extracellular environment by cells. These two pathways are of importance for drug delivery agents with a large size, including liposomes and microparticles.

In addition to determining the type of endocytosis, it is also important to increase the cell entry of non-viral vectors for improved drug delivery efficacy. It has been recognised that if the rate of vesicle formation is unchanged, affinity for the cell surface dominates the rate of cell uptake [71]. A variety of factors influencing the affinity for the plasma membrane include hydrophobicity, charge, and size of polymeric carriers. Cell internalisation can be enhanced by increased exposure and availability of hydrophobic segments for the interaction with the cell surface [72]. Due to the negative charge of the lipid membrane, cationic polymers often exhibit faster and increased cell uptake than anionic polymers [71]. Internalisation of materials into cells is believed to be size-dependent, and therefore polymers with higher molecular weight enter cells more slowly than those with lower molecular weight [69, 72, 73]. In the case of receptor-mediated endocytosis, cell uptake and cell specificity can be increased by the interaction between ligands and receptors [7, 58].

2.2.2 Endosomal Escape

Release of non-viral vectors from endosomal pathways is a key rate-limiting barrier in intracellular delivery of therapeutic agents, especially fragile bioactive macromolecules [7, 74, 75]. During the maturation of vesicles in endosomal pathways, endocytosed materials transport through early endosomes and late endosomes to lysosomes. This maturation process results in the acidification of the vesicle lumen. In early endosomes, the luminal pH value is decreased from physiological 7.4 to the range of 5.9–6.8 due to the effect of ATP-driven proton pumps. The period of transit to early endosomes is short, typically lasting only 2–3 min [63, 76, 77]. The materials are then trafficked to late endosomes, where the action of an ATPase proton-pump enzyme in the vesicle membrane further lowers the luminal pH value gradually to 5.0–6.0 [7, 76]. Once the late endosomes fuse with lysosomes by default, the materials encounter a further decreased luminal pH to as low as $\sim$4.5 [7, 63]. It has been reported that it typically takes 30–60 min [63, 78] to several hours [7] for endocytosed materials to reach lysosomes. The lysosomal environment is usually hostile to intracellular drug delivery. In addition to the low luminal pH environment and the lysosomal membrane barrier, lysosomes contain over 50 digestive enzymes, which are normally used by cells to degrade proteins, lipids and other molecules into smaller moieties that can be recycled for intracellular biosynthesis [79]. In the case of delivery of enzyme-susceptible biological drugs, entrapment within lysosomes presents a major challenge. Even in the case of delivery of drug payload that is not degraded, compartmentalisation within endosomes and/or lysosomes can prevent drugs from reaching their specific intracellular sites of action; therefore the release of drugs from endosomal pathways is needed to avoid non-productive vesicular accumulation [80].

Many strategies have been developed to enable effective cytoplasmic drug delivery using polymers. Initial work with cytoplasmic delivery of small-molecule drugs was conducted to covalently conjugate drugs to polymeric scaffolds through the linkers that can be cleaved by pH-controlled hydrolysis in the acidic endosomal/lysosomal compartments, or by enzymolysis in lysosomes. The drugs, which were cleaved from the polymeric carriers, diffused through the endosomal/lysosomal membrane into the cytoplasm [15, 39, 81]. However, those strategies are limited because it may be difficult to control the rate of linker cleavage to avoid eventual lysosomal accumulation of the formulation [7]. In addition, large, polar biological therapeutics must escape from endosomes through a different mechanism other than diffusion though the membrane. As a result, increasing interest has been focused on designing polymeric delivery agents, which enter endosomal pathways that acidify and can release drug payload into the cytoplasm by disrupting endosomes in response to acidification [7, 78, 82].

2.2.3 Nuclear Localisation

Some drugs, including glucocorticoids and small interfering RNAs (siRNA), act on molecular targets in the cytoplasm [83], while many other drugs act inside membrane-bound intracellular organelles, including nucleus and mitochondria [84]. The nuclear membrane has tightly regulated pores that allow import and export of a specific set of biomolecules, and therefore represents a significant obstacle to delivery of therapeutic agents, which need nuclear localisation.

Once released from endosomes, drug molecules exhibit a relatively high concentration in the cytoplasm, which drives them towards nucleus [85]. Drug molecules smaller than 10–20 kD can diffuse across the nuclear membrane through the molecular sieve function of the nuclear pore complex (NPC). Larger drug molecules can interact with certain protein receptors in the cytosol, and the resultant drug-receptor complexes are transported into the nucleus [7, 86, 87]. Nuclear entry of large drug molecules can also rely on the enhanced permeability of the nuclear membrane during cell division, which was reported to increase the transfection efficiency by 30–500-fold [88–90].

More efforts have been made to explore new methods for nuclear localisation of bioactive macromolecules considering the diffusional and metabolic barriers of the cytoplasm that they encounter. Some non-viral delivery agents have been developed to mimic the mechanisms of viruses for carrying genetic materials into the nucleus [91, 92]. It was argued that nuclear transport of complexes of DNA and polymers might involve the electrostatic interaction between the positively charged complexes and negatively charged phospholipids on the nuclear membrane. The attachment of polymer-DNA complexes with the endosomal membrane debris after disruption of endosomes might also facilitate the fusion of the complexes with the nuclear envelope, which as a result leads to nuclear translocation [93]. The active nuclear transport through the NPC can be realised by the use of nuclear localisation signals (NLS), which are short cationic peptide sequences that are recognised by importins (nuclear import protein) [7].

3 Polymer-Based Non-viral Delivery Agents

A wide variety of delivery agents using polymers have been developed to deliver small-molecule drugs or biopharmaceuticals. Polymer-drug conjugates, polymeric micelles, multi-component polyplexes, polymer-modified liposomes, and polymersomes are reviewed in this section.

3.1 Polymer-Drug Conjugates

Ringsdorf was the first who proposed the concept of polymer-drug conjugates in 1975 [19]. In his general model the water-soluble polymer can be used as a carrier for drug payloads, targeting moieties and solubilisers [94]. The polymer backbone, biostable or biodegradable, is designed to have good water-solubility, biocompatibility, and long-term circulation in the body. Drugs, including small-molecule drugs and biologically active macromolecules, can be covalently conjugated to the water-soluble polymeric scaffold via cleavable linkers, which are stable during transport but susceptible to enzymatic (e.g. peptidyl and ester linkages) or hydrolytic degradation (e.g. pH-sensitive cis-aconityl, hydrazone and acetal linkages) for controlled drug release at an optimum rate upon arrival at the pharmacological target site [15, 52, 95–97]. Attachment of specific targeting moieties, such as sugars, peptides and antibodies, can enable specific delivery of the polymer-drug conjugate to target cells [21, 23, 96].

Due to its narrow distribution of molecular weights and superior biological properties, the highly hydrated polyethylene glycol (PEG) has been widely used as a carrier enabling efficient drug delivery. Abuchowski et al. first covalently attached PEG to bovine serum albumin in 1977, which resulted in a conjugate with improved water solubility, enhanced serum stability, increased circulation time and reduced immunogenicity and antigenicity [25]. Since then PEG has become probably the best candidate among other synthetic hydrophilic polymers tested as drug carriers, especially for developments of the conjugates with bioactive proteins and glycoproteins [98–100]. A number of PEG-based conjugates have already been on the market or undergoing advanced stages of clinical evaluation, including PEG conjugates of adenosine deaminase [101], asparaginase [29], interferon-α2a [102], interferon-α2b [103], antibody anti-TNFα Fab [104, 105] or antibody anti-VEGF [106].

Copolymers of N-(2-hydroxypropyl)methacrylamide (HPMA), first synthesised by Kopeček in the early 1970s, rank among the most intensively investigated water-soluble polymers in synthesis of polymer-drug conjugates. Early work with the HPMA copolymer-drug conjugates used a linear, non-biodegradable HPMA copolymer backbone of molecular weight ∼25 kDa, which was less than the renal excretion threshold of ∼45 kDa to enable the renal elimination of the carrier following drug delivery [49]. One of the most recently described structures of the HPMA copolymers is represented by a branched, high-molecular-weight graft copolymer, wherein the polymer grafts were attached to the main polymer chain through enzymatically or reductively degradable spacers. Following controlled release of doxorubicin, intracellular degradation of the graft copolymer carrier was able to form short low-molecular-weight fragments excretable from the organism by glomerular filtration [107]. HPMA copolymer conjugates with anticancer drugs are one of the few polymer therapeutics, which have already been explored clinically. Since the first HPMA copolymer-based anticancer drug, i.e., sarcolysin bound to the HPMA copolymer with oligopeptide spacer, was described and patented in 1981 [108], there have been at least ten conjugates in early clinical development [15, 24].

3.2 Polymeric Micelles

Since Ringsdorf and co-workers first proposed the use of polymeric micelles for drug delivery in 1984 [109], they have become a promising nanomedicine platform for delivery of small-molecule drugs or nucleic acids [110, 111]. Polymeric micelles present a core-shell type architecture with a narrowly-distributed small size (usually 10–100 nm in diameter) resulting from the self-assembly of amphiphilic block or graft copolymers in aqueous milieu [110, 112]. The hydrophobic core of the polymeric micelles serves as a nano-reservoir for the incorporation of hydrophobic therapeutic compounds, while the hydrophilic shell forms a protective layer, improving the pharmacokinetics and distribution of those drugs [40, 113]. The commonly used shell-forming components include naturally existing polysaccharides, such as chitosan [114] and pullulan [115], and synthetic polymers, such as poly(ethylene glycol) (PEG) [113], poly(L-glutamic acid) (PLGA) [116], poly(acrylic acid) [117], poly(N-isopropylacrylamide) (PNIPAM) [118], poly(N-vinylpyrrolidone) (PVP) [119] and poly(2-ethyl-2-oxazoline) [120]. PEG is the most widely used hydrophilic segment of polymer micelles due to its neutrality, hydrophilicity and flexibility. The presence of PEG can enhance serum stability, increase circulation time in the body and reduce toxicity and immunogenicity, therefore making the PEG-containing micelles attractive for drug delivery applications [25–28]. The commonly used core-forming segments consist of biodegradable hydrophobic polymers, such as poly(ε-caprolactone) (PCL) [121], poly-(D,L-lactide) (PDLLA) [119], poly(N-hexyl stearate L-aspartamide) [122] and poly(γ-benzyl-L-glutamate) (PBLG) [123], and non-biodegradable hydrophobic polymers, such as poly(propylene oxide) (PPO) [124] and poly(vinyl) derivatives [125].

Alternatively, complexation and neutralisation of charged polymers with oppositely charged drug payloads can yield water-insoluble moieties that self-assemble in the same fashion as copolymers featuring a hydrophobic block. The resultant polyion complex micelles are promising for delivery of charged therapeutic agents including nucleic acids [111]. Polycations, such as poly(ethylenimine) (PEI) [126] and poly(L-lysine) (PLL) [127], and polyanions, such as poly(aspartic acid) [128], can be covalently attached to the shell-forming hydrophilic polymer and interact with oppositely charged biological molecules, such as siRNA, oligonucleotide and enzymes, to form this type of micelles [111, 126, 129]. Other works demonstrated an alternative option to develop polyion complex micelles by complexing PEGylated anionic siRNA [130] or antisense oligonucleotide [131] with cationic polymers, such as PEI.

Polymeric micelles are thermodynamically stable, with fairly low critical micelle concentrations [132]. The more relevant property concerning drug delivery is probably their high kinetic stability resulting from the hydrophobic interaction and van der Waals cohesion between hydrophobic segments in the core, as well as the hydrogen bond network between the shell and the surrounding water. The structure of polymeric micelles needs to be properly designed to retain a slow dissociation rate upon the large in vivo dilution, which enables sustained release of drug molecules at the sites of action specifically [16, 30, 111, 113].

Polymeric micelles can be used to deliver therapeutic molecules to tumours either passively through the EPR effect [133] or actively through the attachment of targeting ligands to the corona of micelles [126]. Factors influencing the drug release are crucial since they determine where and how drug molecules are released and, ultimately the effectiveness of the delivery systems. Small-molecule drugs can be released from polymeric micelles upon arrival in the mildly acidic tumour microenvironment and then enter cells by diffusion. Alternatively, cells internalise polymeric micelles by endocytosis and small-molecule drugs are released in endosomes/lysosomes, followed by diffusion to the intracellular sites of action. Compartmentalisation of polymeric micelles in the endosomal compartments, however, can prevent the drug payloads (particularly bioactive macromolecules) from reaching their cytoplasmic or nuclear targets. In such case, it becomes crucial to effectively destabilise the endosomal membrane through the use of amine-containing cationic polymers with pKa values between physiological and lysosomal pHs [134] or pH-responsive membrane-disruptive anionic polymers [135] in the preparation of polyion complex micelles. Moreover, there are different strategies to achieve sustained drug release from polymeric micelles. For instance, release of simply encapsulated small-molecule drugs can be controlled by either the rate of drug diffusion out of the hydrophobic core or the disintegration of assemblies. As for drug molecules covalently bound to the polymers, release is dependent on the collective effects of micelle disassociation to single polymeric chains and cleavage of the liable covalent bond between the drug molecule and the polymer [16]. Some polymeric micelles have been adapted to trigger drug release in response to a change in pH [116, 136], temperature [137] or the redox potential [130]. An example of a polymeric micelle under clinical evaluation is NK105, which consists of modified polyasparate as the hydrophobic core-forming segment and PEG as the hydrophilic shell-forming segment. The positive results obtained from the Phase I and II studies of the micellar nanoparticle formulation with physical entrapment of paclitaxel will lead to a Phase III trial [138–140]. Another agent SP1049C, a novel P-glycoprotein targeting anticancer formulation consisting of doxrorubicin and micelle-forming pluronic copolymers, have entered Phase I and II clinical trials and showed superior anticancer activity in patients with advanced adenocarcinoma of the esophagus and gastroesophageal junction [37, 141].

3.3 Multi-component Polyplexes

In the recent two decades, great interest has been focused on designing non-viral gene-delivery strategies using pH-responsive polymers that form complexes with genetic materials including DNA and RNA, known as multi-component polyplexes [7]. The polyplexes typically consist of: (i) genetic materials, (ii) targeting ligands for receptor-mediated delivery; (iii) amine-containing polycations acting as a nucleic acid packaging and endosomal membrane disruption module, and (iv) in some cases endosomolytic polyanions or peptides that are attached on the surface of complexes [87, 142–144].

A variety of cationic polymers, including poly(L-lysine) (PLL) [145], polyethyleneimine (PEI) [146], poly(2-(dimethylamino)ethyl methacrylate) (PDMAEMA) [147] and polyamidoamine (PAMAM) dendrimers [148], have been developed for gene delivery. Those polymers can be designed to present linear, branched or dendritic structures, and contain primary, secondary, tertiary or quaternary amines that are used to electrostatically complex DNA or RNA into particles with a size of a few tens to several hundred nanometres in diameter. The formed nanoparticles enter cells by endocytosis. The cationic polymers can mediate delivery of genetic materials from endosomes to the cytoplasm or nuclear targets, therefore avoiding lysosomal degradation [146–148]. Many cationic polymers suffered from unsatisfactory gene-transfection efficiency and cytotoxicity, which would limit their clinical applications. One of the strategies to address these problems is to attach pH-responsive endosomolytic polyanions or peptides on the surface of complexes via direct conjugation [148] or electrostatic interaction [143]. The presence of those polyanions or peptides can substantially enhance destabilisation of the endosomal membrane and lead to significant increases of therapeutic potential [149]. The efficiency of the multi-component polyplexes depends on the structure and properties of polymeric carriers, as well as the mechanism used in the formation of complexes, including the dose and time at which polymers are incorporated [149, 150]. The pH-responsive membrane-disruptive polymers will be discussed in more detail in the Section 4.

3.4 Polymer-Modified Liposomes

Liposomes are spherical, concentric lipid bilayers formed by self-assembling of amphiphilic phospholipid molecules in aqueous solution. Their potential as drug carriers was first recognised in 1976 [151]. Typically hydrophilic therapeutic agents can be entrapped within the aqueous interior and hydrophobic drugs incorporated into the lipid bilayer membrane. Since Felgner first reported the use of N-[1-(2,3-dioleyloxy)propyl]-N,N,N-trimethylammonium chloride (DOTMA)-containing unilamellar liposomes for delivery of DNA in 1987 [152], cationic lipids have been widely used for gene delivery [153]. In such case, anionic biomacromolecular drugs, such as DNA and RNA, can be complexed with cationic lipids on the liposomal surface [42]. Through alteration of lipid structures and the surface chemistry, properties of liposomes, including their size, charge and surface functionality, can be easily modified for efficient gene delivery [154]. Liposomes have other attractive properties, such as biocompatibility, biodegradability and relatively high transfection efficiency. Although liposomes have a number of limitations including lack of controlled release mechanisms, poor colloidal stability and major pharmacokinetic issues, they are often recognised as one of the dominant non-viral vectors [7, 155]. Today a number of liposome-based anticancer agents have already been on market, including DaunoXome® (daunorubicin-containing liposomes), Myocet® (doxorubicin-containing liposomes), Doxil®/Caelyx® (doxorubicin-containing PEG-liposomes), and onco TCS® (vincristine-containing liposomes) [4].

Since the concept of pH-responsive liposomes was first proposed in 1980 [156], much attention has been paid to the modified liposomes with covalently attached, adherent, complexed, or entrapped pH-responsive membrane-disruptive polymers [157, 158]. Liposomes can be further modified with protective polymers, such as PEG, to achieve a long-circulating property, and with ligands, such as antibodies, to increase liposomal drug accumulation at desired targets [42]. It has been recognised that liposomes enter cells primarily by endocytosis and encounter acidic environment in endosomes [159]. Upon acidification, pH-responsive polymers can facilitate destabilisation and leakage of liposomes and/or disruption of endosomes. This enables the release of liposomal contents into the cytoplasm, where the released drugs can be trafficked to their cytosolic or nuclear targets [160].

3.5 *Polymersomes*

Polymersomes, a class of artificial vesicles with an architecture similar to that of liposomes, are formed by self-assembling of synthetic, amphiphilic block copolymers. Hammer and Discher first developed polymersomes in aqueous solution using poly(ethylene oxide)-block-poly(ethylethylene) (PEO-b-PEE) diblock copolymers in 1999 [161]. Since then, polymersomes have been widely studied for a variety of tailored applications including drug delivery [155, 162–167]. Polymersomes are engineered to mimic some functions of viral capsids composed of polypeptides, and can be used to deliver various small-molecule drugs, such as doxorubicin (water-soluble drug) and paclitaxel (hydrophobic drug) [163, 164], or bioactive macromolecules, such as nucleic acids [166] and proteins [167], to target cells.

A wide range of synthetic polymer amphiphiles can be used to make polymersomes, some of which can be designed to have triggered release of encapsulated therapeutic agents. An example is polyester-based degradable polymersomes formed from PEG-polylactic acid (PEG-PLA) or PEG-polycaprolactone (PEG-PCL). The release of encapsulated drug payload can be controlled by pH-dependent polyester chain hydrolysis in endolysosomes [163, 168]. Another polymersome from PEO-(propylene sulphide)-PEO triblock copolymers was designed to be responsive to oxidation with glucose oxidase as a trigger [169]. Much work has also been focused on attaching targeting ligands to the surface of polymersomes for specific delivery [170–172]. Most polymersomes are currently used in in vitro and in vivo studies [4], and there are still no clinically approved strategies.

4 pH-responsive Membrane-disruptive Delivery Agents

One of the main aims of drug delivery research is the efficient intracellular delivery of therapeutics, particularly macrodrugs such as proteins and nucleic acids. This delivery is often limited by the default endosomal pathway to lysosomes, where drug molecules may be degraded by enzymes [63, 76], or non-productively accumulated [80]. One strategy to address this issue is to release endocytosed materials efficiently from endosomes into the cytoplasm using pH-responsive endosomolytic delivery agents. Considering the typical pH ranges of early endosomes (pH 5.9–6.8) and late endosomes (pH 5.0–6.0) [63, 76, 77], such delivery agents should be designed to destabilise the phospholipid membrane at mildly acidic pH.

4.1 Viral Vectors

In nature, viruses have evolved over millennia as sophisticated, efficient gene delivery systems. They enter specific cells by receptor-mediated endocytosis. The amphiphilic peptides in the protein coats of some viruses can be transformed to a membrane-disruptive state upon acidification in endosomes, resulting in release of the nucleic acid cargo from endosomes into the cytoplasm [143, 144].

The efficient delivery machinery of viruses has been exploited for drug delivery [14, 173]. Recombinant viruses, such as retrovirus [174, 175], lentivirus [176, 177], adenovirus [178, 179], can be designed as gene-delivery vehicles by replacing part of the viral genome with a therapeutic gene. Coupling of adenovirus particles to transferrin-polylysine-DNA complexes has also been reported to greatly enhance the receptor-mediated gene delivery and expression efficiency [180]. In addition, virosomes, which are the liposome-like reconstituted envelopes of viruses loaded with therapeutic agents, have been found to be an effective approach for the delivery of anticancer drugs, proteins or DNA [181, 182]. Viruses currently dominate the types of vectors being tested in gene therapy clinical trials (67% of ongoing clinical trials) [183]. For clinical purposes, however, enthusiasm for viral delivery systems is dampened by concerns regarding serious safety issues and difficulties in large-scale production [14, 184].

4.2 Viral and Synthetic Peptides as Delivery Agents

Development of viral and synthetic peptides as pH-responsive membrane-disruptive delivery agents has yielded promising results. One of the most commonly studied viral peptides is the fusogenic peptide derived from the N-terminus of the influenza virus hemagglutinin subunit (HA-2), which has been shown to successfully disrupt liposomal membranes and enhance the gene transfer efficiency of transferrin-polylysine-DNA complexes [142]. GALA, a 30 amino acid synthetic peptide with a glutamic acid-alanine-leucine-alanine (EALA) repeat, was rationally designed and synthesised by Szoka's group to mimic the structure and membrane-destabilising behaviour of viral fusogenic peptides [185–187]. Synthetic peptides share some similarities with viral peptides, such as a length of about 15–30 amino acids and an alternating pattern of hydrophobic amino acids (about 1–3 amino acids) interrupted by weakly-charged hydrophilic amino acids [143, 144]. Those viral and synthetic peptides undergo a change of conformation from random coils at

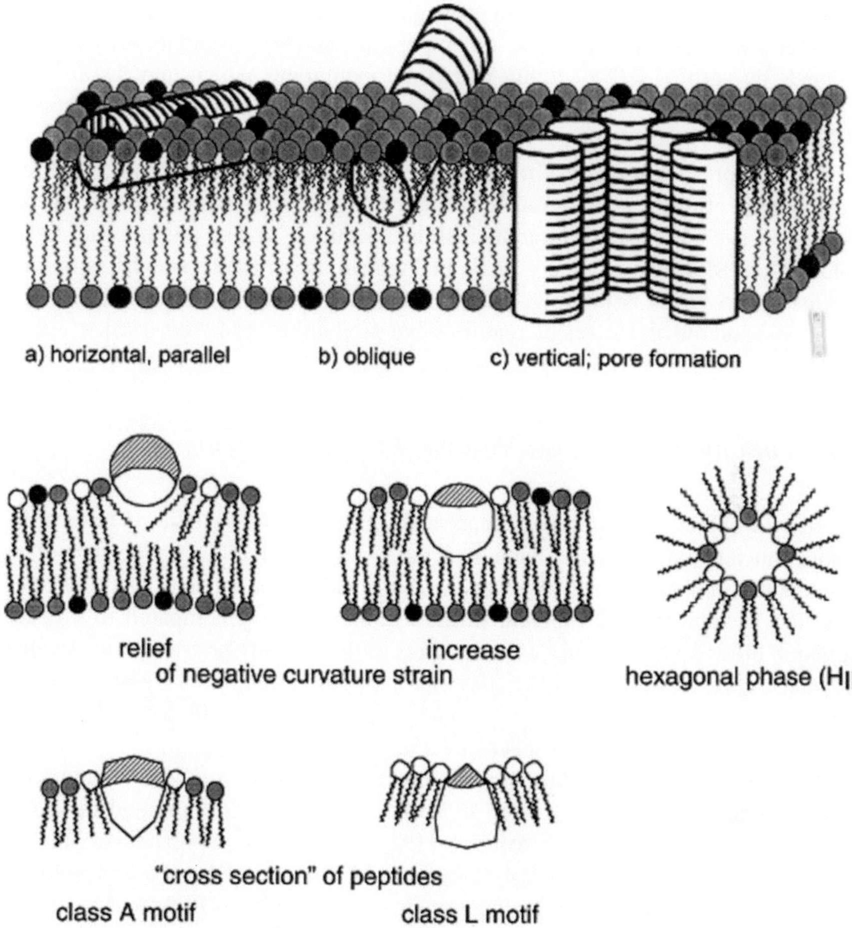

Fig. 3 Mechanisms for membrane destabilisation by amphiphilic peptides. *Top*: interaction of amphiphilic peptides with lipid membrane in horizontal, oblique, or vertical orientation. *Bottom*: peptides with class A motif stabilise membranes by relief of negative intrinsic curvature strain of phospholipid bilayer; peptides with class L destabilise membranes by increase of negative curvature strain, facilitating transition to the hexagonal phase (reproduced from Plank et al. [144])

neutral pH to α-helical structures at mildly acidic pH as a result of protonation of carboxylic acid moieties of the peptides in endosomes [142, 143, 187]. The α-helix is amphiphilic with two opposite faces, one of which is hydrophilic, composed of polar amino acids, and the other hydrophobic, composed of apolar amino acids. The hydrophobic face of the α-helix can associate with the endosomal membrane and cause membrane destabilisation, resulting in release of the endocytosed materials into the cytoplasm [142, 143, 187]. There are at least two different mechanisms (and combinations thereof) reported so far on membrane destabilisation by peptides. As shown in Fig. 3, some peptides bind to the membrane surface in a horizontal or oblique orientation, and disrupt lipid packaging by increasing the negative curvature strain of the lipid bilayer. Some other peptides form transmembrane pores or channels by vertical self-assembly within the membrane, which may, as a result, lead to membrane lysis due to osmotic swelling and collapse of transmembrane potential [144, 185, 187, 188].

The membrane-disruptive viral and synthetic peptides have demonstrated relatively high efficiency in intracellular delivery of small molecules or macromolecules, such as calcein, oligonucleotides, peptides, toxins, and plasmid DNA [143, 144, 148]. However, like viruses, the clinical use of those peptides is limited by the negative safety issues (e.g. potential immunogenicity, toxicity and insertional mutations) and the high costs of synthesis and production [144, 189, 190].

4.3 Cationic Polymers as Non-viral Delivery Agents

In order to obtain drug delivery agents with improved safety, ease of large-scale production and rather controllable structures, much attention has been paid to development of non-viral delivery vectors using synthetic polymers. Generally those non-viral vectors are less efficient in intracellular delivery compared to viruses. Research efforts have been focused on optimisation of polymer design to develop polymeric carriers with high efficiency and minimal toxicity [191]. pH-responsive polymers, which can destabilise the endosomal membrane in a pH-dependent manner, are of particular interest [48].

The capacity of synthetic cationic polymers to form complexes with nucleic acids (known as polyplexes, as discussed in Section 3.3) has been exploited for intracellular delivery of the genetic materials [192]. A variety of pH-responsive amine-containing polymers, including poly(L-lysine) (PLL) [145, 193], polyethyleneimine (PEI) [147, 195], poly(2-(dimethylamino)ethyl methacrylate) (PDMAEMA) [148, 196] and polyamidoamine (PAMAM) dendrimers [148, 196], have shown potential for gene delivery applications. The complexes of these polycations and DNA usually exhibit pronounced adsorption at the cell surface and the subsequent fast rate of internalisation into endosomal vesicles, as a result of the electrostatic attractions between the negatively charged membrane surface and cationic nitrogen [93]. Polycations can enable the escape of genetic materials from endosomes into the cytoplasm through the "proton sponge" effect. As shown in Fig. 4, protonation of the cationic polymers results in an increased influx of protons into endosomes, by the action of the ATPase proton-pump enzyme, to retain their desired luminal pH. The subsequent influx of counter ions ultimately causes increased osmotic pressure, which leads to mechanical swelling and rupture of the endosomal membrane [7, 146, 197]. Moreover, it was argued that the electrostatic interaction between protonated amines and the negatively changed membrane as well as hydrophobic interaction might also contribute to destabilisation of the endosomal membrane [48, 198]. The conformational flexibility [199], molecular weight [146, 200] and pKa value [147] of the cationic polymers can influence their intracellular delivery efficiency.

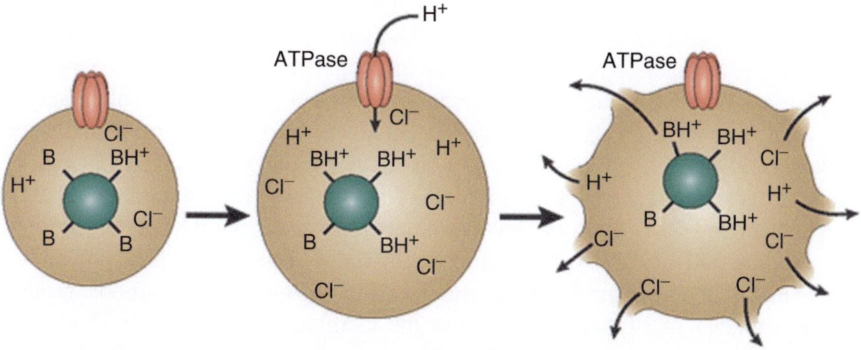

Fig. 4 Schematic illustration of the proton-sponge mechanism of the cationic polymer (reproduced from Pack et al. [7])

Cationic polymers have become promising non-viral gene delivery agents. However, their potential clinical use may be limited by relatively low transfection efficiency and high cytotoxicity. The electrostatic interaction between cationic polymers and the negatively charged cell membrane may result in the inherent high cytotoxicity. Other factors, such as the structure, conformational flexibility and molecular weight of those polymers can also contribute to their cytotoxicity [200, 201].

4.4 Anionic Polymers as Non-viral Delivery Agents

Synthetic anionic polymers are of particular interest as non-viral vectors because they can be designed to mimic the structure and pH-mediated membrane-disruptive properties of viral fusogenic peptides. Like viral fusogenic peptides, those polymers contain ionisable carboxyl groups for pH-responsive functionalities and hydrophobic segments for interacting with the endosomal membrane. They display a pH-triggered coil-to-globule conformational change upon neutralisation at acidic luminal pH within endosomes, resulting in the destabilisation of the endosomal membrane [202] through pore formation [158] or membrane solubilisation [203]. Today, a variety of pH-responsive anionic polymers, with non-biodegradable or biodegradable polymer chains, have been developed for intracellular drug delivery, such as in the form of direct conjugates with biological drugs [97], polymeric micelles [204–206], multi-component polyplexes [149, 207], and pH-responsive liposomes [35, 157].

4.4.1 Anionic Polymers with Non-biodegradable Polymer Chains

Extensive research has been carried out on vinyl poly(α-alkylacrylic acid) polymers with non-biodegradable polymer chains. Early on, Tirrell et al. showed the pH-dependent membrane activity of poly(α-ethylacrylic acid) (PEAA) using phospholipid vesicles [157, 158, 203, 208–210]. Since then, Hoffman, Stayton and other researchers have designed and investigated a family of pH-responsive poly(carboxylic acid)s, including homopolymers and copolymers of α-alkyl acrylic acids (such as methacrylic acid, ethylacrylic acid, propylacrylic acid, and butylacrylic acid) and their copolymers with alkyl acrylates [189, 202, 211–214]. Moreover, the pH-responsive copolymers of maleic anhydride with styrene, alkyl vinyl ethers and other hydrophobic monomers have also been developed for intracellular drug delivery [215]. The membrane disruptive capacity of those polyanions can be modulated by controlling the ratio of carboxyl groups to hydrophobic groups.

The capacity of PEAA to destabilise the membrane of phospholipid vesicles has motivated investigations of the ability of vinyl-based synthetic polymers to disrupt the cell membrane in mildly acidic environment. PEAA and poly (α-propylacrylic acid) (PPAA) were demonstrated to be essentially non membrane-lytic at physiological pH, but significantly enhance destabilisation of red blood cell (RBC) membranes at acidic pH characteristic of the endosome [211]. The cell-membrane activity can be modulated by manipulating the hydrophobic moieties of the polymers [189, 202]. Resulting from the increase of just one methylene unit in the pendant alkyl group, PPAA was found to be 15 times more efficient than PEAA at haemolysis [211, 212]. However, excessively hydrophobic poly(α-butylacrylic acid) (PBAA) may cause undesirable membrane disruption at physiological pH [189]. Further studies showed that by fine-tuning the ratio of carboxyl groups to hydrophobic segments, acrylic acid-alkyl acrylate copolymers and maleic anhydride copolymers could be engineered to exhibit efficient membrane disruption at endosomal pHs [189, 215].

The pH-dependent cell-membrane activity of these vinyl-based polymers has prompted extensive studies into their applications in cytoplasmic delivery of therapeutic agents in vitro and in vivo. Ternary complexes of biotinylated anti-CD3 antibody, streptavidin, and biotinylated PPAA entered Jurkat cells by receptor-mediated endocytosis, and PPAA enhanced translocation of the protein payload to the cytoplasm [216]. PPAA was also designed to aid in endosomal escape of cationic lipids or polymers for improved gene delivery and transfection efficiency with these vectors. Ternary complexes of the cationic lipid DOTAP, pCMVbeta plasmid DNA and PPAA (known as lipoplexes) showed marked enhancement of transfection efficiency in NIH3T3 fibroblasts and significant improvement of stability in serum compared to binary control lipoplexes of DOTAP and DNA [207]. Another example is the incorporation of PPAA into the complexes of chitosan and plasmid DNA, which enhanced gene expression in both HEK293 and HeLa cells [149]. In addition, covalent attachment of therapeutic agents, including antisense oligonucleotides and peptides, to the endosomolytic polyanions via cleavable linkages has been demonstrated to markedly enhance cytoplasmic drug delivery [78, 97]. The successful in vitro cell culture work has been extended to an in vivo murine excisional wound healing model. Positive results showed that the addition of PPAA to lipoplex formulations could considerably increase in vivo transfection efficiency and the inhibition of thrombospondin-2 (TSP2) expression led to improved would healing [59]. A recent in vivo study in an EG.7-OVA mouse tumour protection model showed that direct conjugation of PPAA to ovalbumin (Ova) via reducible disulfide linkages could result in significantly enhanced cytosolic delivery of the protein antigen and the subsequent robust CD8+ T-cell response, demonstrating the potential of PPAA-based vectors for therapeutic vaccine applications [217].

4.4.2 Anionic Polymers with Biodegradable Polymer Chains

The pH-responsive vinyl-based polymers are composed of a carbon-carbon backbone that is non-biodegradable. Their molecular weights should be less than the renal excretion threshold to avoid bioaccumulation [15], therefore limiting their potential clinical applications. To overcome this limitation, hydrolytically or enzymatically cleavable linkages, such as amides, anhydrides and esters, can be incorporated within the polymer backbone [218].

The frequently studied biodegradable poly (amino acid)s-based drug carriers include poly(aspartic acid) (PAA), poly(glutamic acid) (PGA) and their derivatives. The anionic polymers are composed of amino acids bearing ionisable pendant groups such as aspartic acid ($pKa = 3.9$), glutamic acid ($pKa = 4.1$) [219]. Like viral fusogenic peptides, PGA and PAA contain carboxyl groups in its side chains and present a coli-to-helix conformational transition upon reduction of pH [220]. This pH-responsive conformational transition results in increased binding of the poly(amino acid)s to phospholipid membranes and subsequent membrane disruption. The biocompatible, biodegradable poly(amino acid)s have been widely used in targeted intracellular delivery of therapeutic agents, including small-molecule drugs (such as paclitaxel and doxorubicin) [221–224], nucleic acids (such as plasmid DNA) [225, 226], proteins [227, 228].

Chen et al. [205, 229–232] have developed a class of pH-responsive, endosomolytic pseudo-peptidic polymers with a biodegradable polymer backbone. The metabolite-derived synthetic polymers were designed to mimic factors that enable efficient viral transfection. The parent polymer is a polyamide, poly(L-lysine *iso*-phthalamide) [198, 218], and natural hydrophobic amino acids were grafted onto its pendant carboxyl groups to manipulate its amphiphilicity and structure [229, 230, 233]. A large number of pendant carboxyl moieties also provide an opportunity for functionalisation with PEG [205, 206, 232]. The biomimetic polyanions undergo a change of conformation from extended polymer chains to collapsed, hydrophobically-stabilised globular structures with progressive protonation of their pendant carboxyl groups. The interactions between the resultant hydrophobic structures and lipid bilayer membranes lead to enhanced membrane binding of the polymers and subsequent membrane disruption. The pH-responsive conformational transition and cell membrane disruption were examined as a function of pH, concentration, time of incubation, type of side chains and degree of grafting. It has been reported that the balance between their hydrophobic and hydrophilic components, the number of grafts, and aromaticity of the polymers can be manipulated to have highly adaptable conformational change and membrane activity [205, 229–234]. PP-75, the L-phenylalanine grafted polymer with the vastly superior property for intracellular drug delivery, was 35-fold more lytic (on a molar basis) than the membrane-lytic peptide melittin at pH 6.5 (a very early stage of endosomal acidification), but essentially non membrane-lytic at pH 7.4 [231]. After the efficient endosomal release of the model drug calcein was demonstrated [231], PP-75 was used to successfully direct intracellular delivery of the novel anticancer protein apoptin to human osteogenic sarcoma Saos-2 cells [235]. Multicellular spheroids were then developed as an in vitro tumour model [236, 237]. PP-75 was shown to efficiently penetrate from the exterior to the interior of these spheroids and was internalised by almost all the cells (93%) in the spheroids after 1 h uptake of 5 mg mL^{-1} FITC-conjugated PP-75. This, as a result, led to efficient PP-75-mediated endosomal release of the endocytosed model drug calcein within cells in the tumour spheroids [238]. Most recently PP-75 was used for cytoplasmic delivery of stathmin siRNA via cleavable disulfide linkage, which successfully silenced stathmin gene expression both in vitro and in vivo. In a subcutaneous malignant glioma tumour model, the combination therapy through intraperitoneal administration of carmustine (a nitrosourea chemotherapeutic agent) and intratumoral injection of PP-75-stathmin siRNA conjugates has successfully inhibited tumour growth [239]. In addition, the pseudo-peptidic polymers have been fine-tuned to increase the permeability of the cell membrane reversibly at mildly acidic pH ($\sim$7) without causing cell damage. This has proved very useful for efficient delivery of trehalose, a bioprotectant normally impermeable to the phospholipid bilayer, into cells to a level well beyond thresholds for protection of cells during cryopreservation, freeze-drying or desiccation [240–242].

5 Conclusions

The lack of safe and efficient drug delivery agents is a limiting obstacle to drug development and effective treatment of various diseases including cancer. Recombinant viruses have excellent transfection, however, their clinical use can be limited due to serious safety issues and difficulties in large-scale production. This chapter covers the use of polymers for targeted delivery of therapeutic agents, including small-molecule drugs and biopharmaceuticals. The non-viral polymeric delivery agents, although safer than recombinant viruses, generally do not present the desirable efficacy. A variety of extracellular and intracellular barriers to effective polymeric drug delivery have been highlighted in this chapter for rational design of a polymeric carrier that may tackle multiple barriers to achieve optimal efficacy. Today, a wide range of effective polymeric delivery agents have been designed for targeted delivery of small-molecule drugs and/or bioactive macromolecules. Much has been learned about their structure-function relationships from in vitro and in vivo studies, and in some cases, clinical trials. Polymer-drug conjugates, polymeric micelles, multi-component polyplexes, polymer-modified liposomes, and polymersomes have been critically reviewed. A particular focus has been made on pH-responsive, membrane-disruptive polymers as carriers for intracellular delivery of therapeutic agents, particularly biopharmaceuticals, which need precise delivery to their intracellular sites of biological action. Development of promising nanomedicine platforms will depend on "rational design of nanotechnology materials and tools based around a detailed and thorough understanding of biological processes rather than forcing applications for some materials currently in vogue" [243]. With a growing understanding of various extracellular and intracellular barriers, and polymeric drug delivery mechanisms, it is promising that polymer-based delivery agents will play an important role in targeted drug delivery and therapy.

Acknowledgements R.C. wishes to thank Professor Nigel Slater (Department of Chemical Engineering and Biotechnology, University of Cambridge, UK) for support and valuable discussion. R.C. would also like to acknowledge the Biomedical and Health Research Centre (BHRC) in Leeds for the Senior Translational Research Fellowship.

References

1. Glass G (2004) Pharmaceutical patent challenges – time for reassessment. Nat Rev Drug Discov 3:1057–1062
2. Ferrari M (2005) Cancer nanotechnology: opportunities and challenges. Nat Rev Cancer 5:161–171
3. Zhang L, Gu FX, Chan JM, Wang AZ, Langer RS, Farokhzad OC (2008) Nanoparticles in medicine: therapeutic applications and developments. Clin Pharmacol Ther 83:761–769
4. Peer D, Karp JM, Hong S, Farokhzad OC, Margalit R, Langer R (2007) Nanocarriers as an emerging platform for cancer therapy. Nat Nanotechnol 2:751–760
5. Mulligan R (1993) The basic science of gene therapy. Science 260:926–932
6. Pouton CW, Seymour LW (2001) Key issues in non-viral gene delivery. Adv Drug Deliv Rev 46:187–203
7. Pack DW, Hoffman AS, Pun S, Stayton PS (2005) Design and development of polymers for gene delivery. Nat Rev Drug Discov 4:581–593
8. Liechty WB, Kryscio DR, Slaughter BV, Peppas NA (2010) Polymers for drug delivery systems. Annu Rev Chem Biomol Eng 1:149–173
9. Farokhzad OC, Langer R (2009) Impact of nanotechnology on drug delivery. ACS Nano 3:16–20
10. Wong TK, Neumann E (1982) Electric field mediated gene transfer. Biochem Biophys Res Commun 107:584–587
11. Yang NS, Sun WH (1995) Gene gun and other non-viral approaches for cancer gene therapy. Nat Med 1:481–483
12. Seiler C, Pohl T, Wustmann K, Hutter D, Nicolet PA, Windecker S, Eberli FR, Meier B (2001) Promotion of collateral growth by granulocyte-macrophage colony-stimulating factor in patients with coronary artery disease: a randomized, double-blind, placebo-controlled study. Circulation 104:2012–2017
13. Pillai R, Petrak K, Blezinger P, Deshpande D, Florack V, Freimark B, Padmabandu G, Rolland A (1998) Ultrasonic nebulization of catonic lipid-based gene delivery systems for airway administration. Pharm Res 15:1743–1747
14. Thomas CE, Ehrhardt A, Kay MA (2003) Progress and problems with the use of viral vectors for gene therapy. Nat Rev Genet 4:346–358

15. Duncan R (2006) Polymer conjugates as anticancer nanomedicines. Nat Rev Cancer 6:688–701
16. Lavasanifar A, Samuel J, Kwon GS (2002) Poly(ethylene oxide)-block-poly(L-amino acid) micelles for drug delivery. Adv Drug Deliv Rev 54:169–190
17. Ehrlich P (1906) Collected studies on immunity. Wiley, New York
18. de Duve C, Debarsy T, Poole B, Trouet A, Tulkens P, Vanhoof F (1974) Lyosomotropic agents. Biochem Pharmacol 23:2495–2531
19. Ringsdorf H (1975) Structure and properties of pharmacologically active polymers. J Polym Sci Pol Sym 51:135–153
20. Vasey PA, Kaye SB, Morrison R, Twelves C, Wilson P, Duncan R, Thomson AH, Murray LS, Hilditch TE, Murry T, Burtles S, Fraier D, Frigerio E, Cassidy J (1999) Phase I clinical and pharmacokinetic study of PK1 [N-(2-hydroxypropyl)methacrylamide copolymer doxorubicin]: first member of a new class of chemotherapeutic agents—drug-polymer conjugates. Clin Cancer Res 5:83–94
21. Kopeček J, Kopeckova P, Minko T, Lu ZR (2000) HPMA copolymer-anticancer drug conjugates: design, activity, and mechanism of action. Euro J Pharm Biopharm 50:61–81
22. Duncan R (2003) Polymer-drug conjugates. In: Budman D, Calvert H, Rowinsky R (eds) Handbook of anticancer drug development. Lippincott Williams and Wilkins, Baltimore
23. Duncan R (2005) N-(2-Hydroxypropyl)methacrylamide copolymer conjugates. In: Kwon GS (ed) Polymeric drug delivery systems. Marcel Dekker, New York
24. Duncan R, Gac-Breton S, Keane R, Musila R, Sat YN, Satchi R, Searle F (2001) Polymer-drug conjugates, PDEPT and PELT: basic principles for design and transfer from the laboratory to clinic. J Control Release 74:135–146
25. Abuchowski A, McCoy JR, Palczuk NC, van Es T, Davis FF (1977) Effect of covalent attachment of polyethylene glycol on immunogenicity and circulating life of bovine liver catalase. J Biol Chem 252:3582–3586
26. Kozlowski A, Harris JM (2001) Improvements in protein PEGylation: PEGylated interferons for treatment of hepatitis C. J Control Release 72:217–224
27. Davis FF (2002) The origin of pegnology. Adv Drug Deliv Rev 54:457–458
28. Harris JM, Chess RB (2003) Effect of PEGylation on pharmaceuticals. Nat Rev Drug Discov 2:214–221
29. Graham ML (2003) Pegaspargase: a review of clinical studies. Adv Drug Deliv Rev 55:1293–1302
30. Lukyanov AN, Torchilin VP (2004) Micelles from lipid derivatives of water-soluble polymers as delivery systems for poorly soluble drugs. Adv Drug Deliv Rev 56:1273–1289
31. Hrubý M, Konak C, Ulbrich K (2005) Polymeric micellar pH-sensitive drug delivery system for doxorubicin. J Control Release 103:137–148
32. Wang CH, Wang CH, Hsiue GH (2005) Polymeric micelles with a pH-responsive structure as intracellular drug carriers. J Control Release 108:140–149
33. Lee WK, Park JY, Jung S, Yang CW, Kim WU, Kim HY, Park JH, Park JS (2005) Preparation and characterization of biodegradable nanoparticles entrapping immunodominant peptide conjugated with PEG for oral tolerance induction. J Control Release 105:77–88
34. Liang HF, Yang TF, Huang CT, Chen MC, Sung HW (2005) Preparation of nanoparticles composed of poly(gamma-glutamic acid)-poly(lactide) block copolymers and evaluation of their uptake by HepG2 cells. J Control Release 105:213–225
35. Osanai S, Nakamu K (2000) Effects of complexation between liposome and poly(malic acid) on aggregation and leakage behaviour. Biomaterials 21:867–876
36. Leroux JC, Roux E, Le Garrec D, Hong KL, Drummond DC (2001) N-isopropylacrylamide copolymers for the preparation of pH-sensitive liposomes and polymeric micelles. J Control Release 72:71–84
37. Danson S, Ferry D, Alakhov V, Margison J, Kerr D, Jowle D, Brampton M, Halbert G, Ranson M (2004) Phase I dose escalation and pharmacokinetic study of pluronic polymer-bound doxorubicin (SP 1049C) in patients with advanced cancer. Br J Cancer 90:2085–2091

38. Felgner PL, Ringold GM (2003) Cationic liposome-mediated transfection. Nature 337:387–388
39. Christie RJ, Grainger DW (2003) Design strategies to improve soluble macromolecular delivery constructs. Adv Drug Deliv Rev 55:421–437
40. Adams ML, Lavasanifar A, Kwon GS (2003) Amphiphilic block copolymers for drug delivery. J Pharm Sci 92:1343–1355
41. Kaul G, Potineni A, Lynn DM, Langer R, Amiji MM (2002) Surface-modified polymeric nanoparticles for tumor-targeted delivery. SurFacts in Biomaterials 7:1–6
42. Torchilin VP (2005) Recent advances with liposomes as pharmaceutical carriers. Nat Rev Drug Discov 4:145–160
43. Litzinger DC, Buiting AMF, Vanrooijen N, Huang L (1994) Effect of liposome size on the circulation time and intraorgan distribution of amphiphatic poly(ethylene glycol)-containing liposomes. Biochim Biophys Acta 1190:99–107
44. Gref R, Minamitake Y, Peracchina MT, Trubetskoy V, Torchilin V, Langer R (1994) Biodegradable long-circulating polymeric nanospheres. Science 263:1600–1603
45. Düzgüneş N, Nir S (1999) Mechanisms and kinetics of liposome-cell interactions. Adv Drug Deliv Rev 40:3–18
46. Deshpande MC, Davies MC, Garnett MC, Williams PM, Armitage D, Bailey L, Vamvakaki M, Armes SP, Stolnik S (2004) The effect of poly(ethylene glycol) molecular weight architecture on cellular interaction and uptake of DNA complexes. J Control Release 97:143–156
47. Katre NV (1993) The conjugation of proteins with polyethylene-glycol and other polymers—altering properties of proteins to enhance their therapeutic potential. Adv Drug Deliv Rev 10:91–114
48. Yessine MA, Lafleur M, Meier C, Petereit HU, Leroux JC (2003) Characterisation of the membrane-destabilization properties of different pH-sensitive methacrylic acid copolymers. Biochim Biophys Acta 1613:28–38
49. Seymour LW, Duncan R, Strohalm J, Kopeček J (1987) Effect of molecular weight (MW) of N-(2-hydroxypropyl) methacrylamide copolymers on body distribution and rate of excretion after subcutaneous, intraperitoneal, and intravenous administration to rats. J Biomed Mater Res 21:1341–1358
50. Maddox DA, Deen WM, Brenner BM (1992) Glomerular filtration. In: Windhager EE (ed) Handbook of physiology. Section 8: renal physiology, vol 1. Oxford University Press, New York
51. Matsumura Y, Maeda H (1986) A new concept for macromolecular therapeutics in cancer therapy; mechanism of tumoritropic accumulation of proteins and the antitumour agent SMANCS. Cancer Res 46:6387–6392
52. Ulbrich K, Šubr V (2004) Polymeric anticancer drugs with pH-controlled activation. Adv Drug Deliv Rev 56:1023–1050
53. Hobbs SK, Monsky WL, Yuan F, Roberts WG, Griffith L, Torchilin VP, Jain RK (1998) Regulation of transport pathways in tumor vessels: role of tumor type and microenvironment. Proc Natl Acad Sci USA 95:4607–4612
54. Gianasi E, Wasil M, Evagorou EG, Keddle A, Wilson G, Duncan R (1999) HPMA copolymer platinates as novel antitumour agents: In vitro properties, pharmacokinetics and antitumour activity in vivo. Eur J Cancer 35:994–1002
55. Noguchi Y, Wu J, Duncan R, Strohalm J, Ulbrich K, Akaike T, Maeda H (1998) Early phase tumour accumulation of macromolecules: a great difference in clearance rate between tumour and normal tissue. Jpn J Cancer Res 89:307–314
56. Jones AT, Gumbleton M, Duncan R (2003) Understanding endocytic pathways and intracellular trafficking: a prerequisite for effective design of advanced drug delivery systems. Adv Drug Deliv Rev 55:1353–1357
57. Gregoriadis G (1977) Targeting of drugs. Nature 265:407–411

58. Funhoff AM, Van Nostrum CF, Lok MC, Kruijtzer JAW, Crommelin DJA, Hennink WE (2005) Cationic polyacrylates with covalently linked membrane destabilizing peptides as gene delivery vectors. J Control Release 101:233–246
59. Kyriakides TR, Cheung CY, Murthy N, Bornstein P, Stayton PS, Hoffman AS (2002) pH-sensitive polymers that enhance intracellular drug delivery in vivo. J Control Release 78:295–303
60. Luo D, Saltzman WM (2000) Synthetic DNA delivery systems. Nat Biotechnol 18:33–37
61. Rodal SK, Shretting G, Garred O, Vilhardt F, van Deurs B, Sandvig K (1999) Extraction of cholesterol with methyl-beta-cyclodextrin perturbs formation of clathrin-coated endocytic vesicles. Mol Biol Cell 10:961–974
62. Besterman JM, Low RB (1983) Endocytosis – a review of mechanisms and plasma membrane dynamics. Biochem J 210:1–13
63. Mellman I (1996) Endocytosis and molecular sorting. Annu Rev Cell Dev Biol 12:575–625
64. Parton RG, Simons K (2007) The multiple faces of caveolae. Nat Rev Mol Cell Biol 8:185–194
65. Pelkmans L, Kartenbeck J, Helenius A (2001) Caveolar endocytosis of simian virus 40 reveals a new two-step vesicular-transport pathway to the ER. Nat Cell Biol 3:473–483
66. Marsh M, Helenius A (2006) Virus entry: open sesame. Cell 124:729–740
67. Mayor S, Pagano RE (2007) Pathways of clathrin-independent endocytosis. Nat Rev Mol Cell Biol 8:603–612
68. Damm EM, Pelkmans L, Kartenbeck J, Mezzacasa A, Kurzchalia T, Helenius A (2005) Clathrin and caveolin-1-independent endocytosis: entry of simian virus 40 into cells devoid of caveolae. J Cell Biol 168:477–488
69. Pratten MK, Lloyd JB (1986) Pinocytosis and phagocytosis—the effect of size of a particulate substrate on its mode of capture by rat peritoneal—macrophages cultured in vitro. Biochim Biophys Acta 881:307–313
70. Kirkham M, Parton RG (2005) Clathrin-independent endocytosis: new insights into caveolae and non-caveolar lipid raft carriers. Biochim Biophys Acta 1745:273–286
71. Matthews SE, Pouton CW, Threadgill MD (1996) Macromolecular systems for chemotherapy and magnetic resonance imaging. Adv Drug Deliv Rev 18:219–267
72. Lloyd JB, Pratten MK, Duncan R, Kooistra T, Cartlidge SA (1984) Substrate selection and processing in endocytosis. Biochem Soc Trans 12:977–978
73. Pratten MK, Cable HC, Ringsdorf H, Lloyd JB (1982) Adsorptive pinocytosis of polycationic co-polymers of vinylpropyrrolidone with vinylamine by rat yolk-sac and rat peritoneal macrophage. Biochim Biophys Acta 719:424–430
74. Rémy-Kristensen A, Clamme JP, Vuilleumier C, Kuhry JG, Mély Y (2001) Role of endocytosis in the transfection of L929 fibroblasts by polyethylenimine/DNA complexes. Biochim Biophys Acta 1514:21–32
75. Bergen JM, Park IK, Horner PJ, Pun SH (2008) Nonviral approaches for neuronal delivery of nucleic acids. Pharm Res 25:983–998
76. Mukherjee S, Ghosh RN, Maxfield FR (1997) Endocytosis. Physiol Rev 77:759–803
77. Mellman I, Warren G (2000) The road taken: past and future foundations of membrane traffic. Cell 100:99–112
78. Murthy N, Campbell J, Fausto N, Hoffman AS, Stayton PS (2003) Bioinspired pH-responsive polymers for the intracellular delivery of biomolecular drugs. Bioconjug Chem 14:412–419
79. Luzio JP, Pryor PR, Bright NA (2007) Lysosomes: fusion and function. Nat Rev Mol Cell Biol 8:622–632
80. Sheff D (2004) Endosomes as a route for drug delivery in the real world. Adv Drug Deliv Rev 56:927–930
81. Rejmanová P, Kopeček J, Pohl J, Baudyš M, Kostka V (1983) Polymers containing enzymatically degradable bonds, 8. Degradation of oligopeptide sequences in N-(2-hydroxypropyl)methacrylamide copolymers by bovine spleen cathepsin B. Macromol Chem Phys 184:2009–2020

82. Rozema DB, Lewis DL, Wakefield DH, Wong SC, Klein JJ, Roesch PL, Bertin SL, Reppen TW, Chu Q, Blokhin AV, Hagstrom JE, Wolff JA (2007) Dynamic PolyConjugates for targeted in vivo delivery of siRNA to hepatocytes. Proc Natl Acad Sci USA 104:12982–12987
83. Panyam J, Labhasetwar V (2004) Targeting intracellular targets. Curr Drug Deliv 1:235–247
84. Desai MP, Labhasetwar V, Walter E, Levy RJ, Amidon GL (1997) The mechanism of uptake of biodegradable microparticles in Caco-2 cells is size dependent. Pharm Res 14:1568–1573
85. Leonetti JP, Mechti N, Degols G, Gagnor C, Lebleu B (1991) Intracellular-distribution of microinjection antisense oligonucleotides. Proc Natl Acad Sci USA 88:2702–2706
86. Rexach M, Blobel G (1995) Protein import into nuclei – association and dissociation reactions involving transport substrate, transport factors and nucleoporins. Cell 83:683–692
87. Labhasetwar V (2005) Nanotechnology for drug and gene therapy: the importance of understanding molecular mechanisms of delivery. Curr Opin Biotechnol 16:674–689
88. Subramanian A, Ranganathan P, Diamond SL (1999) Nuclear targeting peptide scaffolds for lipofection of nondividing mammalian cells. Nat Biotechnol 17:873–877
89. Brunner S, Sauer T, Carotta S, Cotton M, Saltik M, Wagner E (2000) Cell cycle dependence of gene transfer by lipoplex, polyplex and recombinant adenovirus. Gene Ther 7:401–407
90. Liu G, Li DS, Pasumarthy MK, Kowalczyk TH, Gedeon CR, Hyatt SL, Payne JM, Miller TJ, Brunovskis P, Fink TL, Muhammad O, Moen RC, Hanson RW, Cooper MJ (2003) Nanoparticles of compacted DNA transfect postmitotic cells. J Biol Chem 278:32578–32586
91. Lechardeur D, Sohn KJ, Haardt M, Joshi PB, Monck M, Graham RW, Beatty B, Squire J, O'Brodovich H, Lukacs GL (1999) Metabolic instability of plasmid DNA in the cytosol: a potential barrier to gene transfer. Gene Ther 6:482–497
92. Lechardeur D, Lukacs GL (2002) Intracellular barriers to non-viral gene transfer. Curr Gene Ther 2:183–194
93. Godbey WT, Wu KK, Mikos AG (1999) Tracking the intracellular path of poly(ethylenimine)/DNA complexes for gene delivery. Proc Natl Acad Sci USA 96:5177–5181
94. Cassidy J, Schätzlein AG (2004) Tumour-targeted drug and gene delivery: principles and concepts. Expert Rev Mol Med 6:1–17
95. Soyez H, Schacht E, Vanderkerken S (1996) The crucial role of spacer groups in macromolecular prodrug design. Adv Drug Deliv Rev 21:81–106
96. Duncan R (1992) Drug polymer conjugates – potential for improved chemotherapy. Anticancer Drugs 3:175–210
97. Bulmus V, Woodward M, Lin L, Murthy N, Stayton P, Hoffman A (2003) A new pH-responsive and glutathione-reactive, endosomal membrane-disruptive polymeric carrier for intracellular delivery of biomolecular drugs. J Control Release 93:105–120
98. Greenwald RB, Choe YH, McGuire J, Conover CD (2003) Effective drug delivery by PEGylated drug conjugates. Adv Drug Deliv Rev 55:217–250
99. Pasut G, Veronese F (2006) PEGylation of proteins as tailored chemistry for optimized bioconjugates. In: Satchi-Fainaro R, Duncan R (eds) Advances in polymer science: polymer therapeutics I. Springer, Berlin
100. Pasut G, Veronese FM (2007) Polymer-drug conjugation, recent achievements and general strategies. Prog Polym Sci 32:933–961
101. Levy Y, Hershfield MS, Fernandezmejia C, Polmar SH, Scudiery D, Berger M, Sorensen RU (1988) Adenosine-deaminase deficiency with late onset of recurrent infections response to treatment with polyethylene-glycol modified adenosine-deaminase. J Pediatr 113:312–317
102. Pockros PJ, Carithers R, Desmond P, Dhumeaux D, Fried MW, Marcellin P, Shiffman ML, Minuk G, Reddy KR, Reindollar RW, Lin A, Brunda MJ (2004) Efficacy and safety of two-dose regimens of peginterferon α-2a compared with interferon α-2a in chronic hepatitis C: a multicenter, randomized controlled trial. Am J Gastroenterol 99:1298–1305
103. Wang YS, Youngster S, Grace M, Bausch J, Bordens R, Wyss DF (2002) Structural and biological characterization of pegylated recombinant interferon α-2b and its therapeutic implications. Adv Drug Deliv Rev 54:547–570

104. Chapman AP, Antoniw P, Spitali M, West S, Stephens S, King DJ (1999) Therapeutic antibody fragments with prolonged in vivo half-lives. Nat Biotechnol 17:780–783
105. Chapman AP (2002) PEGylated antibodies and antibody fragments for improved therapy: a review. Adv Drug Deliv Rev 54:531–545
106. Veronese FM, Pasut G (2007) Drug-polymers conjugates. In: Triggle DJ, Taylor JB (eds) Comprehensive medicinal chemistry II. Elsevier, Oxford
107. Etrych T, Chytil P, Mrkvan P, Sírová M, Říhová B, Ulbrich K (2008) Conjugates of doxorubicin with graft HPMA copolymers for passive tumor targeting. J Control Release 132:184–192
108. Kopeček J, Ulbrich K, Zacharieva EI, Georgieva MP, Kabaivanov (1986) Polymers with anticancer activity and methods of their preparation, (PV 5354–81), CZ 223 417, Czechoslovakia
109. Bader H, Ringsdorf H, Schmidt B (1984) Watersoluble polymers in medicine. Macromol Mater Eng 123:457–485
110. Allen C, Maysinger D, Eisenberg A (1999) Nano-engineering block copolymer aggregates for drug delivery. Colloid Surf B 16:3–27
111. Kakizawa Y, Kataoka K (2002) Block copolymer micelles for delivery of gene and related compounds. Adv Drug Deliv Rev 54:203–222
112. van Nostrum CF (2004) Polymeric micelles to deliver photosensitizers for photodynamic therapy. Adv Drug Deliv Rev 56:9–16
113. Kwon GS, Kataoka K (1995) Block copolymer micelles as long-circulating drug vehicles. Adv Drug Deliv Rev 16:295–309
114. Miwa A, Ishibe A, Nakano M, Yamahira T, Itai S, Jinno S, Kawahara H (1998) Development of novel chitosan derivatives as micellar carriers of taxol. Pharm Res 15:1844–1850
115. Akiyoshi K, Kobayashi S, Shichibe S, Mix D, Baudys M, Kim SW, Sunamoto J (1998) Self-assembled hydrogel nanoparticle of cholesterol-bearing pullulan as a carrier of protein drugs: com- plexation and stabilization of insulin. J Control Release 54:313–320
116. Lin J, Zhu J, Chen T, Lin S, Cai C, Zhang L, Zhuang Y, Wang XS (2009) Drug releasing behavior of hybrid micelles containing polypeptide triblock copolymer. Biomaterials 30:108–117
117. Inoue T, Chen G, Nakamae K, Hoffman AS (1998) An AB block copolymer of oligo(methyl methacrylate) and poly(acrylic acid) for micellar delivery of hydrophobic drugs. J Control Release 51:221–229
118. Kim IS, Jeong YI, Cho CS, Kim SH (2000) Core-shell type polymeric nanoparticles composed of poly(L-lactic acid) and poly(N-isopropylacrylamide). Int J Pharm 211:1–8
119. Luo L, Ranger M, Lessard DG, Garrec DL, Gori S, Leroux JC, Rimmer S, Smith D (2004) Novel amphiphilic diblock copolymer of low molecular weight poly(N-vinylpyrrolidone)-block- poly(D, L-lactide): synthesis, characterization and micellization. Macromolecules 37:4008–4013
120. Kim C, Lee SC, Kwon IC, Chung H, Jeong SY (2002) Complexation of poly(2-ethyl-2-oxazoline)-block-poly(ε-caprolactone) micelles with multifunctional carboxylic acids. Macromolecules 35:193–200
121. Savić R, Luo L, Eisenberg A, Maysinger D (2003) Micellar nanocontainers distribute to defined cytoplasmic organelles. Science 300:615–618
122. Lavasanifar A, Samuel J, Kwon GS (2001) Micelles self-assembled from poly(ethylene oxide)-block-poly(N-hexyl stearate L-aspartamide) by a solvent evaporation method: effect on the solubilization and haemolytic activity of amphotericin B. J Control Release 77:155–160
123. Oh I, Lee K, Kwon HY, Lee YB, Shin SC, Cho CS, Kim CK (1999) Release of adriamycin from poly(γ-benzyl-L-glutamate)/poly(ethylene oxide) nanoparticles. Int J Pharm 181:107–115
124. Kabanov A, Lemieux P, Vinogradov S, Alakhov V (2002) Pluronic® block copolymers: Novel functional molecules for gene therapy. Adv Drug Deliv Rev 54:223–233
125. Ranger M, Jones MC, Yessine MA, Leroux JC (2001) From well-defined diblock copolymers prepared by a versatile atom transfer radical polymerization method to supramolecular assemblies. J Polym Sci Polym Chem 39:3861–3874

126. Vinogradov S, Batrakova E, Li S, Kabanov A (1999) Polyion complex micelles with protein-modified corona for receptor-mediated delivery of oligonucleotides into cells. Bioconjug Chem 10:851–860
127. Harada A, Togawa H, Kataoka K (2001) Physicochemical properties and nuclease resistance of antisense-oligodeoxynucleotides entrapped in the core of polyion complex micelles composed of poly(ethylene glycol)-poly(L-lysine) block copolymers. Eur J Pharm Sci 13:35–42
128. Zhang GD, Nishiyama N, Harada A, Jiang DL, Aida T, Kataoka K (2003) pH-sensitive assembly of light-harvesting dendrimer zinc porphyrin bearing peripheral groups of primary amine with poly(ethylene glycol)-b-poly(aspartic acid) in aqueous solution. Macromolecules 36:1304–1309
129. Harada A, Kataoka K (1998) Novel polyion complex micelles entrapping enzyme molecules in the core: preparation of narrowly-distributed micelles from lysozyme and poly(ethylene glycol)-poly(aspartic acid) block copolymer in aqueous medium. Macromolecules 31:288–294
130. Kim SH, Jeong JH, Lee SH, Kim SW, Park TG (2006) PEG conjugated VEGF siRNA for anti-angiogenic gene therapy. J Control Release 116:123–129
131. Jeong JH, Kim SH, Kim SW, Park TG (2005) Polyelectrolyte complex micelles composed of c-raf antisense oligodeoxynucleotide-poly(ethylene glycol) conjugate and poly(ethylenimine): effect of systemic administration on tumor growth. Bioconjug Chem 16:1034–1037
132. Kwon G, Naito M, Yokoyama M, Okano T, Sakurai Y, Kataoka K (1993) Micelles based on AB block copolymers of poly(ethylene oxide) and poly(beta-benzyl L-aspartate). Langmuir 9:945–949
133. Maeda H, Wu J, Sawa T, Matsumura Y, Hori K (2000) Tumor vascular permeability and the EPR effect in macromolecular therapeutics: a review. J Control Release 65:271–284
134. Sonawane ND, Szoka FC, Verkman AS (2003) Chloride accumulation and swelling in endosomes enhances DNA transfer by polyamine-DNA polyplexes. J Biol Chem 278:44826–44831
135. Yessine MA, Dufresne MH, Meier C, Petereit HU, Leroux JC (2007) Proton-actuated membrane-destabilizing polyion complex micelles. Bioconjug Chem 18:1010–1014
136. Schild HG, Tirrell DA (1991) Microheterogeneous solutions of amphiphilic copolymers of N-isopropylacrylamide. An investigation via fluorescence methods. Langmuir 7:1319–1324
137. Chung JE, Yokoyama M, Yamato M, Aoyagi T, Sakurai Y, Okano T (1999) Thermo-responsive drug delivery from polymeric micelles constructed using block copolymers of poly(N-isopropylacrylamide) and poly(butylmethacrylate). J Control Release 62:115–127
138. Kato K, Hamaguchi T, Yasui H, Okusaka T, Ueno H, Ikeda M, Shirao K, Shimada Y, Nakahama H, Muro K, Matsumura Y (2006) Phase I study of NK105, a paclitaxel-incorporating micellar nanoparticle, in patients with advanced cancer. J Clin Oncol 24(Suppl):2018
139. Hamaguchi T, Kato K, Yasui H, Morizane C, Ikeda M, Ueno H, Muro K, Yamada Y, Okusaka T, Shirao K, Shimada Y, Nakahama H, Matsumura Y (2007) A phase I and pharmacokinetic study of NK105, a paclitaxel-incorporating micellar nanoparticle formulation. Br J Cancer 97:170–176
140. Kato K, Chin K, Yoshikawa T, Yamaguchi K, Tsuji Y, Esaki T, Sakai K, Kimura M, Hamaguchi T, Shimada Y, Matsumura Y, Ikeda R (2012) Phase II study of NK105, a paclitaxel-incorporating micellar nanoparticle, for previously treated advanced or recurrent gastric cancer. Invest New Drugs 30:1621–1627
141. Valle JW, Armstrong A, Newman C, Alakhov V, Pietrzynski G, Brewer J, Campbell S, Corrie P, Rowinsky EK, Ranson M (2011) A phase 2 study of SP1049C, doxorubicin in P-glycoprotein-targeting pluronics, in patients with advanced adenocarcinoma of the esophagus and gastroesophageal junction. Invest New Drugs 29:1029–1037
142. Wagner E, Plank C, Zatloukal K, Cotten M, Birnstiel ML (1992) Influenza virus hemagglutinin HA-2 N-terminal fusogenic peptides augment gene transfer by transferrin-polylysine-DNA complexes: toward a synthetic virus-like gene-transfer vehicle. Proc Natl Acad Sci USA 89:7934–7938

143. Plank C, Oberhauser B, Mechtler K, Koch C, Wagner E (1994) The influence of endosome-disruptive peptides on gene transfer using synthetic virus-like gene transfer systems. J Biol Chem 269:12918–12924
144. Plank C, Zauner W, Wagner E (1998) Application of membrane-active peptides for drug and gene delivery across cellular membranes. Adv Drug Deliv Rev 34:21–35
145. Zauner W, Ogris M, Wagner E (1998) Polylysine-based transfection systems utilizing receptor-mediated delivery. Adv Drug Deliv Rev 30:97–113
146. Boussif O, Lezoualc'h F, Zanta MA, Mergny MD, Scherman D, Demeneix B, Behr JP (1995) A versatile vector for gene and oligonucleotide transfer into cells in culture and in vivo: polyethylenimine. Proc Natl Acad Sci USA 92:7297–7301
147. van de Wetering P, Moret EE, Schuurmans-Nieuwenbroek NME, van Steenbergen MJ, Hennink WE (1999) Structure-activity relationships of water-soluble cationic methacrylate/methacrylamide polymers for nonviral gene delivery. Bioconjug Chem 10:589–597
148. Haensler J, Szoka FC Jr (1993) Polyamidoamine cascade polymers mediate efficient transfection of cells in culture. Bioconjug Chem 4:372–379
149. Kiang T, Bright C, Cheung CY, Stayton PS, Hoffman AS, Leong KW (2004) Formation of chitosan-DNA nanoparticles with poly(propyl acrylic acid) enhances gene expression. J Biomater Sci Polym Ed 15:1405–1421
150. Tang MX, Szoka FC (1997) The influence of polymer structure on the interactions of cationic polymers with DNA and morphology of the resulting complexes. Gene Ther 4:823–832
151. Poste G, Papahadjopoulos D (1976) Drug-containing lipid vesicles render drug-resistant tumor-cells sensitive to Actinomycin-D. Nature 261:699–701
152. Felgner PL, Gadek TR, Holm M, Roman R, Chan HW, Wenz M, Northrop JP, Ringold GM, Danielsen M (1987) Lipofection: a highly efficient, lipid-mediated DNA-transfection procedure. Proc Natl Acad Sci USA 84:7413–7417
153. Zabner J (1997) Cationic lipids used in gene transfer. Adv Drug Deliv Rev 27:17–28
154. Koltover I, Salditt T, Radler JO, Safinya CR (1998) An inverted hexagonal phase of cationic liposome–DNA complexes related to DNA release and delivery. Science 281:78–81
155. Discher DE, Ahmed F (2006) Polymersomes. Annu Rev Biomed Eng 8:323–341
156. Yatvin MB, Kreutz W, Horwitz BA, Shinitzky M (1980) pH-sensitive liposomes: possible clinical implications. Science 210:1253–1254
157. Tirrell DA, Takigawa DY, Seki K (1985) Interactions of synthetic-polymers with cell-membranes and model membrane systems. 7. pH-sensitization of phospholipid vesicles via complexation with synthetic poly(carboxylic acid)s. Ann N Y Acad Sci 446:237–248
158. Thomas JL, Tirrell DA (2000) Polymer induced leakage of cations from dioleoyl phosphatidylcholine and phosphatidylglycerol liposomes. J Control Release 67:203–209
159. Straubinger RM, Hong K, Friend DS, Papahadjopoulos D (1983) Endocytosis of liposomes and intracellular fate of encapsulated molecules—encounter with a low pH compartment after internalization in coated vesicles. Cell 32:1069–1079
160. Hafez IM, Cullis PR (2001) Roles of lipid polymorphism in intracellular delivery. Adv Drug Deliv Rev 47:139–148
161. Discher BM, Won YY, Ege DS, Lee JC, Bates FS, Discher DE, Hammer DA (1999) Polymersomes: tough vesicles made from diblock copolymers. Science 284:1143–1146
162. Mabrouk E, Cuvelier D, Brochard-Wyart F, Nassoy P, Li MH (2009) Bursting of sensitive polymersomes induced by curling. Proc Natl Acad Sci USA 106:7294–7298
163. Ahmed F, Pakunlu RI, Srinivas G, Brannan A, Bates F, Klein ML, Minko T, Discher DE (2006) Shrinkage of a rapidly growing tumor by drug-loaded polymersomes: pH-triggered release through copolymer degradation. Mol Pharm 3:340–350
164. Discher DE, Ortiz V, Srinivas G, Klein ML, Kim Y, Christian D, Cai S, Photos P, Ahmed F (2007) Emerging applications of polymersomes in delivery: from molecular dynamics to shrinkage of tumors. Prog Polym Sci 32:838–857
165. Levine DH, Ghoroghchian PP, Freudenberg J, Zhang G, Therien MJ, Greene MI, Hammer DA, Murali R (2008) Polymersomes: a new multi-functional tool for cancer diagnosis and therapy. Methods 46:25–32

166. Lomas H, Canton I, MacNeil S, Du J, Armes SP, Ryan AJ, Lewis AL, Battaglia G (2007) Biomimetic pH sensitive polymersomes for efficient DNA encapsulation and delivery. Adv Mater 19:4238–4243
167. Lee JCM, Bermudez H, Discher BM, Sheehan MA, Won YY, Bates FS, Discher DE (2001) Preparation, stability, and in vitro performance of vesicles made with diblock copolymers. Biotechnol Bioeng 73:135–145
168. Ahmed F, Discher DE (2004) Self-porating polymersomes of PEG-PLA and PEG-PCL: hydrolysis-triggered controlled release vesicles. J Control Release 96:37–53
169. Napoli A, Boerakker MJ, Tirelli N, Nolte RJM, Sommerdijk NAJM, Hubbell JA (2004) Glucose-oxidase based self-destructing polymeric vesicles. Langmuir 20:3487–3491
170. Dalhaimer P, Engler A, Parthasarthy R, Discher DE (2004) Targeted worm micelles. Biomacromolecules 5:1714–1719
171. Meng F, Engbers GHM, Feijen J (2005) Biodegradable polymersomes as a basis for artificial cells: encapsulation, release and targeting. J Control Release 101:187–198
172. Broz P, Benito SM, Saw CL, Burger P, Heider H, Pfisterer M, Marsch S, Meier W, Hunziker P (2005) Cell targeting by a generic receptor-targeted polymer nanocontainer platform. J Control Release 102:475–488
173. Cotten M, Wagner E, Birnstiel ML (1993) Receptor-mediated transport of DNA into eukaryotic cells. Methods Enzymol 217:618–643
174. Keller G, Paige C, Gilboa E, Wagner EF (1985) Expression of a foreign gene in myeloid and lymphoid cells derived from multipotent haematopoietic precursors. Nature 318:149–154
175. Nagano M, Shinohara T, Avarbock MR, Brinster RL (2000) Retrovirus-mediated gene delivery into male germ line stem cells. FEBS Lett 475:7–10
176. May C, Rivella S, Callegari J, Heller G, Gaensler KM, Luzzatto L, Sadelain M (2000) Therapeutic haemoglobin synthesis in beta-thalassaemic mice expressing lentivirus-encoded human beta-globin. Nature 406:82–86
177. Wong LF, Goodhead L, Prat C, Mitrophanous KA, Kingsman SM, Mazarakis ND (2006) Lentivirus-mediated gene transfer to the central nervous system: therapeutic and research applications. Hum Gene Ther 17:1–9
178. Le Gal La Salle G, Robert JJ, Berrard S, Ridoux V, Stratford-Perricaudet LD, Perricaudet M, Mallet J (1993) An adenovirus vector for gene transfer into neurons and glia in the brain. Science 259:988–990
179. Fulton D, Gratton JP, McCabe TJ, Fontana J, Fujio Y, Walsh K, Franke TF, Papapetropoulos A, Sessa WC (1999) Regulation of endothelium-derived nitric oxide production by the protein kinase Akt. Nature 399:597–601
180. Wagner E, Zatloukal K, Cotten M, Kirlappos H, Mechtler K, Curiel DT, Birnsteil ML (1992) Coupling of adenovirus to transferrin polylysine DNA complexes greatly enhances receptor-mediated gene delivery and expression of transfected genes. Proc Natl Acad Sci USA 89:6099–6103
181. Ramani K, Hassan Q, Venkaiah B, Hasnain SE, Sarkar DP (1998) Site-specific gene delivery in vivo through engineered Sendai viral envelopes. Proc Natl Acad Sci USA 95:11886–11890
182. de Jonge J, Leenhouts JM, Holtrop M, Schoen P, Scherrer P, Cullis PR, Wilschut J, Huchriede A (2007) Cellular gene transfer mediated by influenza virosomes with encapsulated plasmid DNA. Biochem J 405:41–49
183. Gene Therapy Clinical Trials Worldwide [online]. J Gene Med (2012). http://www.abedia.com/wiley/vectors.php. Accessed 26 Apr 2012
184. Chen R, Folarin N, Ho VH, McNally D, Darling D, Farzaneh F, Slater NKH (2010) Affinity recovery of lentivirus by diaminopelargonic acid mediated desthiobiotin labeling. J Chromatogr B Analyt Technol Biomed Life Sci 878:1939–1945
185. Subbarao NK, Parente RA, Szoka FC, Nadasdi L, Pongracz K (1987) pH-dependent bilayer destabilization by an amphipathic peptide. Biochemistry 26:2964–2972
186. Parente RA, Nir S, Szoka FC (1988) pH-dependent fusion of phosphatidylcholine small vesicles—induction by a synthetic amphipathic peptide. J Biol Chem 263:4724–4730

187. Li WJ, Nicol F, Szoka FC (2004) GALA: a designed synthetic pH-responsive amphiphatic peptide with applications in drug and gene delivery. Adv Drug Deliv Rev 56:967–985
188. Epand RM, Shai YC, Segrest JP, Anantharamaiah GM (1995) Mechanisms for the modulation of membrane bilayer properties by amphipathic helical peptides. Biopolymers 37:319–338
189. Murthy N, Chang I, Stayton P, Hoffman A (2001) pH-sensitive hemolysis by random copolymers of alkyl acrylates and acrylic acid. Macromol Symp 172:49–55
190. Lam JKM, Ma Y, Armes SP, Lewis AL, Baldwin T, Stolnik S (2004) Phosphorylcholine-polycation diblock copolymers as synthetic vectors for gene therapy. J Control Release 100:293–312
191. Dinçer S, Türk M, Pişkin E (2005) Intelligent polymers as nonviral vectors. Gene Ther 12(Suppl 1):S139–S145
192. Merdan T, Kopeček J, Kissel T (2002) Prospects for cationic polymers in gene and oligonucleotide therapy against cancer. Adv Drug Deliv Rev 54:715–758
193. Midoux P, Monsigny M (1999) Efficient gene transfer by histidylated polylysine pDNA complexes. Bioconjug Chem 10:406–411
194. Merlin JL, Dolivet G, Dubessy C, Festor E, Parache RM, Verneuil L, Erbacher P, Behr JP, Guillemin F (2001) Improvement of nonviral p53 gene transfer in human carcinoma cells using glucosylated polyethylenimine derivatives. Cancer Gene Ther 8:203–210
195. Gary DJ, Lee H, Sharma R, Lee JS, Kim Y, Cui ZY, Jia D, Bowman VD, Chipman PR, Wan L, Zou Y, Mao G, Park K, Herbert BS, Konieczny SF, Won YY (2011) Influence of nano-carrier architecture on in vitro siRNA delivery performance and in vivo biodistribution: polyplexes vs micelleplexes. ACS Nano 5:3493–3505
196. Peng SF, Su CJ, Wei MC, Chen CY, Liao ZX, Lee PW, Chen HL, Sung HW (2010) Effects of the nanostructure of dendrimer/DNA complexes on their endocytosis and gene expression. Biomaterials 31:5660–5670
197. Hubbell JA (2003) Enhancing drug function. Science 300:595–596
198. Eccleston ME, Kuiper M, Gilchrist FM, Slater NKH (2000) pH-responsive pseudo-peptides for cell membrane disruption. J Control Release 69:297–307
199. Tang MX, Redemann CT, Szoka FC (1996) In vitro gene delivery by degraded polyamidoamine dendrimers. Bioconjug Chem 7:703–714
200. van de Wetering P, Cherng JY, Talsma H, Crommelin DJA, Hennink WE (1998) 2-(dimethylamino)ethyl methacrylate based (co)polymers as gene transfer agents. J Control Release 53:145–153
201. Fischer D, Li YX, Ahlemeyer B, Krieglstein J, Kissel T (2003) In vitro cytotoxicity testing of polycations: influence of polymer structure on cell viability and hemolysis. Biomaterials 24:1121–1131
202. Yessine MA, Leroux JC (2004) Membrane-destabilizing polyanions: interaction with lipid bilayers and endosomal escape of biomacromolecules. Adv Drug Deliv Rev 56:999–1021
203. Thomas JL, Barton SW, Tirrell DA (1994) Membrane solubilisation by a hydrophobic polyelectrolyte: surface activity and membrane binding. Biophys J 67:1101–1106
204. Yue Z, Eccleston ME, Slater NKH (2005) PEGylation and aqueous solution behaviour of pH-responsive poly(L-lysine iso-phthalamide). Polymer 46:2497–2505
205. Chen R, Yue Z, Eccleston ME, Slater NKH (2008) Aqueous solution behaviour and membrane disruptive activity of pH-responsive PEGylated pseudo-peptides and their intracellular distribution. Biomaterials 29:4333–4340
206. Eccleston ME, Williams S, Yue Z, Chen R, Lee CK, Anikina E, Pawlyn C, Barrand MA, Slater NKH (2005) Design and in-vitro testing of effective poly(L-lysine iso-phthalamide) based drug targeting systems for solid tumours. Food Bioprod Process 83:141–146
207. Cheung CY, Murthy N, Stayton PS, Hoffman AS (2001) A pH-sensitive polymer that enhances cationic lipid-mediated gene transfer. Bioconjug Chem 12:906–910
208. Thomas JL, Tirrell DA (1992) Polyelecctrolyte-sensitized phospholipids-vesicles. Accounts Chem Res 25:336–342
209. Thomas JL, You H, Tirrell DA (1995) Tuning the response of a pH-sensitive membrane switch. J Am Chem Soc 117:2949–2950

210. Thomas JL, Borden KA, Tirrell DA (1996) Modulation of mobilities of fluorescent membrane probes by adsorption of a hydrophobic polyelectrolyte. Macromolecules 29:2570–2576
211. Murthy N, Robichaud JR, Tirrell DA, Stayton PS, Hoffman AS (1999) The design and synthesis of polymers for eukaryotic membrane disruption. J Control Release 61:137–143
212. Stayton PS, Hoffman AS, Murthy N, Lackey C, Cheung C, Tan P, Klumb LA, Chilkoti A, Wilbur FS, Press OW (2000) Molecular engineering of proteins and polymers for targeting and intracellular delivery of therapeutics. J Control Release 65:203–220
213. Kusonwiriyawong C, van de Wetering P, Hubbell JA, Merkle HP, Walter E (2003) Evaluation of pH-dependent membrane-disruptive properties of poly(acrylic acid) derived polymers. Eur J Pharm Biopharm 56:237–246
214. Jones RA, Cheung CY, Black FE, Zia JK, Stayton PS, Hoffman AS, Wilson MR (2003) Poly (2-alkylacrylic acid) polymers deliver molecules to the cytosol by pH-sensitive disruption of endosomal vesicles. Biochem J 372:65–75
215. Henry SM, El-Sayed MEH, Pirie CM, Hoffman AS, Stayton PS (2006) pH-responsive poly(styrene-alt-maleic anhydride) alkylamide copolymers for intracellular drug delivery. Biomacromolecules 7:2407–2414
216. Lackey CA, Press OW, Hoffman AS, Stayton PS (2002) A biomimetic pH-responsive polymer directs endosomal release and intracellular delivery of an endocytosed antibody complex. Bioconjug Chem 13:996–1001
217. Foster S, Duvall CL, Crownover EF, Hoffman AS, Stayton PS (2010) Intracellular delivery of a protein antigen with an endosomal-releasing polymer enhances CD8 T-cell production and prophylactic vaccine efficacy. Bioconjug Chem 21:2205–2212
218. Eccleston ME, Slater NKH, Tighe BJ (1999) Synthetic routes to responsive polymers; co-polycondensation of tri-functional amino acids with diacylchlorides. React Funct Polym 42:147–161
219. Horton HR, Moran LA, Ochs RS, Rawn JD, Scrimgeour KG (2002) Principles of biochemistry, 3rd edn. Pearson Education, Upper Saddle River
220. Ito Y, Ochiai Y, Park YS, Imanishi Y (1997) pH-sensitive gating by conformational change of a polypeptide brush grafted onto a porous polymer membrane. J Am Chem Soc 119:1619–1623
221. Eldar-Boock A, Miller K, Sanchis J, Lupu R, Vincent MJ, Satchi-Fainaro R (2011) Integrin-assisted drug delivery of nano-scaled polymer therapeutics bearing paclitaxel. Biomaterials 32:3862–3874
222. Guan H, McGuire MJ, Li S, Brown KC (2008) Peptide-targeted polyglutamic acid doxorubicin conjugates for the treatment of alpha(v)beta(6)-positive cancers. Bioconjug Chem 19:1813–1821
223. Prompruk K, Govender T, Zhang S, Xiong CD, Stolnik S (2005) Synthesis of a novel PEG-block-poly(aspartic acid-stat-phenylalanine) copolymer shows potential for formation of a micellar drug carrier. Int J Pharm 297:242–253
224. Han S, Liu Y, Nie X, Xu Q, Jiao F, Li W, Zhao Y, Wu Y, Chen C (2012) Efficient delivery of antitumor drug to the nuclei of tumor cells by amphiphilic biodegradable poly(L-aspartic acid-co-lactic acid)/DPPE co-polymer nanoparticles. Small 8:1596–1606
225. Guo S, Huang Y, Zhang W, Wang W, Wei T, Lin D, Xing J, Deng J, Du Q, Liang Z, Liang XJ, Dong A (2011) Ternary complexes of amphiphilic polycaprolactone-graft-poly (N, N-dimethylaminoethyl methacrylate), DNA and polyglutamic acid-graft-poly(ethylene glycol) for gene delivery. Biomaterials 32:4238–4292
226. Kurosaki T, Kitahara T, Fumoto S, Nishida K, Nakamura J, Niidome T, Kodama Y, Nakagawa H, To H, Sasaki H (2009) Ternary complexes of pDNA, polyethylenimine, and gamma-polyglutamic acid for gene delivery systems. Biomaterials 30:2846–2853
227. Kurosaki T, Kitahara T, Nakamura T, Nishida K, Fumoto S, Kodama Y, Nakagawa H, Higuchi N, Sasaki H (2012) Development of effective cancer vaccine using targeting system of antigen protein to APCs. Pharm Res 29:483–489
228. Shu S, Zhang X, Teng D, Wang Z, Li C (2009) Polyelectrolyte nanoparticles based on water-soluble chitosan-poly(L-aspartic acid)-polyethylene glycol for controlled protein release. Carbohydr Res 344:1197–1204

229. Chen R, Eccleston ME, Yue Z, Slater NKH (2009) Synthesis and pH-responsive properties of pseudo-peptides containing hydrophobic amino acid grafts. J Mater Chem 19:4217–4224
230. Chen R, Khormaee S, Eccleston ME, Slater NKH (2009) Effect of L-leucine graft content on aqueous solution behavior and membrane-lytic activity of a pH-responsive pseudopeptide. Biomacromolecules 10:2601–2608
231. Chen R, Khormaee S, Eccleston ME, Slater NKH (2009) The role of hydrophobic amino acid grafts in the enhancement of membrane-disruptive activity of pH-responsive pseudo-peptides. Biomaterials 30:1954–1961
232. Chen R, Yue Z, Eccleston ME, Williams S, Slater NKH (2005) Modulation of cell membrane disruption by pH-responsive pseudo-peptides through grafting with hydrophilic side chains. J Control Release 108:63–72
233. Khormaee S, Chen R, Park JK, Slater NKH (2010) The Influence of aromatic side-chains on the aqueous properties of pH-sensitive poly(L-lysine iso-phthalamide) derivatives. J Biomater Sci Polym Ed 21:1573–1588
234. Zhang S, Nelson A, Coldrick Z, Chen R (2011) The effects of substituent grafting on the interaction of pH-responsive polymers with phospholipid monolayers. Langmuir 27:8530–8539
235. Liechty WB, Chen R, Farzaneh F, Tavassoli M, Slater NKH (2009) Synthetic pH-responsive polymers for protein transduction. Adv Mater 21:3910–3914
236. Ho VHB, Barcza A, Chen R, Müller KH, Darton NJ, Slater NKH (2009) The precise control of cell labelling with streptavidin paramagnetic particles. Biomaterials 30:6548–6555
237. Ho VHB, Müller KH, Barcza A, Chen R, Slater NKH (2010) Generation and manipulation of magnetic multicellular spheroids. Biomaterials 31:3095–3102
238. Ho VHB, Slater NKH, Chen R (2011) pH-responsive endosomolytic pseudo-peptides for drug delivery to multicellular spheroids tumour models. Biomaterials 32:2953–2958
239. Khormaee S, Choi Y, Shen MJ, Xu B, Wu H, Griffiths GL, Chen R, Slater NKH, Park JK (2013) Endosomolytic anionic polymer for the cytoplasmic delivery of siRNAs in localized in vivo applications. Adv Funct Mater 23:565–574
240. Slater NKH, Lynch A, Chen R (2011) Mammalian cell preservation methods. Patent No. WO/2011/089391, International application No. PCT/GB2011/000072
241. Lynch AL, Chen R, Dominowski PJ, Shalaev EY, Yancey RJ Jr, Slater NKH (2010) Biopolymer mediated trehalose uptake for enhanced erythrocyte cryosurvival. Biomaterials 31:6096–6103
242. Lynch AL, Chen R, Slater NKH (2011) pH-responsive polymers for trehalose loading and desiccation protection of human red blood cells. Biomaterials 32:4443–4449
243. Moghimi SM, Hunter AC, Murray JC (2005) Nanomedicine: current status and future prospects. FASEB J 19:311–330

Amphiphilic Molecules in Drug Delivery Systems

Salomé dos Santos, Bruno Medronho, Tiago dos Santos, and Filipe E. Antunes

Abstract Numerous drug delivery colloidal systems are formulated using polymers or surfactants or a mixture of both, typically due to their self-assembly properties. Molecular self-assembly creates the possibility to dissolve and protect drugs from adverse external environments. Therefore, it is important to understand the interactions behind the self-assembly phenomena of surfactant and polymer molecules, polymer-polymer and polymer-surfactant mixtures. A number of colloidal structures used in drug delivery formulations such as micelles, vesicles, liquid crystalline phases, microemulsions, polymer gels, aerosols, polymer-polymer and polymer-surfactant complexes will be illustrated in this chapter and their main physicochemical properties will be highlighted, keeping in mind their relevance to the drug delivery research field.

Keywords Self-assembly • Amphiphilic • Nanoaggregates • Phase diagrams • Drug delivery systems • Personalized medicine

S. dos Santos • F.E. Antunes (✉)
Department of Chemistry, University of Coimbra, Rua Larga, 3004-535 Coimbra, Portugal
e-mail: filipe.antunes@ci.uc.pt

B. Medronho
Institute for Biotechnology and Bioengineering, Centre of Genomics and Biotechnology (IBB/CGB), Faculty of Sciences and Technology, University of Algarve, Campus de Gambelas, Ed. 8, 8005-139 Faro, Portugal

T. dos Santos
Centre for BioNano Interactions, School of Chemistry and Chemical Biology, University College Dublin, Belfield, Dublin 4, Ireland

J. Coelho (ed.), *Drug Delivery Systems: Advanced Technologies Potentially Applicable in Personalised Treatment*, Advances in Predictive, Preventive and Personalised Medicine 4, DOI 10.1007/978-94-007-6010-3_2,

1 Introduction

The states of matter extend well beyond atomic solids, liquids and gases. Matter organizes itself at many different length scales and in many distinct forms, each distinguished by its microscopic symmetries and dynamics. The properties of most materials result from disorder or heterogeneity at some length scale much larger than the atomic scale. More and more the details of the interactions at the atomic scale need to be understood in order to explain the properties of everyday materials.

The field of soft matter is broad and extremely interesting. There are, for instance, non-crystalline states with various degrees of order (liquid crystals) and there are some states (glasses and gels) that are disordered but which behave as solids. Polymers (biopolymers), surfactants, emulsions, microemulsions and biomembranes are some examples that belong to the complex field of soft matter. The organization within such soft structures, at a certain length scale, brings the potential for encapsulating drugs, turning the structure into a drug delivery vehicle. That is to say, for instance, that the hydrophobic core of a self-assembled structure can dissolve large amounts of water-insoluble drugs.

In applications such as drug therapy, soft systems are generally preferred due to their flexibility and biocompatibility. It is possible to tailor the properties of soft systems such as internal structure and surface activity, in a relative easy way. Furthermore, specific environments (within the body) play their role in tuning such properties and allowing the most desired effect. Also, drugs may be amphiphilic and surface active, altering the organization/structure and the stability of the drug delivery vehicle.

Together with the organization of matter at different length scales, the size of the drug vehicle is in many cases very important. On one hand, the small size of the vehicle creates the possibility for intravenous administration; on the other hand, nanoscale devices and/or nanoscale components of larger devices, of the same size as many biological entities (e.g. proteins and DNA) and structures (e.g. viruses and bacteria), create the potential for crossing many barriers within the body and engage with the cellular machinery.

The "internal" organization and the size of the drug vehicle need to be carefully chosen regarding some crucial aspects of the therapy. For instance, for an efficient and safe therapy, the concentration of the drug should be both sufficiently high at the site of action and constant within the therapeutic window over the period of action. Usually, drugs are randomly distributed to the entire body resulting in high drug concentration in non-target sites, leading to detrimental side effects. Also here, the use of sub-micron drug delivery vehicles is highly advantageous when compared to conventional drug formulations due to the possibility for drug targeting. Other important aspects to take into account when choosing the drug delivery vehicle are: shape, stability, susceptibility to breakdown/degradation, the tendency to undergo self-aggregation, drug selectivity, rate and extent of the drug release, drug adsorption and solubilization capacity, preservation of drug activity and integrity, reduction of drug toxicity, site of action, sustainability and route of administration.

Many complex interactions are present in complex mixtures, such as pharmaceutical formulations. To create appropriate and efficient drug delivery vehicles, one needs to be aware and possess the knowledge on molecular biology and surface/colloid chemistry. The advances in these fields does not always correlate with the development in drug formulations. In part, this is due to the lack of knowledge on the physicochemical and surface properties of the formulation components. In many cases, one component may play more than one role in the system. The "deconstruction" of the formulation recipe and the understanding of the phase behavior of the mixture may be a crucial step to engineer novel ways to deliver drugs. Furthermore, understanding the physicochemical and surface behavior of polymers and, particularly, biopolymers, which play a crucial role in the regulation and integration of life processes and act with high specificity and effectiveness, is very important in the design of physical or chemical modifications that may increase the life-time of the biopolymer and improve its bioavailability.

Polymers and surfactants can be used individually or as mixtures bringing new and strong advantages into the field of drug delivery. The characteristics of these drug vehicles may be tuned varying different parameters such as size and type of the hydrophobic alkyl chain of the surfactant, the nature and size of the polar head group of the surfactant, concentration, salt content, temperature, pH and presence of co-solutes.

Polymers are used in drug delivery due to their efficiency as stabilizers, their capacity to form gels and to control the rheology, even at low concentrations, and also, in special cases, analogous to surfactants, their capacity to form self-assembled stable structures. In some cases, another advantage is their biodegradability potential.

Polymer-surfactant and polymer-polymer associative mixtures present several different properties from the individual behavior of polymer or surfactant systems. Of particular interest, it is the fact that polymer-surfactant and polymer-polymer associative mixtures are capable of forming concentrated complexes/nanoparticles upon dilution. If on one hand the degradation or disruption of surfactant and polymer systems has particular interest in some cases, on the other hand dilution or degradation of the drug vehicle in the body fluids is not desired before a particular site of action is reached, keeping particle integrity. The delay of vehicle degradation and drug release may be achieved by using polymer-surfactant or polymer-polymer complexes as drug vehicles. The polymer and/or the surfactant can be the active component (i.e. drug) and, in this case, the drug is said to be complexed.

The aim of this chapter is to go through the relevant physicochemical features of surfactants and polymers, both individually and in mixtures, to the field of drug delivery. First, surfactant and polymer systems will be analysed individually, followed by the discussion on their synergetic interactions. Finally, a brief practical overview on drug delivery systems/formulations and the in vitro and in vivo applications will be presented.

2 Surfactants

2.1 Introduction

Surfactant systems play an important role in modern drug delivery since they allow, for instance, the control of drug uptake and release rate and minimization of drug degradation and toxicity. An effective synergism between surfactant systems and drugs is nowadays recognized as a key issue to assure therapeutic efficiency. Thus, it becomes important to understand the physicochemical properties and behavior of surfactants in formulations.

2.2 Surfactant Properties

Surfactants, or surface active agents, are very exotic molecules due to their amphiphilic behavior. This means that a surfactant molecule contains both a hydrophobic part (lipophilic) and a hydrophilic part (lipophobic). The non-polar hydrophobic part is typically referred to as tail (composed by one or more hydrocarbon chains, although fluorocarbon and dimethylsiloxane chains can be used) and the polar hydrophilic part is referred to as head group which might be either charged or uncharged.

Surfactants exist in many different forms in nature [1–3]; typically, these molecules are classified according to the chemical nature of their polar head group, i.e. surfactants with a negatively charged head group are referred to as anionic, whereas cationic surfactants contain a positively charged head group. Uncharged surfactants are generally referred to as non-ionic, while zwitterionic surfactants contain both a negatively and a positively charged group. Zwitterionic phospholipids such as phosphatidylcholine and phosphatidylethanolamine are lipids (naturally occurring surfactants) extensively used in drug delivery since they can form a variety of interesting self-assembled structures (liposomes, in particular), frequently presenting low toxicity and good biocompatibility [4]. Lately, new surfactants of low toxicity and high biodegradability, particularly from renewable resources, have been developed. Among them, surfactants with carbohydrate or amino acid polar head groups have been found to be interesting in that respect [5]. Figure 1 shows the structure of surfactant molecules.

Surfactants are found everywhere [6]: in detergency and cleaning, cosmetics and personal care products, plant protection and pest control, paints, lacquers and other coating products, foods and packaging, paper and cellulose products, plastics and composite products, metal processing, textiles and fibbers, oilfield chemicals, leather and furs, mining and flotation, foams and finally in pharmaceuticals, medicine and biochemical research. Surfactants are also responsible for compartmentalization which is fundamental for all living forms.

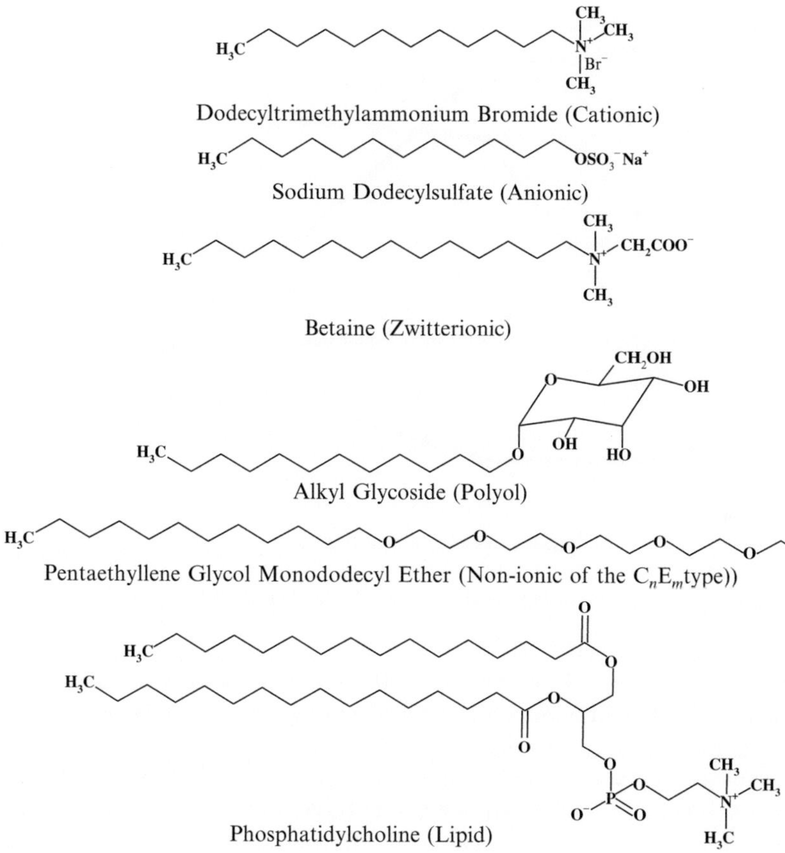

Fig. 1 Examples of surfactant molecules

2.3 *Self-Assembly and Phase Behavior*

Due to the amphiphilic nature, surfactants molecules display two very interesting and useful properties; they reduce the surface tension when adsorbing at a specific interface (i.e. air-water or oil-water) and they have the ability to self-associate and self-organize.

At low temperatures, the solubility is low with surfactant molecules in equilibrium with the surfactant solution. There is a critical point, known as the Krafft temperature, above which the solubility appears to increase rapidly and the solution consists of surfactant aggregates as well as single molecules. Below the Krafft temperature, surfactant aggregates are not formed. Many surfactant molecules aggregate spontaneously in aqueous media generally by starting to form normal micelles with aggregation numbers (number of molecules constituting the aggregate) ranging from 50–100. These micelles are in most cases spherical units, resulting in an isotropic solution with low viscosity.

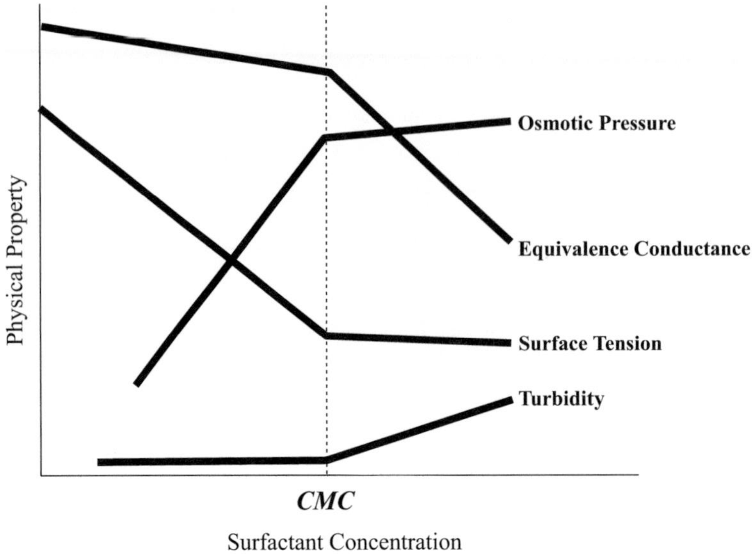

Fig. 2 Variation of different physical properties of a surfactant solution before and after CMC

Micellization is a strongly cooperative self-association process occurring at a particular narrow concentration, critical micellar concentration (CMC). CMC is thus an important parameter to characterize the self-association and may depend on the chemical nature of the surfactant and solvent as well as other factors such as the number and size of the hydrophobic tails. At CMC, the fraction of free monomers in bulk solution is the same as the fraction of molecules building up the aggregate. A further increase of the concentration results in an increase of the number of molecules in the aggregate, while the concentration of monomers in solution remains unchanged. The process is dynamic and, therefore, there is a constant exchange of molecules between the aggregates and the bulk solution [7].

Surface tension measurements are commonly used for CMC determinations. Figure 2 shows that as the surfactant concentration in solution increases the surface tension steadily decreases. This happens due to an increasing adsorption of surfactant molecules at the air/water interface disrupting the local water hydrogen bonding. At CMC, the slope of the surface tension curve decreases to almost zero. There are other physical properties that can be used to monitor micelle formation and CMC determination as represented in Fig. 2. For instance, at CMC, the rate of increase in osmotic pressure falls into a plateau. A sharp increase in turbidity is also observed by light scattering techniques. In conductance measurements, a marked decrease in the slope is observed after crossing CMC indicating that there are much less mobile charged units than expected from the individual surfactant molecules.

The so-called hydrophobic effect is believed to be the main driving force in self-association [8, 9]. It is an entropic driven process; the free energy of a process is

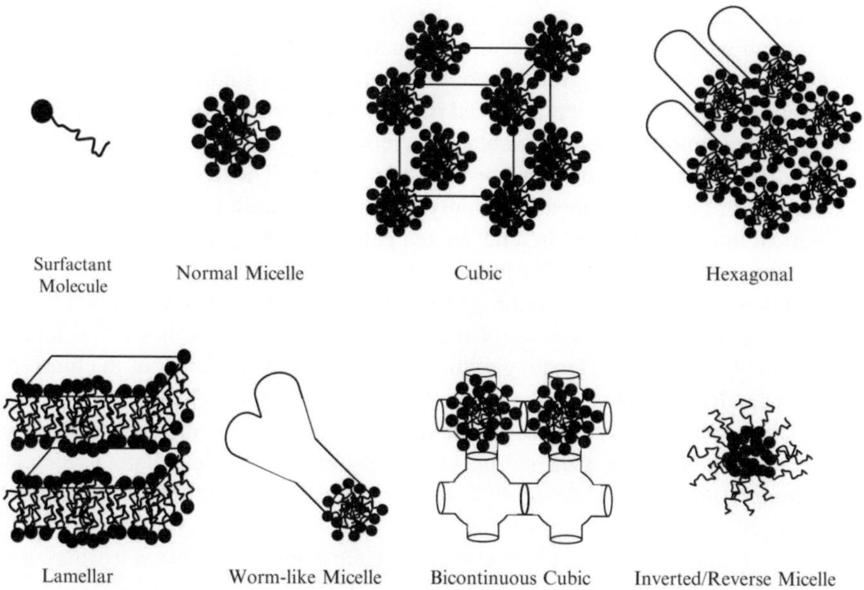

Fig. 3 Examples of different self-assembled surfactant structures

composed both by enthalpic and entropic terms. At room temperature, the enthalpy associated with the transfer of a hydrophobic molecule to an aqueous environment is negligible since the interaction enthalpies are practically the same in both cases. The main contribution comes from the loss of entropy associated with the formation of ordered water cages around the hydrophobe since it implies the disruption of hydrogen bonds between water molecules. As a consequence, non-polar molecules, which decrease the entropy of water, tend to be expelled from the aqueous media triggering the self-aggregation phenomena.

A delicate balance between opposing forces is the key aspect in surfactant self-assembly. It is affected by a range of factors, such as the size of the hydrophobic moiety, surfactant concentration, nature of the polar head group and counterion, salt concentration, pH, temperature and presence of co-solutes [5, 6].

A simple spherical micelle may grow forming cylindrical structures that are anisotropic and show features of macroscopic scale, e.g. flow birefringence. Even in this case, the solution appears as a single phase. Increasing concentration, linear growth can also lead to branched structures that may lead to interconnected structures (normally referred to as bicontinuous), since the solutions are not continuous only in the solvent but also in the surfactant. As concentration increases further amphiphiles can self-assemble to form a great variety of structures as the ones represented in Fig. 3.

Tuning some of the above mentioned parameters may allow the transition of one structure into another, offering interesting opportunities and strategies for drug delivery [4].

Micelles and bilayers are said to be the building blocks of most of the self-assembled structures. Surfactant aggregates (some presented in Fig. 3) can be divided into two main groups: those that are built of limited or discrete self-assemblies, which may be characterized roughly as spherical, prolate or cylindrical; and infinite or unlimited self-assemblies whereby the aggregates are connected over macroscopic distances in one, two or three dimensions. The hexagonal structure is an example of one-dimensional continuity. This phase is built up of (infinitely) long cylindrical micelles arranged in a hexagonal pattern, with each micelle surrounded by six other micelles (Fig. 3). The radius of the circular cross-section (which may be somewhat deformed) is close to the surfactant molecule length. On the other hand, planar lamellae show two-dimensional continuity. This structure is built of layers of surfactant molecules alternating with water layers. The thickness of the bilayers is somewhat lower than twice the surfactant molecule length. The thickness of the water layer can vary considerably depending on the nature of the surfactant. The surfactant bilayer can be stiff and planar or very flexible and undulating whereas the bicontinuous cubic and the sponge structures are examples of three-dimensional continuity.

These supramolecular surfactant structures are considered soft since they are fluid-like, flexible and easily affected by weak external forces. This is due to the nature of the self-assembly where molecules are not held together by covalent bonds but rather by physical forces, such as Van der Waals, hydrogen bonding and hydrophobic associations.

At this point, it becomes important to mention the role of the surfactant molecular geometry in predicting the surfactant structure that is formed [10]. This is of special relevance since physical properties can be quantitatively understood without the need of detailed knowledge of, for instance, the complex short-range forces between surfactant molecules.

2.4 Critical Packing Parameter and Mean Curvature

The driving force for all processes occurring in a non-specific system is the minimization of free energy. Self-association is no exception. As stated above, the balance between favorable and unfavorable interactions between solvent molecules and the particular sites of the surfactant molecule, i.e. minimization of energy penalty by exposing the hydrophobic moiety to water, is crucial for self-assembly. However, two other contributions to the total free energy have to be taken into account; the opposing force to self-assembly, due to head group steric repulsions and the geometric term which requires exclusion of water and head groups from the hydrophobic region occupied by the hydrocarbon tails [11]. These terms can be conveniently expressed by the surfactant critical packing parameter, CPP, which describes how the amphiphile geometry determines the aggregate structure (see Eq. 1).

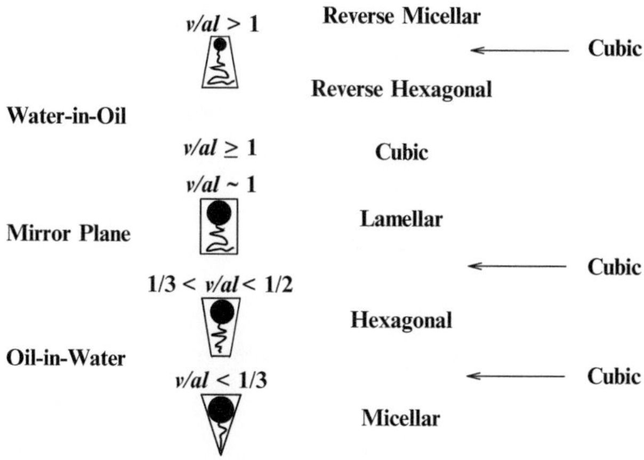

Fig. 4 CPP and preferred surfactant aggregate structures

$$CPP = \frac{a_s}{vl_c} \tag{1}$$

In Eq. 1, a_s is the effective area per head group, v the volume of the hydrocarbon chain and l_c is the maximum effective length that the hydrocarbon chain can assume.

There is a direct correlation between the value of the CPP and the type of aggregate formed [2, 5]. Spherical micelles are formed when CPP<1/3. The spherical aggregates are extremely small and their radius is approximately equal to the maximum stretched out length of the surfactant molecule. Cylindrical micelles are formed when 1/3<CPP<1/2 (single chain surfactants with small head group areas such as non-ionic surfactants and ionic surfactants in high salt concentration). Vesicles, liposomes and flexible bilayers are formed when 1/2<CPP<1 (double chain surfactants with large head group areas such as phospholipids, surfactants with bulky or branched tail groups and the mixture of anionic and cationic surfactants with single chain at nearly equimolar concentration). These types of surfactants cannot pack themselves into a micelle and they form bilayers (lamellar structure). Finally, inverted or reverse structures are formed when CPP>1 (surfactants with small head groups or large tail groups such as double tailed anionic surfactants). These structures are normally formed in non-polar solvents. In these inverted structures, polar head groups are clustered together and hydrophobic tails are extended towards the solvent, as exemplified in Fig. 3. As one will see later, this makes possible the formation of microemulsions, special water-in-oil "containers". Fig. 4 summarizes the expected sequence of surfactant structures regarding the direct dependence on the CPP.

CPP increases by addition of salt to ionic surfactants, since electrostatic repulsion among the head groups is screened. This effect can also be achieved by increasing the surfactant alkyl chain or by adding a co-surfactant (e.g. a long chain alcohol). Other parameters such as concentration, pH or temperature may play a role and contribute to tune CPP and, thus, the structure [3].

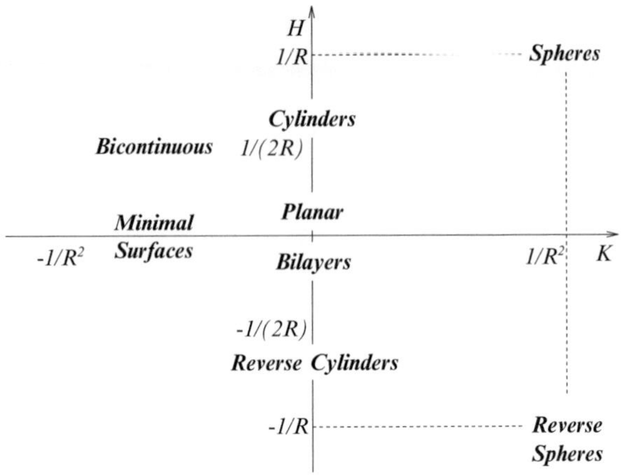

Fig. 5 The various membranc topologies modelled in terms of mean (H) and Gaussian (K) curvature

A related approach to the geometric surfactant parameter, CPP, is the concept of mean curvature, H, of the aggregate which is defined as follows

$$H = \frac{1}{2}\left(\frac{1}{R_1} + \frac{1}{R_2}\right) \tag{2}$$

R_1 and R_2 are the radii of curvature (reciprocal of the principle radii of curvature, c_1 and c_2, respectively) in two perpendicular directions. For a sphere $R_1 = R_2$ and thus $H = 1/R_1$. For a cylinder $R_1 = R$ and $R_2 = \infty$ making $H = 1/2R$. On the other hand, for a planar bilayer $R_1 = R_2 = \infty$, $H = 0$. By convention, the curvature is positive when curved towards the hydrophobic part of the aggregate.

The Gaussian curvature, K, is another useful parameter defined by

$$K = \frac{1}{R_1} \cdot \frac{1}{R_2} = c_1 \cdot c_2 \tag{3}$$

Figure 5 shows a schematic map of the predicted membrane geometries modelled in terms of mean (H) and Gaussian (K) curvature.

The abovementioned observations can be summarized as follows: the more the surfactant aggregate curves towards the oil, the smaller the value of CPP is, i.e. the larger the head group area in relation to the surfactant volume is. For reverse structures, the CPP increases following the order lamellar→hexagonal→ micellar.

For non-ionic surfactants, the surface curvature properties are even more relevant in the sense that they are strongly temperature dependent [12–16]. When the temperature is increased, the water becomes a poor solvent for the ethyleneoxide head group. In this case, the surfactant-surfactant interactions are more attractive and stronger than the surfactant-solvent ones, leading to a release of water molecules from the head group and to a change of the surfactant shape. At low temperatures, water is a good solvent for the ethyleneoxide head group, leading the aggregate to adopt a curvature towards the hydrophobic core. These opposite

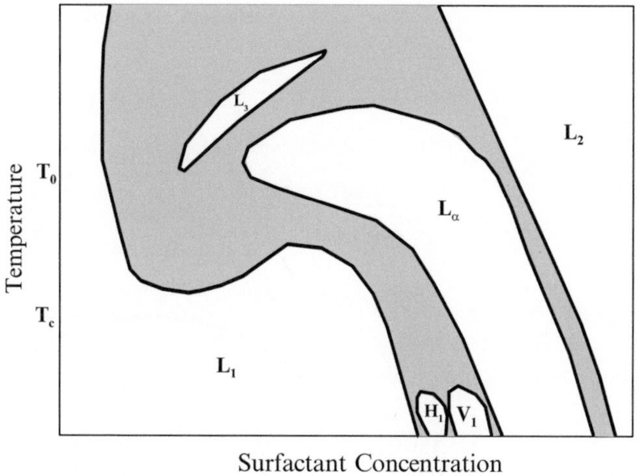

Fig. 6 A general phase diagram for non-ionic surfactants of the C_nE_m type. L_1 – normal micellar; L_α – lamellar; L_2 – reverse micellar; L_3 – sponge; H_1 – hexagonal; and V_1 – cubic. The *grey area* represents multi-phase samples

effects are explained by a change in conformation of the ethyleneoxide; while at low temperatures the lowest energy conformation for the $-OCH_2CH_2O-$ segment (see Fig. 1) corresponds to a gauche conformation around the C–C bond and anti around the C–O bond, which has a low statistical weight but highest dipole moment, at higher temperatures the conformations with low and zero dipole moments become more populated (trans conformations). This leads to an increase of the ordering in the ethyleneoxide chains making the solute–solute interactions more favorable than the solute–solvent ones. Note that the spontaneous curvature, H_0, temperature dependence can be described by

$$H_0 \approx -\beta (T - T_0) \tag{4}$$

where β is the system specific temperature independent constant and T_0 is the phase inversion temperature where the surfactant film neither prefers to curve towards water nor towards oil.

2.5 Phase Diagrams

Phase diagrams are extremely useful in the soft matter field. Often, a binary phase diagram shows the temperature along the y axis and surfactant composition along the x axis. Typically, the phase diagrams of aqueous non-ionic surfactant systems present a great variety of structures [17, 18]. Figure 6 shows a schematic and rather general phase diagram for non-ionic surfactants of the C_nE_m type.

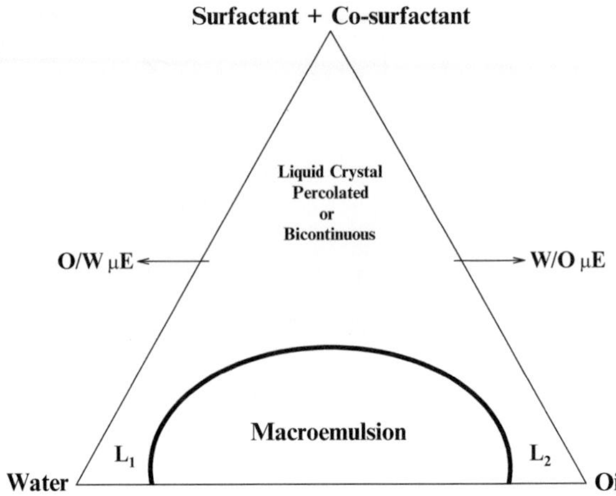

Fig. 7 A hypothetical pseudo-ternary phase diagram of an oil/surfactant/co-surfactant/water system. L_1 – normal micelles; L_2 – reverse micelles; W/O μE – water-in-oil microemulsions; O/W μE – oil-in-water microemulsions

C_n represents the number of carbons in the alkyl chain and E_m represents the number of ethyleneoxide units of the surfactant head group. It is possible to change systematically the hydrophilic head group, m, or the hydrophobic tail, n. For low concentrations and low temperatures, the micellar phase, L_1, dominates. Further increase in concentration leads to bilayer-like phases. As mentioned above, two topologically different organizations of the bilayers are observed: the anisotropic lamellar, L_α, phase characterized by stacks of bilayers with long range order and the sponge, L_3 phase that consists of a network of randomly multi-connected membranes and is isotropic [19, 20]. The latter tends to be more stable at reasonably higher temperatures than the former. Lamellar phases often occupy larger areas in phase diagrams compared to the neighbouring hexagonal, H_1, or cubic, V_1, phases. The reason for this is that three- and two-dimensionally, ordered cubic and hexagonal phases respectively, do not allow so many structural variations before losing their symmetry. On the other hand, the one-dimensionally ordered lamellar phase may sustain larger fluctuations without losing its one-dimensional order. To emphasise this, the melting points of the different phases can be compared; for a collection of phase diagrams see, for instance, Mitchell et al. [17].

Typically, systems are more complex than simple binary compositions. Not only surfactant and water, but also oil can be found in such systems. In such cases, the phase diagram used is generally referred to as ternary [5, 6]. A schematic illustration of a ternary phase diagram is given in Fig. 7. At constant temperature, each surfactant/oil/water system displays a unique ternary phase diagram. The composition of the different phases in a two- or three-phase sample is given by the points where the tie-lines intersect the corresponding one-phase areas. It is important to note that ternary phase diagrams are strictly valid only when the system contains three components (e.g. surfactant, oil, and water). There are situations where multi-dimensional phase diagrams might be needed to fully describe a system, as it will be presented later on in this chapter. The construction and interpretation of multi-dimensional phase diagrams is not an easy task and, thus, one usually fixes one or several variables, and changes systematically the other parameters. One can now note that the richness of surfactant phase behavior is due to the combination of packing constraints and free energy changes associated with hydrophobe-water interactions, the oil-water interface and head group interactions.

2.6 *Theoretical Description of Bilayer Characteristics: Simplified Framework of the Elastic Curvature Energy Model*

The surfactant bilayer may be described as a thin plate whose shape and behavior depends only on the competition between its elastic properties, any applied constraint and the entropy. The stability of a phase made up of membranes is governed by, among other factors, the competition between the elastic energy of the membrane and the thermal energy, k_BT [21–23].

When describing membrane structures and the change on the free energy associated to the transformation from one topology to the other, the concept of curvature free energy has proven to be very useful. The Helfrich flexible surface model basically consists in an interfacial description of the curvature free energy density, g_c, of the polar-apolar interface [24]. For each structure or configuration of the surface, the lowest order in curvature is very often expressed as

$$g_c = 2k(H - H_0)^2 + \bar{k}K \quad (5)$$

where k is the bending rigidity modulus and $\bar{k}$ is the saddle-splay modulus. For a given membrane configuration, the total curvature free energy, G_c, is obtained by integration of the energy over the interfacial area

$$G_c = \int g_c dA \quad (6)$$

The two elastic constants, k and $\bar{k}$, describe the elastic properties of the membrane playing different roles. The bending rigidity modulus, k, can be understood as the energy that has to be provided to bend the membrane around its equilibrium position. When k is comparable to the thermal energy, k_BT, thermal fluctuations give rise to significant displacement fluctuations of the membranes around their equilibrium position. These fluctuations have important consequences on the formation of different phases and on their static and dynamic properties. On the other hand, the saddle-splay modulus, $\bar{k}$, is the driving parameter for the membrane topology. Theoretically, if $-2k < \bar{k} < 0$ (assuming k positive), the membrane prefers to be a plane and this has a tendency to favor the lamellar phase. If $\bar{k} < -2k$, the

Fig. 8 Saddle-splay modulus changes for different membrane topologies

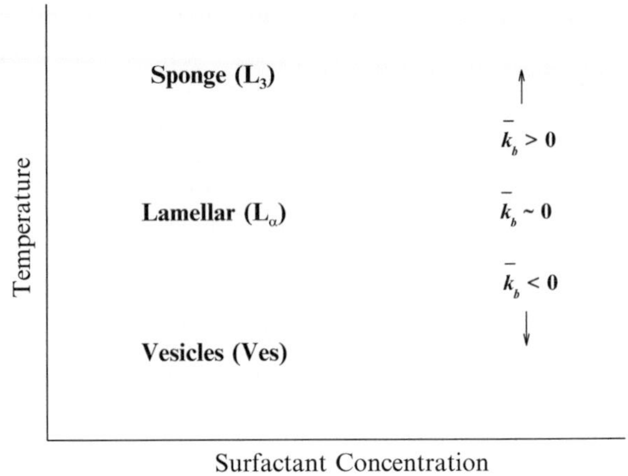

membrane prefers to curve into a spherical shape and the vesicle phase is then favored. If $\bar{k} > 0$ the membrane adopts a saddle-like form, tending to favor cubic or a multi-connected sponge phases. Therefore $\bar{k}$ plays a role each time a structural transformation involves a topological change for the membrane, but it has no effect as long as the curvature fluctuations take place at constant topology or degree of connectivity. Figure 8 shows how the saddle-splay modulus changes for different membrane topologies.

2.7 *Surfactant Structures and their Use in Drug Delivery*

Surfactant systems, such as micellar solutions, liquid crystalline phases or microemulsions are often used in drug delivery [4]. When using these systems in pharmaceutical formulations, generally the goal is to obtain an optimized drug loading and release properties, long shelf-life and low toxicity. Some important features regarding the use of surfactant structures in drug delivery formulations will be mentioned next.

2.7.1 Micelles

Micellar systems are of particular interest in drug delivery [25]. Hydrophobic solutes, which are essentially insoluble in water but readily soluble in oil, can be solubilized in the hydrophobic core of micelles. Such solubilization of hydrophobic compounds is, obviously, achieved when the surfactant concentration is above CMC. Drugs can be physically entrapped in the core of surfactant micelles at concentrations that can exceed their intrinsic water-solubility. Moreover, since the hydrophilic part of the surfactant can interact favorably with the aqueous surroundings, a tight shell around the micellar core is formed. As a consequence, the contents of the hydrophobic core are effectively protected against any adverse effect from the external environment, such as hydrolysis or enzymatic degradation. Thus, the use of micellar systems can effectively reduce the hydrolytic breakdown and degradation rate of a certain compound. This is important because several molecules used for therapeutic ends are moderately stable in aqueous environment. Esters and anhydrides are compounds particularly sensitive to aqueous medium which undergo hydrolysis at high water content, high or low pH and moderately high temperatures [4]. Such hydrolytic degradation may be problematic regarding the storage stability of the formulation. In addition, the degradation products may not be well tolerated or be even toxic.

Another important feature is related to the control of the release rate. The release rate of a drug in a micellar solution depends on its state of solubilization and, hence, on the properties of both the drug and the surfactant system. Since solubilization is nothing more than the partitioning of the drug between the aqueous phase and the micellar microcontainers, both the equilibrium concentration in the aqueous surrounding and the release rate of the solubilized drug is reduced on solubilization. Thus, solubilization may offer an opportunity to control the release rate of certain drugs. In addition, a sustained release over a prolonged time can be obtained. This is particular important in situations where, for instance, the compound is rapidly metabolized.

2.7.2 Liquid Crystalline Phases

Comparable advantages to those obtained using micellar solutions are achieved when solubilizing drugs in liquid crystalline structures [4, 26]. In this case, the release rate is strongly dependent on the localization of the compound in the surfactant self-assembled structure. If the drug contains hydrophobic modifications, an increased partitioning to the hydrophobic domain is observed. This is the case when liquid crystalline phases curve towards the oil (e.g., discrete cubic or hexagonal phases). In principle, liquid crystalline structures are more versatile than micellar solutions since they are able to incorporate rather large amounts of molecules spanning from very hydrophilic to very hydrophobic, and from very small to very large. Apart from the normal hexagonal, reverse hexagonal and lamellar phases, cubic phases are also quite interesting for drug delivery. Liquid crystalline cubic systems can consist of either discrete micelles or reverse micelles close-packed in a cubic symmetry or of a bicontinuous structure. Depending on the structure, the mean curvature can vary considerably and, hence, cubic structures can be found over an entire surfactant concentration range, as shown in Fig. 4. Bicontinuous cubic phases offer the possibility of solubilization of large amounts of both hydrophilic and hydrophobic drugs. In addition, due to their normal high stiffness, these structures allow, for instance, to locate the drug at a desired site. Cubic structures have been shown to deliver small molecule drugs and large proteins by oral and parenteral routes [27]. An interesting feature is the fact that incorporation of a drug in a cubic phase can eventually cause a phase transformation to lamellar or to an inverted hexagonal phase depending on the polarity and concentration of the drug, which can also be used as extra parameter to tune the delivery. The drawback of these systems is very often related to shorter release duration and the extremely high viscosity which may limit their use in certain applications.

2.7.3 Vesicles

Vesicles (sometimes referred to as liposomes when vesicles are made of lipid bilayers) have attracted considerable attention due to their capacity to solubilize oil-soluble substances and to encapsulate water-soluble drugs. Different preparation techniques yield different vesicle type and size, and depending on the requirements or the administration route, a particular preparation method may therefore be necessary or preferred [4]. Once prepared, there are several important properties to be considered. In particular, both the stability of vesicles towards aggregation or fusion, and the leakage rate of solubilized or encapsulated drugs are of major importance.

Channel proteins can be incorporated without loss of their activity within the hydrophobic domain of vesicle membranes acting as size-selective filters, allowing only passive diffusion of small solutes such as ions, nutrients or antibiotics. Thus, drugs that are encapsulated in a nanocage-functionalized with channel proteins are effectively protected from premature degradation by proteolytic enzymes. Multilamellar vesicles, MLVs, are multi-layered liposome-like structures which share similar properties with the planar lamellar state (e.g., interlamellar spacing). Among other advantages, MLVs, show very high encapsulation efficiency for water-soluble and poorly water-soluble molecules [28]. Furthermore, the encapsulation process is devoid of organic solvents and might be less severe when compared to the process for encapsulation into classical liposomes. MLVs have also been used to specifically encapsulate enzymes, small oligonucleotides, and anticancer agents [29]. Since MLVs are only kinetically stable, they can revert to the equilibrium planar lamellar structure with time, working as a multi-shield time-release liposome. Accordingly, this time dependent stability features offer interesting opportunities in drug release.

2.7.4 Microemulsions

As explained above, microemulsions are systems consisting of water, oil and surfactant, which constitute a single optically isotropic, low viscous and thermodynamically stable liquid solution. The stability allows self-emulsification of the system whose properties are not dependent on the preparation process [5]. Microemulsions are easy to prepare and some of them are very versatile for drug delivery since they offer the capacity to solubilize both water-soluble and oil-soluble compounds, frequently in high amounts [30]. Obviously, this is due to the existence of microdomains of different polarity within the same single-phase solution. One can find several administration-specific advantages for these systems.

For instance, the small size of microemulsions droplets, below 100 nm, yields a very large interfacial area, from which the drug can quickly be released into the external medium when absorption (in vitro or in vivo) takes place, maintaining the concentration in the external medium close to initial levels. Microemulsions can improve the efficacy of a drug, allowing the reduction of the initial dose and, consequently, minimizing eventual side effects. The formation of microemulsions is a reversible process, thus these objects may become unstable at low or high temperature, reforming again when the temperature returns to the stability range. This property might be advantageous in some specific situations.

3 Polymers

3.1 Introduction

Polymers are assumed to be giant molecules; they can be more than 1 m long, although most of the investigated polymers are considerable smaller than this. Polymers are common plastics, synthetic fibbers, as well as cellulose, DNA and proteins. They are everywhere and can show different responses. Polymers can adapt to different stimuli. Often, in nature, this stimuli-response is behind many of bioprocesses.

New polymers are being designed to be applied in different fields, from molecular imprinting, artificial tissues, smart clothes, tumor treatment to controlled drug release. Many chemists, physicists, biologists, among others are dedicated to the understanding of different scientific challenges that are behind the design of new efficient materials for specific purposes. In addition, the understanding of polymer features helps to explain many phenomena in nature.

Many applications require natural occurring polymers, which may be chemically modified to reach a specific function. Chemical and physical versatility, and the potential for a broad range of applications, are good reasons to continue the research on the polymer field.

Natural polymers often possess good biocompatibility, making them popular choices for many bio-applications such as tissue engineering scaffolding. There is a lot of research on the drug delivery systems using polymers but there is still a lot to explore and to improve [31].

3.2 Polymer Properties

Polymers are large molecules composed of small chemical repeating units (monomers) covalently bound together. The number of reaping units in a polymer chain is called the degree of polymerization, DP. Synthetic polymers are synthesized by the polymerization of the monomers in any conceivable pattern.

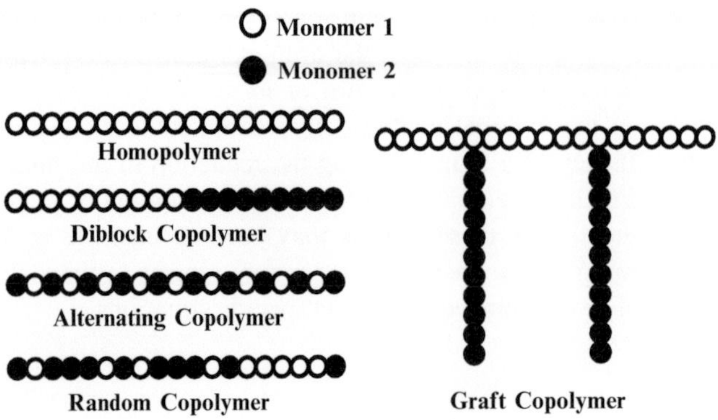

Fig. 9 Different polymers and their monomer organization

Polymers can show a variety of forms. They can be linear, branched or cross-linked and may be composed by monomers of the same type, and consequently, the polymer is referred to as homopolymer, or of different types, and, thus, designated as heteropolymer or copolymer. The latter type of polymers can be further classified by the arrangement of the different monomers along the chain. Therefore, copolymers can be random (randomly distribution of different types of chemical units), block (long segments of the same chemical unit) or graft copolymers (a chain of a type of chemical unit is grafted onto a linear chain composed by another type of chemical unit). Figure 9 shows schematically the different types of monomer organization.

Polymers may also be classified with respect to the charge. When the monomers carry charges the polymer is referred to as a polyelectrolyte. When the monomers are uncharged, the polymer is non-ionic. Figure 10 shows some examples of polymers.

The molecular weight of a polymer is a very important parameter in certain applications. Some physical properties are sensitive to the molecular weight. With the exception of proteins it is virtually impossible to find a polymer batch where all the polymer molecules have the same molecular weight. Since polymers are polydisperse, they are better characterized by the molar mass distribution and the associated molar mass averages, rather than by a unique molar mass. Conceptually, the number average molar mass is defined as

$$M_n = \frac{\sum N_i M_i}{\sum N_i} \tag{7}$$

where N_i is the number of molecules with molecular weight M_i.

Another useful definition of molecular weight average is the weight average molar mass, M_w,

$$M_w = \frac{\sum w_i M_i}{\sum w_i} \tag{8}$$

The ratio of the M_w to the M_n is a measure of the polydispersity of the sample.

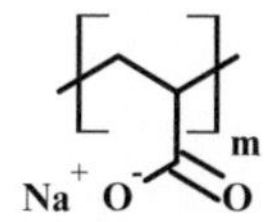

Sodium Polyacrylate (Polyelectrolyte)

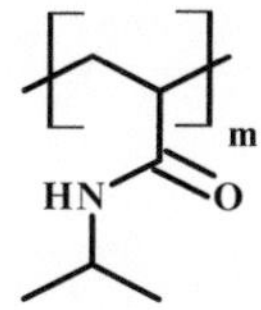

Poly-*N*-Isopropylacrylamide (Non-ionic)

Polyethylene Glycol PEG (Non-ionic)

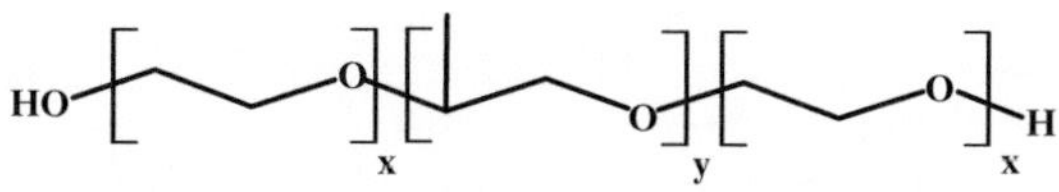

Poloxamer (Pluronic®) (Non-ionic)

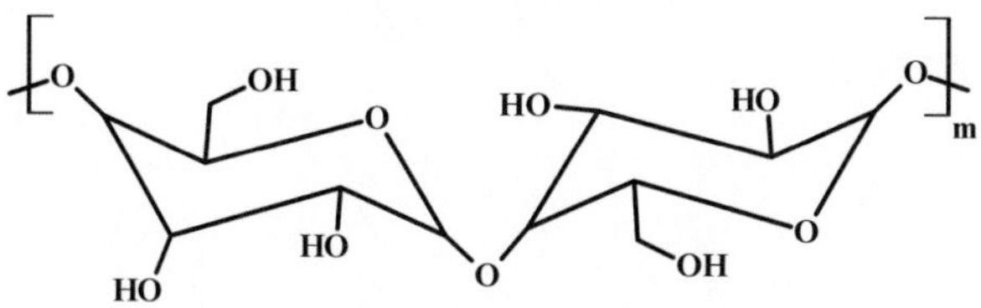

Cellulose (Natural Polymer)

Fig. 10 Examples of different polymer molecules

3.3 *Polymers in Solution*

Water-soluble uncharged polymers dissolve in water due to the gain in conformational entropy of the individual flexible polymer chains when these are diluted by the solvent.

Polymers can present an expanded or compacted conformation; they may be complexed with other substances or be used independently; they may bind active substances and/or release them. Depending on the polymer structure, solution and surface, the properties may differ considerably. To understand these different behaviors, a few considerations about polymers in aqueous solutions will be made. Some of the features are quite general and will be considered independently of the type of polymer. However, some other properties are specific of charged or amphiphilic polymers, and will be analyzed individually.

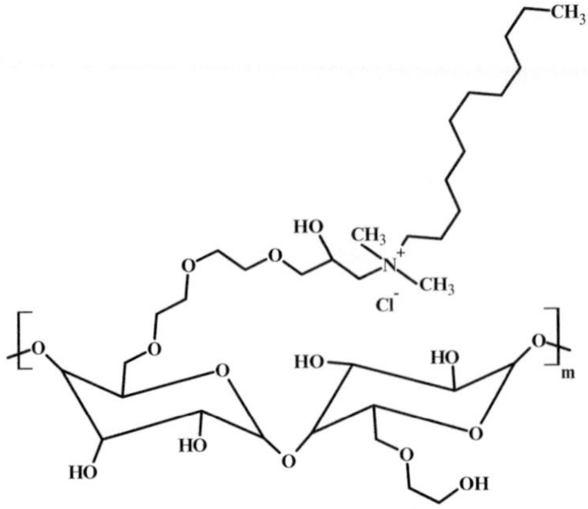

Derivative of Cellulose (Cationic Hydrophobically Modified Polymer - HMP)

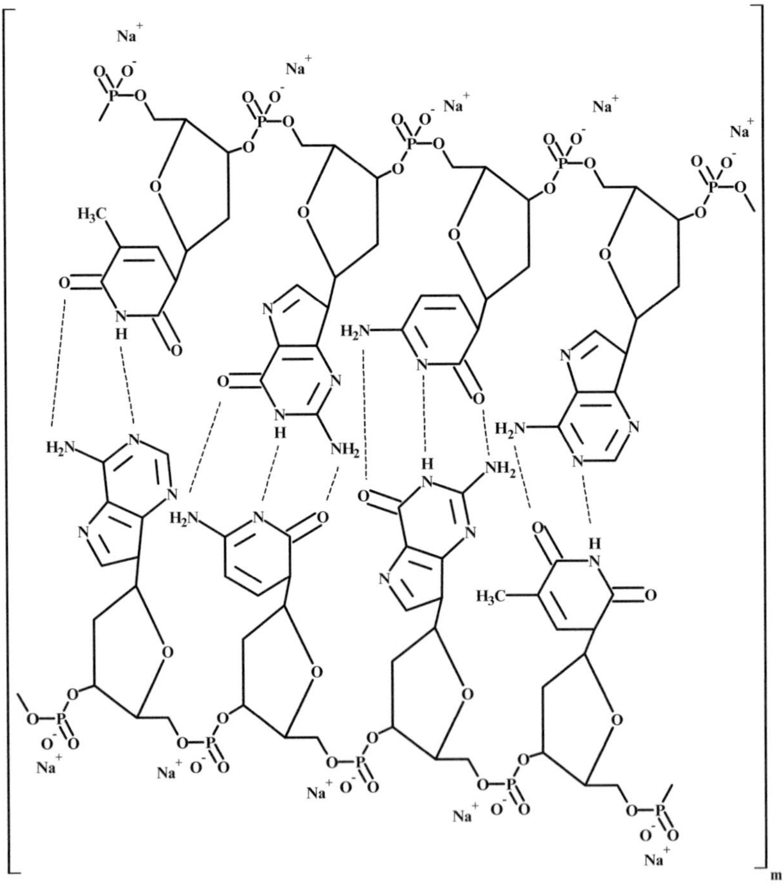

DNA (Natural Polymer)

Fig. 10 (continued)

Polymer coils can adopt different configurations in solution. The general interest is usually on the extension of the polymer chain and its scaling with the change of internal or external parameters. The polymer coil dimension is commonly expressed as the radius of gyration, R_g, which represents the polymer coil radius. R_g is proportional both to the number of polymer segments and the length of the individual segment. The simplest way to express polymer chain dimensions is through the freely jointed chain model, illustrated as:

$$R_g^2 = N\frac{l^2}{6} \tag{9}$$

where R_g is the radius of gyration, N is the number of polymer segments and l is the length of the individual segment. In this model, the interactions among segments are neglected and the polymer chain is allowed to take any bond angle, independently of the positions of the neighbouring segments. This is, however, not seen in real chains, where chain rigidity is present. Flory included this additional parameter for a better representation of real polymer chains: the characteristic ratio, C_∞, seen in Eq. 10 [32]:

$$R_g^2 = C_\infty N\frac{l^2}{6} \tag{10}$$

For flexible polymers, C_∞ is in the range 5–10 [33] while for semiflexible polymers, as the cellulose based, it reaches much higher values [34]. The rigidity of a polymer backbone influences the rheological properties of a polymer solution. The intrinsic viscosity, which is a measure of the capability of a polymer in solution to enhance the viscosity of the solution, is related to the size and molecular weight of a polymer [5, 35]. Flory also corrected the reported model by taking into account the excluded volume, which considers the impossibility of chain segments to overlap. When this effect is present, the number of segments, N, has an exponent of 1.2.

The size of a polymer chain is also strongly affected by the interaction among different molecules in solution. These intermolecular interactions are represented in the Flory-Huggins theory [32, 36, 37]. Here, the interactions between two molecules are quantified by the dimensionless interaction parameter, χ. For $\chi < 0.5$ the solvent is considered to be good and the polymer chain will be extended due to the dominating configurational entropy. It attains a stretched conformation as a rod. When $\chi > 0.5$, the solvent is bad and the monomer-monomer interactions dominate leading to a contracted polymer coil (globule). For $\chi = 0.5$, the steric interactions and the solvent repulsion balance, leading to the formation of a freely jointed chain; this is called the theta condition.

Polymer chains can self-associate in solution. If one chain is composed by different blocks, the general behavior may be extremely rich. When both units of a diblock copolymer are in a good solvent, the copolymer behaves like a homopolymer; however, when placed in a solvent which is good for one of the blocks and bad for the other, there will be segregation between the different blocks and the formation of self-assembled structures, similar to that found in surfactant systems [38–41].

The solubility of some of the polymer blocks can be tuned by temperature. Some non-ionic polymers show a remarkable temperature dependent solubility in water as well as in some other solvents. This behavior is shared with some other substances like cellulose derivatives, poly-N-isopropylacrylamide, poly(ethylene glycol) and poloxamers, commercially known as Pluronics® [16, 42–45]. The temperature effects are associated with conformational changes upon temperature change, as explained previously for the non-ionic surfactants.

3.4 Concentration Regimes

Polymer solutions belong to the dilute, semi-dilute or concentrated regime. For highly soluble polymers, in the dilute regime, the interactions between different polymer molecules are insignificant since the distance between individual polymer coils is considerably longer than the radius of a polymer coil. Thus, the polymer chains move independently from each other. The semi-dilute regime starts at the overlap concentration, c^*, i.e. the concentration where coils start to make entanglements with each other due to the proximity of the polymer coils. Typically the c^* is 0.1–10 wt.% depending on molecular weight and polymer configuration. R_g is larger than the average distance between polymeric chains and they are forced to overlap, resulting in a transient polymeric network, i.e. a network whose structure is based on transient junction points or cross-links arising from interactions between polymer chains; in such networks, as described in Green-Tobolsky model [46], when the old junctions break, new ones are established. Entanglements are topological rather than due to intermolecular forces and the easiest escape for sufficiently long chains will be achieved by sliding along the contorted contour (reptation) [47].

The concentrated regime is reached well above the c^* and the molecules are strongly entangled, behaving more like a melt than like a network.

3.5 Polyelectrolytes

When some or all monomers are charged, the polymer is called polyion. Addition of charges to the polymer chain implies the presence of small mobile counterions to ensure the electroneutrality. The combination of polyion plus its counterions is called polyelectrolyte. Therefore, when dissolved in water, a polyelectrolyte dissociates into a polyion and its counterions. The entropy of mixing increases with the counterion dissociation, since the number of particles increases from 1 to $n+1$ (being n the number of counterions). Thus, when compared with uncharged polymers, polyelectrolytes have higher solubility in water [48]. The distribution of the counterions is enhanced close to the polyion and levels off with increasing the distance from the chain.

The charge distribution along the chain increases R_g and l_p (and thus the rigidity) due to the repulsion between the charges and, for that reason, the polymer adopts a more extended conformation and can overlap at lower concentrations than the uncharged analogue. However, the increase in rigidity may be eliminated with addition of salt. Polyelectrolytes are very sensitive to salt because it screens the electrostatic repulsion among and along the polymer chains. Thus, miscibility can be decreased by the presence of salt since it eliminates the benefit of counterion entropy associated with miscibility. Some polyelectrolytes, as the cellulose derivative presented in Fig. 10, belong to the category of intrinsically insoluble polyelectrolytes because the origin of its solubility is the presence of the charges (these polymers do not dissolve if their charges are screened by the addition of salt).

In non-ionic polymer systems, the presence of salt may either increase or decrease the solubility depending on the type of salt [5]. Iodide and thiocyanate ions increase the solubility of non-ionic polymers while chloride or sulphate decreases the solubility. In some cases the presence of charges is induced by pH; pH can be regulated to obtain a strong polymer network and consequent drug entrapment, or to make the polymer chains collapse and release the drug. This has a tremendous effect when using polymers in drug delivery systems.

3.6 *Amphiphilic Polymers*

When polymers display a small number of hydrophobic side chains, or terminal groups, chemically attached to the backbone, they are called hydrophobically modified polymers (HMPs) [5]. While hydrophilic segments are responsible for the hydration and swelling, the hydrophobic groups tend to minimize the water contact, since their contact with water is energetically unfavorable. HMPs may associate hydrophobically with other amphiphilic or hydrophobic entities, forming aggregates. HMPs can form three-dimensional networks at polymer concentrations even below c^*, triggered by the hydrophobic association among different polymer chains, which largely enhances the viscosity. For this reason, HMPs typically behave as viscosity enhancers in a wide range of applications. The knowledge on the most efficient location, amount and composition of hydrophobes along the polymer chains is a crucial challenge and it has been continuously requested by, for instance, cosmetic and health care companies.

The amphiphilic properties of these polymers should increase compatibility with dispersions by adsorbing onto the surfaces of colloidal particles, at the same time as they give the possibility to form physically transient networks via the hydrophobic moieties of the HMP chains [49].

HMPs may be classified depending on the position of the hydrophobic substitution. They can have grafted modifications (where the hydrophobe is attached along the polymer backbone) or end modifications (hydrophobe at the chain ends) both exhibit intra- and intermolecular association. End-modified block copolymers

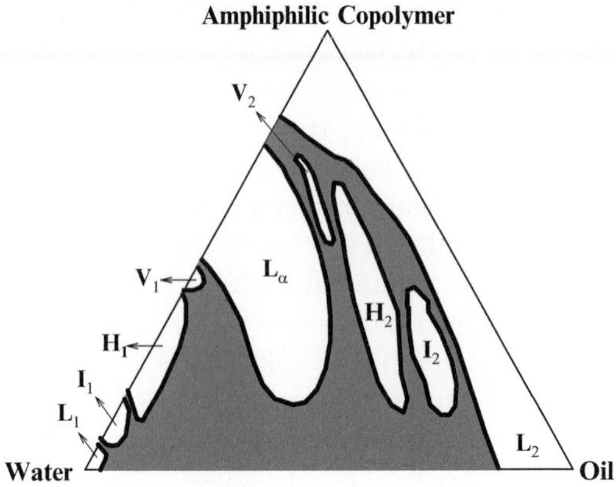

Fig. 11 Phase diagram of a ternary system consisting of an EO-PO-EO copolymer, p-xylene and water. L_1 – normal micellar; I_1 – micellar cubic; L_α – lamellar; L_2 – reverse micellar; L_3 – sponge; H_1 – hexagonal; H_2 – reverse hexagonal; V_1 – bicontinuous; and V_2 – reverse bicontinuous. The *grey* area represents multi-phase samples (reproduced from ref. [5])

have hydrophobic groups on each end. They inter-associate, drawing flower-like micelles in solution, with the hydrophobic groups inserted into the same micelle. At sufficiently high concentrations, the short inter-micelle distance favors polymer bridging between two micelles. The bridges provide connectivity in the system, which gives rise to an attraction between micelles [50, 51] and to strong increase in viscosity.

Poloxamers are also amphiphilic macromolecules. They are triblock copolymers, usually composed by a central hydrophobic chain of polypropyleneoxide and two hydrophilic chains of polyethyleneoxide, as shown in Fig. 10. The amphiphilic nature of these polymers gives rise to complex aggregation, depending on composition, temperature, etc. Each aggregated structure has specific delivery and encapsulation properties. Figure 11 shows the richness of structures of an amphiphilic block copolymer system. Amphiphilic polymers are thus extensively requested as nanoencapsulators in drug delivery systems [52].

3.7 Polymer Microgels

Cross-linked polymers, or microgels, are composed of polymer chemical networks, which have attracted much attention in fundamental studies of soft matter [53, 54] and in applied fields [55, 56]. In particular, they have rapidly gained importance in materials science owing to their applications in drug delivery [57–59] and sensing [60, 61].

The swelling of cross-linked polyelectrolytes can be extremely strong; some microgels may swell in water up to 1,000 times their dry volume [62]. The swelling of microgels has a remarkable effect on thickening, clearly much more pronounced than for linear polymers.

Table 1 Different stimuli used to activate drug release

Stimulus	Mechanism
pH/ionic strength	pH change causes swelling/deswelling and drug release Change in ion concentration inside the gel causes swelling/deswelling and drug release
Chemical	Formation of charge-transfer complex causes swelling/deswelling and drug release
Enzyme-substrate	Product of enzymatic conversion causes swelling/deswelling followed by drug release
Thermal	Change in polymer-polymer and polymer-water interactions cause swelling/deswelling and drug release
Electrical	Change in charge distribution causes swelling/deswelling and drug release
Ultrasound irradiation	Temperature change causes drug release

3.8 *Polymers as Drug Delivery Systems*

Living systems adapt to changes in the environmental conditions. Monomers may be designed to show changes in their behavior upon specific conditions of pH, temperature, ionic strength of the surrounding medium and quality of solvent [53, 63–66]. With a fair knowledge on polymer physical chemistry, it is possible to design versatile materials in the form of shrunk globules or swollen chains. Polyacrylate derivatives that work as efficient thickeners at pH above 7 may behave as liquids at lower pH, to be easily handled. The most commonly used stimuli to activate the release of a drug from the encapsulator are shown in Table 1 (adapted from reference [67]).

The temperature or pH at which the polymer behavior changes can be tuned either by monomer modification, addition of co-solutes (surfactants, second polymer or salt), addition of a second solvent with different polarity, change in concentration or mixture with a second "smart" polymer (synergistic effects were reported elsewhere [68]).

"Smart" polymers can be constituted by different "smart" blocks, covalently linked together. This type of multi-responsive systems used for multi-drug release at various specific conditions or for diagnosis, is currently attracting researchers. These systems can be simultaneously sensitive to temperature, pH, UV radiation and magnetic fields.

4 Polymer-Polymer and Polymer-Surfactant Mixtures

4.1 Introduction

The wide number of studies performed in polymer-polymer and polymer-surfactant aqueous mixtures is a consequence of their importance in many different fields from fundamental science to industrial applications. Mixtures of polymers or polymers and surfactants can be found in the human body, in cleaning products, in constructions materials, in paper, in food, in pharmaceutical products, etc. [69–74]. Early studies on polymer-surfactant systems were connected to the interest in understanding of the interactions between surfactant and proteins or other biological polyelectrolytes, with the aim of comprehending biological processes [75, 76]. In general, the mixtures can be made with several distinct aims. One purpose may be to introduce high viscosity through the polymer and/or to introduce the capacity of solubilizing hydrophobic molecules through the surfactant aggregates. In some other cases, the addition of surfactant to a polymer may intend to increase the polymer solubility [77]. In addition, the mixtures may form very stable nanoparticles [78–81].

Under some particular conditions, some polymers or polymers and surfactants attract each other and form complexes. These complexes can be soluble (homogeneous solution) or insoluble (separating out from the solution), depending on different features such as the chemistry of the molecules, the composition of the mixture and the ionic strength. The insoluble complexes can separate out from the solvent, forming a concentrated phase, which is in equilibrium with a dilute phase [5, 48, 82].

On the other hand, some other compound combinations may result in two separated phases each one enriched of in one of the components.

The type of phase behavior can be adjusted by changing the composition, salt content, etc. Each type may find use in different states of an application [5]. In addition, using mixtures of polymers or polymers and surfactants, different effects can be achieved such as colloidal stability, flocculation, emulsification, structuring or rheological control. In particular, polymers and surfactants are widely used as rheology modifiers or gelling agents and their mixtures can exhibit interesting (sometimes non-monotonic) rheological behavior, different from the individual polymer or surfactant solutions [70, 83–90].

Due to the extensive use of mixtures of polymers and surfactants in drug delivery formulations, it is very important to control and to understand the phase behavior of such mixtures.

4.2 Phase Behavior

The tendency for mixtures of polymers and surfactant self-assembled aggregates to separate into two-phases is much stronger than for mixtures with smaller molecules

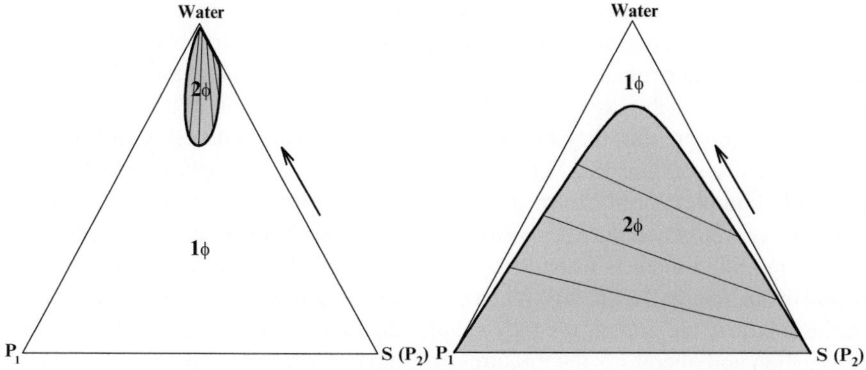

Fig. 12 Schematic phase diagrams for associative (*left*) and segregative (*right*) phase behavior. 1ϕ and 2ϕ stand for one- and two-phase, respectively

due to the lower entropy of mixing of the larger molecules. Thus, phase separation is very common for both polymer and surfactant mixtures. But the behavior of the mixtures is also dictated by the interactions among the macromolecules and between those and the solvent [91]. Two typical phase behaviors may be distinguished considering the distribution of the macromolecules: the associative and the segregative. In the former case, the macromolecules associate into a common concentrated phase in equilibrium with a dilute phase and, in the latter, the macromolecules segregate into different phases of similar total concentration. The two types of behavior were first observed for mixed polymer solutions, but, nevertheless, this is also observed for polymer-surfactant systems [5, 48]. Figure 12 shows schematic phase diagrams for both cases.

In the associative phase separation, the 2ϕ region is anchored in the water corner of the ternary phase diagram and has the shape of a drop, which can be more or less symmetric in relation to the bisector of the water corner. The tie-lines, connecting the equilibrium compositions of the co-existing phases in the 2ϕ region, run from the water corner to the opposite side of the loop, showing that one phase is very dilute and the other is mostly constituted of the complexed molecules. Outside the 2ϕ drop one finds a 1ϕ region, meaning that the mixture of the two associating components may lead to the formation of soluble or insoluble complexes depending on the composition or ratio between the mixed components [5, 77]. This association is quite general for oppositely charged mixtures, but can also be found for non-ionic mixtures [92–100]. The size of the phase separation region increases with polymer molecular weight, micellar size, surfactant alkyl chain length, and polyelectrolyte charge density, while it decreases, for instance, with addition of salt which screens the electrostatic interactions. For certain non-ionic systems, the extension of the phase separation region can also be strongly affected by temperature. Moreover, an associative phase separation is favored by a decreased hydrophilicity of one or both of the solutes.

The oppositely charged polyions interact mainly by electrostatic interactions but if, for instance, the polyions contain hydrophobic moieties, hydrophobic interactions might play a major role and can modulate the thermodynamic phase behavior that may be reflected in the extension of the phase separation region. When hydrophobic interactions are present, they are normally responsible for a decrease in extension of the phase separation [85]. In some cases, due to the hydrophobic interactions, insoluble complexes only form in a narrow composition region (close to charge equivalence), largely reducing phase separation. When both polyions are hydrophobically modified there is a tendency of the hydrophobic groups to associate and reside in one phase [84, 85, 90]. When one of the polyions is in excess (charge-wise) in the mixture, the complexes may acquire an excess charge due to hydrophobic association and, therefore, the mixtures go from two-phase to one-phase.

Segregative phase separation happens when there is no net attraction (hydrophobic or electrostatic) between the macromolecules. The extent of the phase separation is expected to depend on the short-range interactions between the various components of the system, including the solvent, and on the entropy of mixing. Differences in short range interactions generally favor phase separation and the entropy of mixing favors miscibility [77]. As explained earlier in this chapter, the entropy of mixing depends largely on the number of molecules and aggregates present in the mixture at a certain composition. For non-ionic mixtures, this is given by the number of the polymer molecules and the aggregation number of the surfactant aggregates. For charged macromolecules, numerous small ions, the counterions, are present and the entropy of mixing becomes larger when compared to non-ionic.

The segregative behavior may be strongly affected by the introduction of small fractions of ionic moieties; the miscibility increases owing to the fact that confinement of the counterions in one phase is less favorable [77]. This phase behavior can be useful when, in a mixture or formulation, certain components should not associate but kept separated (for instance, to prevent denaturation or precipitation of proteins in a mixture of those with surfactants).

For some systems, at very high salt concentration, the behavior can change from associative to segregative.

4.3 Asymmetry

It is indeed possible to observe that oppositely polyion-polyion (surfactant ion aggregates act as polyions) complexes do not separate out of the solution only at charge equivalence; the mixtures may show a pronounced asymmetry in relation to the charge equivalence. This asymmetry has its origin in some difference between the polyions other than the sign of the charge (a mixture of two polyions differing only in the sign of their charge must necessarily give rise to a symmetric phase diagram) [91]. For instance, a well-established cause of asymmetry in phase behavior in these systems is a pronounced difference in the number of charges per polyion [101]. Also, hydrophobic association can be the cause of asymmetry in the phase separation region in relation to the charge equivalence as can be inferred for the abovementioned discussion [84, 85, 90].

4.4 CAC and Redissolution Phenomena

Uncharged slightly hydrophobic polymers and charged polymers display important similarities in their interaction with ionic surfactants [77]. In both cases, binding of the surfactant to the polymer chain can be observed, resulting in the formation of micelle-like clusters of surfactant molecules adsorbed to the polymer chain. The surfactant binding normally starts at a rather well-defined surfactant concentration, the so-called critical aggregation concentration (CAC) [102], which is always lower than the CMC of the surfactant. The CAC may simply be regarded as the surfactant CMC in the polymer solution. At the CAC there is a strongly cooperative binding. To clarify, the term cooperative binding illustrates a situation where the surfactants prefer to bind in a domain occupied by another surfactant molecule; non-cooperative binding indicates that the molecules have no preference for any domain; and anti-cooperative binding indicates preference by the unoccupied domain.

The binding of the surfactant is normally characterized by its marked cooperativity due to the hydrophobic interaction between the surfactant chains (the binding is actually a micelle formation at the surface of the polymer). But there are other behaviors, for instance, in the case of HMPs the scenario is quite different because the hydrophobic groups have the capacity to associate and the individual surfactant molecules are solubilized by the polymer micelles [77].

Just like the CMC, CAC decreases with increasing surfactant chain length, but in contrast to the CMC for charged systems, CAC increases with increasing polymer concentration or salt concentration, as the entropic gain in releasing the counterions bound to the polyion decreases at higher polymer concentration or salt content. The chemical nature of the surfactant head group may also affect the polymer-surfactant interaction strength. Anionic surfactants show a marked interaction with most homopolymers, while cationic surfactants show a weaker but still significant interaction [5, 77]. Non-ionic surfactants and zwitterionic surfactants only rarely show a noteworthy interaction [5, 77]. When CAC is reached a plateau in the surface tension is observed until all the polymers become saturated with surfactant molecules. Moreover, the plateau increases linearly with polymer concentration [5].

There are different scenarios for the micellization of the surfactant interacting with the polymer chain, involving different types of binding. For ionic surfactants and hydrophilic homopolymers the micellization occurs in the vicinity of the polymer chain. If the attractive interaction between the polymer and the surfactant is strong as for the case of oppositely charged systems, there will be a pronounced cooperative binding of the surfactant molecules to the polymer chain. The micelles sizes are similar with and without the polymer and the aggregation numbers are typically similar or slightly lower than those in the absence of polymer [5].

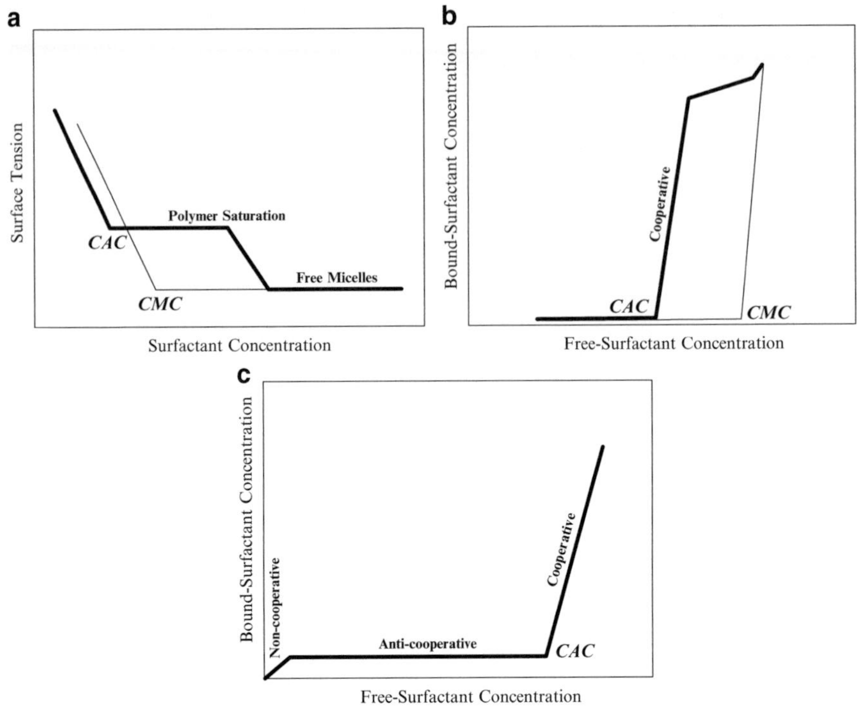

Fig. 13 Schematic trend of the surface tension as a function of the surfactant concentration in the presence and absence of a homopolymer (**a**) binding isotherm for the surfactant in the presence and absence of polymer (**b**) and binding isotherm for the association between an ionic surfactant and a hydrophobically modified polymer (**c**)

For hydrophobically modified polymers the scenario is more complex. There is a pronounced hydrophobic interaction between these polymers and the surfactant molecules. At low surfactant concentrations, there is a low amount of non-cooperative binding of individual surfactant molecules, at intermediate concentrations an anti-cooperative plateau appears and, finally, as the free surfactant concentration equals the CAC, the typically cooperative surfactant binding can be observed [5]. A schematic representation of these phenomena is presented in Fig. 13.

Polyions associate strongly with oppositely charged surfactant ions due to the strong electrostatic attraction between the two multi-charged species together with the large entropic gain (counterions release). Phase separation often occurs, resulting in a concentrated phase enriched in polyions and surfactant ion aggregates in equilibrium with a dilute phase [5, 48, 82]. Phase separation of polyion-surfactant ion complexes appear either at CAC or at concentrations slightly above the CAC, depending on the polyion concentration and the characteristics of the system [82, 103]; in other words, in some systems, it is possible to bind a considerable amount of surfactant ions to the polyion before phase separation occurs whereas in

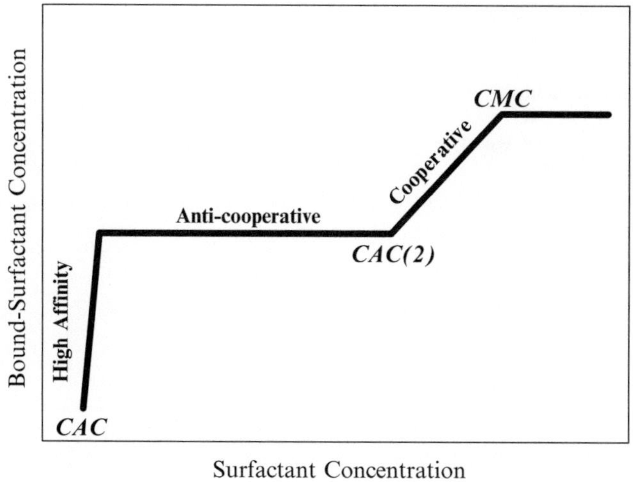

Fig. 14 Schematic trend of the binding isotherm when CAC(2) is present (adapted from Lynch et al. [107])

other mixtures phase separation is observed for very low surfactant concentrations [104–106]. On the other hand, redissolution of the concentrated phase might occur. It occurs when non-stoichiometric complexes are redissolved into a single phase. The redissolution can be accomplished by addition of excess surfactant or excess polyelectrolyte, but the mechanisms behind this phenomenon are not yet fully understood. One possible scenario is the following: when extra surfactant binds to the polyion (above the charge equivalence) the complexes become overcharged. In some systems, a second cooperative binding of surfactant ions can occur at the so-called CAC(2), which is sometimes referred to as the hydrophobic CAC, because cooperative binding happens due to a hydrophobic association of the excess surfactant ions to the neutral complex [107]. This may lead to redissolution. Figure 14 shows a hypothetic representation of CAC(2) type of binding isotherm.

In conclusion, in some systems efficient redissolution occurs; however, there are other systems where efficient redissolution is not achieved by adding excess of surfactant or polyelectrolyte. This is also valid for polymer-polymer systems.

4.5 More Complex Systems

Although some mixtures of oppositely charged systems have been represented using the Gibbs' triangle (ternary plot), as shown in Fig. 12, this is not entirely correct. When the system is composed of oppositely charged macromolecules, the most common type of associative mixtures is a 4-component system and the complete phase diagram would be described by a pyramid as exemplified in Fig. 15 [108]. For multi-component systems, the tie-lines do not generally fall in the mixing

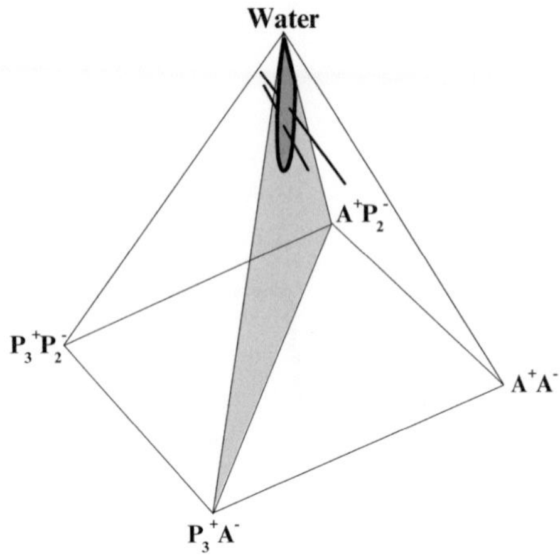

Fig. 15 4-component system represented by a pyramidal phase diagram with an associative phase separation in the $A^+P_2^-/P_3^+A^-/W$ plane. The *tie-lines* are shown as *straight lines* passing through the *grey* mixing plane

plane and the full characterization of the separated phases involves a considerable amount of work [109]. The picture is as follows: when two neutral salts, for instance two polyions of opposite charges ($P_2^-$and P_3^+) with their respective counterions (A^+ and A^-), are mixed, the dissociation of the counterions and the different alternatives to combine the charged species into net neutral (mixed) salts turns the mixture into a 4-component system. Contrary to the non-ionic systems, presented earlier, for oppositely charged systems the classical triangular representation is generally not correct when five species (four neutral salts plus solvent—water) are mixed.

The four apices of the base of the pyramid represent the four different combinations of neutral species. The water content is given by the axis perpendicular to the pyramid base. The conventional triangular representation of polyelectrolyte/surfactant/water system refers to the plane in grey when the two-phase is presented. Despite this representation being not fully correct, it is useful to understand the general features of the system.

Apart from the charged species, there are other factors that may increase the complexity of the mixtures. As discussed, the formation of liquid crystalline phases for individual surfactant or self-assembling polymer systems in water originates multi-phase systems by itself. Furthermore, micellar growth into infinite structures increases the tendency for phase separation. Another factor that clearly increases the complexity of a mixture is the inherent polydispersity [77].

4.6 *Alternative Phase Diagrams for Oppositely Charged Mixtures*

A few number of works have presented the fully characterization of the separated phases including the tie-lines in multi-component polyelectrolyte-surfactant

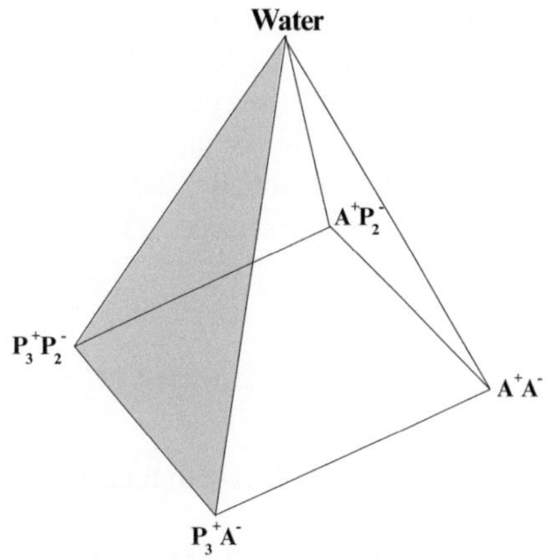

Fig. 16 The pyramid represents all the possible components of a mixture of two charged macromolecules in water. The shadowed $P_3^+P_2^-/P_3^+A^-/H_2O$ area is one of the possible alternative mixing planes

systems [109]. To simplify the study of the complex phase behavior of oppositely charged mixtures, the number of components can be reduced by synthesizing the pure complex salt which results from the complexation of the polyions or polyion and surfactant ions ($P_3^+P_2^-$), and use it as one of the starting components [110–112]. The alternative mixing plane is perpendicular to that corresponding to the conventional mixing plane. This is exemplified in Fig. 16. The complex salt can be then mixed with any other of the components ($P_3^+A^-$, $A^+P_2^-$, A^+A^-). This mixture contains three components and thus represents a truly ternary mixture.

4.7 Rheology of Polymer-Polymer and Polymer-Surfactant Systems

Polymers have a broad range of applications in what concerns to rheological control. Viscosity depends on a number of parameters, such as molecular weight, concentration, solvency conditions and conformation of polymer molecules. The latter is also dependent on the salt concentration and polymer concentration. Associating amphiphilic polymers can give rise to very interesting rheological behavior. Polymers grafted with hydrophobic groups may associate with the hydrophobic groups from other polymers or surfactants. Therefore, for mixed systems, more pronounced changes may be seen.

For associating mixtures of ionic surfactant and non-ionic polymer, the viscosity depends on the binding of the surfactant. As the surfactant binds, the initial non-ionic polymer transforms into a polyelectrolyte and the viscosity rises by orders of magnitude.

The semidilute mixtures of surfactants and hydrophobically modified polymers have been used extensively to modify the rheology of a mixture. The mixed association of the hydrophobes brings physical cross-linking to the complexes and a marked rise in viscosity. However, as the surfactant concentration increases the cross-linking may be lost due to the formation of a 1:1 stoichiometry (i.e. one surfactant micelle per polymer hydrophobe) and viscosity decreases. This non-monotonic viscosity behavior is a very useful phenomenon in many applications [71, 83–88]. Additionally, for non-ionic systems, viscosity can be also tuned using temperature changes, as previously alluded.

4.8 As Drug Delivery Systems

4.8.1 Polyelectrolyte-Surfactant Nanoparticles

In general, the structure of polyelectrolyte-oppositely charged surfactant complexes is dictated by the structure that the surfactant forms in the absence of polyelectrolyte, if the polyelectrolyte is sufficiently flexible to adapt. All the advantages of the self-assemblies formed by the surfactant molecules are present in the mixtures of polymers and surfactants. Many systems, including micelles and (micro)emulsions are very sensitive to dilution. Therefore, the polyelectrolyte-surfactant systems provide a viable alternative to these systems when rupture of the structure due to dilution is undesired.

In dilute solutions, discrete and compact polymer-surfactant nanoparticles form at various charge ratios, with long-range order internal structures [80, 81, 113]. These nanoparticles are regarded as potential drug and gene delivery vehicles. They may be used as carriers of nucleic acids and other biological components into living cells for therapeutic purposes. For instance, the non-viral vectors, formed by DNA and cationic species, may be used to treat cancer and genetic disorders [114, 115].

4.8.2 Polymer Gel-Surfactant Mixtures

Polyelectrolyte gels are swollen due to the osmotic pressure arising from the confinement of the counterions inside the network. When a cross-linked polyelectrolyte gel is exposed to a solution of oppositely charged surfactant, the surfactant molecules will diffuse into the gel, neutralizing the polyion, enabling the release of the counterions as in the case of linear polyions. As the polyion gel is neutralized, it collapses. The concentration of the counterions inside the gel decreases considerably and consequently the internal osmotic pressure decreases as well. The degree of collapsing may be tuned by the surfactant concentration. The collapse can start at the surface of the gel, forming a dense micelle-rich phase ("skin") around the swollen core and as more surfactant is added to the gel, the surface phase grows at the expense of the core until, eventually, the gel fully collapses [116–120].

4.8.3 Protein-Surfactant Mixtures

Drug delivery technology holds several exciting challenges, among which the delivery of intact protein based drugs. Proteins are very susceptible to denaturation under unsuitable pH conditions or metabolizing enzymes. Hence, suitable ways of protecting proteins against denaturation, until the target site has been reached, are necessary.

Reports on interactions between proteins and surfactants comprise both phase separation and redissolution, including formation of physical gels [5, 77]. Systems containing proteins have an important advantage from the beginning; proteins are monodisperse. The differences found between the systems containing proteins and linear polymers are probably related to the exotic nature of the proteins. Proteins are amphiphilic polymers with variable charge density. Additionally, proteins differ very much from one to the other and general considerations are difficult to be drowned. Proteins can be globular or fibrous and the different structures strongly influence the phase behavior. Therefore, studies comprising mixtures of proteins with surfactants or other polymers are limited and scarce general models have been constructed [5, 77].

Proteins and surfactants interact not only because they may have opposite charges but also because they are both amphiphilic. In particular, there are some similarities between protein-surfactant systems and HMP-surfactant systems [5].

Surfactant binding to the protein may lead to irreversible denaturation of the protein and, consequently, the protein may be found in different conformational states in the mixtures. Protein-surfactant complexes involving denatured proteins find some analogies to complexes of surfactants and partially hydrophobic and flexible polymers [5].

4.8.4 Polymer Gels in Protein Delivery

Proteins can be delivered using gels. One possible way to achieve this is to use responsive gels where the drug is loaded through diffusion in the swollen state of the gel and trapped inside it on the collapsed state. Thereafter, the gels are triggered by an external stimuli and the drug released. Prototypes for gel vehicles with the ability to load and release proteins have been designed [121–124]. The interactions between gels and proteins are very complex and simpler models are usually useful to understand general interactions and behaviors. In this regard, the study of the interactions between gels and surfactants is very useful.

4.8.5 DNA-Surfactant Complexes

Many aspects on drug delivery are common to the gene delivery. For gene delivery, DNA, which is a large biopolymer, needs to be compacted in the cells by other biological solutes. Compaction of DNA decreases its radius of gyration and offers protection from nuclease degradation. Positively charged proteins have the capacity to compact DNA in the nucleus of eukaryotic cells. Cationic lipids (natural surfactants) are also widely used to compact DNA. Moreover, the compaction needs to be reversible for transfection and gene expression. The control over the molecular mechanisms influencing the formation of complexes between DNA and lipids or between DNA and other biopolymers, which are very much similar to the synthetic polymers and surfactants, is crucial to obtain the desired formulation.

The strong binding of surfactants to DNA allows compaction and reduction of charges. However, the complexes of DNA and cationic surfactants do not show effective transfection. This is attributed partially to the cytotoxicity of the cationic surfactants and to the low stability of the complexes when the environment is changed [125]. Yet, cationic surfactants can be used in low amounts to charge the complexes of DNA with liposomes and, hence, increase the transfection efficiency [115]. The redissolution or decompaction of the complexes is necessary for DNA to be exposed to the enzymatic machinery in the cell. This can be achieved by the addition of anionic [126] and non-ionic surfactants [127] or even cyclodextrins [128].

DNA chemical gels have been prepared by crosslinking double stranded DNA with ethylene glycol diglycidyl ether (EGDE). The deswelling behavior of the DNA gels has been studied in the presence of different solutes. The deswelling of the gels is reversible when an oppositely charged surfactant is added to the collapsed gel [129].

4.8.6 DNA-Cationic Polymers

Cationic polymers exhibit an extreme efficiency in compacting DNA which may result in DNA-polycation complexes even in very dilute solutions. Phase diagrams of mixtures of DNA and cationic polymers, showing phase separation and redissolution regions, have been studied. The concentration of polycations to induce DNA compaction or phase separation depends on the polycation size and charge density. As expected, the presence of salt may also strongly alter the phase behavior. In general, the maximum precipitation is reached near the point of charge equivalence and typically insoluble complexes are formed already at very low polycation concentration. Often, increasing the concentration of polycation does not lead to redissolution of the complexes [130]. For long DNA molecules, the binding is strongly dependent on salt content and the extent of phase separation decreases with increase of salt concentration until phase separation is suppressed [131]. The size of DNA has also an influence on the phase behavior. While for long DNA molecules, phase separation may occur progressively starting at charge ratios below the charge equivalence, for short DNA molecules the phase separation region may be very narrow and near the charge equivalence [132]. The interaction between DNA and cationic polymers in semidilute regimes have also been studied and redissolution of the complexes was observed with formation of physical DNA-polycation gels [133].

As mentioned earlier for conventional polymer-surfactant systems, nanoparticles also form in DNA-surfactant or -polycation systems. There is a strong interest in controlling the size and shape of the nanoparticles for the development of efficient artificial non-viral gene delivery systems. Single DNA molecules collapse into toroids, rod-like shapes or globules. These morphologies are also found in the complexes with the cationic species.

4.9 Cytotoxicity

Cytotoxicity is an inherent property of many polycations. These cationic polycations can induce membrane damage and initiate apoptosis. In the case of polycations the cytotoxicity could be reduced by lowering the polycation molecular weight or branching [134]. Cytotoxicity can be further reduced by using appropriate type and/or number of side chains [135, 136].

5 Drug Delivery and Cells

5.1 Introduction

The entire concept of having a drug delivery system which allows the body to use the administered drug in the appropriate dosage and site is continuously spreading interest in the scientific community [137–139]. Typically, the drug delivery vehicle is introduced (the process of administration) into the body far from the site of action. The carrier must then move into the bloodstream (the process of absorption) and be transported to the target sites where the drug is needed (the process of distribution). Afterwards, some drugs are chemically altered (the process of metabolism) by the body before they perform their action; others are metabolized afterwards; and others are not metabolized at all. The final step is the removal of the drug and its metabolites from the body (the process of elimination). The main goals for the development of bio-nanotechnologies in drug delivery can be summarized as: specificity of drug targeting and delivery, reduction of toxicity while maintaining the therapeutic activity, biocompatibility and safety.

Currently, many substances are under investigation ranging from the more established systems, e.g. polymeric nanoparticles, surfactant based nanoparticles, solid lipid nanoparticles, liposomes, to the more cutting-edge systems, e.g. dendrimers, nanogels, nanoemulsions.

Nevertheless, the strategy of nanoparticle synthesis will always fade in critical prerequisites that need to be taken into consideration when the appropriate carrier/system is chosen, which includes (i) large carrier capacity (dose of drug to be solubilized), (ii) drug incorporation (increase circulation time) and sustained release rate, (iii) long-term stability and shelf life, (iv) feasibility of incorporation of both hydrophilic and hydrophobic substance, (v) biocompatibility, (vi) bio-distribution and specific targeting, and (vii) functionality.

5.2 Nanoparticle Choice

Numerous factors need to be considered in the choice of the system. Therefore, the main advantages and disadvantages of the different polymer and surfactant systems are summarized in Table 2.

Depending on the preferred drug targeting, the physicochemical properties and stability of the drug, the required time of bioactivity and the patient acceptability, different drug delivery systems tend to have their primary application within a given route of administration, offering different types of advantages and limitations. Amongst all of them, oral route is still the most intensively investigated, which offers the greatest potential for more effective therapeutics and probably collects the highest preference from both the patient and the clinician. However, this type of administration has some critical limitations, due to the way in which a drug typically moves through the digestive tract. For orally administered drugs, during the pathway to the target site, absorption may begin in the mouth and/or stomach. The latter has a quite low pH, influencing drug stability and drug solubility, as well as the properties of the drug carrier system. However, usually most of the drugs are rapidly eliminated due to the hepatic metabolism and the passage through the enzymatic environment of other body tissues, such as those of the gastrointestinal track, where they are easily degraded. Therefore, because the way of drug administration will determine, to some extent, whether or not the patient gains any clinical benefit, and whether he/she suffers any adverse effect from the drug, a list of advantages and disadvantages of some of the main routes of administration are summarized in Table 3.

5.3 Protein Corona

The understanding and the control of the fundamental interactions between the nanovehicles and the bio-systems (cells, tissues) plays a central role in the development and the safe implementation of nanomedicine and drug delivery. It is important to understand how nanoscaled objects interact with living matter, in order to better design drug delivery systems that will interact in pre-determined ways with living matter, increasing the efficiency in terms of targeting and delivering of the drug. Several characteristics of the nanovehicle determine a specific biological response, including size, shape and surface area, which will be considered later. It was found that the effective unit of interest in the cell-nanomaterial interaction is not the vehicle itself, but the associated proteins that are more or less strongly associated with the particle and its corona [140]. In the process of absorption, once

Table 2 Advantages and disadvantages of different structures in drug delivery

System	Advantages	Disadvantages
Micelles	Low surfactant concentrations required	Sensitive to dilution
	Small droplet size	Low solubilization
	Easy preparation	Potential toxicity of surfactant
	Good long-term stability	
	Low viscosity	
Liquid crystalline phases	Good long-term stability	In some cases difficult to prepare generally viscous
	Generally viscous	Short release duration
	Good solubilization capacity	Sensitive to dilution
	Solubilization of both hydrophilic and hydrophobic drugs	
	Solubilization of small and large molecules	
Vesicles and liposomes	Might be made of friendly compound	High viscosity
	Present in the body like lecithin and cholesterol	In some cases, difficult to prepare
	Good solubilization capacity	Often disintegrate once administrated
Microemulsions	Good long-term stability	Requires high surfactant concentrations
	Good solubilization capacity	Drug solubility influenced by the medium conditions
	Small droplet size	Potential toxicity of surfactant
	Easy preparation	
	Clarity	
Emulsions	High capacity of solubilization	Poor stability
		In some cases difficult to prepare
		Poor bioavailability
Polymers	Good stability	Low capacity of solubilization
	Useful controlled release properties	Low reduction in hydrolysis
	Increase stability of drugs/proteins	
Polymer-polymer/polymer-surfactant	Good stability	Too stable
	Reduces molecule size on complexation	Difficult to degrade/dissolve
	Solubilization of both hydrophilic and hydrophobic drugs	
	Solubilization of small and large molecules	

the drug vehicle is dispersed in a biological fluid (e.g. plasma) it acts as a scaffold for biomolecules such as proteins and lipids, which rapidly adsorb onto the "naked" surface of the drug vehicle, conferring a biological identity and resulting in the so called "protein corona". The biomolecular corona is a dynamic entity undergoing continuous exchange of proteins not only between the drug vehicle surface and the

Table 3 Advantages and disadvantages of some different routes of drug administration

Route	Administration	Advantages	Limitations
Oral	Mouth	Most common and convenient	Destruction of the drug
		Safest	Emesis
		Cheaper	Variable absorption
			Patient co-operation
Parenteral	Injection, infusion, or implantation	Active from of the drug is adsorbed	Painful
		Rapid adsorption	Expensive
		Accurate selection of the effective dose	Difficult to self-medicate
			Asepsis is important
Pulmonary	Inhaling orally for local or systemic effect	Low systemic side-effects	Requires significant degree of coordination
		Rapid clinical response	Variability in drug delivery
		Non-invasive needle-free delivery system	Patients with lung disease may not be able to inhale adequately
		Poor gastrointestinal absorption	
		First-pass hepatic metabolism	
Transmucosal	Throw mucous membranes	Rapid drug transport to the systemic circulation	Nasal
			Local toxic effects of the drug
		Poor gastrointestinal absorption	Reduction of systemic effect due to the drug metabolism in the respiratory rack
		First-pass hepatic metabolism	Buccal
			Prolonged exposure to the mucosal surface for significant drug absorption
			Absorption is only favored for drugs with a high pKa (pH of saliva is 6.5–6.9)
			Rectal
			Patient-to-patient irregular uptake
			Administered dose generally higher than intravenously or orally administration
Topical	Outer surface of the body	Effect in a localized area	Irritant local effects
		Less side effects	Local and general allergies
			Can reach the blood circulation and have systemic effects

biological fluid, but also interacting with specific protein receptors that can be found on the cell surface [141]. Very important in this area is the determination of binding rates, affinities, and stoichiometries of protein association with, and dissociation from, drug vehicle in biological fluids, since during their internalization process into the cell, the adsorbed "protein corona" is what the cell "sees". Therefore, nanoparticle-protein complex composition and structure influence not only cell uptake and trafficking, but also particle bio-distribution within the cell [142].

One of the major challenges in the study of drug delivery is to be able to intravenously deliver and target drug vehicles, as much as possible, in cancer-related tissues or cells in order to generate a sufficient effect. This step can only be successful if the drug vehicle is not eliminated from the systemic circulation by blood monocytes and cells of the phagocyte system (associated with the reticuloendothelial system), specialized for uptake of larger particles. Macrophages are designed to eliminate pathogens from the circulation, by a mechanism in which they recognize and bind specific protein at their surface known as opsonins. Well-known opsonins include IgG (Immunoglobulin G) and IgA (Immunoglobulin A) or components of the complement system (innate immune system, especially C3b, C4b, and iC3b). This body defence mechanism obviously leads to a decrease in the bioavailability of the drug in the desired site. In addition, dose-limiting toxic side effects may occur in the reticuloendothelial related tissues/cells. In this context, researchers focused on the development of methods to reduce this undesired effect. One of the proposed methods is related to coating colloidal drug vehicles, e.g. micelles, O/W emulsions or liposomes, with molecules such as polyoxypropylene block copolymers (poloxamers) [143], polyvinyl alcohol [144], PEG [145] and surfactants like sodium oleateor dodecylamine [146], which showed to be quite attractive as carriers in intravenous drug delivery. The reason for this phenomenum is the reduction of non-specific protein binding, which results from a combination of steric hindrance related to the polymer flexibility and hydrophobicity, and also from the increase of surface compatibility with less disruption of protein structure at the interface, if an intermediate layer of a coating agent (e.g. PEG) is presented. Moreover, the coating of the carrier is also needed to prevent agglomeration, keeping the carrier stable in the colloidal suspension.

Significant efforts have been made to identify the most affine proteins of the "protein corona", in order to understand how the protein binding can mediate their bio-interactions, which ultimately would lead to a better prediction on how the drug vehicle would interact with the target site (e.g. tumor cell, tissue) [147]. For example, as previously mentioned, adsorption of opsonins onto large carriers promote their removal from the systemic circulation, while binding of dysopsonins (human serum albumin, apolipoproteins) onto small size carriers leads to a prolonged circulation time. A variety of materials, including biodegradable polymers, polystyrene and silica, have been recently functionalized on the surface with a wide range of biological entities to specifically target cellular receptors [148–150].

5.4 Properties of Nanoparticles

Keeping in mind that nanoparticles enter the human body via various pathways, it is important to understand their interaction with the human body and their health effects, in order to understand both the hazards associated with nanomaterials and the levels of exposure that are likely to occur. Thus, it is easy to understand how important it is for nanomedicine (and all consequences that entails in the strategy of nanoparticle synthesis) to have a broad nanoparticle-cell uptake study, in order to see the impact, and uptake efficiency that different types of nanoparticles have when they are exposed to different cell types. Even if all the details are not yet completely clear, it is evident that new nanoparticle properties will come into play when nanoparticles interact with the human body. For instance, as previously mentioned, it is important to understand the effect of nanoparticles on the immune system and how nanoparticles may contribute to suppress inflammatory responses.

Even with the "protein corona" having a major influence on the cell uptake, other properties of the carrier play a key role in the particle uptake scheme, such as the size, shape, dose and surface properties (hydrophobicity, charge, coating, etc.). Apart from these, also the cell type is of crucial importance in the particle uptake efficiency.

As a result of the variety and complexity in size and scale of materials and components that need to be transported into the cells, a wide range of different cellular uptake mechanisms have evolved. Until now, many distinct uptake pathways have been classified and characterized [151, 152]: (i) phagocytosis is characteristic of specialized mammalian cells such as neutrophils, monocytes and macrophages (particle-dependent); (ii) clathrin-mediated endocytosis (CME) is a mechanism involving specific receptors that recognize and internalize cargo into coated pits formed by the assembly of a cytosolic coat protein (clathrin) which constitutes the main assembly unit (mechanism generally used for nanoparticles of $\sim$120 nm); (iii) caveolae-mediated endocytosis, which involves clustering of lipid "raft" components on the plasma membrane into the so-called caveolae (flask-shaped invaginations) formed due to the interaction of different proteins, mainly caveolin, with the cellular membrane (mechanism generally used for nanoparticles of $\sim$60 nm); (iv) clathrin- and caveolae-independent endocytosis in which other types of cholesterol-rich microdomains (small structures of 40–50 nm in diameter) on the plasma membrane rather than caveolae, diffuse on the cell surface (mechanism generally used for nanoparticles of $\sim$90 nm).

It is well established now that cell uptake leads to a particle-size-dependent biochemical response, with the levels of internalization decreasing when the size of the particle increases. In several studies, the maximum uptake by cells occurs for nanovehicles with 50 nm in diameter [153, 154]. Therefore, these may be the most suitable candidates to be used as carriers for further studies in biological applications. Smaller particles (i.e. less than 50 nm) enter less efficiently, proving the importance of the type of mechanism used by the cell in the process of internalization.

Cellular uptake efficiency is also correlated with the surface charge of the nanoparticles. It has been shown that a positive charge greatly promotes the efficiency of internalization. This occurs due to the electrostatic interactions that are established between the nanoparticle and the negatively charged plasma membrane which promotes their bio-adhesion resulting in an enhanced uptake. However,

Table 4 Toxicity and some adverse effects of different engineered nanoparticles

Adverse effects	Type of nanoparticle
Inhibition of macrophage phagocytosis	TiO_2
	Carbon black
Oxidative stress, (generation of surface radicals and inflammatory action)	Nanoparticle in general
	Carbon nanotubes
Platelet aggregation	Particulate matter
	Single walled carbon nanotubes
	Carboxylic modified nanoparticles
Effect of mitochondria activity	Cationic nanoparticles
Adversely affects cardiac activity	Particulate matter
Access to systemic circulation upon inhalation	Specific coating-nanoparticles

it has been reported that, in some cases, the uptake of positively charged nanoparticles leads to undesirable consequences to the cell. Therefore, one needs to be careful when addressing these questions. After internalization of positive charge nanoparticles, the cell has been shown to undergo distinct processes. The most reported one is cell death via apoptosis [155, 156], which starts with the destabilization of the lysosomal membrane and consequent release of both nanoparticles and proteolytic enzymes such as cathepsins into the cytoplasm [157]. These phenomena result in a series of events that ultimately lead to apoptosis. Other cellular responses have also been described such as the activation of the inflammasome with secretion of interleukin 1β and production of an inflammatory response or activation of the endoplasmic reticulum stress response [158, 159]. Taking the abovementioned into consideration, careful analysis of deleterious effects to the cell needs to be considered when positively charged nanoparticles are to be used. Therefore, this access of nanoparticles to the cellular machinery can lead to unpredictable consequences and biological impacts, especially from those never intended for contact with humans. It is this concern regarding the safety of nanomaterials that opened the door to a careful nanosafety assessment. It is not immediately clear what the consequences of exposure to nanomaterials, even those of materials that are bio-inert in the bulk scale, such as gold, silica and polystyrene, will be for human health and the environment. Some of the toxicological effects caused by nanoparticles are described in Table 4, showing the great importance of the understanding of all the interactions and potential consequences of nanoparticles with living tissues.

In terms of other properties that may influence cell uptake, hydrophobic domains of the particle surface have an important role. Some studies have been performed showing that protein adsorption on the surface of nanoparticles decreases quite significantly the subsequent uptake of these by different cells [160]. In fact, when nanoparticles are exposed in the absence of fetal bovine serum to the culture medium, significantly higher uptake is obtained. The reason behind this phenomenon is explained by non-specific hydrophobic effects. When bare nanoparticles are administrated to cells, their high surface hydrophobicity will cause them to adhere to the already hydrophobic plasma membrane. On the other hand, in the presence of serum, adsorption of proteins screens the non-specific hydrophobic binding of the nanoparticles to the cell surface leading to a decrease of the nanoparticle surface hydrophobicity and a consequently lower, but more specific, uptake of the nanoparticles via protein-corona mediated interactions with cellular receptors. In terms of specificity on the uptake, a balance between the administered particle dose and specific binding needs to be considered. In the case of drug targeting to a cancer cell, non-specific binding should be avoided, since also healthy cells would have higher chance to internalize drug-loaded particles, which would decrease the desired therapeutic effect. As previously mentioned, functionalization of the particle surface with different proteins can result in a more specific uptake. Other strategies should also be considered for this, such as coating the outer particle surface with anionic triblock copolymers, which would reduce non-specific interactions with the negatively charged cells.

Other property receiving increased attention is the effect of particle shape on cell response. Depending on the type of material used, different authors have shown significant different conclusions on how dissimilar shaped nanoparticles (spherical, long rod, short rod, etc.) affect the cell uptake [161, 162]. The conclusions show the importance of careful analysis and individual interpretation for each cell-nanoparticle system, although it is obvious that also the shape of nanoparticles should be considered when these are designed for therapeutic applications.

Studies across all different cell lines from different parts of the human body, representative of the main avenues of exposure to nanoparticles have been widely performed. Different cell types show very different uptake efficiencies for the same materials, with macrophages having the highest uptake rate for all nanoparticle sizes, while cells from protective barriers (epithelial and endothelial) show lower uptake [163]. Once more, this emphasizes the need for engineering nanoparticles that are able to evade the immune system, thereby increasing their circulation time, enhancing the efficiency of drug delivery, while being optimized for uptake in the target cells.

References

1. Schramm LL (2000) Surfactants: fundamentals and applications in the petroleum industry. University Press, Cambridge
2. Goodwin J (2004) Colloids and interfaces with surfactants and polymers—an introduction. Wiley, Chichester
3. Farn RJ (2006) Chemistry and technology of surfactants. Blackwell Publishing Ltd, Oxford
4. Malmsten M (2002) Surfactants and polymers in drug delivery. Marcel Dekker, Inc, New York
5. Holmberg H, Jonsson B, Kronbreg B, Lindman B (2003) Surfactants and polymers in aqueous solution, 2nd edn. Wiley, Hoboken
6. Tadros TF (2005) Applied surfactants—principles and applications. Wiley-VCH Verlag GmbH & Co. KGaA, Weinheim

7. Pfeiffer W, Henkel T, Sackmann E, Knoll W, Richter D (1989) Local dynamics of lipid bilayers studied by incoherent quasi-elastic neutron-scattering. Europhys Lett 8:201–206
8. Tanford C (1980) The hydrophobic effect: formation of micelles and biological membranes. Wiley, New York
9. Hartley G (1936) Aqueous solutions of paraffin-chain salts. Herman & Cie, Paris
10. Israelachvili JN (2002) Intermolecular and surface forces. Academic, London
11. Evans DF, Wennerstrom H (1994) The colloidal domain: where physics, chemistry, biology and technology meet. VCH Publishers Inc, New York
12. Karlström G (1985) A new model for upper and lower critical solution temperatures in poly (ethylene oxide) solutions. J Chem Phys 89:4962–4964
13. Olsson U, Wennerstrom H (1994) Globular and bicontinuous phases of nonionic surfactant films. Adv Colloid Interface Sci 49:113–146
14. Wennerström H, Lindman B (1979) Micelles – physical-chemistry of surfactant association. Phys Rep Rev Sec Phys Lett 52:1–86
15. Strey R (1994) Microemulsion microstructure and interfacial curvature. Colloid Polym Sci 272:1005–1019
16. Lindman B, Karlstrom G (2009) Nonionic polymers and surfactants: temperature anomalies revisited. Comptes Rendus Chimie 12:121–128
17. Mitchell DJ, Tiddy GJT, Waring L, Bostock T, Mcdonald MP (1983) Phase-behavior of polyoxyethylene surfactants with water—mesophase structures and partial miscibility (cloud points). J Chem Soc Faraday Trans 79:975–1000
18. Chernik GG (2000) Phase studies of surfactant – water systems. Curr Opin Colloid Interface Sci 4:381–390
19. Porte G, Appell J, Bassereau P, Marignan J (1989) L-alpha to l3 – a topology driven transition in phases of infinite fluid membranes. J Phys France 50:1335–1347
20. Andersson D, Wennerström H, Olsson U (1989) Isotropic bicontinuous solutions in surfactant solvent systems – the l3 phase. J Phys Chem B 93:4243–4253
21. Golubovic L (1994) Passages and droplets in lamellar fluid membrane phases. Phys Rev E 50:R2419–R2422
22. Nallet F (1991) Membrane fluctuations in dilute lamellar phases. Langmuir 7:1861–1863
23. Skouri M, Marignan J, Appell J, Porte G (1991) Fluid membranes in the semirigid regime – scale-invariance. J Phys II France 1:1121–1132
24. Helfrich W (1973) Elastic properties of lipid bilayers – theory and possible experiments. Z Naturforsch C 28:693–703
25. Torchilin VP (2001) Structure and design of polymeric surfactant-based drug delivery systems. J Control Release 73:137–172
26. Drummond CJ, Fong C (2000) Surfactant self-assembly objects as novel drug delivery vehicles. Curr Opin Colloid Interface Sci 4:449–456
27. Shah JC, Sadhale Y, Chilukuri DM (2001) Cubic phase gels as drug delivery systems. Adv Drug Deliv Rev 47:229–250
28. Medronho B (2009) Shear induced transitions in complex fluids: planar lamellae and multilamellar vesicles. Dissertation, University of Coimbra
29. Olea D, Faure C (2003) Quantitative study of the encapsulation of glucose oxidase into multilamellar vesicles and its effect on enzyme activity. J Chem Phys 119:6111–6118
30. Lawrence MJ, Rees GD (2000) Microemulsion-based media as novel drug delivery systems. Adv Drug Deliv Rev 45:89–121
31. Dang JM, Leong KW (2006) Natural polymers for gene delivery and tissue engineering. Adv Drug Deliv Rev 58:487–499
32. Flory PJ (1953) Principles of polymer chemistry. Cornell University Press, New York
33. Cosgrove T, Ryan K (1990) NMR and neutron-scattering studies on poly(ethylene oxide) terminally attached at the polystyrene/water interface. Langmuir 6:136–142
34. Frank DB (1990) Magnetic resonance of polymers at surfaces. Colloids Surf 45:361–376

35. Hoogendam CW, de Keizer A, Cohen Stuart MA, Bijsterbosch BH, Smit JAM, van Dijk JAPP, van der Horst PM, Batelaan JG (1998) Persistence length of carboxymethyl cellulose as evaluated from size exclusion chromatography and potentiometric titrations. Macromolecules 31:6297–6309
36. Huggins ML (1942) Theory of solutions of high polymers. J Am Chem Soc 64:1712–1719
37. Huggins ML (1942) Some properties of solutions of long-chain compounds. J Phys Chem 46:151–158
38. Torchilin VP (2004) Targeted polymeric micelles for delivery of poorly soluble drugs. Cellular and molecular life sciences. Cell Mol Life Sci 61:2549–2559
39. Zhulina EB, Adam M, LaRue I, Sheiko SS, Rubinstein M (2005) Diblock copolymer micelles in a dilute solution. Macromolecules 38:5330–5351
40. Discher DE, Eisenberg A (2002) Polymer vesicles. Science 297:967–973
41. Howse JR, Jones RA, Battaglia G, Ducker RE, Leggett GJ, Ryan AJ (2009) Templated formation of giant polymer vesicles with controlled size distributions. Nat Mater 8:507–511
42. Lindman B, Karlstrom G, Stigsson L (2010) On the mechanism of dissolution of cellulose. J Mol Liq 156:76–81
43. Antunes FE, Gentile L, Tavano L, Rossi O (2009) Rheological characterization of the thermal gelation of poly(N-isopropylacrylamide) and poly(N-isopropylacrylamide)co-Acrylic Acid. Appl Rheol 19:42064–42072
44. Kjøniksen AL, Zhu K, Pamies R, Nyström B (2008) Temperature-induced formation and contraction of micelle-like aggregates in aqueous solutions of thermoresponsive short-chain copolymers. J Phys Chem B 112:3294–3299
45. Alexandridis P, Zhou D, Khan A (1996) Lyotropic liquid crystallinity in amphiphilic block copolymers: temperature effects on Phase behavior and structure for poly(ethylene oxide)-b-poly(propylene oxide)-b-poly(ethylene oxide) copolymers of different composition. Langmuir 12:2690–2700
46. Green MS, Tobolsky AV (1946) A new approach to the theory of relaxing polymeric media. J Chem Phys 14:80–92
47. Ferry JD (1980) Viscoelastic properties of polymers, 3rd edn. Wiley, New York
48. Piculell L, Lindman B (1992) Association and segregation in aqueous polymer/polymer, polymer/surfactant, and surfactant/surfactant mixtures: similarities and differences. Adv Colloid Interface Sci 41:149–178
49. Tanaka R, Meadows J, Williams PA, Phillips GO (1992) Interaction of hydrophobically modified hydroxyethyl cellulose with various added surfactants. Macromolecules 25:1304–1310
50. Semenov AN, Joanny JF, Khokhlov AR (1995) Associating polymers: equilibrium and linear viscoelasticity. Macromolecules 28:1066–1075
51. Witten TA (1988) Associating polymers and shear thickening. J Phys France 49:1055–1063
52. Cui H, Webber MJ, Stupp SI (2010) Self-assembly of peptide amphiphiles: From molecules to nanostructures to biomaterials. Biopolymers 94:1–18
53. Hubbard A (2002) Encyclopedia of surface and colloid science. Marcel Dekker, New York
54. Senff H, Richtering W (1999) Temperature sensitive microgel suspensions: colloidal phase behavior and rheology of soft spheres. J Chem Phys 111:1705–1711
55. Jeong B, Gutowska A (2002) Lessons from nature: stimuli-responsive polymers and their biomedical applications. Trends Biotechnol 20:305–311
56. Murray MJ, Snowden MJ (1995) The preparation, characterisation and applications of colloidal microgels. Adv Colloid Interface Sci 54:73–91
57. Lopez VC, Snowden MJ (2003) The role of colloidal microgels in drug delivery. Drug Delivery Sys Sci 3:19–23
58. Varma MVS, Kaushal AM, Garg S (2005) Influence of micro-environmental pH on the gel layer behavior and release of a basic drug from various hydrophilic matrices. J Control Release 103:499–510
59. Lopez VC, Hadgraft J, Snowden MJ (2005) The use of colloidal microgels as a (trans)dermal drug delivery system. Int J Pharm 292:137–147

60. Morris GE, Vincent B, Snowden MJ (1997) The interaction of thermosensitive anionic microgel with metal ion solution. Prog Colloid Polymer Sci 105:16–22
61. Byrne R, Benito-Lopez F, Diamond D (2010) Materials science and the sensor revolution. Mater Today 13:16–23
62. Saunders BR, Vincent B (1999) Microgel particles as model colloids: theory, properties and applications. Adv Colloid Interface Sci 80:1–25
63. Barreiro-Iglesias R, Alvarez-Lorenzo C, Concheiro A (2003) Poly(acrylic acid) microgels (carbopol® 934)/surfactant interactions in aqueous media: Part II: ionic surfactants. Int J Pharm 258:179–191
64. Meid J, Friedrich T, Tieke B, Lindner P, Richtering W (2011) Composite hydrogels with temperature sensitive heterogeneities: influence of gel matrix on the volume phase transition of embedded poly-(N-isopropylacrylamide) microgels. Phys Chem Chem Phys 13:3039–3047
65. Berndt I, Popescu C, Wortmann F-J, Richtering W (2006) Mechanics versus thermodynamics: swelling in multiple-temperature-sensitive core-shell microgels. Angew Chemie Int Ed Engl 45:1081–1085
66. Kaneda I, Vincent B (2004) Swelling behavior of PMMA-g-PEO microgel particles by organic solvents. J Colloid Interface Sci 274:49–54
67. Pillai O, Panchagnula R (2001) Polymers in drug delivery. Curr Opin Chem Biol 5:447–451
68. Keerl M, Richtering W (2007) Synergistic depression of volume phase transition temperature in copolymer microgels. Colloid Polym Sci 285:471–474
69. Goddard ED, Hannan RB, Matteson GH (1977) Dye solubilization by a cationic polymer/anionic surfactant system. J Colloid Interface Sci 60:214–215
70. Leung PS, Goddard ED, Han C, Glinka CJ (1985) A study of polycation-anionic-surfactant systems. Colloids Surf 13:47–62
71. Leung PS, Goddard ED (1991) Gels from dilute polymer/surfactant solutions. Langmuir 7:608–609
72. Lee BH, Christian SD, Tucker EE, Scamehorn JF (1991) Effects of an anionic polyelectrolyte on the solubilization of mono- and dichlorophenols by aqueous solutions of N-hexadecylpyridinium chloride. Langmuir 7:1332–1335
73. Magny B, Iliopoulos I, Zana R, Audebert R (1994) Mixed micelles formed by cationic surfactants and anionic hydrophobically modified polyelectrolytes. Langmuir 10:3180–3187
74. Ilias I (1998) Association between hydrophobic polyelectrolytes and surfactants. Curr Opin Colloid Interface Sci 3:493–498
75. Goddard ED, Pethica BA (1951) On detergent-protein interactions. J Chem Soc 2659–2663
76. Laurent TC, Scott JE (1964) Molecular weight fractionation of polyanions by cetylpyridinium chloride in salt solutions. Nature 16:661–662
77. Kwak JCT (1998) Polymer-surfactant systems. Marcel Dekker, New York
78. Zhao D, Feng J, Huo Q, Melosh N, Fredrickson GH, Chmelka BF, Stucky GD (1998) Triblock copolymer syntheses of mesoporous silica with periodic 50 to 300 angstrom pores. Science 279:548–552
79. Thomas A, Goettmann F, Antonietti M (2008) Hard templates for soft materials: creating nanostructured organic materials. Chem Mater 20:738–755
80. Nizri G, Makarsky A, Magdassi S, Talmon Y (2009) Nanostructures formed by self-assembly of negatively charged polymer and cationic surfactants. Langmuir 25:1980–1985
81. Nizri G, Magdassi S, Schmidt J, Cohen Y, Talmon Y (2004) Microstructural characterization of micro- and nanoparticles formed by polymer-surfactant interactions. Langmuir 20:4380–4385
82. Goddard ED, Ananthapadmanabhan KP (1993) Interactions of surfactants with polymers and proteins. Taylor & Francis, Boca Raton
83. Dualeh AJ, Steiner CA (1990) Hydrophobic microphase formation in surfactant solutions containing an amphiphilic graft copolymer. Macromolecules 23:251–255
84. Guillemet F, Piculell L (1995) Interactions in aqueous mixtures of hydrophobically modified polyelectrolyte and oppositely charged surfactant. mixed micelle formation and associative phase separation. J Phys Chem 99:9201–9209

85. Thuresson K, Nilsson S, Lindman B (1996) Effect of hydrophobic modification on phase-behavior and rheology in mixtures of oppositely charged polyelectrolytes. Langmuir 12:530–537
86. Thuresson K, Lindman B, Nyström B (1997) Effect of hydrophobic modification of a nonionic cellulose derivative on the interaction with surfactants. Rheology. J Phys Chem B 101:6450–6459
87. Karlberg M, Thuresson K, Piculell L, Lindman B (2004) Mixed solutions of hydrophobically modified graft and block copolymers. Colloids Surf A 236:159–164
88. Piculell L, Egermayer M, Sjöström J (2003) Rheology of mixed solutions of an associating polymer with a surfactant. Why are different surfactants different? Langmuir 19:3643–3649
89. Antunes FE, Marques EF, Miguel MG, Lindman B (2009) Polymer-vesicle association. Adv Colloid Interface Sci 147–148:18–35
90. Antunes FE, Lindman B, Miguel MG (2005) Mixed systems of hydrophobically modified polyelectrolytes: controlling rheology by charge and hydrophobe stoichiometry and interaction strength. Langmuir 21:10188–10196
91. Bergfeldt K, Piculell L, Linse P (1996) Segregation and association in mixed polymer solutions from flory-huggins model calculations. J Phys Chem 100:3680–3687
92. Hefford RJ (1984) Polymer mixing in aqueous solution. Polymer 25:979–984
93. Chun M-K, Cho C-S, Choi H-K (2001) A novel mucoadhesive polymer prepared by template polymerization of acrylic acid in the presence of poloxamer. J Appl Polymer Sci 79:1525–1530
94. Dubin P, Bock J, Davis R, Schulz ND, Thies C (1994) Macromolecular complexes in chemistry and biology. Springer, Berlin
95. Shenkov S, Baranovsky VY (1994) Complex formation between poly(methacrylic acid) and poly(propylene glycol) in aqueous solutions. J Polymer Sci A 32:1385–1387
96. Cole ML, Whateley TL (1996) Interaction of nonionic block copolymeric (Poloxamer) surfactants with poly (Acrylic Acid), studied by photon correlation spectroscopy. J Colloid Interface Sci 180:421–427
97. Wang Y, Goethals EJ, Du Prez FE (2004) Association behavior between end-functionalized block copolymers PEO-PPO-PEO and poly(acrylic acid). Macromol Chem Phys 205:1774–1781
98. Barreiro-Iglesias R, Alvarez-Lorenzo C, Concheiro A (2003) Poly(acrylic acid) microgels (carbopol® 934)/surfactant interactions in aqueous media: Part I: nonionic surfactants. Int J Pharm 258:165–177
99. Costa T, Schillen K, Miguel MG, Lindman B, Seixas de Melo J (2009) Association of a hydrophobically modified polyelectrolyte and a block copolymer followed by fluorescence techniques. J Phys Chem B 113:6194–6204
100. dos Santos S, Luigjes B, Piculell L (2010) Associative phase behaviour and disintegration of copolymer aggregates on adding poly(acrylic acid) to aqueous solutions of a PEO-PPO-PEO triblock copolymer. Soft Matter 6:4756–4767
101. Kabanov AV, Zezin AB (1984) A new class of complex water-soluble polyelectrolytes. Makroma Chem Suppl 6:259–276
102. Chu DY, Thomas JK (1986) Effect of cationic surfactants on the conformation transition of poly(methacrylic acid). J Am Chem Soc 108:6270–6276
103. Goddard ED (1986) Polymer-surfactant interaction part II. Polymer and surfactant of opposite charge. Colloids Surf 19:301–329
104. Thalberg K, Lindman B (1989) Interaction between hyaluronan and cationic surfactants. J Phys Chem 93:1478–1483
105. Thalberg K, Lindman B, Bergfeldt K (1991) Phase behavior of systems of polyacrylate and cationic surfactants. Langmuir 7:2893–2898
106. Dias R, Mel'nikov S, Lindman B, Miguel MG (2000) DNA phase behavior in the presence of oppositely charged surfactants. Langmuir 16:9577–9583
107. Lynch I, Sjöström J, Piculell L (2005) Reswelling of polyelectrolyte hydrogels by oppositely charged surfactants. J Phys Chem B 109:4258–4262

108. Thalberg K, Lindman B, Karlström G (1991) Phase behavior of a system of cationic surfactant and anionic polyelectrolyte: the effect of salt. J Phys Chem 95:6004–6011
109. Ilekti P, Martin T, Cabane B, Piculell L (1999) Effects of polyelectrolytes on the structures and interactions of surfactant aggregates. J Phys Chem B 103:9831–9840
110. Svensson A, Norrman J, Piculell L (2006) Phase behavior of polyion-surfactant ion complex salts: effects of surfactant chain length and polyion length. J Phys Chem B 110:10332–10340
111. Svensson A, Piculell L, Cabane B, Ilekti P (2002) A new approach to the phase behavior of oppositely charged polymers and surfactants. J Phys Chem B 106:1013–1018
112. Svensson A, Piculell L, Karlsson L, Cabane B, Jonsson B (2003) Phase behavior of an ionic surfactant with mixed monovalent/polymeric counterions. J Phys Chem B 107:8119–8130
113. dos Santos S, Lundberg D, Piculell L (2011) Responsive and evolving mixtures of a hydrolyzing cationic surfactant and oppositely charged polyelectrolytes. Soft Matter 7:5540–5544
114. Fanun M (2010) Colloids in drug delivery. CRC Press/Taylor & Francis, Boca Raton
115. Dias RS, Lindman B (2008) DNA interactions with polymers and surfactants. Wiley, Hoboken
116. Costa D, Hansson P, Schneider S, Miguel MG, Lindman B (2006) Interaction between covalent DNA gels and a cationic surfactant. Biomacromolecules 7:1090–1095
117. Hansson P, Schneider S, Lindman B (2002) Phase separation in polyelectrolyte gels interacting with surfactants of opposite charge. J Phys Chem B 106:9777–9793
118. Nilsson P, Hansson P (2005) Ion-exchange controls the kinetics of deswelling of polyelectrolyte microgels in solutions of oppositely charged surfactant. J Phys Chem B 109:23843–23856
119. Nilsson P, Hansson P (2007) Deswelling kinetics of polyacrylate gels in solutions of cetyltrimethylammonium bromide. J Phys Chem B 111:9770–9778
120. Nilsson P, Unga J, Hansson P (2007) Effect of salt and surfactant concentration on the structure of polyacrylate gel/surfactant complexes. J Phys Chem B 111:10959–10964
121. Morimoto N, Endo T, Ohtomi M, Iwasaki Y, Akiyoshi K (2005) Hybrid nanogels with physical and chemical cross-linking structures as nanocarriers. Macromol Biosci 5:710–716
122. Zhang Y, Zhu W, Wang B, Ding J (2005) A novel microgel and associated post-fabrication encapsulation technique of proteins. J Control Release 105:260–268
123. Murthy N, Thng YX, Schuck S, Xu MC, Frechet JM (2002) A novel strategy for encapsulation and release of proteins: hydrogels and microgels with acid-labile acetal cross-linkers. J Am Chem Soc 124:12398–12399
124. Bysell H, Mansson R, Hansson P, Malmsten M (2011) Microgels and microcapsules in peptide and protein drug delivery. Adv Drug Deliv Rev 63:1172–1185
125. Huang C-I, Olvera de la Cruz M (2002) Polyelectrolytes in multivalent salt solutions: monomolecular versus multimolecular aggregation. Macromol 35:976–986
126. Dias RS, Svingen R, Gustavsson B, Lindman B, Miguel MG (2005) Electrophoretic properties of complexes between DNA and the cationic surfactant cetyltrimethylammonium bromide. Electrophoresis 26:2908–2917
127. Dias RS, Innerlohinger J, Glatter O, Miguel MG, Lindman B (2005) Coil-globule transition of DNA molecules induced by cationic surfactants: a dynamic light scattering study. J Phys Chem B 109:10458–10463
128. Gonzalez-Perez A, Carlstedt J, Dias RS, Lindman B (2010) Cyclodextrins in DNA decompaction. Colloids Surf B Biointerfaces 76:20–27
129. Costa D, Miguel MG, Lindman B (2007) Effect of additives on swelling of covalent DNA gels. J Phys Chem B 111:8444–8452
130. Izumrudov VA, Wahlund PO, Gustavsson PE, Larsson PO, Galaev IY (2003) Factors controlling phase separation in water-salt solutions of DNA and polycations. Langmuir 19:4733–4739
131. Willmitzer L, Bode J, Wagner KG (1977) Phosphorylated protamines. II. Circular dichroism of complexes with DNA, dependency on ionic strength. Nucleic Acids Res 4:163–176
132. Raspaud E, Pelta J, de Frutos M, Livolant F (2006) Solubility and charge inversion of complexes of DNA and basic proteins. Phys Rev Lett 97:068103

133. Costa D, dos Santos S, Antunes F, Miguel MG, Lindman B (2006) Some novel aspects of DNA physical and chemical gels. ARKIVOC iv:161–172
134. Moghimi SM, Symonds P, Murray JC, Hunter AC, Debska G, Szewczyk A (2005) A two-stage poly(ethylenimine)-mediated cytotoxicity: implications for gene transfer/therapy. Mol Ther 11:990–995
135. Putnam D, Gentry CA, Pack DW, Langer R (2001) Polymer-based gene delivery with low cytotoxicity by a unique balance of side-chain termini. Proc Natl Acad Sci USA 98:1200–1205
136. Chen DJ, Majors BS, Zelikin A, Putnam D (2005) Structure-function relationships of gene delivery vectors in a limited polycation library. J Control Release 103:273–283
137. Youan BB (2008) Impact of nanoscience and nanotechnology on controlled drug delivery. Nanomedicine (Lond) 3:401–406
138. Silva GA (2009) Nanotechnology applications and approaches for neuroregeneration and drug delivery to the central nervous system. Ann N Y Acad Sci 1199:221–230
139. Ferrari M (2005) Cancer nanotechnology: opportunities and challenges. Nat Rev Cancer 5:161–171
140. Lynch I, Salvati A, Dawson KA (2009) Protein-nanoparticle interactions: what does the cell see? Nat Nanotechnol 4:546–547
141. Lundqvist M, Stigler J, Elia G, Lynch I, Cedervall T, Dawson KA (2008) Nanoparticle size and surface properties determine the protein corona with possible implications for biological impacts. Proc Natl Acad Sci USA 105:14265–14270
142. Walczyk D, Bombelli FB, Monopoli MP, Lynch I, Dawson KA (2010) What the cell "sees" in bionanoscience. J Am Chem Soc 132:5761–5768
143. Goppert TM, Muller RH (2005) Adsorption kinetics of plasma proteins on solid lipid nanoparticles for drug targeting. Int J Pharm 302:172–186
144. Pandey R, Khuller GK (2005) Solid lipid particle-based inhalable sustained drug delivery system against experimental tuberculosis. Tuberculosis (Edinb) 85:227–234
145. De Campos AM, Sanchez A, Gref R, Calvo P, Alonso MJ (2003) The effect of a PEG versus a chitosan coating on the interaction of drug colloidal carriers with the ocular mucosa. Eur J Pharm Sci 20:73–81
146. Gupta AK, Gupta M (2005) Synthesis and surface engineering of iron oxide nanoparticles for biomedical applications. Biomaterials 26:3995–4021
147. Cedervall T, Lynch I, Lindman S, Berggard T, Thulin E, Nilsson H, Dawson KA, Linse S (2007) Understanding the nanoparticle-protein corona using methods to quantify exchange rates and affinities of proteins for nanoparticles. Proc Natl Acad Sci USA 104:2050–2055
148. Liu GY, Li LY, Yang XL, Dai Z (2008) Preparation of silica/polymer hybrid microspheres and the corresponding hollow polymer microspheres with functional groups. Polymers Adv Technol 19:1922–1930
149. Li W, Chen CY, Ye C, Wei TT, Zhao YL, Lao F, Chen Z, Meng H, Gao YX, Yuan H, Xing GM, Zhao F, Chai ZF, Zhang XJ, Yang FY, Han D, Tang XH, Zhang YG (2008) The translocation of fullerenic nanoparticles into lysosome via the pathway of clathrin-mediated endocytosis. Nanotechnology 19:145102–145114
150. dos Santos T, Varela J, Lynch I, Salvati A, Dawson KA (2011) Effects of transport inhibitors on the cellular uptake of carboxylated polystyrene nanoparticles in different cell lines. PLoS One 6:e24438
151. Doherty GJ, McMahon HT (2009) Mechanisms of endocytosis. Ann Rev Biochem 78:857–902
152. Parton RG, Richards AA (2003) Lipid rafts and caveolae as portals for endocytosis: new insights and common mechanisms. Traffic 4:724–738
153. Osaki F, Kanamori T, Sando S, Sera T, Aoyama Y (2004) A quantum dot conjugated sugar ball and its cellular uptake on the size effects of endocytosis in the subviral region. J Am Chem Soc 126:6520–6521
154. Lu F, Wu SH, Hung Y, Mou CY (2009) Size effect on cell uptake in well-suspended, uniform mesoporous silica nanoparticles. Small 5:1408–1413

155. Bexiga MG, Varela JA, Wang F, Fenaroli F, Salvati A, Lynch I, Simpson JC, Dawson KA (2011) Cationic nanoparticles induce caspase 3-, 7- and 9-mediated cytotoxicity in a human astrocytoma cell line. Nanotoxicology 5:557–567
156. Nel AE, Madler L, Velegol D, Xia T, Hoek EMV, Somasundaran P, Klaessig F, Castranova V, Thompson M (2009) Understanding biophysicochemical interactions at the nano-bio interface. Nat Mater 8:543–557
157. Lunov O, Syrovets T, Loos C, Beil J, Delacher M, Tron K, Nienhaus GU, Musyanovych A, Mailander V, Landfester K, Simmet T (2011) Differential uptake of functionalized polystyrene nanoparticles by human macrophages and a monocytic cell line. ACS Nano 5:1657–1669
158. Fubini B, Ghiazza M, Fenoglio I (2010) Physico-chemical features of engineered nanoparticles relevant to their toxicity. Nanotoxicology 4:347–363
159. Tsai YY, Huang YH, Chao YL, Hu KY, Chin LT, Chou SH, Hour AL, Yao YD, Tu CS, Liang YJ, Tsai CY, Wu HY, Tan SW, Chen HM (2011) Identification of the nanogold particle-induced endoplasmic reticulum stress by omic techniques and systems biology analysis. ACS Nano 5:9354–9369
160. Lesniak A, Campbell A, Monopoli MP, Lynch I, Salvati A, Dawson KA (2010) Serum heat inactivation affects protein corona composition and nanoparticle uptake. Biomaterials 31:9511–9518
161. Vacha R, Martinez-Veracoechea FJ, Frenkel D (2011) Receptor-mediated endocytosis of nanoparticles of various shapes. Nano Lett 11:5391–5395
162. Chithrani BD, Ghazani AA, Chan WCW (2006) Determining the size and shape dependence of gold nanoparticle uptake into mammalian cells. Nano Lett 6:662–668
163. dos Santos T, Varela J, Lynch I, Salvati A, Dawson KA (2011) Quantitative assessment of the comparative nanoparticle-uptake efficiency of a range of cell lines. Small 7:3341–3349

Sizing Nanoparticles in Liquids: An Overview of Methods

Margarida Figueiredo

When you can measure what you are speaking about, you know something about it. But when you cannot measure it, your knowledge is of a meagre and unsatisfactory kind. It may be the beginning of knowledge, but you have scarcely advanced to the stage of science.

Lord Kelvin, 1883

Abstract Measuring nano- and submicron particles suspended in liquids continues to be a difficult task not only because there is no universal sizing technique but, on the contrary, because there are too many alternative methods. These are quite different in their measuring principles and for that may lead to rather different results, especially if the particles under analysis are far from spherical and exhibit broad size distributions.

This chapter is mainly dedicated to the users who are not very acquainted with particle sizing issues but need to select the most adequate method to characterize their suspensions. Having in mind the size range and sample type, seven different methods were selected: besides microscopic methods, that are normally the first choice to "see" the particles, this chapter encompasses techniques based on the measurement of particles Brownian motion (DLS and NTA) and on centrifugal sedimentation (DSC and SdFFF). The operating principles of these techniques as well as their merits and limitations will be discussed.

Keywords Nanoparticles • Microparticles • Light scattering • Microscopy • Sizing • Predictive, preventive and personalized medicine

---

M. Figueiredo (✉)
Department of Chemical Engineering, University of Coimbra, Polo II – Rua Sílvio Lima, 3030-790 Coimbra, Portugal
e-mail: mmf@eq.uc.pt

J. Coelho (ed.), *Drug Delivery Systems: Advanced Technologies Potentially Applicable in Personalised Treatment*, Advances in Predictive, Preventive and Personalised Medicine 4, DOI 10.1007/978-94-007-6010-3_3,

1 Introduction

To determine the average size or size distribution of particle dispersion is a task considerably more complex than it might seem. In the first place because of the ambiguous definition of "size" of an (irregular) particle. In fact, particles are three dimensional objects normally described by three parameters (length, breadth and height) and only spherical particles can be fully described by a single number (its diameter). The strategy usually adopted is to express the particle dimension as an "equivalent spherical diameter", defined as the diameter of the sphere that generates the same result as that of the particle under observation. Volume based, aerodynamic, electrical mobility, optical or Stokes diameters are just a few examples of normally used equivalent diameters (Fig. 1). This concept leads to the fact that different sizing techniques, based on different measuring principles/properties, may generate different equivalent diameter values for the same particles (the less spherical the particle the greater the difference) [1–5]. Conversely the same equivalent diameter may be obtained for different particles. Although this approach is simplistic it is very convenient and for that is employed in almost all particle sizing techniques.

Other issues that arise when trying to assess particle size is the large variety of equipment currently available (several hundreds of different types of instruments) and the fact that, at present, there are no standard techniques or methods to reliably determine this parameter. Indeed the most appropriate method is often selected on a case-by-case basis. In fact, a great variety of techniques are nowadays used for particle size measurement, in particular for nanoparticle sizing, based

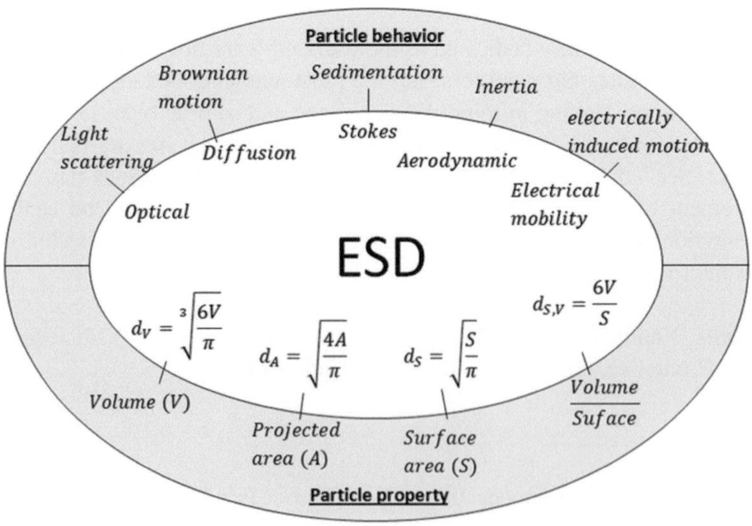

Fig. 1 Common equivalent spherical diameters (ESD) based on different particle properties/behavior (adapted from [2])

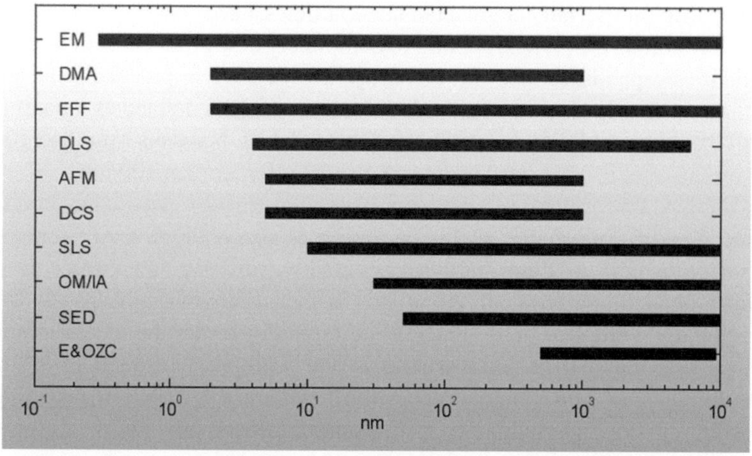

Fig. 2 Commercially available particle sizing techniques (mostly for liquid suspensions) and corresponding measuring ranges: *EM* electron microscopy, *DMA* differential mobility analyzer, *FFF* field flow fractionation, *DLS* dynamic light scattering, *AFM* atomic force microscope, *DCS* differential centrifugal sedimentation, *SLS* static light scattering, *OM/IA* optical microscope/image analysis, *SED* sedimentation, *E&OZC* electrical & optical zone counters

on the measurement of many different properties/parameters, including transport, geometrical, optical or electrical properties [4, 6–8]. These techniques (Fig. 2) correspond to different size ranges, different metrics and will also suit different types of samples (aerosol, liquid or powder).

This leads to an extreme difficulty in trying to classify or even group these techniques. Nonetheless they can broadly be divided in two main groups: particle ensemble characterization methods and single particle techniques. The former have typically low resolution and low sensitivity but a broad dynamic size range and high statistical accuracy whereas, on the contrary, single particle techniques, also referred to as counting techniques, exhibit high sensitivity and resolution but narrow dynamic size ranges. Ensemble analytical techniques can be further divided in classifying and non-classifying techniques.

Examples of single particle analyzers are electric impedance and microscopic/image analysis (SEM, TEM, AFM, etc.). Regarding the ensemble methods, sedimentation, centrifugation and the various forms of particle chromatography are examples of fractionation (or classifying) techniques since the particles may be totally or partially separated in space/time. Light attenuation and static and dynamic light scattering, in which the size distribution is derived from the sum of all the signals resultant from all the particles during the entire measurement, are examples of non-classifying methods. Because these methods normally involve deconvolution procedures, they are more prone to artifacts. Additionally they are fast, easily automated, and can be adapted for on-line and in-line measurements but their resolution is poorer than the classifying methods.

Nanoparticles, specially engineered nanoparticles, have generated a great interest due to their unique properties, leading to completely new products with outstanding potentialities/capabilities, and are currently incorporated in practically all aspects of life. According to the ISO standards [9] nanoparticle is defined as a nano-object with all three external dimensions at the nanoscale, i.e. between 1 nm and 100 nm. However the term is often used to refer to particles larger than 100 nm.

As known, the properties of nanoparticles are heavily dependent upon their size and thus the knowledge of this parameter is absolutely crucial. Nonetheless, the characterization of these particles accurately, robustly and in real time still remains a challenging problem. Although the measuring of nanoaerosols will not be specifically addressed here, it should be highlighted that the recent interest on how to assess the risk of nanoparticles in the workplace resulted in a burst of new techniques, or new applications of old techniques, capable of detecting and measuring very small particles at very low concentrations [10–12].

The difficulties in sizing particles increase as particle size decreases, being the region around one micron a kind of boarder region [6]. In fact, this region is normally considered the dividing line between the techniques adequate for "larger" particles and those particularly suited for nanoparticles. For instance, submicron particles are more conveniently characterized by centrifugation whereas (gravitational) sedimentation is more suited for larger and more dense particles. This region also represents a frontier between static and dynamic light scattering. Moreover regarding zone counting based instruments (electro- and photozone) the detection of particles bellow this value presents technical problems that prevent their adequate characterization. Finally this region also corresponds to the transition between optical and electrical microscopy.

The main purposes of this chapter is to introduce the problematics of particle sizing, specially to those who are not familiar with this topic, and to present a general overview of some of the currently used methods to measure nanoparticles, briefly discussing their operating principles, merits and limitations. However, the techniques discussed will not be restricted to nanoparticles (i.e. <100 nm) but will encompass in general the submicron range. In order to narrow the large number of possibilities, this survey will be mainly focused on **commercial equipment** normally available for **routine analysis** of **nanoparticles suspended in liquids** (or suspensions from dry powders).

Having into consideration the type and the size range of the samples of interest, seven particle characterization techniques were selected for further discussion which can be divided into three major groups that include: (i) microscopic methods, that include Transmission and Scanning Microscopy (**TEM** and **SEM**, respectively) and Atomic Force Microscopy (**AFM**), and that although not applied routinely are normally used (at least once) to really see the particles and get information about their shape, size and size range; (ii) two techniques that although based both on particle Brownian motion correspond to quite distinct working principles: Dynamic Light Scattering (**DLS**), more traditional, and Nanoparticle Tracking Analysis (**NTA**), recently commercialized; and (iii) two fractionation techniques based on centrifugal sedimentation but, again, quite different in their principle of measurement: Differential Centrifugal Sedimentation (**DCS**) and Sedimentation Field Flow Fractionation (**SdFFF**). Finally, it should be pointed out that although this selection was based on an extensive review of the currently available particle sizing instruments and on some personal experience, it does not attempt to provide an exhaustive survey; it is rather intended as a starting point for further investigations.

2 Microscopy-Based Methods

2.1 Electron Microscopy

Microscopy-based instruments are a powerful tool for particle characterization as they involve the direct observation of the particles and thus can provide information about surface features, shape and size. Because of that, they are probably the first choice to investigate unknown particles. However, they should not be considered the primary method of particle size analysis, since, as mentioned, it may be difficult to collect enough data to give a reliable result, and the sample preparation may induce agglomeration. In addition measurements can take long, require skilled operators, being also labor intensive if operated manually.

They are normally coupled to image analysis systems that can record the data of a set of particles to be further processed to generate statistical information. In these systems, particle size is normally expressed as an equivalent diameter, being the most frequent the diameter of a sphere that has the same projected area as the projected image of the particle.

Since the resolution limit of conventional light microscope is about 200 nm (approximately half of the wavelength of the illumination used), in order to see smaller details, smaller wavelengths are needed. Electron microscopes can magnify 100–1,000 times more than a visible light microscope since they use a beam of electrons that, after acceleration, can reach wavelengths of the order of 0.1 nm. This obviously allows a much higher magnification and resolution.

Electron microscopy besides providing a detailed characterization of the particles at sub-nanometer space resolution regarding topography and morphology, may also provide information about composition and crystallography. The extensive sample preparation, the need to analyze samples in vacuum and the necessity for an adequate representative sample are among the main drawbacks of this technique.

Electron microscopes function exactly as their optical counterparts except that they use a focused beam of electrons instead of light to "image" the specimen (Fig. 3). Basically, a beam of electrons is produced at the top of the microscope by heating of a metallic filament. The beam travels through electromagnetic fields and lenses, being focused down towards the sample. Once the beam hits the sample, electrons and X-rays are ejected from the sample (Fig. 4) which are collected in special detectors and converted into a signal that is sent to a screen, producing the final image [13, 14].

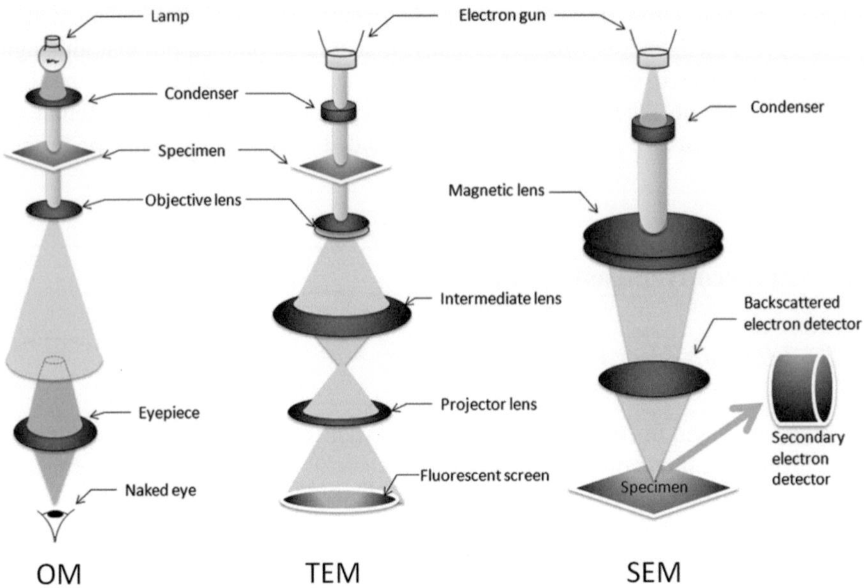

Fig. 3 Comparison of optical microscope (*OM*) with the transmission electron microscope (*TEM*) and the scanning electron microscope (*SEM*), emphasizing their similarities

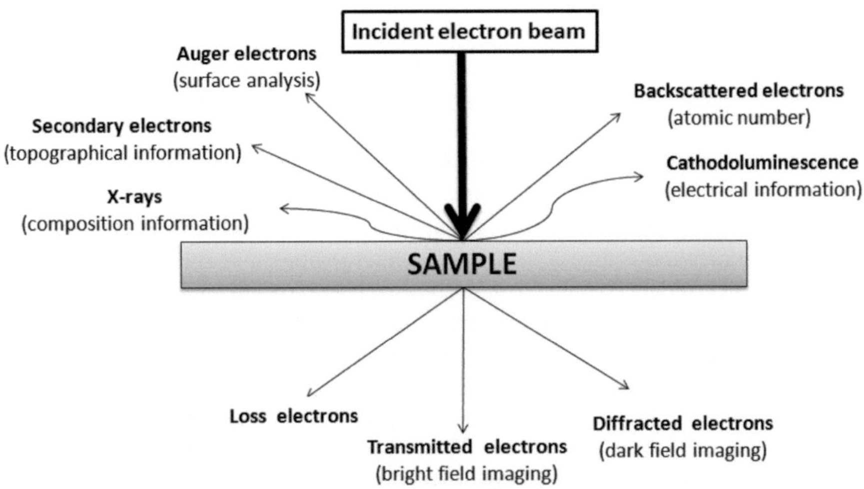

Fig. 4 Electron/sample interaction results in different types of signals (electrons and photons). The interactions on the top side of the diagram are utilized when examining thick or bulk specimens by SEM while on the bottom side are those examined in thin specimens by TEM

There are two common types of electron microscopes: Transmission Electron Microscope (TEM) and Scanning Electron Microscopy (SEM). The latter frequently offers the simplest approach to sample preparation and analysis and allows topographical imaging of particles (giving the impression of three dimensions).

The TEM relies on electrons penetrating through the sample producing a two dimensional image that provides information on the internal particle structure. The use of transmitted electrons requires thin electron transparent samples which limits the maximum diameter of particles analyzed by TEM to a few hundred nanometers. Additionally and due to the necessity of very thin slices of the specimen, TEM usually requires higher level of sample preparation. However, nanoparticles are relatively easy to be investigated by this technique since a single layer will be adequate. TEM is routinely capable of achieving sub-nanometer special resolution depending on the sample and instrument.

In SEM, the electron beam is not projected through the sample. Instead, it is scanned over the sample surface and the back scattering of the electrons emitted from the surface are recorded. The sample must therefore be electrically conductive at the surface, which is normally achieved by sputter coating a non-conductive sample. The bombarding electrons, also referred to as primary electrons, dislodge electrons from specimen, known as secondary electrons, which are attracted and collected by a detector and then translated into a signal. This generates a lower resolution image than TEM, but allows the direct mapping of the surface features. Aside from secondary electrons, the primary electron beam also results in the emission of backscattered (or reflected) electrons from the specimen. Back scattered electrons may also be used in imaging but require a special detector. As back-scattering is related to the atomic number, this imaging mode provides a qualitative information on sample surface composition.

The development of the environmental SEM (ESEM) enabled samples to be observed at low pressure gaseous environment, at high relative humidity (up to 100%) and at higher pressures. The ESEM is ideal for non-metallic surfaces such as biological materials [15].

Finally, electron-specimen interactions may be used to quantitatively characterize particles: characteristic X-rays generated through atomic excitation and decay within the specimen is routinely used in TEM and SEM to detect elemental components through, for instance, Energy Dispersive X-ray Spectroscopy (EDX, sometimes EDS). An EDX detector is used to separate the characteristic X-rays of different elements into an energy spectrum, and is used to determine the abundance of specific elements. EDX can be used to find the chemical composition of materials down to a spot size of a few microns, and to create element composition maps over a much broader raster area [16].

2.2 *Atomic Force Microscopy (AFM)*

Atomic Force Microscopy (AFM), commercially produced in the late 1980s, is part of the field of Scanning Probe Microscopy (SPM) methods which typically use a mechanical probe to "feel" the surface of the sample [17]. AFM consists of scanning a sharp tip on the end of a flexible cantilever over the sample surface, being the forces between the tip and the sample measured from the deflection of the

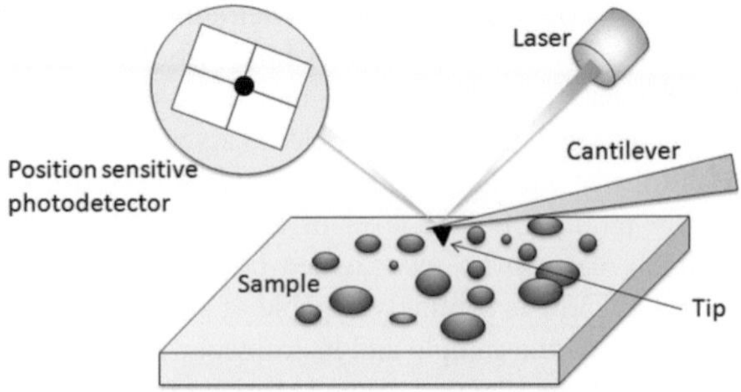

Fig. 5 Schematic illustration of the optical beam deflection system used in AFMs

cantilever. The tips typically have an end radius of 2–20 nm, depending on tip type. The deflection moves a laser spot that reflects into an arrangement of photodiodes, resulting in an extremely high sensitivity detection system (Fig. 5).

One of the greatest advantages of AFM is that it directly produces three dimensional images whereas traditional electron microscopes measure only two dimensional images (electron microscopes cannot measure the vertical dimension (z-direction) of the sample, like the height or depth of the surface features). AFM has been used in a variety of industries from pharmaceuticals to semiconductors with particular relevance in the emerging field of nanotechnology. It should be emphasized that AFM does not only give information on a variety of views of nanoscale structures but, more importantly, can also be used to manipulate those structures.

In addition to measuring the physical topography of samples, other type of information can be extracted as size, morphology, surface texture and roughness. Another major advantage of AFM is the possibility to characterize samples in multiple mediums including ambient air, controlled environments (such as nitrogen or argon) and liquid dispersions with minimal sample preparation. However, it should be pointed out that AFM images are always a convolution of the probe geometry and sample texture. Moreover, the application of AFM to nanoparticle characterization can be limited due to strong tip to surface interactions and applied lateral forces. As a result, the resolution capabilities in the X and Y directions is reduced in comparison with the Z-axis (sub-angstrom resolution in vertical direction against nanometer resolution in XY). Several procedures have been implemented for dimensional characterization of nanoparticles [18–22].

As AFM scanning is done physically with a probe, dry particles must be affixed to a sticky substrate (whose roughness must be less than the size of the nanoparticles). AFM can be operated in contact mode, non-contact mode and intermittent mode (or tapping mode), as illustrated in Fig. 6. The two most common modes are the former and the latter. Contact mode AFM consists of scanning the probe across a sample surface while monitoring the change in cantilever deflection. The distance the scanner moves vertically at each X, Y data point is stored by the computer to form the topographic image of the sample surface. Tapping mode consists of oscillating the cantilever at its resonance frequency and lightly “tapping” on the

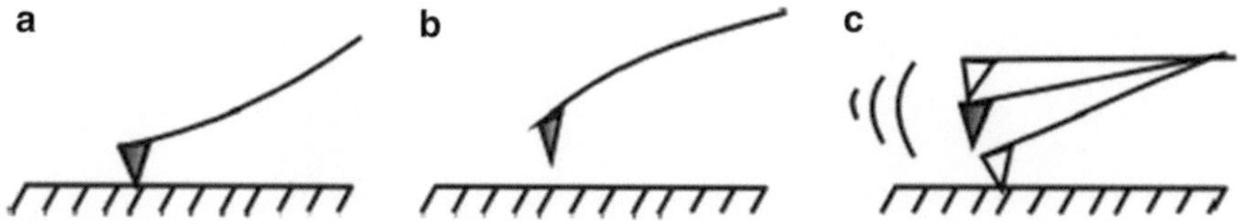

Fig. 6 Modes of operation of AFM: (**a**) contact mode; (**b**) non-contact mode; (**c**) tapping mode

surface during scanning. Contact mode results in higher scan speeds but involves higher forces in air which can damage soft samples. On the other hand, tapping mode overcomes problems associated with friction, adhesion, electrostatic forces. In this mode, lateral forces are virtually eliminated so there is no scrapping, being thus particularly suited for biological materials [23].

Although SEM and AFM appear very different, they both raster a probe across the sample surface to form an image. When possible they should be used together as they provide complementary information [24, 25]. AFM can provide measurements in all three dimensions (including height), whereas SEM is especially suited to very rough samples due to its large depth of field. In fact, and no matter how sharp the tip is, the data collected by AFM can never access the underside of the sample. On the other hand, AFM is more versatile, samples do not need any special treatment, being specially adequate to study biological macromolecules and even living organisms. Nonetheless, AFM cannot scan images as fast as SEM (practically in real time) and the maximum scanning area of AFM is much smaller than that of SEM (SEM can image an area of the order of millimeters whereas the image size of AFM is about $150 \times 150\ \mu m$). Finally, it should be pointed out that samples prepared for TEM and SEM may be directly imaged in an AFM.

3 Dynamic Light Scattering (DLS)

Dynamic Light Scattering (DLS) is an ensemble measuring method that, contrary to the above mentioned techniques, does not permit particle visualization. It has been the method of choice to routinely sizing submicron particles (from a few nanometers to a few micrometers), yielding accurate results in a rather short period of time (of the order of minutes) with little sample preparation requirements.

DLS, also known as Photon Correlation Spectroscopy (PCS), or Quasi-Elastic Light Scattering (QELS) as it was first termed, is a non-invasive user-friendly method widely used for 40 years for measuring the effective size of molecules in solution and particles regarding average size, size distribution and polydispersity [26–30].

The applications of this technique are wide ranging and include many of the fastest growing areas of research—proteins, biomaterials, liposomes and micelles, various polymers, paints, among others [31]. It is based on the determination of the

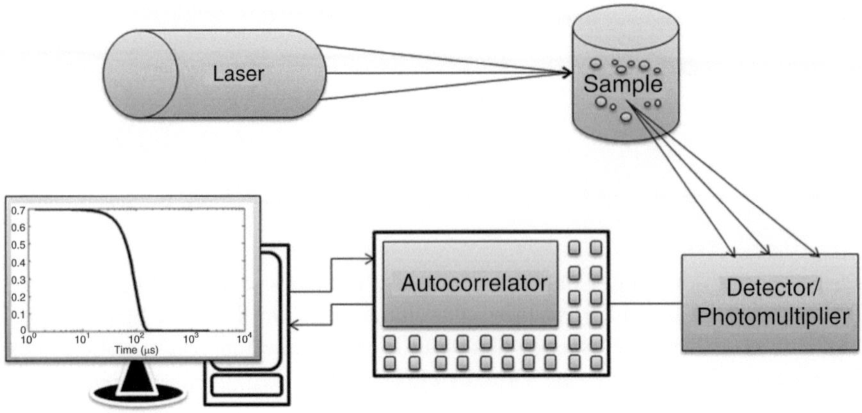

Fig. 7 Typical dynamic light scattering experimental set up

hydrodynamic properties of particles suspended in a liquid medium. If the particles are small, their collisions with the solvent molecules result in an erratic motion, called.

Brownian motion. When the particles are illuminated by a light source such as a laser, the intensity of the scattered light measured by a detector located at some point (Fig. 7) will fluctuate as a result of the different relative positions of the particles caused by the Brownian movements. An important feature of these movements is that small particles move quickly whereas large particles move slowly. These fluctuations can be analyzed as a function of time enabling the evaluation of the average (translational) diffusion coefficient of the particles which in turn can be transformed into an average particle size.

The time dependence of the intensity fluctuations is analyzed using a highly sensitive detector, normally a photomultiplier, that is positioned at a chosen angle (typically 90°or 175°). The output of the photomultiplier resembles a noise signal (Fig. 8) constituted of fluctuations (known as dynamic intensity) around an average value (equivalent to the scattering intensity if the particles were static).

This signal, that contains all the information about particle hydrodynamics, is fed to a digital correlator. This device basically measures the degree of similarity between two signals, or one signal with itself, over a period of time, yielding an intensity autocorrelation function (Fig. 9). Larger particles will result in autocorrelation functions that change slowly and thus the correlation will persist for longer times in comparison to the small, rapidly moving particles for which the correlation will disappear quickly.

Particle size resolution requires a deconvolution of the data contained in the measured correlation curve. A monodisperse sample would generate a single exponential decay from which is relatively simple to calculate the particle size distribution (e.g. cumulants method in which the autocorrelation function is fitted to

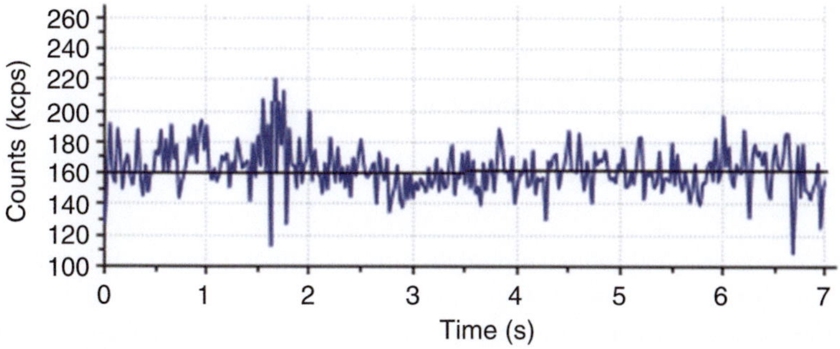

Fig. 8 Scattered light intensity (kilo counts per second, kcps) as a function of time (fluctuations around an average)

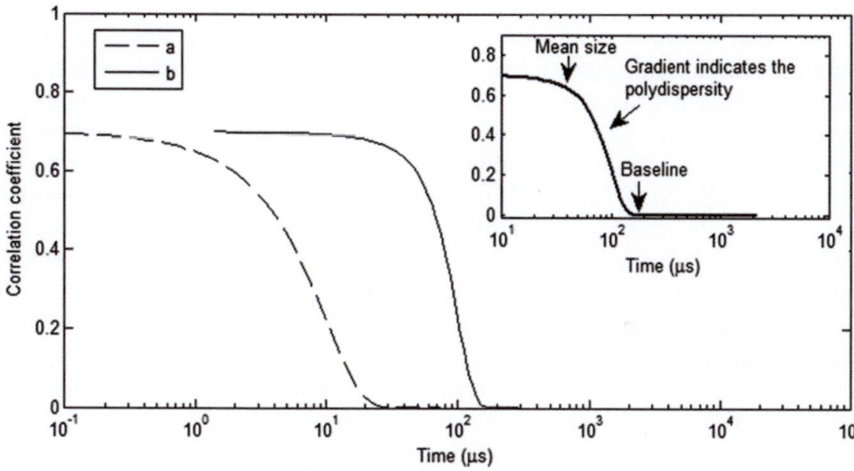

Fig. 9 Examples of the autocorrelation functions of two different particle populations (a-small particles and high polydispersity; b-large particles and medium polydispersity). The shape of curves provides information about sample size and size distribution: the point at which the function starts to decay gives information about particle size; the decay rate is related with the distribution polydispersity and a flat (or noisy) base line generally indicates the absence (or presence) of large particles or agglomerates

a low order polynomial [9, 27]). The decay rate is related to translational diffusion coefficient (D). Knowing the temperature and the viscosity of the suspension, the diffusion coefficient can be transformed into an average diameter, using the Stokes-Einstein Equation:

$$\mathbf{d}\,(\mathbf{H}) = \frac{\mathbf{kT}}{\mathbf{3\pi\eta D}} \quad (1)$$

where

d(H) Hydrodynamic diameter
D Translational diffusion coefficient
k Boltzmann's constant
T Absolute temperature and
η Viscosity.

This diameter refers to how a particle diffuses within a fluid so it is referred to as a hydrodynamic diameter, being defined as the diameter of a sphere that has the same translational diffusion coefficient as the particle. This average size is weighted by the particle scattering intensity and known as the Z average (or intensity average). In addition to the average decay rate (and the corresponding average size), the cumulants method also provides an indication of the spread of the decay rates and hence of the spread of particle size about the average value (called polydispersity index).

The autocorrelation function of polydisperse samples is no longer a single exponential decay function but a summation of single exponential functions (each of which corresponding to a different size class). In this case, the use of the cumulants method is obviously inadequate. Instead, to derive the particle size distribution, quite complex deconvolution algorithms have to be applied for the fitting process (e.g. CONTIN, NNLS) [26, 32–35]. However, most of these algorithms concern a mathematically ill-conditioned problem with no unique solution, assume a functional form for size distribution and select the simplest solution that fits the data. Noise in the scattered intensity signal and the dependence of the scattered intensity on particle size make resolution of broad or multimodal distributions difficult regardless the algorithm used.

Although the fundamental size distribution generated by DLS is an intensity based distribution (i.e. a plot of the relative intensity of light scattered by the particles versus the respective particle size), this can be converted to a volume distribution. For this Mie theory is applied (which requires a priori knowledge of the particles optical properties). Applying further assumptions about particle shape, a number distribution can be estimated but these are generally considered to be inaccurate.

Figure 10 exemplifies the differences between intensity, volume and number distributions. As the intensity of the scattered light of very small particles is proportional to the sixth power of the particle diameter, particles that are ten times larger will scatter one million times more light than the smaller particles and thus the contribution of the latter will be reduced in this type of distribution. On the contrary, a small artifact at low particle size when transformed from intensity—to number-weighting may become predominant. Special care must then be taken in sample preparation to avoid the presence of impurities, being pre filtering normally required. Additionally, systems where agglomerates may be present should be carefully analyzed. On the other hand, and to avoid multiple scattering and/or

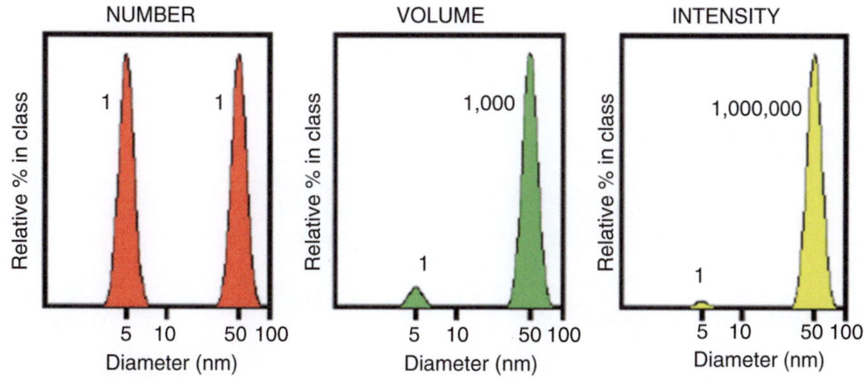

Fig. 10 Number, volume and intensity distributions of a bimodal mixture of 5 and 50 nm lattices in equal numbers (reproduced with the permission of Malvern Instruments)

particle-particle interactions, low particle concentration is normally necessary. Nevertheless new equipment configurations allow to overcome this limitation (e.g. the collection of back scattered light).

The main drawback of DLS is, as mentioned, its limitation to produce accurate particle size distributions from multimodal samples. Moreover, it is generally accepted that this technique can only resolve particle populations that differ in size by a factor larger than 2:1. Nonetheless it is very accurate, reproducible and extremely rapid for sizing narrow distributions, which makes this technique very popular.

4 Nanoparticle Tracking Analysis (NTA)

Nanoparticle Tracking Analysis is a relatively new technique (commercialized in 2006) that measures simultaneously, many individual particles by video tracking [36].

This technique combines laser light scattering microscopy with a CCD camera, which enables the real time visualization of nanoparticles in liquids. Particle motion is analyzed by video, allowing to track particle positional changes, from which particle diffusion and ultimately particle size can be determined.

Like Dynamic Light Scattering, NTA is based on the particles Brownian motion and uses the Stokes-Einstein equation to compute sizes but the operating principle of these two techniques is quite diverse. As discussed below, NTA is not an ensemble method; instead particles are visually analyzed on a particle-by-particle basis. However, it should be pointed out that NTA does not enable a resolved image of the particles, like in microscopic methods, but the particles are only visualized as point scatterers. It covers the range between 30 nm and 1 μm.

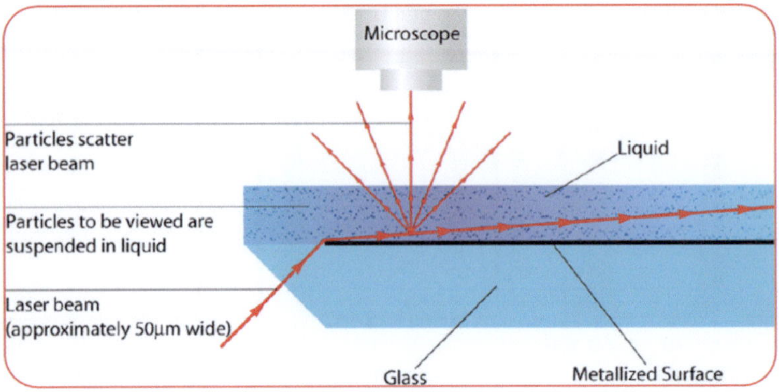

Fig. 11 Schematic illustration of the NTA working principle [37] (reproduced with the permission of NanoSight Ltd)

As depicted in Fig. 11, samples prepared in an appropriate liquid (mainly water) are placed in a sample chamber and illuminated by specially configured laser beam. The dispersed light is captured by a CCD camera, via microscope and thus an idea of the presence, polydispersity and concentration can be immediately obtained. As mentioned, Brownian motion of each particle (separately but simultaneously) is followed in real time via video. Such videos are subsequently analyzed using specially designed software, enabling the estimation of the diffusion coefficients, from which hydrodynamic particle diameters can be derived through Stokes-Einstein equation (Eq. (1)). Furthermore, because NTA tracks particles within a known volume of liquid, the size distribution which is produced is a true and direct number frequency distribution and relative particle concentrations can therefore be determined.

Comparison studies involving both techniques, NTA and DLS, led to the conclusion that they complement each other and thus should be used together [38]. Nonetheless, and as described, their working principle is quite distinct. Whilst DLS measures changes in scattering intensity from a bulk sample, NTA observes individual particle diffusion directly and thus builds the distribution one particle at a time. On the other hand, NTA is not intensity-weighted and hence does not tend to be biased towards larger particles. Moreover, it is claimed to have much higher resolving power than DLS with respect to polydisperse and heterogeneous samples. Finally it provides particle number concentration and does not require information on refractive index. Nonetheless for monodisperse samples, more reproducible results are obtained with DLS since in this technique many more particles are analyzed. Besides, sample concentration range is less critical in DLS which is also less time consuming.

5 Differential Centrifugal Sedimentation (DCS)

Sedimentation of particles in a fluid has long been used to determine particle size distribution [1]. It is based on the determination of the time required for the particles to settle a known distance **h** in a fluid of known viscosity and density. The particle diameter is calculated using the Stokes Equation:

$$\mathbf{D} = \sqrt{\frac{18\mu \mathbf{h}}{\mathbf{g}\left(\rho_p - \rho_m\right)\mathbf{t}}} \tag{2}$$

where

D Diameter of particle
μ Viscosity of medium
h Settling distance during time **t**
g Gravitational acceleration
ρ_p Density of particle and
ρ_m Density of medium.

This equation, valid for single spheres settling under gravity, in a fluid of infinite extent and under laminar flow conditions, clearly shows that the sedimentation velocity (h/t) increases as the square of the particle diameter (hence particles that differ in size by only a few percent settle at significantly different rates, improving the resolving power of sedimentation methods).

Sedimentation can be either gravitational or centrifugal. Gravitational sedimentation is limited to larger particles since in the settling process particles smaller than about a micron, besides resulting in too long settling times, are highly affected by convection, diffusion and Brownian motion, which become more important than the motion due to gravitational settling. By applying a centrifugal force, the size range measured can be extended down to a few nanometers. The relationship between the settling time and particle size when a centrifuge is used is given by a modified Stokes law that takes into account the variation in centrifugal force with distance from the center of rotation:

$$\mathbf{D} = \sqrt{\frac{18\mu \mathbf{l}\,\mathbf{n}\left(\frac{\mathbf{R_f}}{\mathbf{R_0}}\right)}{\left(\rho_p - \rho_f\right)\omega^2 \mathbf{t}}} \tag{3}$$

where $\mathbf{R_0}$ = initial radius of rotation; $\mathbf{R_f}$ = final radius of rotation; $\boldsymbol{\omega}$ = rotational velocity; the other variables as in Eq. (2). This equation is valid for the line start technique, in which all the particles start to settle at the same position.

For a centrifuge running at constant speed and temperature, all the parameters of Eq. (3), except time, are constant during analysis. Their values are either known or

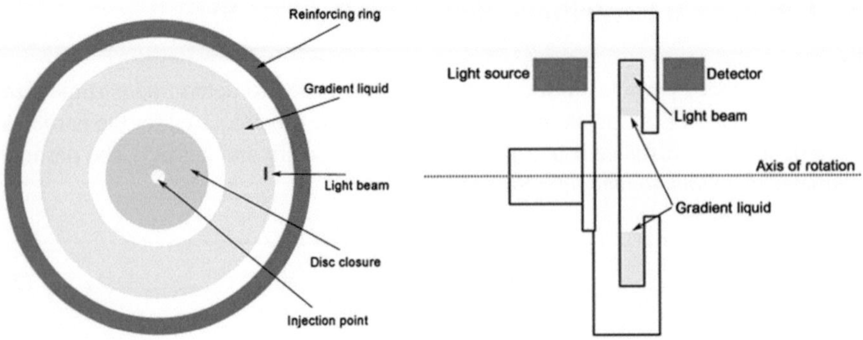

Fig. 12 Typical configuration of DSC disc (reproduced with the permission of Analytic Ltd)

can be accurately measured. To improve the accuracy of these values, it is normal to use a narrow calibration standard of known size to determine the effective combined value for all these parameters, K, defined as:

$$\mathbf{K} = \sqrt{\frac{\mathbf{18\mu l\,n}\left(\frac{\mathbf{R_f}}{\mathbf{R_0}}\right)}{\left(\rho_p - \rho_f\right)\omega^2}} \tag{4}$$

and thus the time scale can be directly converted to particle size as:

$$\mathbf{D} = \mathbf{K}\sqrt{\frac{\mathbf{1}}{\mathbf{t}}} \tag{5}$$

The disc centrifuge operates by forcing particles radially outward through a spin fluid under high centrifugal force. The most common design for centrifugal sedimentation instruments is a hollow, optically clear disc with a central opening on one side. The opposite side of the disc is centrally mounted on a drive shaft that is driven by a variable speed motor (from a few hundreds to several thousands rpm). A typical disc cross section is shown in Fig. 12.

As illustrated in Fig. 12, the empty spinning disc chamber is partially filled with liquid, known as spin disc (normally sucrose solution). Typically it forms a ring inside the chamber with a slight density gradient to prevent hydrodynamic instabilities. A dilute sample (<1% solid content) is injected into the center of the disc at the beginning of the analysis. The time for particles to reach the detector versus beam intensity is converted to size distribution according to Stokes's law (Eq. (4)). For irregular particles this leads to an equivalent spherical diameter, also known as Stokes equivalent diameter, equal to the diameter of a sphere that sediments at the same velocity as the actual particle.

At any time, only particles of one particular size range (a "differential") are being measured by the detector (this is why the disc centrifuge method is named "differential sedimentation"). Indeed all larger particles have reached the detection zone earlier (higher settling rates) and all the smaller ones have not yet arrived (lower settling rates). The detector signal is reduced when particles reach the detector beam, being the reduction in intensity related to particle concentration. The detector beam is usually monochromatic light or X-rays. Shorter wavelength light (400–500 nm) gives better detection sensitivity, being especially suited for nanoparticle measurements. In the case of an X-ray detector, the reduction in intensity is proportional to particle concentration, the only limit is the weak signal produced by materials of low atomic number (<13, aluminium). When a monochromatic light beam is used, Mie theory light scattering (involving extinction efficiency factors) has to be applied to the intensity data to compute particle concentration [39, 40]).

Recent advances in technology allowed to greatly improve Differential Centrifugal Sedimentation overcoming previous limitations and difficulties associated to this technique namely ease of use, speed of analysis, and accuracy. DCS is now considered a powerful tool in measuring particle size distributions down to a few nanometers and up to a few tenths of micron (depending on particle density).

One of the advantages of this technique is its excellent resolution: very close multi-modal particle distributions (modes with a diameter ratio of 5% can be completely resolved) which makes this technique very useful to detect/measure small differences or changes in particle size. Analysis time depend upon the range of sizes that is analyzed as well as the density of the particles, reaching, for very small particles, more than one hour. Special discs and techniques are now available to enable ramping of a disc centrifuges speed during an analysis. This is useful if a sample contains some relatively large particles as well as nano particles. An initial slow centrifuge speed is used to measure the larger particles, being the speed ramped up to the maximum so that the nano particles can still be measured within a reasonable timeframe [41].

Accuracy can be verified using calibration standards. However, absolute weight accuracy of the distribution depends on knowing optical properties (if a light detector is used) and particle shape [4].

6 Sedimentation Field Flow Fractionation (SdFFF)

Sedimentation Field flow Fractionation (SdFFF or SdF3) is one of a number of field flow fractionation (FFF) methods, first introduced by Giddings in 1966, particularly suited for separating and characterizing abroad range of fine particles and molecules, from nanoparticle suspensions to proteins, polysaccharides, synthetic macromolecules, microgels, or environmental pollutants [42–45].

The operating principle of SdFFF is illustrated in Fig. 13. The separation of the sample takes place inside a narrow ribbon-like channel (a few hundred microns

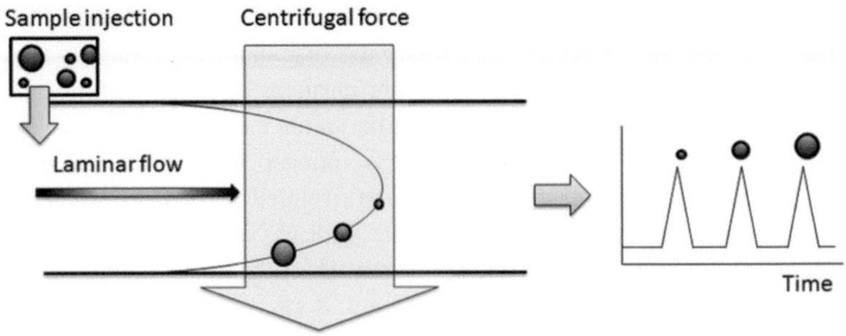

Fig. 13 Schematic illustration of SdFFF operation principle (adapted from [46])

thick and 0.25–1 m in length) which sits within a rotating disc. A carrier liquid is pumped through this channel from the inlet, where the sample is injected, to the outlet, to which a detector is connected. Inside the channel, a parabolic flow profile is established, with the highest velocity at the center of the channel. As the main flow carries the sample particles through the channel they are affected by a centrifugal field applied across the channel perpendicular to the flow direction. The particles are thus driven towards the outside edge of the flow chamber. Particles with a larger effective mass (larger size) are compressed against the chamber wall forming a thin compact layer while small particles, with greater Brownian motion form a cloud of particles and thus a more diffuse layer more close to the center. Interaction with the stream lines of the parabolic profile separates the particles according to size fractions: smaller particles, located in fast stream lines, elute first whereas bigger particles, located in slow stream lines, elute later. The determination of concentration is usually based on light scattering (UV as well as dynamic and static light scattering). The measured concentration versus elution time is converted into a size distribution using instrument software.

Since the field force is proportional to particle volume multiplied by the density difference between particle and liquid, particles are separated according to their buoyant mass, which in turn is proportional to the volume equivalent spherical diameter provided that the particle density is uniform. This working principle allows a very high resolution separation of the particles (peaks as 5% different in relative size can be resolved). Additionally, a very large size range of particles can be separated, from submicron to supramicron size (from 5 to 10 nm nanometers to more than 100 micron, depending on sample characteristics). SdFFF has been shown to have an extraordinary ability to study aggregation phenomena and to track particle size distribution changes caused by aggregation in colloidal samples [47]. For samples which contain a broad size distribution, programs that permit centrifuge speed decays are available to improve resolving power across the size range [48].

The complicated algorithms used for size separation together with the long run times ($\sim$1 h) and the mechanical/maintenance problems are among the main drawbacks of this technique. On the other hand, its unique fractionation principle that not only allows particle separation and characterization but also allows the collection of narrow size fractions for further analysis (e.g. electron microscopy, elemental analysis, dynamic light scattering) constitutes a major advantage. Additionally, as the separation is relatively gentle, SdFFF has gain wide application in the biological field [49].

7 Conclusions

From the above considerations, it is clear that, although limited to the submicron range, the selection of the most convenient particle sizing technique continues to be a challenging task. In fact all the methods present strong and weak points and no universal technique is yet available. Additionally, the lack of written standards, absolutely necessary to ensure consistent sample preparation and measuring procedures, is also a major limitation.

As discussed, imaging techniques are indispensable to visualize the nanoparticles as well as to assess structural parameters, however sample preparation and the operating conditions may mask some of the particles features. Besides, the limited number of particles per image may cause statistical problems. On the other hand, ensemble techniques, like dynamic light scattering, involve deconvolution processes that limit their resolution. Nonetheless they are extremely quick and most reliable for narrow monomodal size distributions, and for that they are widely used. NTA could be combined with PCS to improve resolution. The other alternative for broad size or multi-modal distributions is to use fractionation methods, namely centrifugal sedimentation, that despite providing very good separations may be relatively slow. Field flow fractionation methods still offer the additional advantage of providing monosized fractions for further investigation.

Finally, it should be once again emphasized that each technique provides a different type of equivalent spherical diameter (projected area based, hydrodynamic, Stokes, etc.) and generates distributions weighted in different ways (number-weighted as in single particle methods, mass-weighted as in sedimentation or intensity-weighted as in dynamic light scattering). Furthermore, for most of the methods involving light scattering (either as a working principle or to evaluate particle concentration), additional information about the particle optical properties (not always available) is frequently required.

The overall conclusion is that a comparison of methods, yielding complementary information and eliminating eventual artifacts, is highly recommended to obtain accurate data.

References

1. Allen T (1997) Particle size measurement, vol I: Powder sampling and particle size measurement. Chapman & Hall, London
2. Baron PA, Willeke K (2001) Aerosol measurement: principles, techniques, and applications. Wiley, New York

3. Provder T (1991) Particle size distribution II: assessment and characterization. American Chemical Society, Washington, DC
4. Bowen P (2002) Particle size distribution measurement from millimeters to nanometers and from rods to platelets. Taylor & Francis, Philadelphia
5. Barreiros FM, Ferreira PJ, Figueiredo MM (1996) Calculating shape factors from particle sizing data. Part Part Syst Charact 13:368–373
6. Weiner B (2011) A guide to choosing a particle sizer. In: Focus on microscopy & microtechniques. http://www.labmate-online.com. Accessed Mar 2012
7. Stintz M (2009) Review of nanometrology for engineered nanoparticles within europe – strengths & opportunities. In: Co-nanomet European workshop. Technical University Dresden, DE. http://www.co-nanomet.eu. Accessed Mar 2012
8. Łojkowski W, Turan R (2006) In: Łojkowski W, Turan R, Proykova A, Daniszewska A (eds) Eighth nanoforum report: nanometrology European nanotechnology gateway, Düsseldorf
9. ISO (2008) Particle size analysis—dynamic light scattering (DLS). International Standards for Business, Government and Society
10. Aitken RJ, Creely KS, Tran CL (2004) Nanoparticles: an occupational hygiene review. In: Research report 274. Institute of Occupational Medicine. http://www.hse.gov.uk/research/rrpdf/rr274.pdf. Accessed Mar 2012
11. Kuhlbusch TAJ, Asbach C, Fissan H, Göhler D, Stintz M (2011) Nanoparticle exposure at nanotechnology workplaces: a review. Part Fibre Toxicol 8:8–22
12. ISO (2007) Workplace atmospheres – ultrafine, nanoparticle and nano-structured aerosols – Inhalation exposure characterization and assessment. International Standards for Business, Government and Society
13. Goodhew PJ, Humphreys FJ, Beanland R (2001) Electron microscopy and analysis. Taylor & Francis, London
14. Flegler SL, Heckman JW, Klomparens KL (1993) Scanning and transmission electron microscopy: an introduction. Oxford University Press, Oxford
15. Heath JP (2005) Dictionary of microscopy. Wiley, Chichester
16. Goldstein J, Newbury DE, Echlin P, Joy DC, Romig AD Jr, Lyman CE, Fiori C, Lifshin E (1992) Scanning electron microscopy and X-ray microanalysis: a text for biologists, materials scientists, and geologists. Plenum Press, New York
17. Binnig G, Quate CF, Gerber C (1986) Atomic force microscope. Phys Rev Lett 56:930–933
18. Boyd RD, Cuenat A (2010) New analysis procedure for fast and reliable size measurement of nanoparticles from atomic force microscopy images. J Nanopart Res 13:105–113
19. Starostina N, West P (2006) Quantitative and qualitative nanopowder nanoparticle characterization with AFM. Adv Powder Metall Part Mater 1:02–17
20. Starostina N, Brodsky M, Prikhodko S, Hoo CM, Mecartney ML, West P (2008) AFM capabilities in characterization of particles and surfaces: from angstroms to microns. J Cosmet Sci 59:225–232
21. Gélinas V, Vidal D (2010) Determination of particle shape distribution of clay using an automated AFM image analysis method. Powder Technol 203:254–264
22. Mechler Á, Kopniczky J, Kokavecz J, Hoel A, Granqvist CG, Heszler P (2005) Anomalies in nanostructure size measurements by AFM. Phys Rev B 72:125407–125412
23. Mott D, Cotts B, Lim IS, Luo J, Park HY, Njoki P, Schadt MJ, Zhong CJ (2008) Size determination of nanoparticles based on tapping-mode atomic force microscopy measurements. J Scan Probe Microsc 3:1–8
24. Gupta S, Brouwer P, Bandyopadhyay S, Patil S, Briggs R, Jain J, Seal S (2005) TEM/AFM investigation of size and surface properties of nanocrystalline ceria. J Nanosci Nanotechnol 5:1101–1107
25. Russell P, Batchelor D, Thornton J (2001) Scanning Electron Microscopy (SEM) and Atomic Force Microscopy (AFM): complementary techniques for high resolution surface investigations. Surf Mod Technol 134(10):109–118
26. Berne BJ, Pecora R (2000) Dynamic light scattering: with applications to chemistry, biology, and physics. Dover Publications, Mineola

27. ISO (1996) Particle size analysis – photon correlation spectroscopy. International Standards for Business, Government and Society
28. Pecora R (1985) Dynamic light scattering: applications of photon correlation spectroscopy. Plenum, New York
29. Dahneke BE (1983) Measurement of suspended particles by quasi-elastic light scattering. Wiley, Chichester
30. Brown W (1993) Dynamic light scattering: the method and some applications. Oxford University Press, Oxford
31. Cao A (2003) Light scattering. Recent applications. Anal Lett 36:3185–3225
32. Provencher SW (1982) A constrained regularization method for inverting data represented by linear algebraic or integral equations. Comput Phys Commun 27:213–227
33. Finsy R (1994) Particle sizing by quasi-elastic light scattering. Adv Colloid Interface Sci 52:79–143
34. Stepanek P (1993) Data analysis in dynamic light scattering. In: Brown W (ed) Dynamic light scattering: the method and some applications. Oxford University Press, Oxford
35. Roig AR, Alessandrini JL (2006) Particle size distributions from static light scattering with regularized non-negative least squares constraints. Part Part Syst Charact 23:431–437
36. Malloy A, Carr B (2006) Nanoparticle tracking analysis – the Halo™ system. Part Part Syst Charact 23:197–204
37. Malloy A (2011) Count, size and visualize nanoparticles. Mater Today 14:170–173
38. Filipe V, Hawe A, Jiskoot W (2010) Critical evaluation of Nanoparticle Tracking Analysis (NTA) by NanoSight for the measurement of nanoparticles and protein aggregates. Pharm Res 27:796–810
39. Weiner BB, Fairhurst D, Tscharnuter WW (1991) Particle size analysis with a disc centrifuge: importance of the extinction efficiency. In: Provder T (ed) Particle size distribution II: assessment and characterization. ACS symposium series, Washington, DC
40. Devon MJ, Provder T, Rudin A (1991) Measurement of particle size distributions with a disc centrifuge. Data analysis considerations. In: Provder T (ed) Particle size distribution II: assessment and characterization. ACS symposium series, Washington, DC
41. Wittmer S. CPS disc centrifuges – a well proven method with modern technology. LOT-Oriel, Darmstadt
42. Giddings JC, Myers MN, Moon MH, Barman BN (1991) Particle separation and size characterization by sedimentation field-flow fractionation. In: Provder T (ed) Particle size distribution II: assessment and characterization. American Chemical Society, Washington, DC
43. Cölfen H, Antonietti M (2000) Field-flow fractionation techniques for polymer and colloid analysis. Adv Polym Sci 157(2000):67–187
44. Messaud FA, Sanderson RD, Runyon JR, Otte T, Pasch H, Ratanathanawongs Williams SK (2009) An overview on field-flow fractionation techniques and their applications in the separation and characterization of polymers. Prog Polym Sci 34:351–368
45. Contado C, Argazzi R (2009) Size sorting of citrate reduced gold nanoparticles by sedimentation field-flow fractionation. J Chromatogr A 1216:9088–9098
46. Arakawa T, Philo JS, Ejima D, Sato H, Tsumoto K (2007) Aggregation analysis of therapeutic proteins, part 3: principles and optimization of Field-Flow Fractionation (FFF). Bioproc Int 5:52–70
47. Barman BN, Giddings JC (1991) Overview of colloidal aggregation by sedimentation field-flow fractionation. In: Provder T (ed) Particle size distribution II. American Chemical Society, Washington, DC
48. Murphy DM, Garbarino JR, Taylor HE, Hart BT, Beckett R (1993) Determination of size and element composition distributions of complex colloids by sedimentation field-flow fractionation – inductively coupled plasma mass spectrometry. J Chromatogr A 642:459–467
49. Roda B, Zattoni A, Reschiglian P, Moon MH, Mirasoli M, Michelini E, Roda A (2009) Field-flow fractionation in bioanalysis: a review of recent trends. Anal Chim Acta 635:132–143

Nanoparticles for Oral Delivery of Insulin

Antonio J. Ribeiro, Raquel Seiça, and Francisco Veiga

Abstract The development of insulin nanoparticles (NP) in addition to protecting insulin against a harmful gastric environment can avoid its enzymatic degradation, while controlling its release and enhancing its absorption in the small intestine.

The enhancement of oral bioavailability of nanoencapsulated insulin has been demonstrated in animal studies. However, the question is why they were not followed by industrial development and clinical trials is still open. The variability of the insulin concentration delivered to the blood, the low bioavailability of encapsulated insulin requiring the administration of high doses and the lack of uptake mechanisms evidence, represent handicaps for the development of such delivery systems.

Administration of such high doses of insulin transcellularly may induce mitogenic changes in the gastrointestinal epithelium, as insulin is a growth factor. Current NP systems must be improved both to increase the insulin loading and to lower the insulin dose required to produce a therapeutic glycemia reduction.

Keywords Gastrointestinal epithelium • Insulin • Nanoparticle • Oral bioavailability • Therapeutic dose • Personalized medicine

A.J. Ribeiro (✉) • F. Veiga
Centre for Pharmaceutical Studies, Faculty of Pharmacy, University of Coimbra, Azinhaga de Santa Comba, 3000-548 Coimbra, Portugal
e-mail: aribeiro@ff.uc.pt

R. Seiça
Institute of Biomedical Research in Light and Image (IBILI), Faculty of Medicine, University of Coimbra, 3000-354 Coimbra, Portugal

J. Coelho (ed.), *Drug Delivery Systems: Advanced Technologies Potentially Applicable in Personalised Treatment*, Advances in Predictive, Preventive and Personalised Medicine 4, DOI 10.1007/978-94-007-6010-3_4,

1 Introduction

Diabetes mellitus is a chronic disease that occurs either when the pancreas does not produce enough insulin and/or when the body became resistant to insulin actions. Diabetes and its complications have a significant economic impact on individuals, families, health systems and countries. For example, WHO estimates that 346 million people worldwide have diabetes and that deaths related to it will double between 2005 and 2030 [1]. Type 1 diabetes (previously known as insulin-dependent, juvenile or childhood-onset) is characterized by an almost complete loss of insulin mainly due to selective autoimmune destruction of the pancreatic β-cells and requires daily specific forms of diabetes. The etiology of type 1 diabetes is multifunctional and may involve genetic and environmental factors. Type 2 diabetes (formerly called non-insulin-dependent or adult-onset) results from insulin resistance and deterioration and β-cell function. Type 2 diabetes comprises 90% of people with diabetes around the world, and is largely related to excess body weight and physical inactivity. Gestational diabetes is hyperglycemia with onset or first recognition during pregnancy. Impaired glucose tolerance (IGT) and impaired fasting glucose (IFG) are intermediate conditions in the transition between normality and diabetes. People with IGT or IFG are at high risk of progressing to type 2 diabetes, although this is not inevitable.

Other specific are associated with endocrinopathies, exocrine pancreas disease, genetic deterioration of the β-cell or in insulin action.

There is a need of a timely and reliable diagnosis based on the availability of adequate options for prevention and therapy, which include a healthier life style, such as food restriction and intensified physical exercise. The therapy of diabetes 2 is currently based on approved drug classes encompassing both small molecules (biguanides, sulfonylureas, meglitinide derivatives, alpha-glucosidase inhibitors, thiazolidinediones (TZDs), dipeptidyl peptidase-IV (DPP-IV) inhibitors) and proteins (insulin and analogues, glucagon-like polypeptide-I and analogues). Certainly, this repertoire does not meet the requirements for the individualized treatment of this common disease with its estimated 2,000–3,000 susceptibility genes and even higher numbers of the underlying polymorphisms and combinations thereof [2].

The gene products encoded by susceptibility genes are intrinsically characterized by exquisite potency and, in consequence, safety, in general, as well as by efficient quasi 'inborn' discovery [2]. Therefore, upon their identification they may be used for therapy directly by substitution of defective or missing gene products or indirectly by blockade of over-active or over-produced gene products using neutralizing antibodies directed against them.

As insulin, gene products are potentially administrated by parenteral route, often multiple daily injections are necessary, not contributing to improve the quality of life of diabetics. To eliminate the pain and fear associated with injections and increase diabetics compliance, researchers have actively worked to identify administration routes or alternatives that are painless and non-invasive. Generally, oral administration can improve disease management, enhance patient compliance and reduce the long-term complications of diabetes [3]. However, therapeutic proteins and antibodies have to overcome a number of critical hurdles prior to beneficial systemic action. One of the most important is the passage along the gastrointestinal tract and the transport across intestinal epithelial cells into the circulation.

In recent years, there has been considerable work in preparing materials and finding new uses for nanoscale structures based on biomaterials. Uses such as nanocarriers for controlled and targeted drug delivery, systems for biological recognition, have shown the versatility of these biopolymeric materials [4]. The development of nanoparticulate systems for drug delivery applications, particularly for insulin delivery, has evolved through remarkable advances never before seen in the field of drug delivery [5]. It is a prerequisite of oral insulin delivery systems that nanoparticles provide the hormone the protection in the harmful gastrointestinal environment, to facilitate its absorption through the gastrointestinal epithelium, and that insulin is delivered in its bioactive form to ensure glycemic control. Taking in to account the macromolecule nature of genes and insulin and the aimed oral delivery of both, the aim of chapter is to present a comprehensive review of the role of nanoparticles on insulin absorption through intestinal epithelium.

2 Diabetes Treatment

Type 1 diabetes treatment is a daily task. Lack of insulin production by the pancreas is particularly difficult to control. Treatment requires a regimen that typically includes multiple daily insulin injections and home blood glucose testing a number of times per day. Treatment of type 2 diabetes typically includes medication of oral hypoglycemic agents (biguanides, sulfonylureas, meglitinide derivatives, alpha-glucosidase inhibitors, thiazolidinediones (TZDs), dipeptidyl peptidase-IV (DPP-IV) inhibitors). If the diabetes is not controlled with this approach, combination therapy that is medication with different drug classes like a combination of a biguanide and sulfonylurea may be used. Traditional oral hypoglycemic agents such as sulfonylurea, metformin, the glitazones and gliptins are able to lower hemoglobin A1c (A1c) levels about 0.5–2.0%. Combinations of these agents can under optimum conditions achieve an A1c reduction of 3%. Thus in people close to goal of 7.0% [6] then exercise, diet and a single agent are appropriate. If the A1c is between 7.5 and 9.0%, combinations of oral agents are an appropriate first choice. When A1c is >9.0 unlimited capacity of insulin is recruited to achieve goal [1]. This can be accomplished in a number of ways which include addition to the oral regimen, use of a single long acting insulin analog, use of combinations of different forms of insulin and finally a basal long acting insulin together with a short acting bolus based upon the prevailing blood glucose and the anticipated carbohydrate intake [7, 8]. Approximately 40% of people with type 2 diabetes require insulin injections and data from the United Kingdom Prospective Diabetes Study (UKPDS) indicate that after insulin is introduced either alone or in combination with oral therapy, the long-term outcome is improved glycemic control. Therefore, yet controversial, soon or later the diabetic will be on insulin.

3 Insulin Treatment

The therapeutic goal of the treatment with insulin is to reproduce, as accurately as possible, the physiological pattern of insulin secretion, in order to control glycemia (which generally fluctuates between 3.5 and 7 mmol/l during the course of a day), avoiding the complications arising from hyperglycemia [9].

As we can see in Fig. 1, insulin is secreted along the day not only following meals, as a response to higher blood sugar levels (in order to stimulate glucose uptake by hepatic, muscular and fat cells), but also in lesser quantities, in a permanent way (basal secretion), during night and outside the mealtimes according to the level of counter-regulatory hormones in order to maintain blood glucose levels.

The optimal method of insulin delivery must be safe, provide the hormone in a way that compensates for the metabolic abnormalities of diabetes mellitus, and be psychologically and socially acceptable [10, 11].

Insulin, as in the case of most peptidic drugs, is currently and mostly delivered by parenteral administration. Since the discovery of insulin, it is delivered to diabetic patients exclusively via the subcutaneous route. There are four different types of

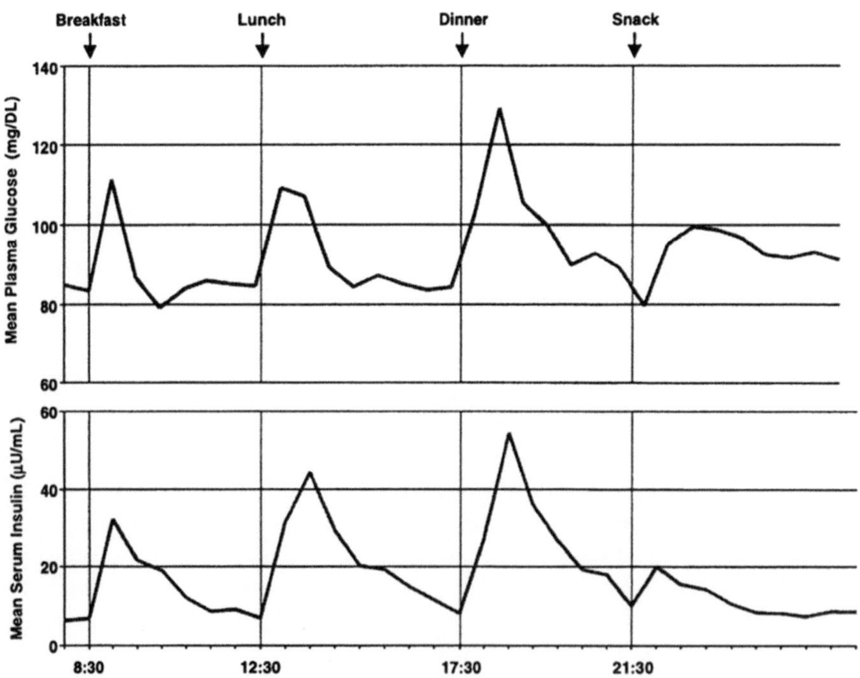

Fig. 1 Mean 24-h physiologic serum insulin and plasma glucose levels in non-diabetic subjects. Insulin outbursts coincide with increases in plasma glucose levels at mealtimes, but it is also secreted in a permanent way, though in lesser quantities [9]

Table 1 Types of insulin formulations available in the market

Category	Insulin formulation	Onset (h)	Peak (h)	Duration (h)
Rapid	Lispro	0.25–0.50	0.50–2.50	3-00–6.00
	Aspart	0.70–0.33	0.50–1.50	3.00–5.00
	Glulisine	1.00–2.00	1.00–3.00	1.00–2.50
Short-acting	Regular	0.50–2.00	1.00–5.00	4.00–10.00
Intermediate	NPH	1.00–4.00	6.00–14.00	12.00–24.00
	Lente	1.00–4.00	4.00–14.00	16.00–24.00
	Detemir	1.00–4.00	6.00–8.00	6.00–24.00
Long-acting	Ultralente	4.00–12.00	8.00–20.00	18.00–30.00
	Glargine	2.00–4.00		20.00–24.00

insulin available in the market as we can see in Table 1. The usual duration of action is relatively short; i.e., 4–8 h and therefore daily two to four injections are required for proper control of severe diabetic condition [12].

Intensive insulin therapy aims to imitate the complex kinetics of physiological insulin secretion. Reproducibility studies of time-action profiles of insulin preparations after subcutaneous injection show considerable intra- and interindividual variations. Furthermore, insulin administered subcutaneously results in peripheral hyperinsulinemia, which is assumed to be involved in vascular disease and cause hypoglycemia episodes. Several approaches are being made to develop other routes of administration, such as oral intake and aerosol inhalation [9, 13], which would constitute much more comfortable, convenient and compliance-encouraging alternatives. Furthermore, these non-parenteral routes of administration would minimize the health hazard implied in repeated injection of these short-acting drugs, surpass complications arising from the need for sterile techniques associated with parenteral formulations, and could provide a more suitable way of controlling their release profile. However, no clinically relevant noninvasive insulin administration route has been found [14, 15].

Assuming that internal absorption takes place following oral administration, insulin enters the portal circulation from the gastrointestinal tract. The physiological route of endogenous insulin release is thus limited, which may improve the pharmacodynamic properties of insulin.

In former times, the emphasis of oral delivery of insulin was more on convenience and avoidance of needles. With the current understanding of the pathogenesis of multifactorial complex diseases, such as type 2 diabetes, additional emphasis has also been placed on the physiological importance of the most appropriate, that is natural, route of delivery of the protein therapeutic to its site(s) of action in the target tissues and cells. In case of therapy of type 2 diabetes this is the hepatic route [3]. Existing insulin therapies including newer, injectable and short-acting analogues introduce insulin at the periphery and do not mimic the normal, physiological portal-to-peripheral gradient. These therapies continue to over-insulinize the periphery resulting in undesirable side effects, such as peripheral hyperinsulinemia with resulting insulin resistance, weight gain and hypoglycemia [2]. With hepatically

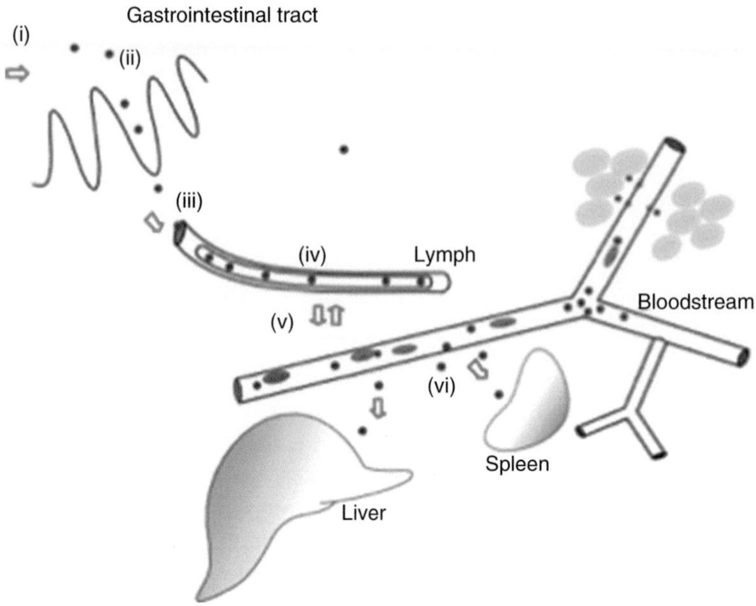

Fig. 2 Schematic representation of the areas of flow and transport of nanoparticles for delivery and targeting (adapted from [36]): (i) flow in the gastrointestinal tract; (ii) uptake and translocation through paracellular, transcellular or lymphatic pathway; (iii) passage into the mesenteric lymph; (iv) flow in the lymph vessels and encapsulation in the lymph nodes; (v) transport between lymph and systemic circulation; (vi) passage to target sites

delivered oral insulin – as we can see in Fig. 2 – there is the possibility of achieving high portal vein concentrations of insulin without sustained peripheral hyperinsulinemia [16, 17]. The physiological rational and therapeutic advantage for oral insulin has meanwhile been widely acknowledged by diabetologists [3].

4 Barriers to Oral Insulin Delivery

After oral ingestion, insulin has to surpass gastric enzymes like pepsin (which is active at pH 2–3) trypsin, chymotrypsin and carboxypeptidase in the duodenum (due to sharp rise in pH there is a risk of protein precipitation if it reaches its isoelectric point (pH 5.3–5.4) [16]. As the proteins reach the brush border of enterocytes they are acted upon by exopeptidases as well as insulin degrading enzymes present in the cytoplasm of absorptive cells. It is estimated that more than 90% of the proteolytic activity of the small intestine occurs at the brush border membrane [18].

Once insulin escapes the chemical and biochemical environment in the GI tract, it has to surpass the epithelial lining of the small intestine, which presents the next obstacle. The epithelial lining is composed of 2 types of cells, columnar epithelial cells (enterocytes) and goblet cells that secrete mucus and Peyer'patches.

Glycocalyx secreted by enterocytes along with mucins hinders the absorption of insulin. Insulin plays a vital role in the regulation of small intestine development because there are insulin receptors in the gut and insulin stimulates the proliferation of epithelial cells expressing its receptors [19].

5 Nanotechnology and Nanoparticles

5.1 Insulin Carrier

A carrier for oral delivery of insulin should provide:

a) A high insulin-loading capacity thereby reduce the therapeutic amount of carrier and consequently the quantity of matrix materials for administration;
b) Resistance against gastrointestinal enzymes and pH gradients;
c) A stable and biocompatible environment to ensure that the main fraction of insulin will be biologically active following encapsulation and stabilize and preserve physiological activity during both particle processing and insulin release;
d) Insulin permeability within the intestinal epithelium;
e) Absorption of insulin, so it can interact with cell-surface receptors of enterocytes or be captured by lymphatic cells, or pass through or be entrapped in the lymph nodes or transfer to the blood, provided that the carrier remain intact and its size are below a certain threshold limit;
f) A long intestinal residence time, reversibly increase the permeability of the mucosal epithelium to enhance the absorption of insulin and provide the intact insulin to the systemic circulation;
g) An accurate amount of insulin fast enough to control the glucose concentration in blood, and the function has to be reproducible each time oral insulin is delivered;
h) Safety after its oral administration.

Thus, the first goal to develop an oral formulation for insulin is to reduce or avoid enzymatic degradation and the second goal is to increase insulin permeability within the intestinal epithelium. Some current strategies being investigated to orally deliver proteins are the use of permeation enhancers, protease inhibitors, polymeric delivery systems, or chemical modification of a protein [20]. Several of these strategies have produced promising results like pH sensitivity, enzymatic inhibition etc., but the bioavailability is still low [21]. Among the possible strategies to improve the absorption of drugs, polymeric micro- and nanoparticles represent an exciting approach to enhance the uptake and transport of administered molecules [22]. However, gastrointestinal uptake of particles depends on the size range and on polymeric delivery systems [23]. Mucoadhesive delivery systems provide intimate contact of insulin with the mucosa, thereby reducing drug degradation between the delivery system and the absorbing membrane and to localize the drug delivery system to a certain site [24].

5.2 *Nanoparticles*

Nanoparticle (NP) delivery systems may provide a solution for oral administration of insulin. In addition to protecting it against a harmful gastric environment (chemical degradation), encapsulating insulin in nanoparticulate carriers can prevent enzymatic degradation, while controlling its release and increasing their residence time in the small intestine. With a higher surface-to-volume ratio than that of microparticulate systems, NPs can significantly increase their cellular contact, thus increasing their absorption.

6 Insulin Nanoparticles

Numerous delivery systems that are based on nanoparticles have been evaluated for the oral administration of insulin. A number of review articles have summarized the variety of nanotechnology-based strategies that have been used to encapsulate and deliver insulin orally [16, 25–29].

Oral insulin-delivery by means of NPs systems has received considerable attention. Using various materials and formulation processes facilitates the modulation of physicochemical properties of NPs (e.g., size, surface charge and mucoadhesivity), extent of insulin loading, insulin release profile as well as their biological behavior. Among these NPs, the ones made of biodegradable hydrophobic or hydrophilic polymers are focused in this chapter. Hydrophobic polymers, including poly(lactide) (PLA), poly(lactide-co-glycolide)(PLGA), and poly(ε-caprolactone) (PCL), have been used to produce NPs for oral insulin delivery. Among the hydrophilic polymers, alginate, chitosan (CS) and its derivatives have attracted the most attention owing to their excellent mucoadhesive properties, permeabilization capacity and biocompatibility. Other polymeric materials, including dextran, solid lipid, polyallylamine and polyacrylic acid (PAA), have also been used as carriers for oral insulin delivery.

While the chemical environment in the gastrointestinal (GI) tract tend to modify insulin by a broad range of pH (from min. pH 1.3 in stomach to max. pH 8.0 in the intestine) the most severe threat for loss of insulin is the enzymatic degradation along the way through the GI tract. Additionally the surface of the intestinal epithelium is covered with the high viscous mucus which is negatively charged and only permeable for nutrients, water and small molecules and prevents efficiently the entrance of nanoparticles [30].

Different mechanisms may coexist to allow NPs to be taken up by the intestinal epithelium, depending on the polymers used to form NPs. Designing NP carriers to enhance the absorption of insulin via the GI tract involves adapting their physicochemical properties to optimize non-specific cell uptake or attaching ligands to the particle surface to specifically target cells and enhance endocytosis.

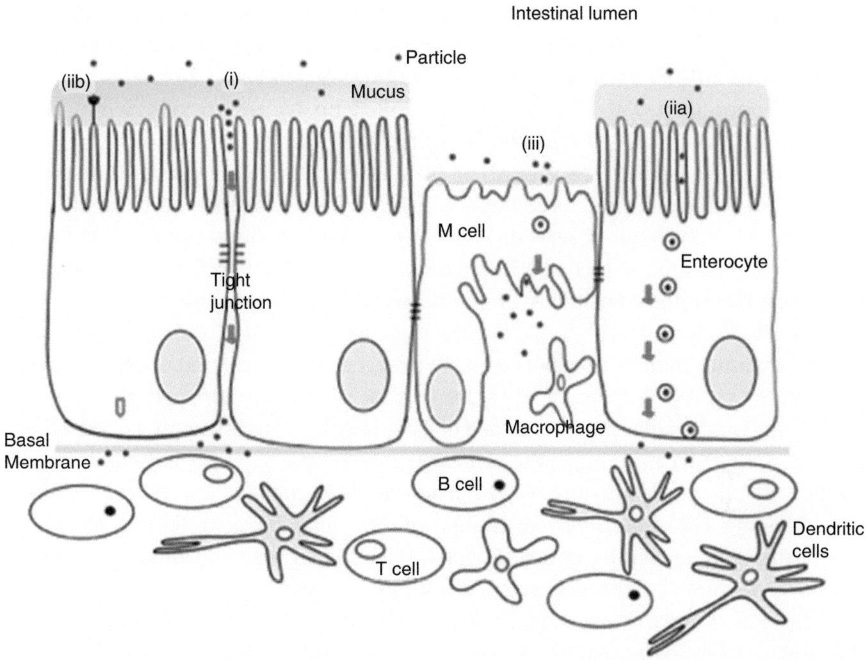

Fig. 3 Schematic representation of the intestinal structures and the route of particle translocation cross the intestinal epithelium: (i) paracellular transport, (iia) transcellular process by passive transcytosis, (iib) transcellular process by receptor-mediated transcystosis and (iii) transport via M cells of Peyer's patches

The mechanisms proposed to explain the translocation of particles through the intestinal epithelium are: (i) paracellular transport; (ii) passive transcytosis or receptor mediated transcystosis; and (iii) transport via M cells of Peyer's patches [31, 32]. Figure 3 schematically illustrates the intestinal structures and the route of particle translocation across the intestinal epithelium.

Here especially the transcellular pathway, in the form of non specific or specific uptake of NP and paracellular pathway are explored. The role of nanoparticulate carrier on insulin uptake by cells is discussed in a different section.

6.1 Transcellular Insulin Pathway

6.1.1 Non-specific Uptake of NPs

Transport of NPs via the transcellular pathway is related to their particle size, surface charge and mucoadhesivity [33]. Related studies have demonstrated that the interaction between NPs and epithelial cells is optimized when their particle size ranges from 50 to 500 nm [34, 35]. Transcytosis of NPs generally increases when the particle size decreases [36]. Additionally, according to a previous study, the uptake of NPs by enterocytes or M cells is size dependent.

Small particles (<50–100 nm) can be taken up by enterocytes, while larger particles are more likely internalized by M cells [37].

Exactly how the surface charge of NPs affects their non-specific uptake by enterocytes or M cells remains contentious [38]. A high variability between different systems has often been observed. A previous study reported that the negatively charged poly(styrene) NPs had a lower affinity to intestinal epithelia, especially M cells, than that of their positively or neutrally charged counterparts [39].

However, Shakweh et al. found that the negatively or neutrally charged NPs had a better capability in their transport across Peyer's patches than that of the positively charged ones [40]. Polyethylene glycol (PEG), a neutrally charged polymer, has been employed for coating the surface of NPs in drug-delivery applications. Dense coating with PEG minimizes the adhesive interaction between NPs and mucins (negatively charged), allowing NPs to penetrate through the mucus layer [41, 42].

Enhancing the mucoadhesive ability of NPs can be an effective means of increasing their uptake by epithelial cells. NPs coated with bioadhesive materials have been designed to develop mucoadhesion.

Adhesion of a carrier system to the mucus can increase its residence time and the contact of the subsequently released drug with the underlying epithelium, thus increasing the drug concentration at the absorption site [43]. Hydrophilic polymers such as chitosan (CS), polyacrylic acid (PAA), thiomers and their derivatives are notable examples of such mucoadhesive materials [44–46].

Additionally, coating particles with polyvinyl alcohol (PVA) or vitamin E succinylated polyethylene glycol 1,000 (Vitamin E TGSP) can significantly enhance their adhesion and absorption capability by the intestinal epithelial cells [38].

Previous studies have also described several insulin derivatives with such properties, including alkylated/acylated insulin, PEGylated insulin and polysialylated insulin [47, 48].

A novel strategy for both oral delivery and tissue targeting is based on a 'natural' co-translational modification of insulin with a glycolipid structure, the covalent coupling to glycosylphosphatidylinositol (GPI) [49].

6.1.2 Specific Uptake of NPs

Via the transcellular pathway, NPs can be taken up by enterocytes or M cells of Peyer's patches [27]. Theoretically, supramolecular structures such as NPs cannot diffuse through cells by passive diffusion owing to their large size. Conversely, active transport, as illustrated by different energy-dependent mechanisms, is largely used to transport drugs encapsulated in NPs [50].

The active transcellular transport of NPs begins with an endocytic process that occurs at the apical cell membrane; the particles then transport through the cells and release at their basolateral pole [51].

A variety of potential ligands such as proteins and vitamins have been conjugated to nanoparticle surfaces to facilitate receptor-mediated endocytosis of the particles, including lectins [52, 53], wheat germ agglutinin (WGA) [54, 55], vitamin B12 [56, 57].

B12-derivatized-nanoparticles were transported into Caco-2 cells with significantly higher efficiency than unmodified particles [58]. The pharmacological availability of B12 (NPs) containing 2, 3, and 4% (w/w) insulin was 1.1-, 1.9-, and 2.6-fold higher than that of NPs without B12 and authors hypothesized that the uptake of the NPs was mediated by B12 [57, 59]. Vitamin B12 can be bound by an intrinsic factor (IF) in the small intestine to form vitamin B12eIF complex; this complex can be recognized by an IF-specific receptor located on the luminal surface of intestinal cells and facilitate transport across the intestinal enterocytes [56].

Wheat germ agglutinin (WGA) is a special binder for N-acetylglucosamine and sialic-acid [54]. WGA has been described to bind not only to the surface of Caco-2 cells and enterocytes, but also to be taken up into cells by receptor-mediated endocytosis involving the epidermal growth factor receptor that is expressed at a considerable density on enterocytes [60].

In contrast to typical transmembrane proteins, glycosylphosphatidylinositol (GPI)-anchored proteins are embedded in the outer leaflet of the plasma membrane by a specific GPI glycolipid which consists of phosphatidylinositol followed by a glycan core and a phosphodiester-ethanolamine bridge [49, 61].

6.2 Paracellular Pathway

Although paracellular space represents <1% of intestinal mucosal surface, paracellular pathway is a preferred route for transporting hydrophilic drugs [62]. However, junctional complexes either restrict or completely block the passage between cells of macromolecules or aggregates larger than approximately 1 nm [63]. Therefore, as is widely recognized, polymeric NPs cannot diffuse through the intestinal barrier via the paracellular route. To improve their paracellular transport, tight-junctions (TJs) must be opened reversibly by using enhancers such as cationic chitosan (CS) and its derivatives and anionic polyacrylic acid (PAA) and its derivatives polymers and calcium chelators [64]. The former (CS and PAA) open TJs by interacting with either extracellular domains of TJs proteins or with a surface receptor, thus activating a cascade leading to disassembly of TJs; the latter (calcium chelators) modulates TJs by chelating calcium inducing disruption of adherens junctions and TJs via activation of protein kinase C [65].

Although the opening of TJs increases paracellular permeability, intercellular spaces between epithelial cells remain limited to transport intact NPs to the bloodstream [50]. Until now, no paracellular transport of intact NPs has been demonstrated. Increased paracellular permeability must be more likely ascribed to the anticipated release of the drug from the disintegrated NPs [50]. In some cases,

NPs or their constituted components play a role in TJ disruption [66–69]. Notably, although reversible, the opening of TJs may have altered physiological functions when they are opened.

6.3 Role of Nanoparticulate on Insulin Uptake

The investigation of the mechanism of nanoparticles pathway through the intestinal epithelium is often not enough to design new formulations to increase the bioavailability / pharmacological action of insulin. Recent experiments have allowed to obtain data on how the insulin is uptaken by enterocytes, free, or anchored within the components of the nanoparticles. A reliable comparison of results obtained with insulin uptake from micro- and nanoparticulate carriers can only be made when experiments were performed using the same experimental conditions. Any assumption about it may raise some speculation even tough is easier to make it for experiments done by the same research group.

Cationic polyelectrolytes such as chitosan or its derivatives (i.e. trimethylated chitosan, TMC have been shown to form nano-complexes with proteins and facilitate protein transport across gut epithelial cells in vitro via both paracellular and transcellular routes [70–73]. They promote paracellular transport by interacting with TJs proteins via electrostatic interaction, thereby causing a temporary change in cell morphology. This leads to TJs opening which in turn allows protein passage between cells [74–76]. These interactions have been shown to be reversible and less likely to result in permanent cellular damage compared to traditional permeation enhancers, i.e. carbomers and surfactants [77].

Cellular uptake of free insulin is dependant on insulin-receptor interaction, while insulin complexed with PAA amphiphilic polyelectrolytes uptake and transport are affected by the polymer architecture. Changes in polymer structure resulted in differences in cellular uptake rate, localization within cells and mechanism of uptake, as well as facilitation of insulin transport across cell layers [78]. Nanosizing insulin-polyelectrolytes (PEC) were taken up as a whole by Caco-2 cells, and PEC did not dissociate into free insulin and polyelectrolytes before being taken up by the cells [78].

Differences in cellular localization and uptake mechanism is likely due to the presence of quaternary ammonium moieties in polyelectrolytes-insulin conjugates [78]. Quaternary ammonium moiety confers a permanent, pH-independent, positive charge, which allows stronger electrostatic interaction with insulin [79].

The effect of nanoparticulate delivery system made of albumin/chitosan coated alginate/dextran sulfate core on enhancing insulin permeation through intestinal membrane was evaluated in different intestinal epithelial models using cell cultures and excised intestinal tissues. Insulin permeation through Caco-2 cell monolayer was enhanced 2.1-fold, facilitated by the nanoparticles compared with insulin alone, 3.7-fold through a mucus-secreting Caco-2/HT29 co-culture, and 3.9-fold through excised intestinal mucosa of Wistar rats.

Transcellular permeation was evidenced by internalization of independently labeled insulin and nanoparticles into enterocytes, in which insulin appeared to remain associated with the nanoparticles [80].

7 Conclusions

Several NP systems have been developed for oral insulin delivery in recent years. To be consequent, their long term efficacy must be demonstrated in larger animals and in humans.

The enhancement of oral bioavailability of encapsulated insulin in nanoparticles has been clearly demonstrated in animal studies. However, the question as why they were not followed by industrial development and clinical trials is still open.

It seems that the high variability of the insulin concentration delivered to the blood, the still low bioavailability of encapsulated insulin requiring the administration of high doses, higher than the typical dose of 1 IU/kg needed to induce the same level of glycemia via the SC injection, as well as the lack of information allowing to adapt these results to humans, represent heavy handicaps for the development of an oral formulation of insulin to prevent or treat diabetes.

Diabetes is a chronic disease and therefore a long-term administration of such high doses of insulin transcellularly may induce mitogenic changes in the GI epithelium, as insulin is also a growth factor. Given this challenge, current NP systems must be improved both to increase the insulin loading and to lower the insulin dose required to produce a therapeutic blood glucose reduction.

Maintaining the physicochemical and biological stability of insulin in the GI tract is of priority concern when using a NP system with prolonged retention in the small intestine. Combining enzyme inhibitors within NPs represents the potential to increase the potency of orally administered insulin.

The technology of glycolipid-based could enable the efficient oral delivery of insulin and its targeting to the sites of action of the corresponding abnormal gene products encoded by polymorphic/mutant susceptibility genes.

This is presumably a prerequisite for achieving the goal of personalized prevention and therapy of multifactorial complex common disease like diabetes in the medium term. Ideally, they will encompass patient-tailored combinations of GPI proteins/antibodies which address as many as feasible of the products affected by the susceptibility genes in a tissue-specific fashion.

Hopefully, effectively addressing the above challenges will lead to the practical implementation of the NP system for oral delivery of insulin.

References

1. WHO (2012) Diabetes. Fact sheet N°312
2. Müller G (2010) Oral protein therapy for the future—transport of glycolipid-modified proteins: vision or fiction? Pharmacology 86:92–116

3. Arbit E (2004) The physiological rationale for oral insulin administration. Diabetes Technol Ther 6:510–517
4. Langer R, Peppas NA (2003) Advances in biomaterials, drug delivery, and bionanotechnology. AIChE J 49:2990–3006
5. Pinto Reis C, Neufeld RJ, Ribeiro AJ, Veiga F (2006) Nanoencapsulation II. Biomedical applications and current status of peptide and protein nanoparticulate delivery systems. Nanomedicine 2:53–65
6. American Diabetes Association (2010) Standards of medical care in diabetes—2010. Diabetes Care 33:S11–S61
7. Meneghini L (2008) Demonstrating strategies for initiation of insulin therapy: matching the right insulin to the right patient. Int J Clin Pract 62:1255–1264
8. Freeman JS (2009) Insulin analog therapy: improving the match with physiologic insulin secretion. J Am Osteopath Assoc 109:26–36
9. Gerich JE (2002) Novel insulins: expanding options in diabetes management. Am J Med 113:308–316
10. Silva CM, Ribeiro AJ, Figueiredo IV, Goncalves AR, Veiga F (2006) Alginate microspheres prepared by internal gelation: development and effect on insulin stability. Int J Pharm 311:1–10
11. Rossetti P, Porcellati F, Fanelli CG, Perriello G, Torlone E, Bolli GB (2008) Superiority of insulin analogues versus human insulin in the treatment of diabetes mellitus. Arch Physiol Biochem 114:3–10
12. Carino GP, Mathiowitz E (1999) Oral insulin delivery. Adv Drug Deliv Rev 35:249–257
13. Patel N, Craddock BL, Staniforth JN, Tobyn MJ, Welham MJ (2001) Spray-dried insulin particles retain biological activity in rapid in-vitro assay. J Pharm Pharmacol 53:1415–1418
14. Heinemann L, Pfutzner A, Heise T (2001) Alternative routes of administration as an approach to improve insulin therapy: update on dermal, oral, nasal and pulmonary insulin delivery. Curr Pharm Des 7:1327–1351
15. Owens DR, Zinman B, Bolli G (2003) Alternative routes of insulin delivery. Diabet Med 20:886–898
16. Damge C, Reis CP, Maincent P (2008) Nanoparticle strategies for the oral delivery of insulin. Expert Opin Drug Deliv 5:45–68
17. Müller G (2010) Personalized prognosis and diagnosis of type 2 diabetes—vision or fiction? Pharmacology 85:168–187
18. Wong TW (2010) Design of oral insulin delivery systems. J Drug Target 18:79–92
19. Saffran M, Pansky B, Budd GC, Williams FE (1997) Insulin and the gastrointestinal tract. J Control Release 46:89–98
20. Mahato RI, Narang AS, Thoma L, Miller DD (2003) Emerging trends in oral delivery of peptide and protein drugs. Crit Rev Ther Drug Carrier Syst 20:153–214
21. Gomez-Orellana I (2005) Strategies to improve oral drug bioavailability. Expert Opin Drug Deliv 2:419–433
22. Allémann E, Leroux J, Gurny R (1998) Polymeric nano- and microparticles for the oral delivery of peptides and peptidomimetics. Adv Drug Deliv Rev 34:171–189
23. Delie F (1998) Evaluation of nano- and microparticle uptake by the gastrointestinal tract. Adv Drug Deliv Rev 34:221–233
24. Varum FJ, McConnell EL, Sousa JJ, Veiga F, Basit AW (2008) Mucoadhesion and the gastrointestinal tract. Crit Rev Ther Drug Carrier Syst 25:207–258
25. Babu VR, Patel P, Mundargi RC, Rangaswamy V, Aminabhavi TM (2008) Developments in polymeric devices for oral insulin delivery. Expert Opin Drug Deliv 5:403–415
26. Gholamipour-Shirazi A (2008) Insulin encapsulation. Expert Opin Drug Deliv 5:1335–1355
27. Ramesan RM, Sharma CP (2009) Challenges and advances in nanoparticle-based oral insulin delivery. Expert Rev Med Devices 6:665–676
28. Woitiski CB, Carvalho RA, Ribeiro AJ, Neufeld RJ, Veiga F (2008) Strategies toward the improved oral delivery of insulin nanoparticles via gastrointestinal uptake and translocation. BioDrugs 22:223–237

29. Chen MC, Sonaje K, Chen KJ, Sung HW (2011) A review of the prospects for polymeric nanoparticle platforms in oral insulin delivery. Biomaterials 32:9826–9838
30. Lai SK, Wang Y-Y, Hanes J (2009) Mucus-penetrating nanoparticles for drug and gene delivery to mucosal tissues. Adv Drug Deliv Rev 61:158–171
31. Aprahamian M, Michel C, Humbert W, Devissaguet JP, Damge C (1987) Transmucosal passage of polyalkylcyanoacrylate nanocapsules as a new drug carrier in the small intestine. Biol Cell 61:69–76
32. Eldridge JH, Hammond CJ, Meulbroek JA, Staas JK, Gilley RM, Tice TR (1990) Controlled vaccine release in the gut-associated lymphoid tissues. I. Orally administered biodegradable microspheres target the peyer's patches. J Control Release 11:205–214
33. Hillaireau H, Couvreur P (2009) Nanocarriers' entry into the cell: relevance to drug delivery. Cell Mol Life Sci 66:2873–2896
34. Desai MP, Labhasetwar V, Amidon GL, Levy RJ (1996) Gastrointestinal uptake of biodegradable microparticles: effect of particle size. Pharm Res 13:1838–1845
35. Jung T, Kamm W, Breitenbach A, Hungerer K-D, Hundt E, Kissel T (2001) Tetanus toxoid loaded nanoparticles from sulfobutylated poly(vinyl alcohol)-graft-poly(lactide-co-glycolide): evaluation of antibody response after oral and nasal application in mice. Pharm Res 18:352–360
36. Florence AT (2005) Nanoparticle uptake by the oral route: fulfilling its potential? Drug Discov Today 2:75–81
37. des Rieux A, Fievez V, Theate I, Mast J, Preat V, Schneider YJ (2007) An improved in vitro model of human intestinal follicle-associated epithelium to study nanoparticle transport by M cells. Eur J Pharm Sci 30:380–391
38. des Rieux A, Fievez V, Garinot M, Schneider Y-J, Préat V (2006) Nanoparticles as potential oral delivery systems of proteins and vaccines: a mechanistic approach. J Control Release 116:1–27
39. Jani P, Halbert GW, Langridge J, Florence AT (1989) The uptake and translocation of latex nanospheres and microspheres after oral administration to rats. J Pharm Pharmacol 41:809–812
40. Shakweh M, Besnard M, Nicolas V, Fattal E (2005) Poly (lactide-co-glycolide) particles of different physicochemical properties and their uptake by peyer's patches in mice. Eur J Pharm Biopharm 61:1–13
41. Wang YY, Lai SK, Suk JS, Pace A, Cone R, Hanes J (2008) Addressing the PEG mucoadhesivity paradox to engineer nanoparticles that "slip" through the human mucus barrier. Angew Chem Int Ed Engl 47:9726–9729
42. Tang BC, Dawson M, Lai SK, Wang YY, Suk JS, Yang M, Zeitlin P, Boyle MP, Fu J, Hanes J (2009) Biodegradable polymer nanoparticles that rapidly penetrate the human mucus barrier. Proc Natl Acad Sci U S A 106:19268–19273
43. Smart JD (2005) The basics and underlying mechanisms of mucoadhesion. Adv Drug Deliv Rev 57:1556–1568
44. Pinto-Alphandary H, Aboubakar M, Jaillard D, Couvreur P, Vauthier C (2003) Visualization of insulin-loaded nanocapsules: in vitro and in vivo studies after oral administration to rats. Pharm Res 20:1071–1084
45. Dünnhaupt S, Barthelmes J, Hombach J, Sakloetsakun D, Arkhipova V, Bernkop-Schnürch A (2011) Distribution of thiolated mucoadhesive nanoparticles on intestinal mucosa. Int J Pharm 408:191–199
46. Khutoryanskiy VV (2011) Advances in mucoadhesion and mucoadhesive polymers. Macromol Biosci 11:748–764
47. Galloway JA, Chance RE (1994) Improving insulin therapy-achievements and challenges. Horm Metab Res 26:591–598
48. Pillai O, Panchagnula R (2001) Insulin therapies—past, present and future. Drug Discov Today 6:1056–1061
49. Müller G (2011) Novel applications for glycosylphosphatidylinositol-anchored proteins in pharmaceutical and industrial biotechnology. Mol Membr Biol 28:187–205
50. Roger E, Lagarce F, Garcion E, Benoit JP (2010) Biopharmaceutical parameters to consider in order to alter the fate of nanocarriers after oral delivery. Nanomedicine (Lond) 5:287–306

51. Shakweh M, Ponchel G, Fattal E (2004) Particle uptake by Peyer's patches: a pathway for drug and vaccine delivery. Expert Opin Drug Deliv 1:141–163
52. Hussain N, Jani PU, Florence AT (1997) Enhanced oral uptake of tomato lectin-conjugated nanoparticles in the rat. Pharm Res 14:613–618
53. Russell-Jones GJ (2004) Use of targeting agents to increase uptake and localization of drugs to the intestinal epithelium. J Drug Target 12:113–123
54. Weissenboeck A, Bogner E, Wirth M, Gabor F (2004) Binding and uptake of wheat germ agglutinin-grafted PLGA-nanospheres by Caco-2 monolayers. Pharm Res 21:1917–1923
55. Lochner N, Lobmaier C, Wirth M, Leitner A, Pittner F, Gabor F (2003) Silver nanoparticle enhanced immunoassays: one step real time kinetic assay for insulin in serum. Eur J Pharm Biopharm 56:469–477
56. Petrus A, Vortherms A, Fairchild T, Doyle R (2007) Vitamin B12 as a carrier for the oral delivery of insulin. ChemMedChem 2:1717–1721
57. Chalasani K, Russell-Jones G, Jain A, Diwan P, Jain S (2007) Effective oral delivery of insulin in animal models using vitamin B12-coated dextran nanoparticles. J Control Release 122:141–150
58. McEwan JF, Veitch HS, Russell-Jones GJ (1999) Synthesis and biological activity of ribose-5′-carbamate derivatives of vitamin B12. Bioconjug Chem 10:1131–1136
59. Chalasani KB, Russell-Jones GJ, Yandrapu SK, Diwan PV, Jain SK (2007) A novel vitamin B12-nanosphere conjugate carrier system for peroral delivery of insulin. J Control Release 117:421–429
60. Lochner N, Pittner F, Wirth M, Gabor F (2003) Wheat germ agglutinin binds to the epidermal growth factor receptor of artificial Caco-2 membranes as detected by silver nanoparticle enhanced fluorescence. Pharm Res 20:833–839
61. Orlean P, Menon AK (2007) Thematic review series: lipid posttranslational modifications. GPI anchoring of protein in yeast and mammalian cells, or: how we learned to stop worrying and love glycophospholipids. J Lipid Res 48:993–1011
62. Pappenheimer JR (1987) Physiological regulation of transepithelial impedance in the intestinal mucosa of rats and hamsters. J Membr Biol 100:137–148
63. Nellans HN (1991) Mechanisms of peptide and protein absorption. 1. Paracellular intestinal transport-modulation of absorption. Adv Drug Deliv Rev 7:339–364
64. Salamat-Miller N, Johnston TP (2005) Current strategies used to enhance the paracellular transport of therapeutic polypeptides across the intestinal epithelium. Int J Pharm 294:201–216
65. Plapied L, Duhem N, des Rieux A, Préat V (2011) Fate of polymeric nanocarriers for oral drug delivery. Curr Opin Colloid Interface Sci 16:228–237
66. Mi FL, Wu YY, Lin YH, Sonaje K, Ho YC, Chen CT, Juang JH, Sung HW (2008) Oral delivery of peptide drugs using nanoparticles self-assembled by poly(gamma-glutamic acid) and a chitosan derivative functionalized by trimethylation. Bioconjug Chem 19:1248–1255
67. Lin YH, Sonaje K, Lin KM, Juang JH, Mi FL, Yang HW, Sung HW (2008) Multi-ion-crosslinked nanoparticles with pH-responsive characteristics for oral delivery of protein drugs. J Control Release 132:141–149
68. Cheng WP, Gray AI, Tetley L, Hang Tle B, Schatzlein AG, Uchegbu IF (2006) Polyelectrolyte nanoparticles with high drug loading enhance the oral uptake of hydrophobic compounds. Biomacromolecules 7:1509–1520
69. Jevprasesphant R, Penny J, Attwood D, McKeown NB, D'Emanuele A (2003) Engineering of dendrimer surfaces to enhance transepithelial transport and reduce cytotoxicity. Pharm Res 20:1543–1550
70. Bayat A, Dorkoosh FA, Dehpour AR, Moezi L, Larijani B, Junginger HE, Rafiee-Tehrani M (2008) Nanoparticles of quaternized chitosan derivatives as a carrier for colon delivery of insulin: ex vivo and in vivo studies. Int J Pharm 356:259–266
71. Sadeghi AM, Dorkoosh FA, Avadi MR, Weinhold M, Bayat A, Delie F, Gurny R, Larijani B, Rafiee-Tehrani M, Junginger HE (2008) Permeation enhancer effect of chitosan and chitosan derivatives: comparison of formulations as soluble polymers and nanoparticulate systems on insulin absorption in Caco-2 cells. Eur J Pharm Biopharm 70:270–278

72. Thanou M, Verhoef JC, Junginger HE (2001) Oral drug absorption enhancement by chitosan and its derivatives. Adv Drug Deliv Rev 52:117–126
73. Artursson P, Lindmark T, Davis SS, Illum L (1994) Effect of chitosan on the permeability of monolayers of intestinal epithelial cells (Caco-2). Pharm Res 11:1358–1361
74. Kotzé AF, de Leeuw BJ, Lueßen HL, de Boer AG, Verhoef JC, Junginger HE (1997) Chitosans for enhanced delivery of therapeutic peptides across intestinal epithelia: in vitro evaluation in Caco-2 cell monolayers. Int J Pharm 159:243–253
75. van der Lubben IM, Verhoef JC, Borchard G, Junginger HE (2001) Chitosan and its derivatives in mucosal drug and vaccine delivery. Eur J Pharm Sci 14:201–207
76. Ranaldi G, Marigliano I, Vespignani I, Perozzi G, Sambuy Y (2002) The effect of chitosan and other polycations on tight junction permeability in the human intestinal Caco-2 cell line. J Nutr Biochem 13:157–167
77. Thanou MM, Kotzé AF, Scharringhausen T, Luessen HL, de Boer AG, Verhoef JC, Junginger HE (2000) Effect of degree of quaternization of N-trimethyl chitosan chloride for enhanced transport of hydrophilic compounds across intestinal Caco-2 cell monolayers. J Control Release 64:15–25
78. Thompson C, Cheng W, Gadad P, Skene K, Smith M, Smith G, McKinnon A, Knott R (2011) Uptake and transport of novel amphiphilic polyelectrolyte-insulin nanocomplexes by Caco-2 cells—towards oral insulin. Pharm Res 28:886–896
79. Thompson CJ, Tetley L, Uchegbu IF, Cheng WP (2009) The complexation between novel comb shaped amphiphilic polyallylamine and insulin-towards oral insulin delivery. Int J Pharm 376:46–55
80. Woitiski CB, Sarmento B, Carvalho RA, Neufeld RJ, Veiga F (2011) Facilitated nanoscale delivery of insulin across intestinal membrane models. Int J Pharm 412:123–131

Bioartificial Pancreas: In the Road to Clinical Application

Joana Crisóstomo, Jorge F.J. Coelho, and Raquel Seiça

Abstract With the increasing incidence of diabetes, it is urgent to develop a reliable and safe source of insulin production. Conventional transplantation of pancreatic tissue has been demonstrated to be an efficacious method of restoring glycaemic control in type 1 diabetes. However, the lack of donors, the use of immunosuppressants and the early β-cell failure limit the use of this approach. Regenerative medicine has focused on the design and application of bioartificial pancreas to treat this disease. Islet microencapsulation technique enables the transplantation in the absence of immunosuppression by protecting the cells through an immunoisolative membrane. This membrane should be able to protect transplanted allo- and xenogenic cells from the host, while facilitating adequate transport of oxygen, nutrients, and secreted hormones. Although many different materials and

J. Crisóstomo (✉)
Laboratory of Physiology, Institute of Biomedical Research in Light and Image,
Faculty of Medicine of University of Coimbra, Sub-unit 1, 1st floor, Azinhaga de Santa Comba,
Celas, 3060-350 Coimbra, Portugal

Research Centre for Chemical Processes Engineering and Forest Products, Chemical Engineering Department, Faculty of Sciences and Technology of University of Coimbra, Polo II – Rua Sílvio Lima, 3030-790 Coimbra, Portugal
e-mail: joanacrisostomo@hotmail.com

J.F.J. Coelho
Department of Chemical Engineering, University of Coimbra, Polo II,
Pinhal de Marrocos 3030-790, Coimbra, Portugal
e-mail: jcoelho@eq.uc.pt

R. Seiça
Laboratory of Physiology, Institute of Biomedical Research in Light and Image,
Faculty of Medicine of University of Coimbra, Sub-unit 1, 1st floor, Azinhaga de Santa Comba,
Celas, 3060-350 Coimbra, Portugal
e-mail: rmfseica@gmail.com

J. Coelho (ed.), *Drug Delivery Systems: Advanced Technologies Potentially Applicable in Personalised Treatment*, Advances in Predictive, Preventive and Personalised Medicine 4, DOI 10.1007/978-94-007-6010-3_5,

immunoisolation devices build-up processes have been studied, their viability in *in vivo* applications has been compromised. With clarification of all obstacles to the clinical application, new strategies will be developed that would increase the islet performance after transplantation.

Keywords Diabetes treatment • Bioartificial pancreas • Islets encapsulation • Personalized medicine

1 The Target Disease: Diabetes

Diabetes mellitus is a metabolic disease characterized by an abnormal increase in blood glucose levels. The high glucose concentration can result from the body incapacity to produce insulin (β-cell dysfunction/failure/mass reduction) and from the cells inability to respond to the produced insulin (peripheral and hepatic insulin resistance) [1] (Fig. 1). Chronic hyperglycemia, through several pathways, leads to the development of diabetes associated complications, such as diabetic neuropathy, retinopathy, nephropathy and cardiovascular disease, which are related to high rates of morbidity and mortality [2–6].

1.1 Epidemiology

In the last decades the incidence of diabetes mellitus highly increased, affecting millions of people worldwide. According to the World Health Organization, at least 171 million people worldwide were diabetic in 2000. Its incidence is increasing rapidly, and it is estimated that by 2030, this number will almost double [7].

1.2 Classification

In 1997, the World Health Organization and the American Diabetes Association agreed on a new classification for diabetes mellitus [8]. The most frequent diabetes forms are type 1 diabetes and type 2 diabetes. Type 1 diabetes is typically detected in young people who become insulin-dependent, due to an autoimmune attack to β-cells present in islets. Type 2 diabetes affects mainly adults and overweight people who usually are non-insulin-dependent and its origin is multifactorial. Other forms of diabetes include gestational diabetes and diabetes resulting from a specific cause. Some common causes include exocrine pancreas damage, endocrinopathies like Cushing syndrome and acromegaly, rare forms of genetic disorders and diabetes induced by drugs as thiazide diuretics and glucocorticoids [9].

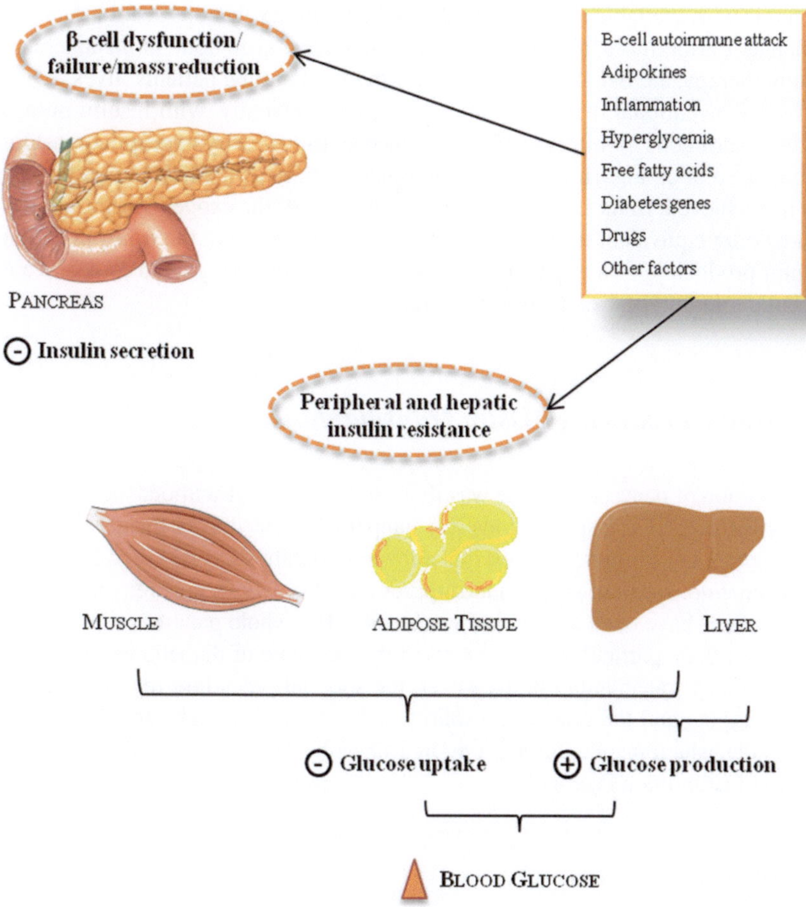

Fig. 1 Diabetes mellitus: β-cell changes and reduction of insulin action

1.3 Conventional Treatments

An appropriate diet and exercise used to be recommended for all diabetic patients [10, 11], but the maintenance of good blood glucose levels is mainly carried out by insulin injections and other appropriate anti-hyperglycemic medications [12, 13]. Commonly used in type 2 diabetic patients, oral anti-diabetic drugs have different mechanisms of action based on their capacity to increase the amount of insulin secretion or to increase the hepatic and peripheral action of insulin [13]. Once there is no insulin production in type 1 diabetes, insulin therapy is critical to control glucose levels.

A good glycemic control has been demonstrated to be vital in delaying or avoiding the diabetes-associated vascular complications [14]. However, the control of glycemia with available anti-diabetic drugs, including exogenous insulin, remains a challenge. Furthermore, although the efforts that have been done in management of insulin therapy, the risk of hypoglycemia still exists putting patients' lives at risk [15]. With conventional insulin injections and, more recently, with insulin pumps involving continuous monitoring of blood glucose, followed by appropriate insulin infusion based in an algorithm, the physiological pattern of the hormone secretion is still not achieved. In fact, the regulatory mechanism of the exogenous hormone is far from being equivalent to that in living cells. In this context, the transplantation of insulin productive tissue, pancreas or islets of Langerhans, gained a significant therapeutic role in type 1 diabetes treatment.

1.4 Insulin-Producing Tissue Transplantation

Transplantation of pancreatic tissue was first tried in 1890 but without success [16]. The improvement of transplantations just happened latter, during the 1960s, with the development of immunosuppression, allowing implant survival [17, 18].

With the improvement of immunosuppression, the whole pancreas and islets transplantations have been carried out worldwide. The whole pancreas transplantation has been done particularly in patients in the end-state of diabetic nephropathy, requiring kidney transplantation. However, the surgical procedure remains technically challenging and the rates of morbidity and mortality are substantial [19]. On the other side, islet transplantation is ongoing since the development and promissory outcomes of Edmonton Protocol.

1.4.1 Edmonton Protocol

The Edmonton Protocol, a method of pancreatic islets implantation for the treatment of type 1 diabetes mellitus, was developed in 1989 in the Canadian city of Edmonton [20]. This protocol involves islets isolation from pancreas of cadaveric donors by digestive process. The immunosuppressed patients receive the islets through infusion into the portal vein that results in islet entrapment within hepatic sinusoids. Since the first two patients transplanted with the Edmonton Protocol in 1989, several hundred people have received islet transplants through this methodology or variants of it. In average, the transplanted patients follow up shows that, although 50–68% of patients maintain insulin-independence 1 year after the transplantation, less than 10% are free of daily insulin supplementation 5 years later [20]. These disappointing results are related to the islets incapacity to cope with surrounding host environment. The development of fibrosis induced by inflammatory reaction deprives cells from oxygen and trigger their death. In addition, the lack of organ donors and the required immunosuppressive drugs limits this practice. The importance of the appropriate immunosuppression regimen is demonstrated in humans by the fact that in autotransplantation 2,500 islet equivalents per kilogram are provided, while in allotransplantation three or four times more islets are required to reach insulin independence [21]. On the other hand, the deleterious effects of the immunosuppressants treatments are well documented. Besides diabetogenic effects, these drugs are associated with increased susceptibility to viral, fungus and bacterial infections, as well as with increased cancer development risk [22]. Taking into account this reality, the main efforts that have been done in this area are aimed to protect the cells from the surrounding host environment, avoiding the use of immunosuppressants.

1.5 Innovative Treatments: The Regenerative Medicine

Intensive research using sophisticated technologies has been done aiming the future clinical application of islets transplantation. In this context, the regenerative medicine has been explored due to its ability to create live and functional tissues to repair/replace tissue or organ function loss due to age, disease, damage, or congenital defects. Hopefully, it might be possible to cure diseases such as Alzheimer, Parkinson and Huntington diseases, kidney, liver and heart damages, cancer, diabetes mellitus and spinal cord injury [23–25]. This new fast-emerging field takes advantage of different techniques such as tissue engineering, cellular therapies and the development of artificial/bioartificial organs. In the context of type 1 diabetes treatment, the most promising technique that enables the transplantation in the absence of immunosuppression, even between different species, is the islets encapsulation, which is the basis of the bioartificial pancreas [23, 26].

As reported before, to face the immune system, islet transplantation therapy requires an inexhaustible source of cells capable of delivering the therapeutic agent in response to physiological cues. Other innovative techniques such as the use of stem cells [27, 28] and genetic therapy [29] can help to overcome this problem. Over the past 5 years, significant advances have been made in understanding the signaling pathways that control lineage development to generate human insulin producing cells. Despite the efforts, the majority of insulin producing cells generated are polyhormonal and non-functional, reflecting the first transition in pancreatic development [27, 28]. A recent study shows the generation of functional β cells from human pluripotent stem cells (hPSC), which arise during the second transition of pancreatic development, detected only following transplantation of progenitor stage cells into immunocompromised mice. The authors concluded and highlighted that besides the improvement of cell expansion and survival it is imperative to immune protect the *in vitro* generated β cells in order to fully realize the goal of hPSC-driven regenerative medicine [27].

Taking advantage of genetic engineering, it is possible to develop cells with the ability to produce and release insulin. However, besides the technical complexity, genetically engineered cells often arise from an allogenic or xenogenic source and will also require protection from the host immune system [29]. In these two therapies, even if cells are from autologous origin, the immunoprotection is crucial in autoimmune type 1 diabetes. All innovative strategies seem to have a common requirement: the immune protection of cells/islets. Consequently, the efforts towards encapsulation are highly valuable in the road to clinical application.

2 The Bioartificial Pancreas: A New Era of Diabetes Treatment

The history of the bioartificial pancreas extends for over 30 years. Since the pioneer work by Chick et al. in the development of a bioartificial pancreas [30] and Lim and Sun's introduction of alginate-encapsulated islets [31], extensive research has been made in this field. Several groups have successfully transplanted encapsulated islets into diabetic animal models and, in several of these experiments, glycemic control was achieved for relatively long periods without immunosuppression.

The principle of bioartificial pancreas is shown in Fig. 2. The basic idea of a bioartificial pancreas is to enclose islets within a semi-permeable membrane that protects the implant from the type 1 diabetes produced auto-antibodies and from the immune responses of the host organisms. On the other way, oxygen, glucose, nutrients and waste products should flow freely through the membrane. Furthermore, the diffusion of insulin secreted by encapsulated islets must be appropriated for a good glycaemic control. The ideal bioartificial pancreas is a therapeutic approach to enable islets' allo- and xenotransplantation without immune suppression. In this way, the lack of islet donors and the harmful use of immunosuppressants would be solved. The bioartificial pancreas requires specific materials' features and

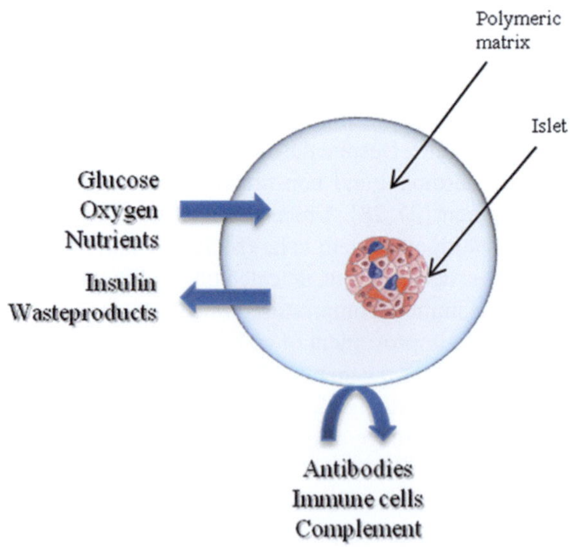

Fig. 2 The principle of bioartificial pancreas

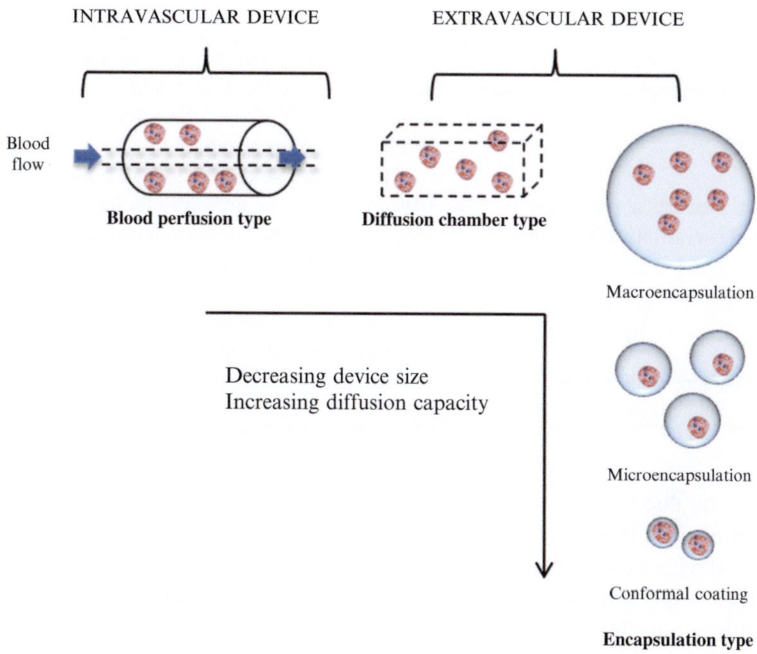

Fig. 3 Different types of bioartificial pancreas

sophisticated islet encapsulation to ensure adequate blood supply, to avoid immune rejection, and to limit fibrous capsule formation. Additionally, from auto- to allo- and to xenotransplantation, these characteristics have a growing importance due to the crescent rejection likelihood.

2.1 *Types of Bioartificial Pancreas*

A variety of polymeric, inorganic matrices and membranes have been tested to produce immunoisolation devices of diverse physicochemical properties and geometries [23–25, 32], namely the type of blood perfusion, diffusion chamber, encapsulation (macro- and microencapsulation), and more recently, conformal coating (Fig. 3). All of these devices are typically comprised of three basic components: a selective membrane, an internal matrix (except in conformal coating) and the living cells, in this case, the islets. Being an intravascular chamber, the required vascular surgery and the risk of thrombosis in blood perfusion type limit the application of this device. Although the advantage of being an extravascular device, the diffusion chamber type also demonstrates some pitfalls that restrict its use, such as the reported islet aggregation and the reduced oxygen/nutrients supply. When islets congregate and form large aggregates, the cells in the center

Table 1 Biomaterials' important factors for developing bioartificial organs (adapted from [32])

Material chemistry	Material must be obtainable in a pure form
Materials formulation and processing	Ease of processing for desired properties
	Reproducibility of desired properties
	Sterilizability of the material
	Aseptic materials processing for cell contacting
Material mechanical properties	Adequate mechanical properties for the application
Material interactions with the host	Adequate biostability for the implant lifetime
	Non-tumorigenic, non-toxic, non-teratological
	Non-thrombogenic
	No complement activation
	Stability of material properties in the host environment
Material interactions with the encapsulated cells	Non-cytotoxic material to the encapsulated cells
	Maintenance of cell viability once implanted

are necrotic due to oxygen shortage. These limitations can somehow be overcome with an encapsulation strategy, especially microencapsulation, making it a more suitable device. The spherical geometry of the capsules provides a high surface-to-volume ratio which is advantageous for diffuse transport between the graft and surrounding host tissue and, adding to this, the implant is introduced with minor surgery. Although microcapsules are indeed smaller than macrocapsules, their volume still limits its potential for transplantation [26, 33]. As a result, recent studies have focused in conformal coating of the surface of transplanted cells with very thin layers of different materials. Conformal coating has been performed using different techniques such as interfacial photopolymerization [34], layer-by-layer polyelectrolyte deposition [35, 36], hydrophobic polymer coating [35, 37–40] and also conformal coating with living cells [41]. However, other studies showed that these types of coatings display limited stability and most of them dissociate from the cell surface after only a few days. Moreover, the mechanical stability of such very thin coatings remains undefined [26, 38].

2.2 *Membrane Materials for Cell Encapsulation*

Success of encapsulated cells often hinges on the proper engineering of the component materials and properties to fulfill the requirements of encapsulated cells and recipient host. The choice of membrane material and formulation method dictates the resulting transport, morphology and strength characteristics for the desired purpose. There are some important issues that have to be considered in order to choose a biomaterial for the development of the devices (Table 1).

The materials used to develop semi-permeable membranes can be categorized as polymeric or non-polymeric. Hydrogels and thermoplastics are the two polymeric major categories used in this application.

2.2.1 Non-polymeric Materials

In non-polymeric materials, such as silicon and ceramics, the potential to better control the pore size overcomes the common unsolved problem of blockage of harmful immune system components observed in many polymeric formulations [33]. Smith and collaborators developed a silicon nanopore membrane for islets immunoisolation, which allows size selectivity [42]. However, the complicated and harsh conditions of production limit their further development as cell encapsulation devices. Moreover, their fabrication techniques are restricted to macrocapsule devices, mostly in the form of hollow fibres. Macrocapsules are mechanically durable and easily retrievable but have a geometry less ideal for diffusive transport compared to microcapsules.

2.2.2 Polymeric Materials

The properties of polymers have been subject of intensive study and many reports highlighted that these materials are suitable to afford an optimal barrier and be successfully used in devices for transplantation. Furthermore, using sophisticated synthesis methods, it is possible to manipulate some of the polymers' properties and to introduce some new functions through functionalization with bioactive substances [25, 26, 32, 43]. The ability to manipulate all of these issues allows the polymers to be tailored with very specific features for the desired application, which turns these methods very promising for clinical application.

Regarding to thermoplastic polymers, they have more chemical and mechanical stability than hydrogels [33], being preferential in cell encapsulation systems. Polyurethane, polysulfone, and dialysis membranes are examples of thermoplastics tested in islets encapsulation [44–46]. Despite their favorable stability properties, thermoplastics exhibit low permeability to water-soluble nutrients and subsequently limited viability of encapsulated cells, limiting their clinical potential [32]. Moreover, they are often selected for incorporation into undesirable vascular devices and macrocapsules [33]. As reported, this type of device limits the choice of transplantation site and, in islets transplantation studies, this device demonstrated to fail little time after its implantation.

The most popular polymers in the development of cell encapsulation systems, and specifically in the bioartificial pancreas, are the hydrogels. These network materials constituted by crosslinking hydrophilic polymers chains have the ability to absorb large amounts of water. They were first introduced into biomedical applications in the late 1950s following the development of poly(2-hydroxyethyl methacrylate) (PHEMA) gels for soft contact lenses [33]. Due to their viscoelasticity and high water content, they possess a degree of flexibility very similar to biological tissues, displaying minimal inflammatory response and easily adjustable permeability properties [33].

A variety of natural and synthetic polymers have been applied to islet encapsulation. The most common natural polymers used are the polysaccharides agarose and alginate [22, 25, 26], which are obtained from marine algae and form biocompatible gels, with low toxicity and with low costs (Fig. 4). Teramura's group has been working on an agarose hydrogel to microencapsulate islets, and some reports showed that agarose membrane can be applied to transplantation of allogenic islets for treating type 1 diabetic patients without immunosuppressive therapy [26]. Comparing with agarose, the alginate has long term immunoisolatory properties [32]. Alginate-based technology was used in the first human trial of this type for allograft transplantation therapy. This preliminary trial was performed by Soon-Shiong and partners on a single patient with type 1 diabetes, who demonstrated insulin independence with tight glycemic control nine months after implantation [47].

Among synthetics polymers, poly(vinylalcohol) (PVA) and poly(ethylene glycol) (PEG) are the most popular [26, 48, 49] (Fig. 4). PEG has been widely used to functionalize solid surfaces for inert and biocompatible surface engineering, as well as to limit immunogenicity and prolong the circulating half-life of proteins and peptides for drug delivery. Recently, it has been suggested that PEG grafted onto the collagen matrix of isolated islets could inhibit pathways leading to islet xenograft rejection [48, 49]. Specifically, PEGylation leads to a reduction in antibody/complement mediated cytotoxicity [48, 49].

Both PVA and PEG show the capacity to have different structures, carrying different groups in their ends, turning these polymers suitable to interact with cell surface in different ways. Their intrinsic characteristics allow the fabrication of particles by coatings through polyion complex formation, chemical reaction and hydrophobic interaction (Figs. 4 and 6).

2.3 Fabrication of Particles

The development of microparticles' build-up processes has been made during the last 30 years of research. In fact, several experiments demonstrated that the microparticles are the type of device that most satisfies the requirements for *in vivo* application. Without compromising the stability of particles and the host reaction, different processes have been applied to maximally reduce the particles' size. The reduction of the transplantation volume facilitates the implantation procedure and removes the constraints with respect to the choice of transplant location (Fig. 5).

Firstly, the particles were generated by a simple method. In alginate formulation, for example, the suspension of islets and alginate was extruded through a needle into an aqueous solution containing a divalent cation that ionically crosslinked with alginate [25, 32, 50]. The resulting alginate beads usually have a diameter of around 700 μm. More sophisticated droplet generators employing coaxial air flow [32], electrostatic droplet generators [50], or laminar jet break up [51] have been used to decrease particle size. Other approach that has been used since its development by Calafiore and co-workers is the emulsion coating technique [52]. Instead of extruding a mixture of pancreatic islets and alginate, islets are enclosed in a thin layer of alginate using an emulsion coating technique. Then, the alginate layer

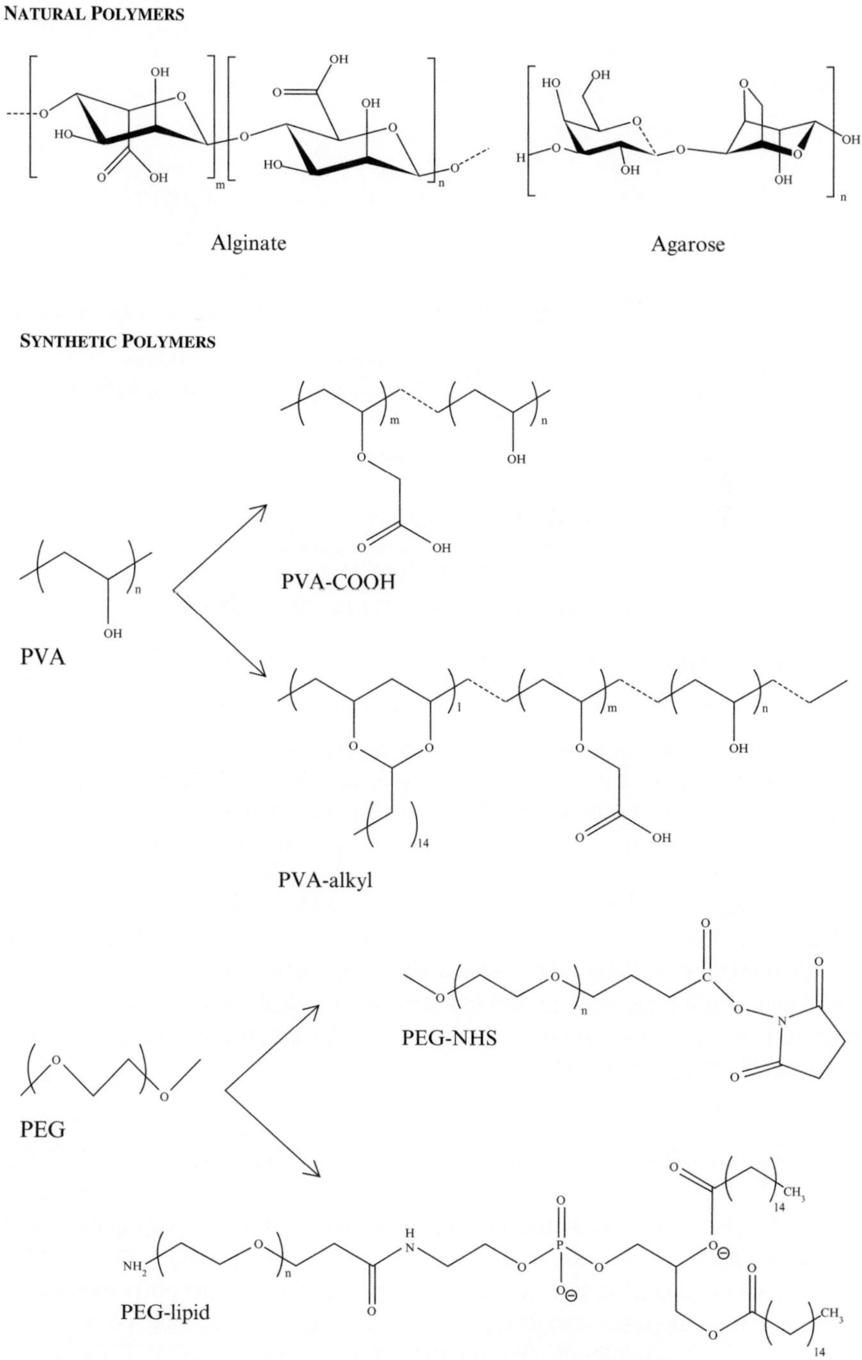

Fig. 4 Chemical structures of natural and synthetic polymers most commonly used in bioartificial pancreas

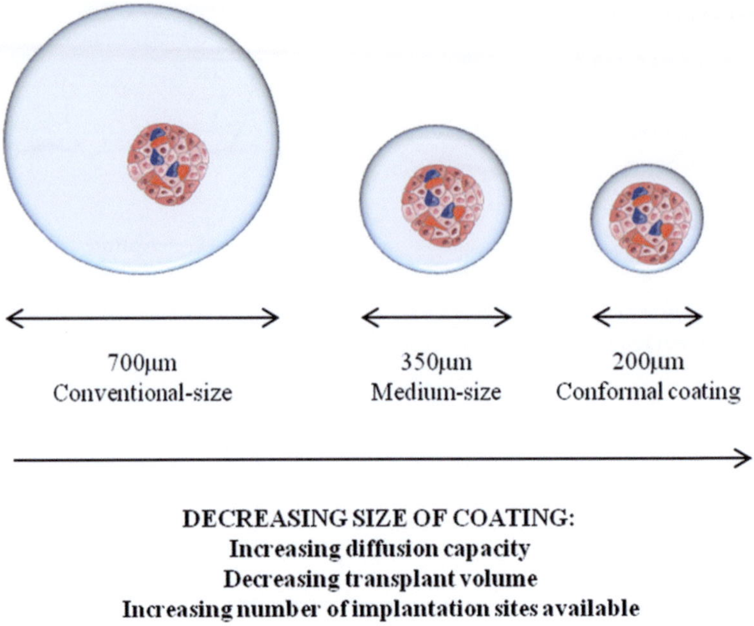

Fig. 5 Size comparison of microparticles generated by different technologies

is cross-linked with calcium and a polyelectrolyte membrane is formed through sequential immersion in poly-L-ornithine (PLO) and alginate. In fact, the techniques of islets coating with a very thin membrane (conformal coating) have been preferred over the technology of conventional biopancreas. Coherent coatings cause only a minimal increase in the volume of the islet itself, resulting, in average, in a particle diameter of 200 μm. Beyond Calafiore's emulsion coating technique, the reduction of the coating's thickness has been carried out by fluid entrainment and interfacial precipitation [50, 53] and photopolymerization [34]. More recently conformal coating has been performed through chemical reaction [26], layer-by-layer polyelectrolyte deposition [35, 36], hydrophobic polymer coating [35, 37–40], and also with living cells [41].

2.3.1 Conformal Coating Through Layer-by-Layer Method

A layer-by-layer method has been reported by some groups to encapsulate β cells and islets [36, 37]. The negatively charged cell surface is treated with a cationic polymer solution, and the surface is further exposed to an anionic polymer solution to form a layer-by-layer membrane. Krol et al. attempted to coat islets with a poly(allylaminehydrochloride) (PAH)/poly(styrene) sulphate (PSS)/PAH multilayer membrane [36] and Miura et al. examined an alginate/PLL/alginate multilayer

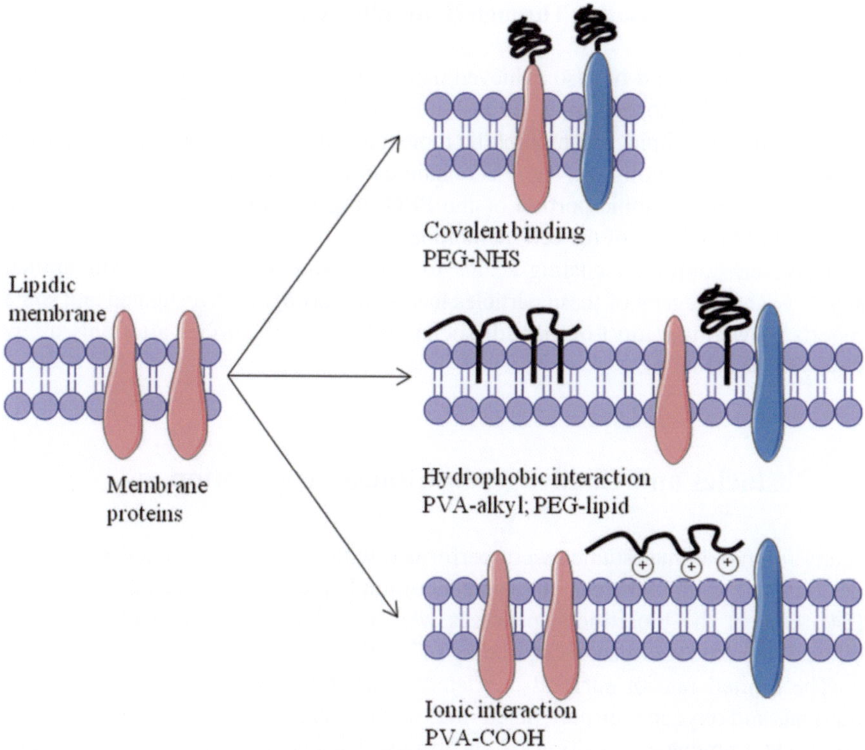

Fig. 6 Examples of interactions between synthetic polymers and islets in conformal coatings

coating on the surface of islets [37]. Although this method achieved small particles, direct interactions between the cationic polymer and cell surface should be avoided because the cell membrane would be gradually destroyed. The problem is that most cationic polymers, such as poly-L-lysine (PLL) and poly(ethyleneimine), are extremely cytotoxic and treated cells are severely damaged [50] (Fig. 6).

2.3.2 Conformal Coating Through Chemical Reaction

Important properties, such as good biocompatibility, nonimmunogenicity, and resistance to protein adsorption, make PEG an important type of hydrophilic polymer for biomedical applications. In some studies, islets were also covered with a PEG layer. PEG carrying an activated ester group (N-hydroxyl-succinimidyl ester—NHS) at one chain end, which is suitable to react with amino groups of the membrane proteins or collagen layer on the islet surface [26] (Fig. 6).

2.3.3 Conformal Coating Through Hydrophobic Interaction

Surface coating could be also achieved using amphiphilic polymers, such as PEG conjugated with phospholipids (PEG-lipid) and PVA, carrying long alkyl chains [35, 37–40]. PEG-lipid, an amphiphilic molecule, has a hydrophilic PEG chain and hydrophobic phospholipids. The PEG chain can be anchored to the β cell surface because the hydrophobic portion of the PEG-lipid is spontaneously incorporated into the lipidic bilayer of the cell membrane.

However, conformal coating seems to have a drawback related with limited stability. The integrity of these particles has been thoroughly investigated and some reports showed that most of them dissociate from the cell surface after only a few days [26, 38] (Fig. 6).

3 Obstacles and Challenges in Clinical Application

Many *in vivo* animal studies were performed with bioartificial pancreas but their performance is limited and a huge percent of implant failure has been demonstrated [54]. Few of the bioartificial pancreas were tested in human patients and their clinical outcome was also disappointing [47, 55].

The limited rate of survival is related to several reasons, such as inadequate nutrients and oxygen transport, deleterious innate inflammatory responses to foreign body and immune recognition of encapsulated islets [25, 54, 56], where many factors are implicated.

3.1 Particle Size

The size of microcapsules is an important factor that influences not only the efficiency of oxygen supply and the insulin secretion, but also the volume to transplant and consequently the implantation site, as described below (see Sect. 3.2).

The efficiency of oxygen supply is negatively influenced by the increase of the microparticles size: large particles have more probability to suffer cellular hypoxia. Secreted insulin diffusion is similarly influenced. The delay of insulin response induced by glucose changes is most clearly understood if taking into account the distance between the capsule surface and the islet outer cell layer. This distance, between of ≈100–400 μm, creates a void space that glucose and insulin must cross prior to transport in or out of the device. Accordingly, capsule formulations seem to influence the response time of bioartificial pancreas [31, 57, 58]. Specifically, the pore size and other physicochemical properties of the immunoisolation membrane and cell immobilization matrix influence the diffusivity of important solutes. Decreasing capsule size has been shown to minimize insulin secretion delay [59, 60]. Much effort has been made in order to reduce the size of the capsules. As reported before, more efficient techniques of particles build-up have been developed. Specifically, recent approaches that directly enclose some polymers through surface islet coating have reported the smallest encapsulated islets [26].

3.2 Transplantation Site

Regarding the size's influence in the volume transplantation, current clinical islet transplantation protocols require ~600–700 thousand islets (average diameter of islets ~150 μm), a volume of approximately 5–10 mL which is affordable in hepatic sinusoids. In contrast, trials using islets entrapped in 500 μm microcapsules (average diameter of conventional capsules is approximately three time as large as the original islets) will require a transplant volume of 50 mL, representing approximately a five to tenfold increase in transplant volume [20, 25, 61]. As a result, the volume to transplant is a constraint of the procedure of intra-hepatic transplantation. It is well documented that the implantation sites should meet important requirements: be well vascularized to supply the graft with nutrient and oxygen, give physical support and allow the implant introduction in a minimally invasive procedure. Although the intra-hepatic transplantation satisfies the referred features, within the hepatic portal environment islets face with instant blood-mediated inflammatory reactions (IBMIR), low oxygen tension and are exposed to high levels of glucose and pharmaceuticals, including the immunosuppressants [62]. Regarding the encapsulated islets, an extravascular and extrahepatic implantation eliminates the IBMIR and the hepatic malignant environment, while a biomaterial platform provides physical support for the islets [63]. Different sites for encapsulated islets implantation have been studied in animal model such as peritoneal cavity [22, 64, 65], subcutaneous [64, 66], intramuscular [66, 67] and intra-epididymis tissues [66], bone marrow cavity [67], kidney [64], liver [68] and anterior chamber of the eye [55]. The subcutaneous and the intraperitoneal spaces were already experienced in human patients [47, 55]. As reported, for large spaces such as the peritoneal cavity and the subcutaneous tissue the microcapsule size can be increased up to approximately 500 μm. In contrast, for the transplantation into the liver via the portal vein, the size should be as small as possible and it is necessary to apply islets conformal coating approaches [26]. Some studies suggest avoiding the peritoneal cavity as an implant site due to the consequent aggregation and subsequent fibrosis and hypoxia, culminating in islets necrosis [25]. Taking advantage of the optimal localization, some groups intended to pre-vascularize a subcutaneous region for posterior implantation of microencapsulated islets [69, 70]. However, the superficial presence of the transplant is associated with high risk of graft mechanical stress and damage.

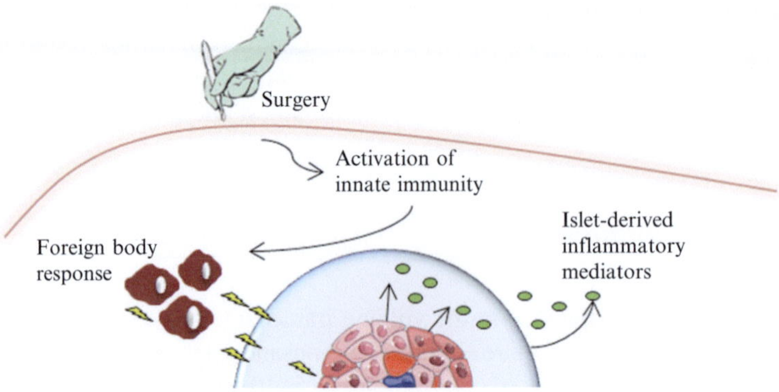

Fig. 7 Cytotoxic inflammatory responses during the islet transplantation

3.3 *The Inflammatory Reaction*

The inflammatory processes are the basis of all problems that lead to islets failure (Fig. 7). The inflammatory response to immunoisolated islets may be mediated by a foreign body response to the device itself, the release of inflammatory mediators from the encapsulated cells and the surgical procedure necessary for implantation of the device (Fig. 7).

3.3.1 The Device

Inflammation has an important role in the protection of the body from foreign agents which, in this case, is the device with non autologous cells. Device-induced inflammation is related with non-specific adsorption of proteins to the material surface, followed by recruitment of neutrophils and macrophages to the implant site, and the subsequent attachment and overgrowth of the macrophages and fibroblasts leading to the graft loss by necrosis [25, 50, 54]. This inflammatory scenario involves the production of many low molecular weight inflammatory mediators which are able to cross the membrane and interact with the islets. As discussed before, the membrane permeability desired for islet encapsulation should block the passage of immune cells and antibodies but still has to allow free insulin diffusion and transit of oxygen, nutrients and metabolic wastes for maintaining cellular function [54, 71]. However, the permeability that accommodates insulin diffusion will permit pro-inflammatory cytokines, such as interleukin-1 alpha (IL-1β), tumor necrosis factor-alpha (TNF-α), Interferon-gamma (IFN-γ), reactive oxygen species (ROS) and other low molecular weight molecules to enter the capsules and exert deleterious effects on β-cell function and islet viability [54].

Besides transplantation site and implant size, the material properties critically influence the biocompatibility of encapsulated islets, namely surface charge, porosity, roughness, surface chemistry and free energy [54]. The capsule size also has an important role in physical irregularities. Usually, the smaller capsules contain more irregularities and therefore the fibrosis is potentiated. After several years of research, the biocompatibility of the devices still compromises their performance *in vivo*. Therefore, many groups focused their efforts in improving the biocompatibility of immunoisolation membranes [50]. Surface PEGylation has been used to improve the biocompatibility of synthetic immunoisolation membranes [35, 48, 49], as well as alginate/PLL/alginate microcapsules [25, 50].

3.3.2 The Encapsulated Cells

As reported before, another source of inflammation is directly related to the islets themselves. They might produce low molecular weight inflammatory mediators such as IL-1β and 6, TNF-α, monocyte chemotactic protein-1 (MCP-1) and ROS, which are able to diffuse across immunoisolation membranes triggering inflammatory cell recruitment and activation [25, 54, 57].

3.3.3 The Surgery

Inevitably, the device implantation surgical procedure, even minimally invasive, generates an inflammatory response [25]. Inflammatory mediators released during this early phase by nearby immune cells are able to induce IBMIR, namely blood coagulation and complement system activation. This kind of reactions gains a great proportion in the vascular delivery, such as the usual clinical infusion into the hepatic portal vein, that can lead to the loss of up to two-thirds of the islets within the first few days post transplantation [62]. Beyond the recruitment of immune cells to the implant site and the subsequent attachment and overgrowth of macrophages and fibroblasts leading to the graft hypoxia, the produced small molecules are, once more, able to cross the membrane and to interact with encapsulated cells. After the interaction with the islets receptors, these molecules trigger, again, inflammatory pathways that culminate in apoptosis and inevitably in implant loss [54, 63].

4 Finding the Optimal Environment for the Implant

In order to overcome the primary obstacle of IBMIR observed in vascular delivery, systemic administration of anticoagulants drugs such as activated protein C [72], low molecular weight dextran sulphate [73] and the thrombin inhibitor melagatran [74], has been performed in the clinical practice. However, the systemic application of these drugs is limited by the increased risk of bleeding. The pitfalls observed in the delivery of this kind of substances and others encouraged the development of controlled drug delivery systems. Other innovative methodology tests the outcome of encapsulated islets whit the particularity of immobilization of substances that prevent IBMIR directly on the capsule surface [39, 40, 75–79]. Other than substances that prevent IBMRI, protein receptors, important peptides or enzymes with vital activity could be also used with these approaches. These methodologies restrict the pharmacological effect of certain drugs only to the transplantation site. Subsequently, it will be possible to locally control the transplant microenvironment, with local immunomodulation and important cells requirements, promoting engraftment and long-term function. Hence, the required islet mass can be reduced and it will be possible to use cell therapies involving alternative β cells sources, such as xenogenic islets or engineering insulin-producing cells (Sect. 1.5).

4.1 Polymers Functionalization with Bioactive Substances: New Features and Abilities

The polymers functionalization is based in the introduction of desired chemical groups or molecules into the polymer to create specific chemical, physical, biological, pharmacological or other properties. In the context of transplantation, this methodology allows the functionalization with substances that protect the implant from the host and that improve the β cell functioning. With the improvement of immune-protection barriers and the developing of an essential environment to cell survival, the implants are more prone to long-term survive. Some approaches to construct functionalized membranes have already been proposed. Specifically in bioartificial pancreas, previous reports show different strategies to immobilize important substances in encapsulated islets, where some *in vitro* tests demonstrated promissory results (Table 2). However, the functionalization of the bioartificial pancreas is still a challenge and the clinical application of this device depends on its success.

4.2 Delivery Systems

The systemic delivery has shown to be problematic due to the required administration doses, short half-live of some substances and the side effects on non-target tissues [63]. The development of controlled drug delivery systems (CDDS) can solve these problems. The delivery can be made from the scaffold used for cells' encapsulation or from separate systems [25, 63]. These structures have been tested to deliver from small drugs to large bioactive proteins and genes encoding these proteins [85, 86]. Multiple factors can be delivered simultaneously and delivery can be tailored to persist for days to weeks [63].

A critical consideration for the islet transplantation is the reestablishment of a functional vasculature to support efficient glucose sensing and the insulin demand of islets [26, 63]. The implantation sites can be poor vascularized, yielding studies with localized production or pre-treatment (injections or treatment in culture) with

Table 2 Polymers functionalization with bioactive substances in bioartificial pancreas

Activity acquired	Biosubstances	Polymer functionalized	Strategy to immobilize the biosubstance	Ref
Anti-coagulant/ Fibrinolytic/ Complement inhibition (IBMRI)	Urokinase	PEG-lipid carrying oligo(dT)$_{20}$	*DNA hybridization* ➔ urokinase carrying oligo(dA)$_{20}$	[79]
		PVA-alkyl carrying thiol groups	*Thiol/maleimide chemistry* ➔ urokinase carrying maleimide groups	[40]
	Urokinase & heparin	PEG-lipid with biotin groups	Multilayered of biotin-BSA/streptavidin by *layer-by-layer* method	[39]
			Schiff base between urokinase and aldehyde groups produced by dextran oxidized at surface	
			Ionic interaction between heparin and protamine anchored to the surface	
	Thrombomodulin	PEG-NHS reacts with collagen's amino groups on islets surface	*Staudinger ligation* between recombinant azido-thrombomodulin and a bifunctional PEG linker	[78]
	Urokinase & thrombomodulin	PEG-lipid carrying maleimide groups	*Thiol/maleimide chemistry* ➔ urokinase and thrombomodulin carrying thiol groups	[77]
	sCR1 & heparin	PEG-lipid carrying maleimide groups	*Thiol/maleimide chemistry* between sCR1 carrying thiol groups and PEG	[76]
			Layers of sCR1 and heparin formed by *layer-by-layer* method	
	sCR1	Agarose carrying maleimide groups	*Thiol/maleimide chemistry* ➔ sCR1 carrying thiol groups	[75]
		PEG-lipid carrying maleimide groups	*Thiol/maleimide chemistry* ➔ sCR1 carrying thiol groups	[80]
Anti-inflammatory	Inhibitory peptide for islet cell surface IL-1 receptor	4-arm-PEG	*Thiol/maleimide chemistry* ➔PEG carrying thiol groups	[71]

(continued)

Table 2 (continued)

Activity acquired	Biosubstances	Polymer functionalized	Strategy to immobilize the biosubstance	Ref
	Inhibitory peptide for islet cell surface TNF-α receptor	PEG-diacrylate	*Photopolymerization* between PEG-diacrylate and TNF-α receptor functionalized with acrylic groups	[81]
Anti-oxidative	Superoxide dismutase (SOD)	PEG-diacrylate	*Photopolymerization* between PEG-diacrylate and SOD containing acrylic groups	[82]
Pro-apoptotic	Anti-Fas monoclonal antibody	PEG-diacrylate	*Photopolymerization* between PEG-diacrylate and NHS-PEG-acrylate Conjugation between anti-Fas MAbs and pendant NHS groups	[83]
Islets function improvement	GLP-1	PEG-diacrylate carrying maleimide groups	*Thiol acrylate photopolymerization* ➔ GLP-1 with thiol groups	[84]

angiogenic factors such as vascular endothelial growth factor and platelet-derived growth factor, improving vascularization and subsequent islet function and engraftment [87, 88]. Therefore, different CDDS have been employed to deliver angiogenic promoting factors [86, 89, 90] and some studies have already demonstrated the improvement of islets implantation in a prevascularized site [70, 91].

4.3 *The Contribution of Cells*

Another approach uses living cells and the benefits from their protection/immunemodulation capacity [26].

Teramura and co-authors propose a new method for the microencapsulation of islets with a layer of living cells derived from a recipient for islet encapsulation, aiming to improve the biocompatibility of islets after transplantation [41, 92]. The outermost surface is composed by the recipient's own cells and the host immune defense system is not provoked. They employed a biotin/streptavidin reaction [41] and DNA hybridization [92] to attach the HEK293 cells (human endoderm kidney cell line) on the surface of PEG encapsulated islets.

Besides the modification of coating surface with living cells, biomaterial scaffolds can also allow islets to be co-localized with other transplanted cell

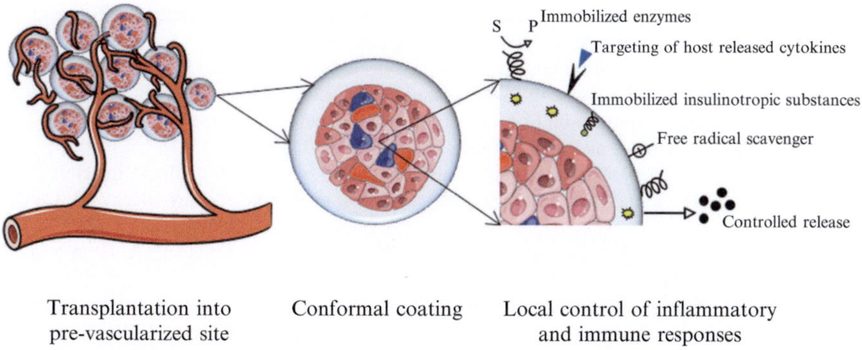

Fig. 8 Emerging paradigms in the islet transplantation

types to create an immunoprivileged environment. Promising results report the co-transplantation of islets with Sertoli cells, demonstrating the creation of an immunoprivileged zone with this kind of cells, resulting in long-term implant survival [93].

5 Conclusions and Future Directions

To allow routine clinical implementation, a range of factors must be properly balanced as for any given application: cost vs. benefit, allowable volume, implant duration, robustness, immunoisolatory function and process scalability. Figure 8 reviews the emerging paradigms in the islets transplantation. Continuous collaborations between engineers, biological and clinical-scientists will be essential for the definition of realistic objectives and to overcome existing obstacles.

References

1. Alsahli M, Gerich JE (2010) Abnormalities of insulin secretion and β-cell defects in type 2 diabetes. In: Holt RIG, Cockram C, Flyvbjerg A, Glodstein BJ (eds) Textbook of diabetes. Wiley-Blackwell, Chichester
2. Giacco F, Brownlee M (2010) Oxidative stress and diabetic complications. Circ Res 107: 1058–1070
3. Marshall SM, Flyvbjerg A (2010) Diabetic nephropathy. In: Holt RIG, Cockram C, Flyvbjerg A, Glodstein BJ (eds) Textbook of diabetes. Wiley-Blackwell, Chichester
4. Nilsson PM, Viljoen A, Wierzbicki AS (2010) Cardiovascular risk factors. In: Holt RIG, Cockram C, Flyvbjerg A, Glodstein BJ (eds) Textbook of diabetes. Wiley-Blackwell, Chichester
5. Scanlon PH (2010) Diabetic retinopathy. In: Holt RIG, Cockram C, Flyvbjerg A, Glodstein BJ (eds) Textbook of diabetes. Wiley-Blackwell, Chichester

6. Ziegler D (2010) Diabetic peripheral neuropathy. In: Holt RIG, Cockram C, Flyvbjerg A, Glodstein BJ (eds) Textbook of diabetes. Wiley-Blackwell, Chichester
7. Rathmann W, Giani G (2004) Global prevalence of diabetes: estimates for the year 2000 and projections for 2030—response to Rathman and Giani. Diabetes Care 27:2568–2569
8. American Diabetes Association (1997) Report of the expert committee on the diagnosis and classification of diabetes mellitus. Diabetes Care 20:1183–1197
9. Alberti KGMM (2010) The classification and diagnosis of diabetes mellitus. In: Holt RIG, Cockram C, Flyvbjerg A, Glodstein BJ (eds) Textbook of diabetes. Wiley-Blackwell, Chichester
10. Toeller M (2010) Lifestyle issues: diet. In: Holt RIG, Cockram C, Flyvbjerg A, Glodstein BJ (eds) Textbook of diabetes. Wiley-Blackwell, Chichester
11. Yardley JE, Alberga A, Kenny GP, Sigal RJ (2010) Lifestyle issues: exercise. In: Holt RIG, Cockram C, Flyvbjerg A, Glodstein BJ (eds) Textbook of diabetes. Wiley-Blackwell, Chichester
12. Gough S, Narendran P (2010) Insulin and insulin treatment. In: Holt RIG, Cockram C, Flyvbjerg A, Glodstein BJ (eds) Textbook of diabetes. Wiley-Blackwell, Chichester
13. Bailey CJ, Krentz AJ (2010) Oral antidiabetic agents. In: Textbook of diabetes. Wiley-Blackwell, Chichester, pp 452–477
14. Bluestone JA, Herold K, Eisenbarth G (2010) Genetics, pathogenesis and clinical interventions in type 1 diabetes. Nature 464:1293–1300
15. Cryer PE (2010) Hypoglycemia in type 1 diabetes mellitus. Endocrinol Metab Clin North Am 39:641–654
16. Williams P (1894) Notes on diabetes treated with extract and by grafts of sheep's pancreas. Br Med J 2:1303–1304
17. Ballinge WF, Lacy PE (1972) Transplantation of intact pancreatic islets in rats. Surgery 72:175–186
18. Pirsch JD, Miller J, Deierhoi MH, Vincenti F, Filo RS (1997) A comparison of tacrolimus (FK506) and cyclosporine for immunosuppression after cadaveric renal transplantation. Transplantation 63:977–983
19. Sutherland DE, Gruessner RW, Dunn DL, Matas AJ, Humar A, Kandaswamy R, Mauer SM, Kennedy WR, Goetz FC, Robertson RP, Gruessner AC, Najarian JS (2001) Lessons learned from more than 1,000 pancreas transplants at a single institution. Ann Surg 233:463–501
20. Shapiro AM, Lakey JR, Ryan EA, Korbutt GS, Toth E, Warnock GL, Kneteman NM, Rajotte RV (2000) Islet transplantation in seven patients with type 1 diabetes mellitus using a glucocorticoid-free immunosuppressive regimen. N Engl J Med 343:230–238
21. Oberholzer J, Triponez F, Mage R, Andereggen E, Bühler L, Crétin N, Fournier B, Goumaz C, Lou J, Philippe J, Morel P (2000) Human islet transplantation – lessons from 13 autologous and 13 allogeneic transplantations. Transplantation 69:1115–1123
22. Agudelo CA, Teramura Y, Iwata H (2009) Cryopreserved agarose-encapsulated islets as bioartificial pancreas: a feasibility study. Transplantation 87:29–34
23. Prokop A, Davidson JM (2005) Bioartificial pancreas: an update. Appl Cell Immobil Biotechnol 8B:39–54
24. Sumi S (2011) Regenerative medicine for insulin deficiency: creation of pancreatic islets and bioartificial pancreas. J Hepatobiliary Pancreat Sci 18:6–12
25. Wilson JT, Chaikof EL (2008) Challenges and emerging technologies in the immunoisolation of cells and tissues. Adv Drug Deliv Rev 60:124–145
26. Teramura Y, Iwata H (2010) Bioartificial pancreas microencapsulation and conformal coating of islet of Langerhans. Adv Drug Deliv Rev 62:827–840
27. Nostro MC, Keller G (2012) Generation of beta cells from human pluripotent stem cells: potential for regenerative medicine. Semin Cell Dev Biol 23:701–710
28. Borowiak M, Melton DA (2009) How to make beta cells? Curr Opin Cell Biol 21:727–732
29. Bowie KM, Chang PL (1998) Development of engineered cells for implantation in gene therapy. Adv Drug Deliv Rev 33:31–43

30. Chick WL, Like AA, Lauris V (1975) Beta cell-culture on synthetic capillaries: artificial endocrine pancreas. Science 187:847–849
31. Lim F, Sun AM (1980) Microencapsulated islets as bioartificial endocrine pancreas. Science 210:908–910
32. Li RH (1998) Materials for immunoisolated cell transplantation. Adv Drug Deliv Rev 33:87–109
33. Nafea EH, Marson A, Poole-Warren LA, Martens PJ (2011) Immunoisolating semi-permeable membranes for cell encapsulation: focus on hydrogels. J Control Release 154:110–122
34. Desmangles AI, Jordan O, Marquis-Weible F (2001) Interfacial photopolymerization of beta-cell clusters: approaches to reduce coating thickness using ionic and lipophilic dyes. Biotechnol Bioeng 72:634–641
35. Teramura Y, Kaneda Y, Iwata H (2007) Islet-encapsulation in ultra-thin layer-by-layer membranes of poly(vinyl alcohol) anchored to poly(ethylene glycol)-lipids in the cell membrane. Biomaterials 28:4818–4825
36. Krol S, del Guerra S, Grupillo M, Diaspro A, Gliozzi A, Marchetti P (2006) Multilayer nanoencapsulation. New approach for immune protection of human pancreatic islets. Nano Lett 6:1933–1939
37. Miura S, Teramura Y, Iwata H (2006) Encapsulation of islets with ultra-thin polyion complex membrane through poly(ethylene glycol)-phospholipids anchored to cell membrane. Biomaterials 27:5828–5835
38. Teramura Y, Kaneda Y, Totani T, Iwata H (2008) Behavior of synthetic polymers immobilized on a cell membrane. Biomaterials 29:1345–1355
39. Teramura Y, Iwata H (2008) Islets surface modification prevents blood-mediated inflammatory responses. Bioconjug Chem 19:1389–1395
40. Totani T, Teramura Y, Iwata H (2008) Immobilization of urokinase on the islet surface by amphiphilic poly(vinyl alcohol) that carries alkyl side chains. Biomaterials 29:2878–2883
41. Teramura Y, Lwata H (2009) Islet encapsulation with living cells for improvement of biocompatibility. Biomaterials 30:2270–2275
42. Smith C, Kirk R, West T, Bratzel M, Cohen M, Martin F, Boiarski A, Rampersaud AA (2005) Diffusion characteristics of microfabricated silicon nanopore membranes as immunoisolation membranes for use in cellular therapeutics. Diabetes Technol Ther 7:151–162
43. van Dijk M, Rijkers DT, Liskamp RM, van Nostrum CF, Hennink WE (2009) Synthesis and applications of biomedical and pharmaceutical polymers via click chemistry methodologies. Bioconjug Chem 20:2001–2016
44. Silva AI, Mateus M (2009) Development of a polysulfone hollow fiber vascular bio-artificial pancreas device for in vitro studies. J Biotechnol 139:236–249
45. Kadam SS, Sudhakar M, Nair PD, Bhonde RR (2010) Reversal of experimental diabetes in mice by transplantation of neo-islets generated from human amnion-derived mesenchymal stromal cells using immuno-isolatory macrocapsules. Cytotherapy 12:982–991
46. Kessler L, Legeay G, Jesser C, Damgé C, Pinget M (1995) Influence of corona surface-treatment on the properties of an artificial membrane used for Langerhans islets encapsulation-permeability and biocompatibility studies. Biomaterials 16:185–191
47. Soon-Shiong P, Heintz RE, Merideth N, Yao QX, Yao Z, Zheng T, Murphy M, Moloney MK, Schmehl M, Harris M, Mendez R, Mendez R, Sandford PA (1994) Insulin independence in a type 1 diabetic patient after encapsulated islet transplantation. Lancet 343:950–951
48. Teramura Y, Iwata H (2009) Surface modification of islets with PEG-lipid for improvement of graft survival in intraportal transplantation. Transplantation 88:624–630
49. Weber LM, He J, Bradley B, Haskins K, Anseth KS (2006) PEG-based hydrogels as an in vitro encapsulation platform for testing controlled beta-cell microenvironments. Acta Biomater 2:1–8
50. Leung A, Nielsen LK, Trau M, Timmins NE (2010) Tissue transplantation by stealth – coherent alginate microcapsules for immunoisolation. Biochem Eng J 48:337–347
51. Brandenberger H, Widmer F (1998) A new multinozzle encapsulation/immobilisation system to produce uniform beads of alginate. J Biotechnol 63:73–80

52. Basta G, Osticioli L, Rossodivita ME, Sarchielli P, Tortoioli C, Brunetti P, Calafiore R (1995) Method for fabrication of coherent microcapsules—a new potential immunoisolatory barrier for pancreatic-islet transplantation. Diabetes Nutr Metab 8:105–112
53. Zekorn T, Siebers U, Horcher A, Schnettler R, Zimmermann U, Bretzel RG, Federlin K (1992) Alginate coating of islets of Langerhans—in vitro studies on a new method for microencapsulation for immunoisolated transplantation. Acta Diabetol 29:41–45
54. de Groot M, Schuurs TA, van Schilfgaarde R (2004) Causes of limited survival of microencapsulated pancreatic islet grafts. J Surg Res 121:141–150
55. Valdés-González RA, Dorantes LM, Garibay GN, Bracho-Blanchet E, Mendez AJ, Dávila-Pérez R, Elliott RB, Terán L, White DJ (2005) Xenotransplantation of porcine neonatal islets of Langerhans and sertoli cells: a 4-year study. Eur J Endocrinol 153:419–427
56. Bretzel RG, Jahr H, Eckhard M, Martin I, Winter D, Brendel MD (2007) Islet cell transplantation today. Langenbecks Arch Surg 392:239–253
57. de Vos P, de Haan BJ, de Haan A, van Zanten J, Faas MM (2004) Factors influencing functional survival of microencapsulated islet grafts. Cell Transplant 13:515–524
58. Trivedi N, Keegan M, Steil GM, Hollister-Lock J, Hasenkamp WM, Colton CK, Bonner-Weir S, Weir GC (2001) Islets in alginate macrobeads reverse diabetes despite minimal acute insulin secretory responses. Transplantation 71:203–211
59. Chicheportiche D, Reach G (1988) In vitro kinetics of insulin release by microencapsulated rat islets—effect of the size of the microcapsules. Diabetologia 31:54–57
60. Canaple L, Rehor A, Hunkeler D (2002) Improving cell encapsulation through size control. J Biomater Sci Polym 13:783–796
61. Calafiore R, Basta G, Luca G, Lemmi A, Montanucci MP, Calabrese G, Racanicchi L, Mancuso F, Brunetti P (2006) Microencapsulated pancreatic islet allografts into nonimmunosuppressed patients with type 1 diabetes: first two cases. Diabetes Care 29:137–138
62. Barshes NR, Wyllie S, Goss JA (2005) Inflammation-mediated dysfunction and apoptosis in pancreatic islet transplantation: implications for intrahepatic grafts. J Leukoc Biol 77:587–597
63. Gibly RF, Graham JG, Luo X, Lowe WL Jr, Hering BJ, Shea LD (2011) Advancing islet transplantation: from engraftment to the immune response. Diabetologia 54:2494–2505
64. Dufrane D, Steenberghe M, Goebbels RM, Saliez A, Guiot Y, Gianello P (2006) The influence of implantation site on the biocompatibility and survival of alginate encapsulated pig islets in rats. Biomaterials 27:3201–3208
65. Sun Y, Ma X, Zhou D, Vacek I, Sun AM (1996) Normalization of diabetes in spontaneously diabetic cynomologus monkeys by xenografts of microencapsulated porcine islets without immunosuppression. J Clin Invest 98:1417–1422
66. Liu XY, Nothias JM, Scavone A, Garfinkel M, Millis JM (2010) Biocompatibility investigation of polyethylene glycol and alginate-poly-L-lysine for islet encapsulation. ASAIO J 56:241–245
67. Yang KC, Wu CC, Qi Z, Chen JC, Sumi S, Lin FH (2010) Comparison of bioartificial pancreas performance in the bone marrow cavity and intramuscular space. Arch Med Res 41:151–153
68. Cabric S, Sanchez J, Lundgren T, Foss A, Felldin M, Källen R, Salmela K, Tibell A, Tufveson G, Larsson R, Korsgren O, Nilsson B (2007) Islet surface heparinization prevents the instant blood-mediated inflammatory reaction in islet transplantation. Diabetes 56:2008–2015
69. De Vos P, Hillebrands JL, De Haan BJ, Strubbe JH, Van Schilfgaarde R (1997) Efficacy of a prevascularized expanded polytetrafluoroethylene solid support system as a transplantation site for pancreatic islets. Transplantation 63:824–830
70. Wang W, Gu Y, Tabata Y, Miyamoto M, Hori H, Nagata N, Touma M, Balamurugan AN, Kawakami Y, Nozawa M, Inoue K (2002) Reversal of diabetes in mice by xenotransplantation of a bioartificial pancreas in a prevascularized subcutaneous site. Transplantation 73:122–129
71. Su J, Hu BH, Lowe WL Jr, Kaufman DB, Messersmith PB (2010) Anti-inflammatory peptide-functionalized hydrogels for insulin-secreting cell encapsulation. Biomaterials 31:308–314
72. Contreras JL, Eckstein C, Smyth CA, Bilbao G, Vilatoba M, Ringland SE, Young C, Thompson JA, Fernández JA, Griffin JH, Eckhoff DE (2004) Activated protein C preserves functional islet mass after intraportal transplantation: a novel link between endothelial cell activation, thrombosis, inflammation, and islet cell death. Diabetes 53:2804–2814

73. Johansson H, Goto M, Dufrane D, Siegbahn A, Elgue G, Gianello P, Korsgren O, Nilsson B (2006) Low molecular weight dextran sulfate: a strong candidate drug to block IBMIR in clinical islet transplantation. Am J Transplant 6:305–312
74. Ozmen L, Ekdahl KN, Elgue G, Larsson R, Korsgren O, Nilsson B (2002) Inhibition of thrombin abrogates the instant blood-mediated inflammatory reaction triggered by isolated human islets: possible application of the thrombin inhibitor melagatran in clinical islet transplantation. Diabetes 51:1779–1784
75. Luan NM, Teramura Y, Iwata H (2010) Immobilization of the soluble domain of human complement receptor 1 on agarose-encapsulated islets for the prevention of complement activation. Biomaterials 31:8847–8853
76. Luan NM, Teramura Y, Iwata H (2011) Layer-by-layer co-immobilization of soluble complement receptor 1 and heparin on islets. Biomaterials 32:6487–6492
77. Chen H, Teramura Y, Iwata H (2011) Co-immobilization of urokinase and thrombomodulin on islet surfaces by poly(ethylene glycol)-conjugated phospholipid. J Control Release 150:229–234
78. Stabler CL, Sun XL, Cui W, Wilson JT, Haller CA, Chaikof EL (2007) Surface re-engineering of pancreatic islets with recombinant azido-thrombomodulin. Bioconjug Chem 18:1713–1715
79. Takemoto N, Teramura Y, Iwata H (2011) Islet surface modification with urokinase through DNA hybridization. Bioconjug Chem 22:673–678
80. Luan NM, Teramura Y, Iwata H (2011) Immobilization of soluble complement receptor 1 on islets. Biomaterials 32:4539–4545
81. Lin CC, Metters AT, Anseth KS (2009) Functional PEG-peptide hydrogels to modulate local inflammation induced by the pro-inflammatory cytokine TNF alpha. Biomaterials 30: 4907–4914
82. Cheung CY, McCartney SJ, Anseth KS (2008) Synthesis of polymerizable superoxide dismutase mimetics to reduce reactive oxygen species damage in transplanted biomedical devices. Adv Funct Mater 18:3119–3126
83. Cheung CY, Anseth KS (2006) Synthesis of immunoisolation barriers that provide localized immunosuppression for encapsulated pancreatic islets. Bioconjug Chem 17:1036–1042
84. Lin CC, Anseth KS (2009) Glucagon-like peptide-1 functionalized PEG hydrogels promote survival and function of encapsulated pancreatic beta-cells. Biomacromolecules 10:2460–2467
85. Rives CB, des Rieux A, Zelivyanskaya M, Stock SR, Lowe WL Jr, Shea LD (2009) Layered PLG scaffolds for in vivo plasmid delivery. Biomaterials 30:394–401
86. Chen RR, Silva EA, Yuen WW, Mooney DJ (2007) Spatio-temporal VEGF and PDGF delivery patterns blood vessel formation and maturation. Pharm Res 24:258–264
87. Lai Y, Schneider D, Kidszun A, Hauck-Schmalenberger I, Breier G, Brandhorst D, Brandhorst H, Iken M, Brendel MD, Bretzel RG, Linn T (2005) Vascular endothelial growth factor increases functional beta-cell mass by improvement of angiogenesis of isolated human and murine pancreatic islets. Transplantation 79:1530–1536
88. Olsson R, Maxhuni A, Carlsson PO (2006) Revascularization of transplanted pancreatic islets following culture with stimulators of angiogenesis. Transplantation 82:340–347
89. Matsui M, Tabata Y (2012) Enhanced angiogenesis by multiple release of platelet-rich plasma contents and basic fibroblast growth factor from gelatin hydrogels. Acta Biomater 8:1792–1801
90. Oliviero O, Ventre M, Netti PA (2012) Functional porous hydrogels to study angiogenesis under the effect of controlled release of vascular endothelial growth factor. Acta Biomater 8:3294–3301
91. Balamurugan AN, Gu Y, Tabata Y, Miyamoto M, Cui W, Hori H, Satake A, Nagata N, Wang W, Inoue K (2003) Bioartificial pancreas transplantation at prevascularized intermuscular space: effect of angiogenesis induction on islet survival. Pancreas 26:279–285
92. Teramura Y, Minh LN, Kawamoto T, Iwata H (2010) Microencapsulation of islets with living cells using polyDNA-PEG-lipid conjugate. Bioconjug Chem 21:792–796
93. Dufour JM, Lord SJ, Kin T, Rayat GR, Dixon DE, Bleackley RC, Korbutt GS, Rajotte RV (2007) Comparison of successful and unsuccessful islet/sertoli cell cotransplant grafts in streptozotocin-induced diabetic mice. Cell Transplant 16:1029–1038

Cationic Liposome-Based Systems for Nucleic Acid Delivery: From the Formulation Development to Therapeutic Applications

Henrique Faneca, Ana Luísa Cardoso, Sara Trabulo, Sónia Duarte, and Maria C. Pedroso de Lima

Abstract Significant progress has been made in the development of different types of nucleic acids, including plasmid DNA, siRNA and oligonucleotides with the potential to form the basis of new treatment options for genetic and acquired diseases. However, the lack of suitable vectors for efficient delivery of nucleic acids into target cells represents a major hurdle for the successful application of gene therapy.

Since they were first described by Felgner et al. in the late 1980s, cationic liposomes have been considered one of the most promising carriers for nucleic acid delivery to mammalian cells in vitro and in vivo. Nevertheless, cationic liposomes suffer from relatively low levels of gene expression, thus the drive to improve these vectors continues.

Here, we describe crucial physicochemical parameters of cationic liposome-based systems that modulate their biological activity. Moreover, we provide an overview of different strategies that have been explored to surpass cellular barriers towards improving the efficacy of these systems to mediate nucleic acid delivery and therapeutic activity. Finally, examples illustrating the application of cationic liposome-based systems in clinical trials are presented.

H. Faneca • A.L. Cardoso • S. Trabulo • S. Duarte
Center for Neuroscience and Cell Biology, University of Coimbra, Coimbra, Portugal

M.C. Pedroso de Lima (✉)
Center for Neuroscience and Cell Biology, University of Coimbra, Coimbra, Portugal

Department of Life Sciences, University of Coimbra, Apartado 3046, 3001-401 Coimbra, Portugal
e-mail: mdelima@ci.uc.pt

J. Coelho (ed.), *Drug Delivery Systems: Advanced Technologies Potentially Applicable in Personalised Treatment*, Advances in Predictive, Preventive and Personalised Medicine 4, DOI 10.1007/978-94-007-6010-3_6,

Graphical Abstract

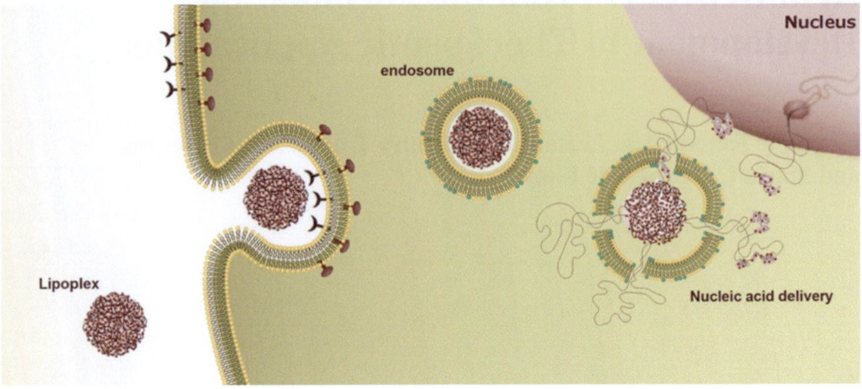

Following binding to the plasma membrane of the target cell, lipoplexes are internalized by the endocytotic pathway. Once inside the endosome, lipoplexes promote endosomal membrane destabilization allowing nucleic acid delivery into the cytoplasm and subsequent nuclear translocation. All these steps are crucial for achieving efficient therapeutic activity

Keywords Nucleic acid delivery • Cationic liposomes • Lipoplexes • Gene therapy • Targeting ligands • Personalized medicine

Abbreviations

5-FC	5-fluorocytosine
AD	Atopic dermatitis
Apo B	Apolipoprotein B
ASO	Antisense oligonucleotide
BBB	Blood brain barrier
BCR-ABL	Break cluster region-Abelson
CD	Cytosine deaminase
$CD4^+$ T lymphocytes	T lymphocytes expressing the cluster differentiation 4
$CD8^+$ T lymphocytes	T lymphocytes expressing the cluster differentiation 8
CF	Cystic fibrosis
CFTR	Cystic fibrosis transmembrane conductance regulator
Chol	Cholesterol
c-myc	Myelocytomotosis oncogene
CNS	Central nervous system
CpG	Cytosine-phosphate-guanine motifs
CPP	Cell-penetrating peptide
ctDNA	Calf timus DNA

DC-Chol	3ß-[N-(N′,N′-dimethylaminoethane)-carbamoyl] cholesterol
DLS	Delivery liposomal system
DMPE-PEG_{5000}	Dimyristoylphosphatidylethanolamine-polyethylene $glycol_{5000}$
DMRIE	1,2-dimyristyloxypropyl-3-dimethyl-hydroxy ethyl ammonium bromide
DMTAP	Dimyristoyl 1,2-diacyl-3-trimethylammonium-propane
DOGS	Dioctadecyl amino glycyl spermine
DOPE	1,2-dioleoyl-sn-glycero-3-phosphoethanolamine
DOSPA	2,3 dioleyloxy-*N*-[2[sperminecarboxaminino]ethyl]-*N*,*N*-dimethyl-1-propanaminium trifluroacetate
DOTAP	1,2-dioleoyl-3-trimethylammonium-propane
DOTIM	Octadecenoyloxy[ethyl-2-heptadecenyl-3-hydroxyethyl] imidazolinium chloride
DOTMA	2,3-*bis*[oleyl]oxipropyltrimethylammoniumchloride
DSPE	1,2-distearoyl-sn-glycero-3-phosphoethanolamine
EBV	Epstein-Barr virus
EGFR	Endothelial growth factor receptor
ePC	Egg yolk phosphatidylcholine
EPOPC	1-palmitoyl-2-oleoyl-sn-glycero-3-ethylphosphocholine
eSph	Egg yolk sphingomyelin
FA	Folic acid
FR	Folate receptor
GCV	Ganciclovir
HA2	Hemagglutinin subunit 2
HBV	Hepatitis B virus
HER-2/neu	Human epidermal growth factor receptor 2 proto-oncogene
HIV-1	Human immunodeficiency virus type 1
HSA	Human serum albumin
HSV-tk	Herpes simplex virus-tymidine kinase
hTR	Human telomerase RNA component
HVJ	Hemagglutinating virus of Japan
IC_{50}	Half maximal inhibitory concentration
IFN-α	Interferon-alpha
IFN-β	Interferon-beta
IFN-γ	Interferon-gamma
IL-12	Interleukin-12
IL-13	Interleukin-13
IL-18	Interleukin-18
IL-4	Interleukin-4
LMO2	LIM domain only 2
LPS	Lipopolysaccharide
MAbs	Monoclonal antibodies

Man-liposomes	Mannosylated liposomes
md-LErafAON	Modified formulation of antisense oligonucleotide targeting Raf-1
MDM2	Murine double minute oncogene
MEND	Multifunctional envelope-type nanodevices
MIP-1β	Macrophage inflammatory protein-1β
miRNA	microRNA
mRNA	Messenger RNA
NFκB	Nuclear factor kappa-light-chain-enhancer of activated B cells
NK	Natural killer
ON	Oligonucleotide
PC	Phosphatidylcholine
pDNA	Plasmid DNA
PEG	Polyethylene glycol
Raf-1	Proto-oncogene serine/threonine-protein kinase
RNAi	RNA interference
shRNA	Short hairpin RNA
siRNA	Small interference RNA
SNALPs	Stabilized nucleic acid lipid particles
SSO	Splice-switching oligonucleotide
SV40	Simian vacuolating virus 40
Tf-lipoplexes	Transferrin-associated lipoplexes
TfR	Transferrin receptor
THL	Trojan Horse liposomes
TIL	Tumor infiltrating lymphocytes
TNF-α	Tumor necrosis factor-alpha
TSA	Mouse mammary adenocarcinoma
UCL	Ultradeformable cationic liposomes
VEGF	Vascular endothelial growth factor

1 Introduction

The advances in the field of gene therapy have significantly increased the prospect for nucleic acids as highly promising agents towards the treatment of a large number of human diseases. However, despite the enormous pharmacological potential of these molecules, successful clinical application of nucleic acid-based strategies remains dependent on the development of safe delivery systems capable of overcoming numerous biological barriers, so that efficient uptake of nucleic acids into target cells is achieved.

Starting at the formulation level, cost, sterilization and shelf-live are parameters that have to be considered. Then, for the molecules to reach the target cells in the body, stability in biological fluids must be guaranteed, as well as the ability

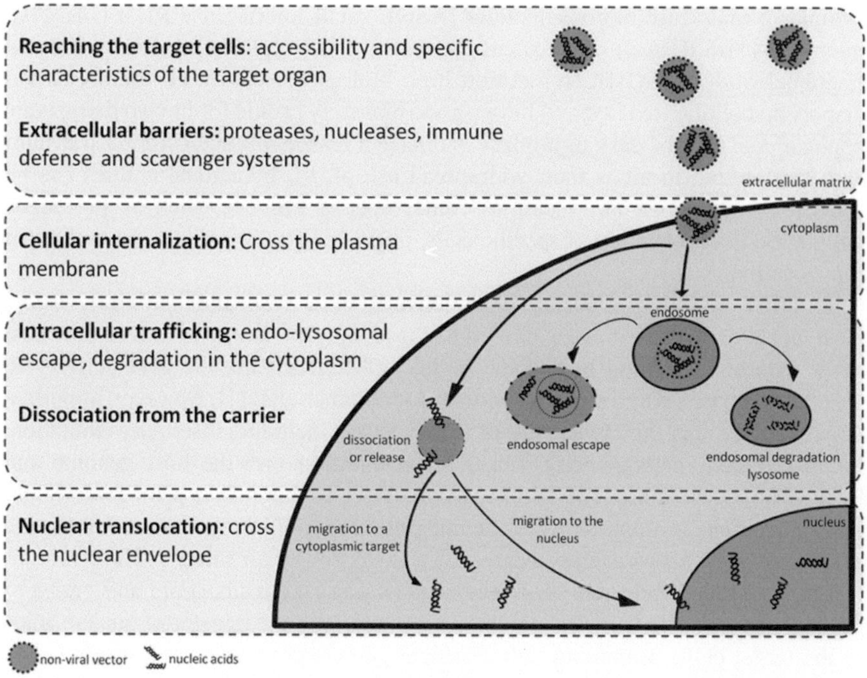

Fig. 1 Physiological/cellular barriers to transfection. A successful nucleic acid delivery system should reach the target tissue/organ, avoid enzymatic degradation and diffuse through the extracellular matrix, reach the target cell, cross the plasma membrane, escape the endosome and/or avoid degradation in the cytoplasm, release the nucleic acids into the cytoplasm and ensure that the nucleic acids reach their target, either in the cytoplasm or the nucleus

to escape the immunological surveillance and clearance mechanisms. Cellular barriers, such as the plasma and nuclear membranes, the endocytic pathway and potential degradation by cytoplasmic enzymes, constitute further obstacles that must be overcome. Therefore, appropriate systems must be devised to ensure that the various physiological/cellular barriers to transfection are surpassed, thus allowing the correct targeting in a wide variety of cells, tissues and whole organs (Fig. 1) [1–7].

When a systemic administration route is used, the first barrier consists in crossing the bloodstream which is, in general, restricted to molecules or particles smaller than 5 nm [8]. However, certain tissues, including the liver, spleen and also some tumors allow the passage of molecules up to 200 nm in diameter [8]. While local application of compounds allows the administration of larger molecules, and surpassing the initial barriers, there is still the need to diffuse through the extracellular matrix so that the plasma membrane of target cells is reached [8].

Ideally, nucleic acid delivery systems should: (i) protect nucleic acids from degradation; (ii) be effectively internalized in specific target tissues/organs/cells, including non-dividing cells; (iii) promote release of the carried cargos in the cytoplasm (antisense oligonucleotides [ASO], small interference RNA [siRNAs], microRNA [miRNAs]) or nucleus (plasmid DNA [pDNA], splice-switching oligonucleotides [SSOs]); (iv) exhibit high biological activity at low doses; (v) display no cellular toxicity; (vi) have a good biosafety profile for in vivo therapeutic applications; (vii) be easy to produce and have a reasonable shelf-life for transport and distribution, to allow their widespread use [4, 9]. Functional entities can be further introduced in the vector to confer special abilities, such as prolonged circulation time, targeting of specific cells, improved endosomal release or nuclear translocation.

In order to improve nucleic acid delivery, both viral and non-viral vectors have been explored. Viruses have developed elegant strategies to invade host cells, naturally evolving to become highly efficient for nucleic acid delivery to specific cell types, while avoiding immunosurveillance [10, 11]. However, important drawbacks hamper the clinical use of viral vectors, including insert-size limitation, immunogenicity/pathogenicity, unspecific integration into the host genome and large-scale manufacture [12]. In this regard, strategies that reduce pathogenicity and immunogenicity promote site-specific integration and allow regulation of transgene expression have been developed and shown to improve their safety [10].

On the other hand, non-viral vectors are clearly nonpathogenic and easier to produce, but are also less efficient than viral vectors in the transfer of nucleic acids to the target cells, promoting only transient gene expression [3, 11, 13–16]. The non-viral delivery systems can be divided into two groups: physical methods, such as microinjection, electroporation and hydrodynamic injection, and chemical-based vehicles, which are usually cationic in nature, and include liposomes, polymers and cell-penetrating peptides [13, 14, 17]. Although physical methods offer a simple protocol to analyze gene function, the application of such nucleic acid delivery strategies in the clinic is unlikely [17]. On the other hand, the developed chemical-based non-viral vectors have been shown to improve transfection efficiency and biocompatibility are, therefore, appropriate for clinical applications [17]. Among these, cationic liposomes constitute promising systems for the bench-to-bedside transposition of nucleic acid-based therapeutics.

2 Cationic Liposome-Based Systems: Physicochemical Parameters That Modulate Their Biological Activity

Since cationic liposomes were first described by Felgner el al. [18], innumerous cationic liposome formulations have been developed, several of them presenting high biological activity both in vitro and in animal models of disease [19]. However, the results obtained from the application of most of the therapeutic strategies in clinical trials using cationic liposomes have shown a limited therapeutic efficiency [20]. In this context, an enormous effort is currently being made to increase the efficiency of cationic liposomes as gene delivery systems, including the synthesis of new cationic lipids and the design of new cationic liposome formulations.

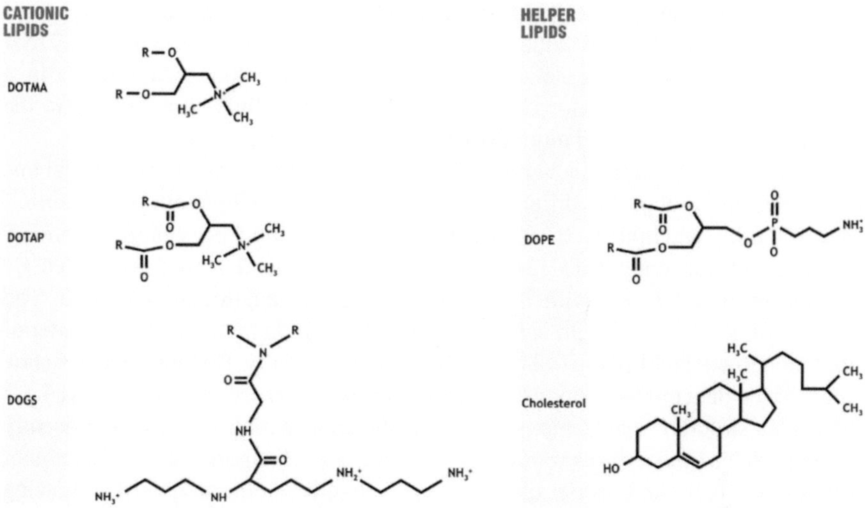

Fig. 2 Chemical structures of common cationic and helper lipids. Cationic lipids, such as DOTMA and DOTAP (monovalent lipids) and DOGS (multivalent lipid), are used for the preparation of cationic liposomes usually in the presence of a helper lipid, such as DOPE and cholesterol. Cationic lipids present amphipathic nature, with a hydrophobic core bridged by a linker group to a hydrophilic head group, which can vary significantly in structure and in the number of positive charges

Cationic lipids include a group of amphiphiles that exhibit a positive charge which triggers their interaction with negatively charged nucleic acids leading to the formation of cationic liposome/DNA complexes (lipoplexes). Cationic lipids are frequently composed of a positively charged headgroup bridged by a linker group to a hydrophobic lipid anchor, a specific role being attributed to each of these components.

The headgroup, which interact electrostatically with the genetic material, can be categorized as follows: quaternary ammonium; primary, secondary or tertiary amines; guanidinium; heterocyclics; amino acids and peptides. The positive charge of cationic lipids described so far is usually associated with amine groups with different degrees of substitution, but other functions like amidine, guanidinium or pyridinium groups have been reported [19, 21, 22]. The headgroup may be single or multiple positively charged (Fig. 2). Multivalent headgroups (e.g., 2,3 dioleyloxy-*N*-[2[sperminecarboxaminino]ethyl]-*N*, *N*-dimethyl-1-propanaminium trifluroacetate [DOSPA] and dioctadecyl amino glycyl spermine [DOGS]) are more efficient in condensing DNA and more active than monovalent lipids (e.g., 2,3-*bis*[oleyl]oxipropyltrimethylammoniumchloride [DOTMA], 1,2-dioleoyl-3-trimethylammonium propane [DOTAP], 3 β[*N*-[*N*′,*N*′-dimethyl amino ethane]-carbamoyl]cholesterol [DC-Chol] and 1,2-dimyristyloxypropyl-3-dimethyl-hydroxy ethyl ammonium bromide [DMRIE]) which may be related to the greater ability of the former to condense and protect DNA as well as to promote cell binding [21, 23]. However, the higher number of positive charges and the consequent high DNA condensation, observed for some cationic lipid/DNA ratios, may hamper the genetic material dissociation, which could result in a lower biological activity. Moreover, multivalent cationic lipids usually lead to the formation of less stable and more toxic lipoplexes [21, 23].

The hydrophobic lipid anchor is typically composed of either a diacyl backbone or a cholesterol group. This different composition affects the lipid bilayer properties, such as fluidity and lipid exchange, which consequently influences the gene delivery efficiency of the complexes [24]. Usually, the biological activity mediated by cationic lipids decreases with increasing alkyl chain length and saturation. The increase of alkyl chain length and saturation reduces the intermembrane transfer of lipid monomers and lipid membrane mixing [21]. Recently, Koynova and Tenchov summarized data on the relationship between transfection activity and cationic lipid molecular structure and demonstrated that the transfection activity significantly depends on the lipid hydrocarbon chain structure. These authors presented a number of examples, including a series of cationic phosphatidylcholine derivatives, showing that optimum transfection is displayed by lipids with chain length of approximately 14 carbon atoms and that the transfection efficiency strongly increases with increase of chain unsaturation, specifically upon replacement of saturated with monounsaturated chains [25].

The linker group of cationic lipids has also been considered to have a crucial influence on the biological activity and cytotoxicity of lipoplexes [24]. Such effect has been associated to the degree of stability and biodegradability of the linker. In order to reduce the cytotoxicity of cationic lipids without compromising the biological activity, some biodegradable lipids have been used, which can be metabolized by various enzymes (esterases and peptidases) [26]. Usually, cationic lipids with stable linker groups, like ether linkage (e.g., DOTMA) are more toxic than lipids containing less stable linker groups, like ester linkage (e.g., DOTAP). However, the lower stability of cationic lipids could reduce the biological activity of lipoplexes and therefore, the linkers of cationic lipids should present a balance between stability and biodegradability. On the other hand, some authors claim that the increase in the linker length could be correlated with an increase in the biological activity [19].

Overall, small modifications in the composition/structure of cationic lipids could result in a significant change in the biological activity of the lipoplexes. Therefore, a complete understanding of the relationships between cationic lipids composition/structure and their biological activity, as well as their cytotoxicity, is crucial for the development of efficient gene delivery systems.

Besides the cationic lipids, cationic liposomes are usually composed of helper lipids, like dioleoylphosphatidylethanolamine (DOPE) and cholesterol (Chol) (Fig. 2), since it has been proven that the presence of these co-lipids can increase their transfection activity. Transfection studies performed in different cell lines clearly show that complexes prepared from cationic liposomes containing DOPE or Chol usually present a much higher biological activity than those produced with cationic liposomes containing only the cationic lipid [27, 28]. The enhanced effect of DOPE is due to its conic structure and consequently to its ability to promote a transition from a bilayer to a hexagonal configuration, which may facilitate fusion with the target membranes and/or their destabilization [24, 29, 30]. Moreover, DOPE can facilitate the dissociation of the genetic material from the cationic lipid component of the complexes after their internalization [31]. This could occur due to the interaction of the amine group of DOPE with the DNA phosphate groups, which weakens the interaction between DNA and cationic lipids [31]. On the other hand, the incorporation of cholesterol in the cationic liposomes also results in an increase in the biological activity of the complexes, this being particularly evident for transfection carried out in the presence of serum or in vivo [27, 28, 32]. This is most probably due to the effect of cholesterol on controlling the bilayer fluidity, which in in vivo applications as well as in the presence of serum could result in an increase in the stability of the complexes and consequently in enhanced biological activity [28, 32–34]. These observations clearly show that the presence of a co-lipid can modulate the transfection capacity of the complexes, this effect being dependent on the type and percentage of the co-lipid in the cationic liposome formulation.

Besides the lipid composition, other factors can influence the gene delivery capacity of the lipoplexes. One of the most important factors is the ratio between cationic lipids and DNA. Our studies, in agreement with those obtained by other authors, have shown that complexes prepared at a low cationic lipid/DNA charge ratio present a homogenous size distribution (mean diameter ranging from 150 to 300 nm) and a negative superfice charge, which is due to the presence of DNA that is not completely condensed by the cationic lipid [28, 35, 36]. Complexes that are prepared at a lipid/DNA charge ratio close to neutrality present a neutral or approximately neutral zeta potential and a large size distribution, with a mean diameter that can easily exceed 1,000 nm. In the case of the lipoplexes prepared with an excess of cationic liposomes (complexes with a positive lipid/DNA charge ratio) a homogeneous size distribution (mean diameter near 200–300 nm) and a positive surface charge were observed. Complexes with an excess of negative or positive charge exhibit a higher colloidal stability than those prepared at a charge ratio close to neutrality. This is most likely due to the electrostatic repulsive forces that occur for the highly positively and negatively charged complexes, which would prevent their aggregation. In the case of complexes close to neutrality, the absence of electrostatic repulsive forces promotes aggregation and consequently an increase in their mean diameter [27, 28, 35–38].

The surface charge of the complexes was demonstrated to play a relevant role in the biological activity. In this regard, lipoplexes with a positive zeta potential usually mediate a much higher transfection activity than those with a negative zeta potential, which is most probably due to the interaction of the positively charged complexes with the negatively charged cell membrane components, such as proteoglycans [16, 21]. However, the high positive charge of the complexes can be responsible for a decrease in their biological activity, since the more extensive condensation of the genetic material may prevent its cytoplasmic release. Moreover, the highly charged lipoplexes usually present a very small mean diameter and consequently carrying low amounts of nucleic acids, which could also contribute for a lower transfection activity than those presenting only a slightly positive charge [16, 27, 35]. In agreement with these arguments, it has been found that, in most of the cases, lipoplexes that exhibit a slight positive charge and a mean diameter higher than 300 nm are more efficient in mediating transfection than those with the same lipid composition but with an excess of negative or positive charge [28, 33, 39]. However, in accordance with previous reports, our results indicated that the composition of the complexes plays a very important role in their biological activity, since complexes with similar zeta potential and mean diameter presented a substantial difference in their transfection activity as a consequence of their different compositions [27, 28, 40, 41].

Besides the lipid/DNA charge ratio, other factors could influence the mean diameter and zeta potential of the lipoplexes and consequently their biological activity, such as lipoplex concentration, ionic strength and temperature of the medium in which the complexes are prepared and their maturation time. The increase of these parameters usually leads to an increase in the mean diameter of the complexes and in the possible occurrence of slight changes in their surface charge [3, 21]. Altogether, these changes in the lipoplex properties frequently result in alterations in their biological activity, as evidenced by the decreased transfection activity observed upon a drastic increase of concentration and maturation time of the complexes due to the high increase of their mean diameter.

3 Strategies to Enhance Biological Activity of Cationic Liposome-Based Systems

Although significant progress has been made regarding the development of lipoplex formulations, their efficiency is still below that required for their successful clinical application. In this regard, functional devices have been introduced in lipoplex formulations to help overcome the biological barriers associated to the transfection process, such as targeting a specific tissue or cell type, transversing the plasma membrane, escaping lysosomal degradation and overcoming the nuclear envelope. These devices include, for example, the use of targeting ligands to increase the specificity of cellular uptake, proteins or fusogenic peptides to enhance endosomal release and peptides to promote nuclear translocation.

3.1 Targeting Ligands

It has been demonstrated that adding cell-specific ligands to cationic liposome-based systems allows the use of lower doses of these systems and their cargo, while simultaneously facilitating tissue targeting, improving transfection efficiency and reducing side-effects. Figure 3 shows several examples of cell-surface molecules and ligands which have already been explored in targeting strategies.

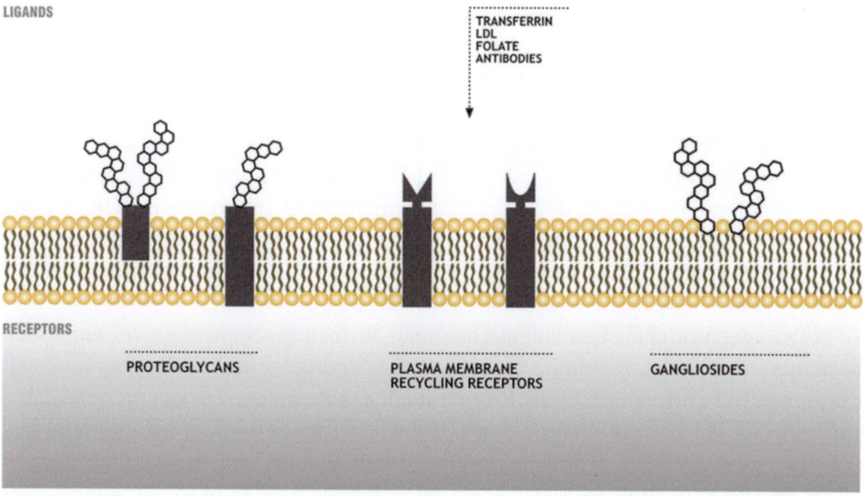

Fig. 3 Cell-surface receptors and ligands used in gene therapy targeting strategies. By coupling specific ligands to cationic liposome-based systems it is possible to improve cell internalization and, in certain cases, achieve tissue and cell specificity

One example of a widely explored targeting ligand is transferrin. The transferrin receptor (TfR) is a glycoprotein abundantly expressed in cancer cells and in certain tissues, such as the nervous tissue and the endothelial tissue that composes the blood brain barrier (BBB). The mechanism of TfR internalization and recycling to the cell surface has been well characterized and several studies have demonstrated that modifying cationic liposomes to improve their binding to this receptor can enhance both DNA [42] and siRNA [43] internalization. The approaches used to target TfR typically involve the electrostatic or covalent association of transferrin, or the chemical conjugation of anti-TfR antibodies to the surface of the liposomes. Several studies have been published demonstrating the ability of Trojan Horse liposomes (THL), which consist of cationic immunoliposomes coated with polyethylene glycol (PEG), to mediate the transvascular delivery of therapeutic genes or short hairpin RNA (shRNAs) to the brain. The targeting of THL is accomplished by the conjugation of 1% of PEG residues to peptidomimetic monoclonal antibodies (MAbs) against TfR. These MAbs are able to mediate both receptor-mediated transcytosis of THL through the BBB and endocytosis into brain cells. An example of the application of this technology involved the use of TfR-targeted THL to silence the endothelial growth factor receptor (EGFR) in a human glioma model established in immunocompromised mice [44]. Although antibody conjugation has proven to be an effective strategy to target the TfR, transferrin itself is much more affordable and is commercially available in larger quantities. Thus, direct conjugation or electrostatic binding of this ligand to cationic liposomes has also been widely explored as a targeting strategy. Our group has developed transferrin-associated lipoplexes (Tf-lipoplexes), composed of DOTAP and Chol, which have proven to mediate efficient delivery of plasmid DNA encoding the nerve growth factor [42] and anti-c-jun siRNAs to the mice brain [43].

Folic acid (FA) is another targeting ligand that has been explored to enhance nucleic acid delivery. This molecule is a crucial vitamin for proliferating cells, being taken up in large quantities, due to its involvement in the biosynthesis of nucleotide bases [45]. The choice of this vitamin is very attractive because of high binding affinity to its receptor, low immunogenicity, small size, stability during storage and low cost [46].

Folate receptor (FR) is a single-chain cell surface glycoprotein over expressed in numerous cancers, including breast, ovarian and head and neck squamous cell carcinoma [47–50], which binds FA and mediates its uptake by receptor-mediated endocytosis. The ease of conjugation of FA to lipoplexes without losing biological activity makes the FR a promising target for tumor-selective delivery. Conjugation of FA by covalent coupling to cationic liposomes has been used to deliver different nucleic acid molecules, such as DNA [51–53], antisense oligodeoxyribonucleotides [54, 55] and siRNA [56].

Work performed by Hofland et al. demonstrated that lipoplexes prepared from combination of FA-PEG–1,2-distearoyl-sn-glycero-3-phosphoethanolamine (DSPE) or FA-PEG–Chol with cationic liposomes, composed of a cationic lipid RPR209120 and DOPE, greatly reduced normal tissue gene transfer and efficiently increased gene delivery to tumors [53]. A completely different cationic liposome formulation was developed by Laine et al. In this study, the authors synthesized novel di- and tetraether-type archaeal lipid analogues with FA covalently coupled to the end of the PEG chain. These lipids were then combined with the conventional glycine betaine-based cationic lipid to form archaeosomes. The fact that the activity of FA-modified lipids was inhibited when free FA was added to the transfection medium indicated that a ligand/receptor-based targeted transfection was involved [51]. Recently, Li et al. developed a lipoplex formulation involving the complexation of calf timus DNA (ctDNA) with FA-covalently coupled cationic liposomes (ctDNA-FA-CLs). The authors demonstrated a selective uptake of ctDNA-FA-CLs in FR-positive L1210JF cancer cells, which were able to mediate a higher antitumoral effect in these cells, both in culture and in an animal model of disease compared to non-targeted complexes [57].

An alternative strategy to associate FA to cationic liposomes is based on the electrostatic interaction between the FA and liposomes. Our group generated a new gene delivery system by non-covalent association of FA to DOTAP:Chol or EPOPC:Chol cationic liposomes (FA-associated lipoplexes). This approach led to considerably higher levels of biological activity than that based on covalent coupling of FA. Moreover, FA-associated lipoplexes were shown to mediate significantly higher transfection efficiency and antitumoral activity than plain lipoplexes (lacking FA) [33].

3.2 Endosomolytic Agents

Aiming at achieving efficient biological activity, the cationic liposome-based gene delivery systems should exhibit the capacity to promote endosomal release of nucleic acids to avoid lysosomal degradation. Although cationic lipids can promote the destabilization of the endosomal membrane, several endosomolytic agents have been developed based on known fusogenic or pore-forming proteins and peptides, which significantly improve the cytoplasmic availability of nucleic acid molecules [58–60]. A typical example of an endosomolytic agent is the hemagglutinin subunit 2 (HA2) fusion peptide of the Influenza virus. This peptide undergoes a conformational change under acidic pH and assumes fusogenic properties that potentiate the destabilization of the endosomal membrane, and has been, therefore, associated with cationic liposomes to induce DNA release from the endocytic compartment and increase transfection efficiency [59]. Another example of a pH-dependent fusogenic peptide is GALA, a peptide composed of repeating sequences of Glu-Ala-Leu-Ala, which was designed to mimic the function of viral fusion protein sequences and shown to enhance transfection efficiency of lipoplexes [59]. Recently, this peptide was also applied as a functionalizing agent of multifunctional envelope-type nanodevices (MEND). Association of GALA to particles containing a cationic polymer/nucleic acid complexation core encapsulated in cationic liposomes resulted in improved release of both DNA [61] and siRNAs [62]. Lundberg et al. have reported a similar strategy, which uses a modified penetratin peptide called EB1, containing specially placed histidine insertions and capable of inducing the formation of a membrane-destabilizing alpha-helix upon protonation. Because the EB1 peptide induced higher luciferase silencing levels when compared to unmodified penetratin, it was hypothesized that this effect was due to enhancement of endosomal siRNA release [63].

On the other hand, our results on the enhancement of transfection promoted by association of human serum albumin (HSA) to EPOPC:Chol/DNA lipoplexes led us to conclude that HSA may act at two different stages of the endocytotic pathway: by promoting binding/endocytosis of the lipoplexes and by facilitating escape of DNA from the endosome [34]. Such effect of albumin to promote DNA release from the endocytotic pathway may be attributed to its capacity to undergo a low pH-induced conformational change, thereby acquiring fusogenic properties [64–66]. In this regard, it should be noted that, similarly to HSA, the enhancing effect of transferrin on transfection may be also partially attributed to the capacity of this protein to acquire fusogenic properties under acid conditions, as those observed in the endosome [60].

3.3 Nuclear Translocation

Since the nuclear membrane represents one the main barriers associated to the transfection process, different strategies have been applied to promote the nuclear entry of DNA. Most of these strategies have been based on the observation that proteins bearing Nuclear Localization Signal (NLS) peptides have the ability to be translocated into the nucleus by a receptor-mediated process. In this regard, peptide sequences composed of a short stretch of five to ten basic amino acids based on the simian vacuolating virus 40 (SV40) core nuclear localization signal are those that have been associated most frequently with non-viral vectors aiming at enhancing transfection. Different versions of this strategy have been reported, including the simple association of NLS peptides with lipoplexes [67] or their conjugation with polycations [68]. Behr and collaborators reported a procedure consisting of covalent coupling of a single NLS sequence to capped DNA constructs containing a hairpin oligonucleotide enriched with amino groups, which resulted in a significant enhancement of transfection [69]. A simpler but equally effective technique to couple NLS sequences to plasmids and to target them to the nucleus involved the combination of a peptide nucleic acid (PNA) with the SV40 NLS, creating a bi-functional PNA-NLS peptide [70]. The use of a nonclassical NLS containing the M9 sequence of the heterogeneous nuclear ribonucleoprotein (hnRNP) A1 was reported as being a promising strategy to enhance transfection of non-dividing cells mediated by cationic liposomes [71]. Alternative strategies have been reported involving the use of intermediate length cationic peptides to pre-condense DNA and increase its nuclear entry. Recently, Chen and coworkers have employed protamine, a cationic and naturally occurring polypeptide bearing an arginine-rich NLS sequence, to form homogenous and tight DNA particles that could be further complexed with cationic liposomes [72]. The presence of this polypeptide enhanced the membrane translocation and nuclear localization of a plasmid DNA, thus contributing to an increase in transfection efficiency. Karyophilic cell-penetrating peptides (CPPs) including that derived from human immunodeficiency virus type 1 (HIV-1) tat protein [73, 74] and the $S4_{13}$-PV peptide, a chimeric peptide resulting from the combination of a cell-penetrating sequence derived from the Dermaseptin S4 peptide with the SV40 large T antigen NLS, have also been shown to enhance transfection mediated by cationic liposomes [75, 76], suggesting that CPPs can also be of great value to enhance nuclear translocation.

4 Design of Nucleic Acid-Based Therapeutic Strategies Mediated by Cationic Liposomes

The original concept of gene therapy consisted in the replacement of a nonfunctional gene with its normal version or introduction of a gene that is dysfunctional or absent in the host, aiming at treating the causes of diseases rather than the symptoms. While initially gene therapy was focused on the delivery of DNA to treat monogenic diseases, its application was soon extended to include the transfer of other types of nucleic acid molecules to treat or prevent all human genetic and acquired diseases.

4.1 Cationic Liposomes and DNA Delivery

The replacement of a nonfunctional gene is technically very difficult due to the need of controlled recombination between DNA fragments to achieve the insertion of the functional form of the gene and, at the same time, the elimination of the dysfunctional gene from the genome. Therefore, the correction of a genetic defect is usually accomplished by the introduction of a plasmid DNA containing the functional gene in the target cells or tissue. The introduction of a therapeutic gene (not involved in monogenic diseases) can be performed by a similar procedure, where the administered gene has an intrinsic therapeutic function, such as an immune-stimulating gene, suicide gene, or a tumor-suppressor gene. The strategy involving the introduction of plasmid DNA into target cells offers innumerous therapeutic possibilities, as exemplified in the studies described below. In this regard, cationic liposomes have been extensively used to deliver plasmid DNA containing different therapeutic genes towards application in gene therapy approaches.

In 2001, Miyata et al. prepared hemagglutinating virus of Japan (HVJ) cationic liposomes containing various lipids such as egg yolk phosphatidylcholine (ePC), egg yolk sphingomyelin (eSph), DOPE, Chol, and DC-Chol. Following injection of HVJ cationic liposomes carrying a plasmid DNA coding for macrophage inflammatory protein-1β (MIP-1β) into subcutaneous tumor masses from colon, local expression of MIP-1β and accumulation of T lymphocytes expressing the cluster differentiation 4 ($CD4^+$ T lymphocytes) were detected. Moreover, intraperitoneal injection of these lipoplexes in a mouse model of peritoneal dissemination of colon tumor significantly increased the animal survival, as compared to control lipoplexes (containing a non-coding plasmid DNA) [77]. In another study, an autologous tumor cell vaccine based on the Epstein-Barr virus (EBV)/lipoplex was developed by Asada et al. EBV/lipoplex containing TFL2-3, a novel cationic lipid, and an EBV-based plasmid was designed to express the interleukin-12 (IL-12) and/or interleukin-18 (IL-18). These authors used a melanoma mouse model, generated by subcutaneous injection of B16 cells, to test the therapeutic effect resulting from repetitive immunization with irradiated B16 cells previously transfected with the EBV/lipoplex. Strong tumor suppression, high interferon-gamma (IFN-γ) production and elevated levels of cytotoxic T lymphocyte and natural killer activation were obtained in mice vaccinated with B16 cells transfected with EBV/lipoplex encoding IL-12 or both (IL-12/IL-18) but not in mice vaccinated with IL-18 alone. Moreover, metastatic tumor outgrowth was suppressed in a pulmonary metastatic mouse model vaccinated with B16 cells transfected with EBV/lipoplex coding for IL-12 [78]. Recently, we reported the use of a lipoplex formulation prepared from EPOPC:Chol cationic liposomes associated to HSA to deliver IL-12 and/or herpes simplex virus-tymidine kinase (HSV-tk) encoding genes. Intratumoral administration of lipoplexes containing plasmids encoding IL-12 or HSV-tk induced a strong antitumoral effect in mouse model bearing a subcutaneous TSA (mouse mammary adenocarcinoma) tumor. Smaller tumor size, higher T-lymphocyte tumor infiltration and more extensive tumor necrotic and hemorrhagic areas were observed in mice treated with the referred lipoplexes, when compared to animals treated with control complexes (lacking the therapeutic genes) [79]. The same lipoplex formulation was also applied to deliver the HSV-tk gene to TSA cells in combination with vinblastine. Such combined strategy resulted in a great enhancement of the antitumoral activity, a significant synergistic effect being observed [34]. We also tested another lipoplex formulation generated from Tf-associated DOTAP:Chol cationic liposomes in suicide therapy strategies using a syngeneic, orthotopic murine model for head and neck squamous cell carcinoma. A significant tumor reduction was achieved upon intratumoral delivery of HSV-tk or cytosine deaminase (CD) genes, followed by intraperitoneal injection of ganciclovir (GCV) or 5-fluorocytosine (5-FC), respectively. Moreover, tumor microenvironment was modified following suicide gene therapy and enhanced apoptosis, recruitment of natural killer (NK) cells, $CD4^{+}$ and $CD8^{+}$ T-lymphocytes and an increase in the levels of several cytokines/chemokines were observed within the tumors [80].

Besides cancer, gene therapy mediated by cationic liposomes has been applied to treat other types of diseases. Ultradeformable cationic liposomes (UCL), containing DOTAP and DOPE, were used to deliver the interleukin-4 (IL-4) gene in the K14-VEGF transgenic mouse model for psoriasis. In these studies, lipoplexes were administered by topical transdermal penetration and DNA expression was detected in ear skin. The results showed that the animal group treated with UCL/mIL-4 exhibited mild psoriasis phenotype, an antipsoriatic effect being confirmed by histological analysis [81]. In a different study, cationic liposomes composed of octadecenoyloxy[ethyl-2-heptadecenyl-3-hydroxyethyl] imidazolinium chloride (DOTIM) and Chol were complexed with non-coding DNA and administered intranasally to BALB/c mice in order to evaluate the efficacy of protection against the lethal bacterial pathogen *Francisella tularensis*. The authors found that pretreatment with lipoplexes 24 h prior to bacterial challenge conferred nearly total protection of mice against the lethal bacteria. The mucosal treatment with the lipoplexes resulted in a significant increase of survival time in mice challenged with a virulent strain of the bacteria and was associated to a significant increase in the expression of IL-12, IFN-γ, tumor necrosis factor-alpha (TNF-α), interferon-beta (IFN-β) and interferon-alpha (IFN-α) genes in the lung [82].

4.2 *Cationic Liposomes and RNA Interference*

The discovery of the RNA interference (RNAi) pathway can be considered as one of the most remarkable biological breakthroughs of the last decade. Since its initial description by Fire and Mello [83] in 1998, RNAi has transformed the genomic field and introduced scientists to a series of new and complex gene regulation pathways, mediated by small RNA molecules previously discarded as irrelevant. RNAi is a highly conserved pathway present in eukaryotic cells, by which small RNA molecules, known as siRNAs, are able to bind and mediate cleavage of complementary mRNA molecules [84]. During the last decade, scientists have developed different methods to manipulate RNAi in the laboratory, exploiting its potential as both a research tool and a novel approach to treat human diseases, including viral infections [85], cancer [86], neurodegenerative diseases and genetic metabolic pathologies [87].

Several research groups have demonstrated successful and highly efficient siRNA internalization and siRNA-induced gene knockdown mediated by cationic liposome-based systems, both in vitro and in vivo.

Several studies have shown efficient silencing of liver-related protein targets in vivo, such as the GB virus [88] protein and the endogenous gene gp46, the rat homolog of the human heat shock protein 47 [89], using cationic liposome-based systems. However, the most promising results were achieved with a cationic liposome-based formulation of stable nucleic acid-lipid particles (SNALPs), which were able to dramatically reduce apolipoprotein B (ApoB) levels in the liver of mice and non-human primates following intravenous injection [90]. A single siRNA injection resulted in a dose-dependent silencing of ApoB mRNA expression in the liver 48 h after administration. As a consequence of an impressive 90% reduction of ApoB, serum cholesterol and low-density lipoproteins were reduced 65 and 85%, respectively. The applicability of the SNALP formulation was also demonstrated in animal models of Hepatitis B [85] and Ebola infection [91]. A study by Morrissey and co-workers has shown that Hepatitis B virus (HBV) replication could be inhibited through the delivery of a siRNA-SNALP complex targeting the HBV RNA. Three daily injections reduced serum HBV levels by at least one order of magnitude, and this effect was found to be specific, dose-dependent and to last for at least 7 days [85]. Concerning the Ebola virus, another study showed that, when contained within the SNALP formulation, siRNAs targeting the polymerase gene of the virus completely protected guinea pigs and prevented animal death when administered shortly after an Ebola virus challenge [91].

Significant progress in the development of cationic lipid-based RNAi delivery systems has also been made in the field of cancer therapeutics. A cationic cardiolipin analog carrying c-raf siRNAs was shown to inhibit the growth of breast tumor xenographs in mice [92], while cationic liposomes containing anisamide-conjugated PEG effectively penetrated the lung metastasis of melanoma tumors in mice and resulted in 80% gene silencing after a single intravenous injection of a mixture of siRNAs against the cellular myelocytomotosis oncogene (c-myc), murine double minute oncogene (MDM2) and vascular endothelial growth factor (VEGF) [93]. A more recent study has demonstrated that SNALPs targeted to TfRs were able to deliver encapsulated siRNAs against break cluster region-Abelson (BCR-ABL) and mediate efficient and specific silencing of the BCR-ABL oncogenes in cultured human chronic myeloid leukemia cells [94]. This cationic liposome-based formulation was also used to co-encapsulate anti-BCR-ABL siRNAs with the drug imatinib, resulting in a 3.84-fold reduction of imatinib half maximal inhibitory concentration (IC_{50}) and promoting significant cytotoxicity [95].

An interesting example of efficient local delivery to the brain was reported by Kumar in 2006. In this study, intrathecal administration of siRNA-i-FECT complexes protected mice from fatal Japanese encephalitis virus and West Nile virus [96]. Our own studies have also shown efficient delivery of Tf-associated lipoplexes, prepared from DOTAP: Chol, to the central nervous system (CNS). This lipoplex formulation was able to mediate successful knockdown of the transcription factor c-Jun in an animal model of excitotoxic brain injury, achieving significant neuroprotection and decreasing brain inflammation [43]. A number of studies have also shown promising results after systemic delivery of lipoplexes containing siRNAs against IL12p40 [97, 98] and TNF-α [99], in models of LPS-induced inflammation or collagen-induced arthritis.

Finally, mucosal administration of cationic lipid-based formulations has been considered of active interest, due to the ease of access, favorable tolerability and large number of potential target disorders associated with these tissues. Successful intravaginal delivery of siRNAs against Herpes simplex virus 2 complexed with cationic lipids was recently achieved by Palliser and co-workers [100]. Vaginal instillation of siRNAs complexed with Oligofectamine led to significant uptake by the epithelial and lamina propria cells and protected mice against lethal infection for up to 9 days. TNF-α silencing by Lipofectamine 2,000/siRNA lipoplexes was also reported by Zhang and colleagues in a model of colonic inflammation [101].

4.3 Cationic Liposomes and Oligonucleotide Delivery

The capacity of the antisense technology to target any desired gene and thus modulate a variety of cellular functions is of paramount pharmacological interest [102, 103]. This technology is based on the use of sequence specific oligonucleotides (ONs) which, once inside the cells, can hybridize with complementary mRNA strands, causing translational arrest or mRNA degradation through activation of the cellular enzymes of the RNase H family and consequently blocking gene expression [102]. These ONs can disrupt protein production through three main mechanisms: (i) the formation of an ON/RNA duplex which is a substrate for endogenous RNase H, leading to mRNA cleavage; (ii) the formation of an ON/mRNA duplex that sterically hinders the assembly of the ribosomal complex or arrests a ribosomal complex already engaged in translation, in both cases affecting protein biosynthesis; (iii) the formation of an ON/mRNA duplex in the nucleus which alters pre-mRNA splicing in the nucleus through a steric-blocking mechanism [6, 104]. Regarding this latter mechanisms, the importance of modulating pre-mRNA splicing is underscored by the finding that the frequency of alternative splicing, the process by which a single primary transcript gives rise to several mature mRNAs and proteins, is higher than previously anticipated, with most studies reporting alternative splice forms in more than 70% of human genes [105, 106]. Moreover, aberrant splicing is often related to human diseases, such as β-thalassemia, cystic fibrosis, neurofibromatosis type 1, ataxia-telangiectasia, Alzheimer disease, myotonic dystrophy and cancer [105, 106]. The control of pre-mRNA splicing pattern can be accomplished by the use of antisense oligonucleotides, termed splice-switching oligonucleotides (SSOs) [105, 107, 108].

Cationic liposomes have been successfully applied to deliver ONs into cells in vitro [94, 109–113], and in vivo [114, 115]. Resina et al. [110] used a cationic lipid-based formulation with a low positive net charge and high particle size homogeneity, the DLS (delivery liposomal system, DOGS/DOPE) liposomes, for the delivery of negatively charged ONs. The 2′-O-methyl ON analogues delivered as complexes with DLS liposomes were shown to correct splicing more efficiently than PNA oligomers delivered as CPP conjugates [110]. Our own studies demonstrated that the optimal splice correcting activity of 2′-O-methyl-RNA molecules was exhibited when both cationic liposomes and the $S4_{13}$-PV CPP were combined in the same formulation [76].

Another group used 2′-O-methyl-RNA antisense molecules directed against the human telomerase RNA component (hTR), a potent and specific telomerase inhibitor, as a potential candidate for anticancer therapy [116]. These studies showed that the cationic lipid DOTAP was the most suitable transfection reagent for the delivery of 2′-O-methyl-RNA in human lung cancer cells, showing relatively low cytotoxicity and considerable ability to promote efficient uptake and consequent inhibition of telomerase [116].

In another study, cationic liposomes were loaded with a pro-apoptotic peptide, D-(KLAKLAK)(2) and the Bcl-2 antisense oligodeoxynucleotide, G3139 [109]. First, the cationic peptide was condensed with the anionic ON G3139, forming negatively charged peptide/ON complexes which were then entrapped into DOTAP/DOPE cationic liposomes [109]. This sequential charge interaction led to a very efficient entrapment of both the pro-apoptotic peptide and the ON [109]. Overall, it was demonstrated that the combination of the pro-apoptotic peptide D-(KLAKLAK)(2) and the antisense ON G3139 in a cationic liposome enhanced apoptotic/antitumor efficacy and could thus provide a promising tool for cancer treatment [109].

A cationic liposomal formulation suitable for systemic delivery was developed for the transfer of either antisense ONs or siRNAs [115]. In this case, a modified formulation of antisense ON targeting the proto-oncogene serine/threonine-protein kinase (Raf-1) containing dimyristoyl 1,2-diacyl-3-trimethylammonium-propane (DMTAP), phosphatidylcholine (PC) and Chol (md-LErafAON), was tested in vivo, in human tumor xenografts grown in athymic mice [115]. It was demonstrated that md-LErafAON constitutes a promising system for the systemic delivery of antisense therapeutics, being well tolerated and suitable for clinical evaluation [115].

Interleukin-13 (IL-13) is overproduced in the skin of atopic dermatitis (AD), playing an essential role in the pathogenesis of this disease. In a recent study, IL-13 antisense ON was formulated with cationic liposomes composed of DOTAP, which improved its cellular uptake in vitro and transdermal delivery in vivo [114]. The results suggested that this system constitutes a potential therapeutic tool for the treatment of AD, which could additionally be applied to other immune diseases associated with the production of IL-13 [114].

As a novel therapy for preventing bone loss, Dinh et al. developed macrophage-targeted carriers, consisting of mannosylated cationic liposomes (Man-liposomes), to deliver a decoy (a double-stranded ON) to the nuclear factor kappa-light-chain-enhancer of activated B cells (NFκB) [117]. It was verified that cellular accumulation of NFκB decoy was enhanced by Man-liposomes and NFκB activation in macrophages was suppressed when NFκB decoy complexes were delivered by Man-liposomes, indicating a promising role for this formulation for bone loss therapy [117].

5 Cationic Liposome-Based Systems in Clinical Trials

The starting point for gene therapy clinical trials was in 1989, upon approval of a clinical trial involving the use of tumor infiltrating lymphocytes (TIL), containing a marker gene (coding for resistance to neomycin) introduced by a retroviral vector, with the purpose of optimizing the treatment approach, including the analysis of bio-distribution and survival of the modified TIL. This treatment was well tolerated by all the patients and no side effects due to gene transduction were observed [118]. This clinical trial was the first of a period, in the early 1990s, of great enthusiasm for gene therapy. However, by the mid-1990s, it became clear that efficient gene therapy strategies would be slow to come, due to the obtained unsatisfactory results [119].

Gene therapy faced its first serious drawback in the late 1990s, when a patient enrolled in a phase I clinical trial involving intravascular gene transfer mediated by an adenovirus type 5-based vector, that contained the human ornithine transcarbamylase cDNA, experienced lethal complications [120]. This case led to a considerable examination of the safety and toxicity associated to the use of viral vectors [119]. Nevertheless, some time later Cavazzana-Calvo et al. reported the first clinical success of gene therapy [121]. These scientists successfully treated children suffering from Severe Combined Immunodeficiency-X1, through the application of an ex vivo gene therapy protocol involving the autologous infusion of hematopoietic stem cells transduced with a Moloney-retrovirus derived vector containing the gene encoding the gamma-c cytokine receptor subunit of interleukin-2, -4, -7, -9 and -15 receptors [121, 122]. Unfortunately, the enthusiasm gave way to alarm at the end of 2002, when two of the ten treated children presented a leukemia-like lymphocyte proliferation. This abnormality was attributed to retroviral vector integration near the LIM domain only 2 (LMO2) proto-oncogene promoter, leading to aberrant transcription and expression of LMO2 [123]. In the beginning of 2005 a third child developed a proliferative condition, involving a different oncogene. Although this and one of the earlier two patients responded well to chemotherapy treatment, one of the three affected patients did not respond to the chemotherapy treatment for the leukemia-like condition and died [124].

All the above safety problems associated to the application of viral vectors prompted the scientists to develop new gene delivery systems. Cationic liposomes have been one of the most used non-viral vectors in worldwide human clinical trials of gene therapy, so far being applied in 109 protocols. As illustrated in Table 1, clinical application of cationic liposomes in gene therapy has been essentially performed in cancer (88 protocols) and monogenic diseases (14 protocols), cystic fibrosis (CF) being the most frequently addressed hereditary disease [20].

Table 1 Representative gene therapy clinical trials involving cationic liposomes

Disease	Clinical trial	Clinical phase	Reference
Cystic fibrosis	Delivery of the cystic fibrosis transmembrane conductance regulator gene to nasal epithelium and lung by nebulisation of cationic liposomes (GL-67:DOPE:DMPE-PEG_{5000})/DNA complexes.	I	[125]
Cystic fibrosis	Repeated administration of the cystic fibrosis transmembrane conductance regulator gene to nasal epithelium by cationic liposomes (DC-Chol:DOPE)/DNA complexes.	I	[126]
Glioblastoma multiforme	Delivery of interferon beta gene via cationic liposomes (composition not revealed).	I	[127]
Metastatic melanoma	Intratumoral injection of Allovectin-7, which consists of a plasmid, encoding human leukocyte antigen-B7 and beta-2 microglobulin, complexed with a cationic liposome formulation (DMRIE/DOPE), as an immunotherapeutic agent.	II	[128]
Breast and ovarian cancer	Intracavitary injection of E1A gene complexed with cationic liposomes (DC-Chol).	I	[129]
Glioblastoma multiforme	Stereotactically guided intratumoral delivery of an HSV-tk gene by a cationic liposomal formulation (composition not revealed) and systemic ganciclovir.	I/II	[130]

CF is caused by mutation of the cystic fibrosis transmembrane conductance regulator (CFTR) gene. The normal CFTR protein is localized at the apical membrane of epithelial cells and plays a vital role in regulating transepithelial ion transport and water balance. In CF, the airway secretions dehydrate, mucociliary clearance is impaired and opportunistic bacterial infections take hold. This causes inflammation and progressive lung damage which is usually fatal by early adulthood [131]. Studies performed in transgenic CF mutant mice showed that the electrophysiological defect in the trachea of the animals can be partially or completely corrected by instillation or aerosolization of human CFTR cDNA expression constructs complexed with cationic liposomes [132, 133]. However, in most of the human clinical trials performed, the gene transfer efficiency was low, and most likely insufficient to achieve clinical benefit. Such low efficiency is probably due to the extracellular barriers such as mucus, the glycocalyx, tight junctions and mucociliary clearance, which a major function is to prevent uptake of foreign materials, including gene transfer agents [134]. Hyde et al. demonstrated for the first time that lipoplexes, prepared with the DC-Chol:DOPE cationic liposome formulation and a plasmid containing the human CFTR cDNA, could be successfully readministered to the nose of CF patients. These authors showed that repeated administration of these complexes was safe, with no activation of the immune system, and resulted in gene transfer with some changes in CFTR-dependent cellular function, after each of the three doses (administered 4 weeks apart) in six out of ten treated patients [126]. In another study, Alton et al. showed that the administration by nebulization of the lung of the patients with lipoplexes, prepared with a plasmid containing the human CFTR cDNA (pCF-1-CFTR) and the cationic liposome formulation GL-67/DOPE/dimyristoylphosphatidylethanolamine-polyethylene glycol$_{5000}$ (DMPE-PEG$_{5000}$), resulted in a significant degree of correction of the chloride abnormality in all the eight patients treated with the complexes. Although no alterations in the sodium transport abnormality were detected, the authors observed influenza-like symptoms in seven of the eight treated patients, which are likely to be triggered by cytosine-phosphate-guanine (CpG) motifs in the plasmid DNA [125]. Recently, through extensive nucleotide optimization, the UK CF Gene Therapy Consortium has generated a CFTR expression plasmid that is completely depleted of all CpG sequences, allowing reduced inflammation and sustained pulmonary gene expression [135]. Unfortunately, in the last decade, no significant therapeutic improvement was achieved, independently of the type of gene delivery system used, showing that gene transfer to the airway epithelial cells is a difficult issue, most probably because lung has evolved to keep foreign particles out. However, technologies to improve non-viral gene transfer to the lung continue to be developed, in order to increase the level and persistence of transgene expression. In parallel, studies on how plasmid DNA is taken up by primary airway epithelial cells, its movement through the cytoplasm and subsequently into the nucleus are being addressed and this knowledge may lead to significant improvements in non-viral gene transfer to CF patients [136].

Most of the approved protocols for gene therapy clinical trials (approximately 65%), including those involving cationic liposomes, are focused on cancer diseases. These protocols include the use of various gene types in different antitumoral strategies, such as immunopotentiation, oncogene inactivation and suicide gene therapy. In 2004, Yoshida et al. published a clinical study, involving five patients with malignant glioma (glioblastoma multiforme or anaplastic astrocytoma), where the safety and effectiveness of IFN-β gene transfer via cationic liposomes were evaluated. These authors detected transgene expression and antitumour activity in four out of five treated patients. Two patients showed a partial response (>50% tumour reduction) and two others had stable disease 10 weeks after beginning the therapy [127]. Significant changes in gene expression related to immunoresponse, apoptosis and inhibition of neovascularization were observed. Moreover, many tumor cells showed necrotic changes, and numerous CD8^{+} lymphocytes and macrophages infiltrating the tumor and surrounding tissues were identified by immunohistochemistry [137]. These results suggest that this antitumoral strategy is safe and results in a significant therapeutic effect induced by the activation of the immune system [127, 137]. On the other hand, several clinical trials in phase II and two in phase III have been performed using the Allovectin-7 formulation (Vical Inc, San Diego, California) as immunotherapeutic agent against cancer [20]. Allovectin-7 consists of a bicistronic plasmid, encoding the human leukocyte antigen-B7 and beta-2 microglobulin, complexed with a cationic liposome formulation (DMRIE/DOPE), which was designed to express the allogeneic major histocompatibility complex class I antigen upon intralesional administration. Recently, Bedikian et al. performed a phase II dose-escalation study to evaluate the safety and efficacy of high-dose Allovectin-7 in patients with metastatic melanoma [128]. Eligible patients had stage III or IV metastatic melanoma recurrent or unresponsive to prior therapy. Patients received six weekly intratumoral injections followed by 3 weeks of observation and evaluation, and those with stable or responding disease were submitted to additional cycles of Allovectin-7. All 133 patients were evaluated for safety while 127 patients were evaluated for efficacy (2 mg, high dose), 15 patients achieving an objective response with median duration of response of 13.8 months. Moreover, a histological examination of tissue from two responding patients who had their lesions resected has shown no evidence of melanoma. In conclusion, these results indicate that high-dose of Allovectin-7 seems to be an active and well-tolerated treatment for selected stage III/IV metastatic melanoma patients [128].

The human epidermal growth factor receptor 2 proto-oncogene (HER-2/neu) encodes an epidermal growth factor receptor-related transmembrane protein and is over expressed in many types of human neoplastic diseases. Therefore, this oncogene constitutes an excellent molecular target for therapeutic strategies against HER-2/neu-overexpressing cancers, such as human breast and ovarian cancers [138]. In this context, Hortobagyi et al. evaluated the feasibility of intracavitary injection of E1A gene complexed with DC-Chol cationic liposomes in patients with breast and ovarian cancers in a phase I clinical trial. The adenovirus type 5 E1A gene is associated with antitumor activities by transcriptional repression of HER-2/neu and induction of apoptosis. The authors observed E1A gene expression which was accompanied by HER-2/neu down-regulation, increased apoptosis, and reduced proliferation [129].

Vilarret et al. performed a clinical trial involving 24 patients, with recurrent, unresectable head and neck cancer, which were submitted to intratumoral injections of tgDCC-E1A lipoplexes (E1A gene complexed with DC-cholesterol:DOPE cationic liposomes) over 8 weeks. The authors showed that intratumoral administration of tgDCC-E1A was well tolerated in all patients, no significant toxicities related to tgDCC-E1A being reported. One patient had a complete response to the therapeutic strategy, two patients had minor response, and seven patients had stable disease. The median time to progression was 8.6 weeks, and median survival was 4.6 months. Although intratumoral injections of tgDCC-E1A were safe and well tolerated, a modest tumor response was observed, showing the need for further development of therapeutic strategies involving tgDCC-E1A [139].

In a prospective phase I/II clinical study, Voges et al. treated eight patients, suffering from recurrent glioblastoma multiforme, with stereotactically guided intratumoral convection-enhanced delivery of an HSV-1-tk gene-bearing cationic liposomal vector and systemic ganciclovir. In two of the eight patients, the authors observed a greater than 50% reduction of tumour volume, and in six of the eight patients focal treatment effects were reported. The results of this phase I/II study demonstrated the feasibility and safety of this therapeutic strategy [130].

6 Concluding Remarks

Significant progress has been made regarding the development of cationic liposome-based formulations for nucleic acid delivery into target cells, including the synthesis of numerous novel cationic lipids, optimization of physicochemical properties of the lipoplexes and incorporation of new components to enhance their biological activity. The demonstrated efficacy of such systems to mediate delivery of different types of nucleic acid molecules such as plasmid DNA, siRNAs and oligonucleotides and the achievement of a therapeutic activity in animal models have made them promising candidates towards application in gene therapy. However, the results obtained in most of the therapeutic strategies using cationic liposomes and other nucleic acid delivery systems in gene therapy clinical trials have shown a limited therapeutic efficiency. It should be noted that gene therapy protocols have been applied to patients with highly developed cancers or suffering from a disease for which no alternative treatment exists, which has undoubtedly contributed to the reported limited success. Taking this into consideration and the observed complications experienced with viral vectors in gene therapy clinical trials, an enormous effort is currently being made to further enhance the efficiency of cationic liposomes as nucleic acid delivery systems. It is expected that in a near future new-generation cationic liposome-based systems will find their place in successful gene therapies.

Acknowledgments The authors wish to thank Pedro Carvalho for the artwork included in this chapter.

References

1. Meade BR, Dowdy SF (2007) Exogenous siRNA delivery using peptide transduction domains/cell penetrating peptides. Adv Drug Deliv Rev 59:134–140
2. Opalinska JB, Gewirtz AM (2002) Nucleic-acid therapeutics: basic principles and recent applications. Nat Rev Drug Discov 1:503–514
3. Pedroso de Lima MC, Simoes S, Pires P, Faneca H, Duzgunes N (2001) Cationic lipid-DNA complexes in gene delivery: from biophysics to biological applications. Adv Drug Deliv Rev 47:277–294
4. Somia N, Verma IM (2000) Gene therapy: trials and tribulations. Nat Rev Genet 1:91–99
5. Torchilin VP (2008) Cell penetrating peptide-modified pharmaceutical nanocarriers for intracellular drug and gene delivery. Biopolymers 90:604–610
6. Veldhoen S, Laufer SD, Restle T (2008) Recent developments in peptide-based nucleic acid delivery. Int J Mol Sci 9:1276–1320

7. Wiethoff CM, Middaugh CR (2003) Barriers to nonviral gene delivery. J Pharm Sci 92:203–217
8. Whitehead KA, Langer R, Anderson DG (2009) Knocking down barriers: advances in siRNA delivery. Nat Rev Drug Discov 8:129–138
9. Heitz F, Morris MC, Divita G (2009) Twenty years of cell-penetrating peptides: from molecular mechanisms to therapeutics. Br J Pharmacol 157:195–206
10. Pfeifer A, Verma IM (2001) Gene therapy: promises and problems. Annu Rev Genomics Hum Genet 2:177–211
11. Robbins PD, Tahara H, Ghivizzani SC (1998) Viral vectors for gene therapy. Trends Biotechnol 16:35–40
12. Mountain A (2000) Gene therapy: the first decade. Trends Biotechnol 18:119–128
13. Pedroso de Lima MC, Simoes S, Pires P, Gaspar R, Slepushkin V, Duzgunes N (1999) Gene delivery mediated by cationic liposomes: from biophysical aspects to enhancement of transfection. Mol Membr Biol 16:103–109
14. El-Aneed A (2004) An overview of current delivery systems in cancer gene therapy. J Control Release 94:1–14
15. Lechardeur D, Verkman AS, Lukacs GL (2005) Intracellular routing of plasmid DNA during non-viral gene transfer. Adv Drug Deliv Rev 57:755–767
16. Simoes S, Filipe A, Faneca H, Mano M, Penacho N, Duzgunes N, Pedroso de Lima MC (2005) Cationic liposomes for gene delivery. Expert Opin Drug Deliv 2:237–254
17. Li S, Huang L (2000) Nonviral gene therapy: promises and challenges. Gene Ther 7:31–34
18. Felgner PL, Gadek TR, Holm M, Roman R, Chan HW, Wenz M, Northrop JP, Ringold GM, Danielsen M (1987) Lipofection: a highly efficient, lipid-mediated DNA-transfection procedure. Proc Natl Acad Sci USA 84:7413–7417
19. Montier T, Benvegnu T, Jaffres PA, Yaouanc JJ, Lehn P (2008) Progress in cationic lipid-mediated gene transfection: a series of bio-inspired lipids as an example. Curr Gene Ther 8:296–312
20. Gene Therapy Clinical Trials Worldwide (2011) [cited; The journal of gene medicine website for Gene Therapy Clinical Trials Worldwide]. Available from: http://www.wiley.co.uk/genmed/clinical/. Accessed 7 Jan 2013
21. Pedroso de Lima MC, Neves S, Filipe A, Duzgunes N, Simoes S (2003) Cationic liposomes for gene delivery: from biophysics to biological applications. Curr Med Chem 10:1221–1231
22. Tros de Ilarduya C, Sun Y, Duzgunes N (2010) Gene delivery by lipoplexes and polyplexes. Eur J Pharm Sci 40:159–170
23. Marshall J, Yew NS, Eastman SJ, Jian C, Schelue RK, Cheng SH (1999) Cationic lipid-mediated gene delivery to the airways. In: Huang L, Hung M-C, Wagner E (eds) Nonviral vectors for gene therapy. Academic Press, San Diego
24. Felgner JH, Kumar R, Sridhar CN, Wheeler CJ, Tsai YJ, Border R, Ramsey P, Martin M, Felgner PL (1994) Enhanced gene delivery and mechanism studies with a novel series of cationic lipid formulations. J Biol Chem 269:2550–2561
25. Koynova R, Tenchov B (2010) Cationic lipids: molecular structure/transfection activity relationships and interactions with biomembranes. Top Curr Chem 296:51–93
26. MacDonald RC, Rakhmanova VA, Choi KL, Rosenzweig HS, Lahiri MK (1999) O-ethylphosphatidylcholine: a metabolizable cationic phospholipid which is a serum-compatible DNA transfection agent. J Pharm Sci 88:896–904
27. Faneca H, Simoes S, Pedroso de Lima MC (2002) Evaluation of lipid-based reagents to mediate intracellular gene delivery. Biochim Biophys Acta 1567:23–33
28. Faneca H, Simoes S, Pedroso de Lima MC (2004) Association of albumin or protamine to lipoplexes: enhancement of transfection and resistance to serum. J Gene Med 6:681–692
29. Koltover I, Salditt T, Radler JO, Safinya CR (1998) An inverted hexagonal phase of cationic liposome-DNA complexes related to DNA release and delivery. Science 281:78–81
30. Zuidam NJ, Barenholz Y (1998) Electrostatic and structural properties of complexes involving plasmid DNA and cationic lipids commonly used for gene delivery. Biochim Biophys Acta 1368:115–128

31. Harvie P, Wong FM, Bally MB (1998) Characterization of lipid DNA interactions. I. Destabilization of bound lipids and DNA dissociation. Biophys J 75:1040–1051
32. Smith JG, Wedeking T, Vernachio JH, Way H, Niven RW (1998) Characterization and in vivo testing of a heterogeneous cationic lipid-DNA formulation. Pharm Res 15:1356–1363
33. Duarte S, Faneca H, Pedroso de Lima MC (2011) Non-covalent association of folate to lipoplexes: a promising strategy to improve gene delivery in the presence of serum. J Control Release 149:264–272
34. Faneca H, Faustino A, Pedroso de Lima MC (2008) Synergistic antitumoral effect of vinblastine and HSV-Tk/GCV gene therapy mediated by albumin-associated cationic liposomes. J Control Release 126:175–184
35. Pires P, Simoes S, Nir S, Gaspar R, Duzgunes N, Pedroso de Lima MC (1999) Interaction of cationic liposomes and their DNA complexes with monocytic leukemia cells. Biochim Biophys Acta 1418:71–84
36. Radler JO, Koltover I, Salditt T, Safinya CR (1997) Structure of DNA-cationic liposome complexes: DNA intercalation in multilamellar membranes in distinct interhelical packing regimes. Science 275:810–814
37. Lasic DD (1996) Doxorubicin in sterically stabilized liposomes. Nature 380:561–562
38. Zelphati O, Nguyen C, Ferrari M, Felgner J, Tsai Y, Felgner PL (1998) Stable and monodisperse lipoplex formulations for gene delivery. Gene Ther 5:1272–1282
39. Lundberg D, Faneca H, Moran Mdel C, Pedroso de Lima MC, Miguel Mda G, Lindman B (2011) Inclusion of a single-tail amino acid-based amphiphile in a lipoplex formulation: effects on transfection efficiency and physicochemical properties. Mol Membr Biol 28:42–53
40. Li S, Tseng WC, Stolz DB, Wu SP, Watkins SC, Huang L (1999) Dynamic changes in the characteristics of cationic lipidic vectors after exposure to mouse serum: implications for intravenous lipofection. Gene Ther 6:585–594
41. Sternberg B, Hong K, Zheng W, Papahadjopoulos D (1998) Ultrastructural characterization of cationic liposome-DNA complexes showing enhanced stability in serum and high transfection activity in vivo. Biochim Biophys Acta 1375:23–35
42. da Cruz MT, Cardoso AL, de Almeida LP, Simoes S, Pedroso de Lima MC (2005) Tf-lipoplex-mediated NGF gene transfer to the CNS: neuronal protection and recovery in an excitotoxic model of brain injury. Gene Ther 12:1242–1252
43. Cardoso AL, Costa P, de Almeida LP, Simoes S, Plesnila N, Culmsee C, Wagner E, Pedroso de Lima MC (2010) Tf-lipoplex-mediated c-Jun silencing improves neuronal survival following excitotoxic damage in vivo. J Control Release 142:392–403
44. Zhang Y, Zhang YF, Bryant J, Charles A, Boado RJ, Pardridge WM (2004) Intravenous RNA interference gene therapy targeting the human epidermal growth factor receptor prolongs survival in intracranial brain cancer. Clin Cancer Res 10:3667–3677
45. Lu Y, Low PS (2002) Folate-mediated delivery of macromolecular anticancer therapeutic agents. Adv Drug Deliv Rev 54:675–693
46. Reddy JA, Low PS (1998) Folate-mediated targeting of therapeutic and imaging agents to cancers. Crit Rev Ther Drug Carrier Syst 15:587–627
47. Saba NF, Wang X, Muller S, Tighiouart M, Cho K, Nie S, Chen Z, Shin DM (2009) Examining expression of folate receptor in squamous cell carcinoma of the head and neck as a target for a novel nanotherapeutic drug. Head Neck 31:475–481
48. Sudimack J, Lee RJ (2000) Targeted drug delivery via the folate receptor. Adv Drug Deliv Rev 41:147–162
49. Meier R, Henning TD, Boddington S, Tavri S, Arora S, Piontek G, Rudelius M, Corot C, Daldrup-Link HE (2010) Breast cancers: MR imaging of folate-receptor expression with the folate-specific nanoparticle P1133. Radiology 255:527–535
50. Campbell IG, Jones TA, Foulkes WD, Trowsdale J (1991) Folate-binding protein is a marker for ovarian cancer. Cancer Res 51:5329–5338
51. Laine C, Mornet E, Lemiegre L, Montier T, Cammas-Marion S, Neveu C, Carmoy N, Lehn P, Benvegnu T (2008) Folate-equipped pegylated archaeal lipid derivatives: synthesis and transfection properties. Chemistry 14:8330–8340

52. Li H, Piao L, Yu B, Yung BC, Zhang W, Wang PG, Lee JL, Lee RJ (2011) Delivery of calf thymus DNA to tumor by folate receptor targeted cationic liposomes. Biomaterials 32:6614–6620
53. Hofland HE, Masson C, Iginla S, Osetinsky I, Reddy JA, Leamon CP, Scherman D, Bessodes M, Wils P (2002) Folate-targeted gene transfer in vivo. Mol Ther 5:739–744
54. Rait AS, Pirollo KF, Xiang L, Ulick D, Chang EH (2002) Tumor-targeting, systemically delivered antisense HER-2 chemosensitizes human breast cancer xenografts irrespective of HER-2 levels. Mol Med 8:475–486
55. Chiu SJ, Marcucci G, Lee RJ (2006) Efficient delivery of an antisense oligodeoxyribonucleotide formulated in folate receptor-targeted liposomes. Anticancer Res 26:1049–1056
56. Feng C, Wang T, Tang R, Wang J, Long H, Gao X, Tang S (2010) Silencing of the MYCN gene by siRNA delivered by folate receptor-targeted liposomes in LA-N-5 cells. Pediatr Surg Int 26:1185–1191
57. Li P, Liu D, Sun X, Liu C, Liu Y, Zhang N (2011) A novel cationic liposome formulation for efficient gene delivery via a pulmonary route. Nanotechnology 22:245104
58. Medina-Kauwe LK, Xie J, Hamm-Alvarez S (2005) Intracellular trafficking of nonviral vectors. Gene Ther 12:1734–1751
59. Simoes S, Slepushkin V, Gaspar R, Pedroso de Lima MC, Duzgunes N (1998) Gene delivery by negatively charged ternary complexes of DNA, cationic liposomes and transferrin or fusigenic peptides. Gene Ther 5:955–964
60. Simoes S, Slepushkin V, Pires P, Gaspar R, Pedroso de Lima MC, Duzgunes N (1999) Mechanisms of gene transfer mediated by lipoplexes associated with targeting ligands or pH-sensitive peptides. Gene Ther 6:1798–1807
61. Yamauchi J, Hayashi Y, Kajimoto K, Akita H, Harashima H (2010) Comparison between a multifunctional envelope-type nano device and lipoplex for delivery to the liver. Biol Pharm Bull 33:926–929
62. Hatakeyama H, Ito E, Akita H, Oishi M, Nagasaki Y, Futaki S, Harashima H (2009) A pH-sensitive fusogenic peptide facilitates endosomal escape and greatly enhances the gene silencing of siRNA-containing nanoparticles in vitro and in vivo. J Control Release 139:127–132
63. Lundberg P, El-Andaloussi S, Sutlu T, Johansson H, Langel U (2007) Delivery of short interfering RNA using endosomolytic cell-penetrating peptides. FASEB J 21:2664–2671
64. Shenkman S, Araujo PS, Dijkman R, Quina FH, Chaimovich H (1981) Effects of temperature and lipid composition on the serum albumin-induced aggregation and fusion of small unilamellar vesicles. Biochim Biophys Acta 649:633–641
65. Garcia LAM, Shenkman S, Araujo PS, Chaimovich H (1983) Fusion of small unilamellar vesicles induced by bovine serum albumin fragments. Brazil J Med Res 16:89–96
66. Simões S, Slepushkin V, Pires P, Gaspar R, Pedroso de Lima MC, Düzgüneş N (2000) Human serum albumin enhances DNA transfection by lipoplexes and confers resistance to inhibition by serum. Biochimica et Biophysica Acta 1463:459–469
67. Aronsohn AI, Hughes JA (1998) Nuclear localization signal peptides enhance cationic liposome-mediated gene therapy. J Drug Target 5:163–169
68. Chan CK, Jans DA (1999) Enhancement of polylysine-mediated transfer infection by nuclear localization sequences: polylysine does not function as a nuclear localization sequence. Hum Gene Ther 10:1695–1702
69. Zanta MA, Belguise-Valladier P, Behr JP (1999) Gene delivery: a single nuclear localization signal peptide is sufficient to carry DNA to the cell nucleus. Proc Natl Acad Sci USA 96:91–96
70. Branden LJ, Mohamed AJ, Smith CI (1999) A peptide nucleic acid-nuclear localization signal fusion that mediates nuclear transport of DNA. Nat Biotechnol 17:784–787
71. Subramanian A, Ranganathan P, Diamond SL (1999) Nuclear targeting peptide scaffolds for lipofection of nondividing mammalian cells. Nat Biotechnol 17:873–877
72. Chen J, Yu Z, Chen H, Gao J, Liang W (2011) Transfection efficiency and intracellular fate of polycation liposomes combined with protamine. Biomaterials 32:1412–1418

73. Hyndman L, Lemoine JL, Huang L, Porteous DJ, Boyd AC, Nan X (2004) HIV-1 Tat protein transduction domain peptide facilitates gene transfer in combination with cationic liposomes. J Control Release 99:435–444
74. Rudolph C, Plank C, Lausier J, Schillinger U, Muller RH, Rosenecker J (2003) Oligomers of the arginine-rich motif of the HIV-1 TAT protein are capable of transferring plasmid DNA into cells. J Biol Chem 278:11411–11418
75. Trabulo S, Mano M, Faneca H, Cardoso AL, Duarte S, Henriques A, Paiva A, Gomes P, Simoes S, Pedroso de Lima MC (2008) S4(13)-PV cell penetrating peptide and cationic liposomes act synergistically to mediate intracellular delivery of plasmid DNA. J Gene Med 10:1210–1222
76. Trabulo S, Resina S, Simoes S, Lebleu B, Pedroso de Lima MC (2010) A non-covalent strategy combining cationic lipids and CPPs to enhance the delivery of splice correcting oligonucleotides. J Control Release 145:149–158
77. Miyata T, Yamamoto S, Sakamoto K, Morishita R, Kaneda Y (2001) Novel immunotherapy for peritoneal dissemination of murine colon cancer with macrophage inflammatory protein-1beta mediated by a tumor-specific vector, HVJ cationic liposomes. Cancer Gene Ther 8:852–860
78. Asada H, Kishida T, Hirai H, Satoh E, Ohashi S, Takeuchi M, Kubo T, Kita M, Iwakura Y, Imanishi J, Mazda O (2002) Significant antitumor effects obtained by autologous tumor cell vaccine engineered to secrete interleukin (IL)-12 and IL-18 by means of the EBV/lipoplex. Mol Ther 5:609–616
79. Faneca H, Cabrita AS, Simoes S, Pedroso de Lima MC (2007) Evaluation of the antitumoral effect mediated by IL-12 and HSV-tk genes when delivered by a novel lipid-based system. Biochim Biophys Acta 1768:1093–1102
80. Neves S, Faneca H, Bertin S, Konopka K, Duzgunes N, Pierrefite-Carle V, Simoes S, Pedroso de Lima MC (2009) Transferrin lipoplex-mediated suicide gene therapy of oral squamous cell carcinoma in an immunocompetent murine model and mechanisms involved in the antitumoral response. Cancer Gene Ther 16:91–101
81. Li J, Li X, Zhang Y, Zhou XK, Yang HS, Chen XC, Wang YS, Wei YQ, Chen LJ, Hu HZ, Liu CY (2010) Gene therapy for psoriasis in the K14-VEGF transgenic mouse model by topical transdermal delivery of interleukin-4 using ultradeformable cationic liposome. J Gene Med 12:481–490
82. Troyer RM, Propst KL, Fairman J, Bosio CM, Dow SW (2009) Mucosal immunotherapy for protection from pneumonic infection with francisella tularensis. Vaccine 27:4424–4433
83. Fire A, Xu S, Montgomery MK, Kostas SA, Driver SE, Mello CC (1998) Potent and specific genetic interference by double-stranded RNA in caenorhabditis elegans. Nature 391:806–811
84. Elbashir SM, Harborth J, Lendeckel W, Yalcin A, Weber K, Tuschl T (2001) Duplexes of 21-nucleotide RNAs mediate RNA interference in cultured mammalian cells. Nature 411:494–498
85. Morrissey DV, Blanchard K, Shaw L, Jensen K, Lockridge JA, Dickinson B, McSwiggen JA, Vargeese C, Bowman K, Shaffer CS, Polisky BA, Zinnen S (2005) Activity of stabilized short interfering RNA in a mouse model of hepatitis B virus replication. Hepatology 41:1349–1356
86. Halder J, Kamat AA, Landen CN Jr, Han LY, Lutgendorf SK, Lin YG, Merritt WM, Jennings NB, Chavez-Reyes A, Coleman RL, Gershenson DM, Schmandt R, Cole SW, Lopez-Berestein G, Sood AK (2006) Focal adhesion kinase targeting using in vivo short interfering RNA delivery in neutral liposomes for ovarian carcinoma therapy. Clin Cancer Res 12:4916–4924
87. Frank-Kamenetsky M, Grefhorst A, Anderson NN, Racie TS, Bramlage B, Akinc A, Butler D, Charisse K, Dorkin R, Fan Y, Gamba-Vitalo C, Hadwiger P, Jayaraman M, John M, Jayaprakash KN, Maier M, Nechev L, Rajeev KG, Read T, Rohl I, Soutschek J, Tan P, Wong J, Wang G, Zimmermann T, de Fougerolles A, Vornlocher HP, Langer R, Anderson DG, Manoharan M, Koteliansky V, Horton JD, Fitzgerald K (2008) Therapeutic RNAi targeting PCSK9 acutely lowers plasma cholesterol in rodents and LDL cholesterol in nonhuman primates. Proc Natl Acad Sci USA 105:11915–11920

88. Yokota T, Iijima S, Kubodera T, Ishii K, Katakai Y, Ageyama N, Chen Y, Lee YJ, Unno T, Nishina K, Iwasaki Y, Maki N, Mizusawa H, Akari H (2007) Efficient regulation of viral replication by siRNA in a non-human primate surrogate model for hepatitis C. Biochem Biophys Res Commun 361:294–300
89. Sato Y, Murase K, Kato J, Kobune M, Sato T, Kawano Y, Takimoto R, Takada K, Miyanishi K, Matsunaga T, Takayama T, Niitsu Y (2008) Resolution of liver cirrhosis using vitamin A-coupled liposomes to deliver siRNA against a collagen-specific chaperone. Nat Biotechnol 26:431–442
90. Zimmermann TS, Lee AC, Akinc A, Bramlage B, Bumcrot D, Fedoruk MN, Harborth J, Heyes JA, Jeffs LB, John M, Judge AD, Lam K, McClintock K, Nechev LV, Palmer LR, Racie T, Rohl I, Seiffert S, Shanmugam S, Sood V, Soutschek J, Toudjarska I, Wheat AJ, Yaworski E, Zedalis W, Koteliansky V, Manoharan M, Vornlocher HP, MacLachlan I (2006) RNAi-mediated gene silencing in non-human primates. Nature 441:111–114
91. Geisbert TW, Hensley LE, Kagan E, Yu EZ, Geisbert JB, Daddario-DiCaprio K, Fritz EA, Jahrling PB, McClintock K, Phelps JR, Lee AC, Judge A, Jeffs LB, MacLachlan I (2006) Post exposure protection of guinea pigs against a lethal ebola virus challenge is conferred by RNA interference. J Infect Dis 193:1650–1657
92. Chien PY, Wang J, Carbonaro D, Lei S, Miller B, Sheikh S, Ali SM, Ahmad MU, Ahmad I (2005) Novel cationic cardiolipin analogue-based liposome for efficient DNA and small interfering RNA delivery in vitro and in vivo. Cancer Gene Ther 12:321–328
93. Li SD, Chono S, Huang L (2008) Efficient oncogene silencing and metastasis inhibition via systemic delivery of siRNA. Mol Ther 16:942–946
94. Mendonca LS, Firmino F, Moreira JN, Pedroso de Lima MC, Simoes S (2010) Transferrin receptor-targeted liposomes encapsulating anti-BCR-ABL siRNA or as ODN for chronic myeloid leukemia treatment. Bioconjug Chem 21:157–168
95. Mendonca LS, Moreira JN, Pedroso de Lima MC, Simoes S (2010) Co-encapsulation of anti-BCR-ABL siRNA and imatinib mesylate in transferrin receptor-targeted sterically stabilized liposomes for chronic myeloid leukemia treatment. Biotechnol Bioeng 107:884–893
96. Kumar P, Lee SK, Shankar P, Manjunath N (2006) A single siRNA suppresses fatal encephalitis induced by two different flaviviruses. PLoS Med 3:e96
97. Flynn MA, Casey DG, Todryk SM, Mahon BP (2004) Efficient delivery of small interfering RNA for inhibition of IL-12p40 expression in vivo. J Inflamm 1:4
98. Sorensen DR, Leirdal M, Sioud M (2003) Gene silencing by systemic delivery of synthetic siRNAs in adult mice. J Mol Biol 327:761–766
99. Khoury M, Louis-Plence P, Escriou V, Noel D, Largeau C, Cantos C, Scherman D, Jorgensen C, Apparailly F (2006) Efficient new cationic liposome formulation for systemic delivery of small interfering RNA silencing tumor necrosis factor alpha in experimental arthritis. Arthritis Rheum 54:1867–1877
100. Palliser D, Chowdhury D, Wang QY, Lee SJ, Bronson RT, Knipe DM, Lieberman J (2006) An siRNA-based microbicide protects mice from lethal herpes simplex virus 2 infection. Nature 439:89–94
101. Zhang Y, Cristofaro P, Silbermann R, Pusch O, Boden D, Konkin T, Hovanesian V, Monfils PR, Resnick M, Moss SF, Ramratnam B (2006) Engineering mucosal RNA interference in vivo. Mol Ther 14:336–342
102. Jarver P, Langel U (2004) The use of cell-penetrating peptides as a tool for gene regulation. Drug Discov Today 9:395–402
103. Laufer SD, Restle T (2008) Peptide-mediated cellular delivery of oligonucleotide-based therapeutics in vitro: quantitative evaluation of overall efficacy employing easy to handle reporter systems. Curr Pharm Des 14:3637–3655
104. Gleave ME, Monia BP (2005) Antisense therapy for cancer. Nat Rev Cancer 5:468–479
105. Sazani P, Kole R (2003) Modulation of alternative splicing by antisense oligonucleotides. Prog Mol Subcell Biol 31:217–239
106. Tazi J, Bakkour N, Stamm S (2009) Alternative splicing and disease. Biochim Biophys Acta 1792:14–26

107. Kole R, Williams T, Cohen L (2004) RNA modulation, repair and remodeling by splice switching oligonucleotides. Acta Biochim Pol 51:373–378
108. Mercatante DR, Mohler JL, Kole R (2002) Cellular response to an antisense-mediated shift of Bcl-x pre-mRNA splicing and antineoplastic agents. J Biol Chem 277:49374–49382
109. Ko YT, Falcao C, Torchilin VP (2009) Cationic liposomes loaded with proapoptotic peptide D-(KLAKLAK)(2) and Bcl-2 antisense oligodeoxynucleotide G3139 for enhanced anticancer therapy. Mol Pharm 6:971–977
110. Resina S, Abes S, Turner JJ, Prevot P, Travo A, Clair P, Gait MJ, Thierry AR, Lebleu B (2007) Lipoplex and peptide-based strategies for the delivery of steric-block oligonucleotides. Int J Pharm 344:96–102
111. Resina S, Kole R, Travo A, Lebleu B, Thierry AR (2007) Switching on transgene expression by correcting aberrant splicing using multi-targeting steric-blocking oligonucleotides. J Gene Med 9:498–510
112. Thierry AR, Abes S, Resina S, Travo A, Richard JP, Prevot P, Lebleu B (2006) Comparison of basic peptides- and lipid-based strategies for the delivery of splice correcting oligonucleotides. Biochim Biophys Acta 1758:364–374
113. Thierry AR, Vives E, Richard JP, Prevot P, Martinand-Mari C, Robbins I, Lebleu B (2003) Cellular uptake and intracellular fate of antisense oligonucleotides. Curr Opin Mol Ther 5:133–138
114. Kim ST, Lee KM, Park HJ, Jin SE, Ahn WS, Kim CK (2009) Topical delivery of interleukin-13 antisense oligonucleotides with cationic elastic liposome for the treatment of atopic dermatitis. J Gene Med 11:26–37
115. Zhang C, Newsome JT, Mewani R, Pei J, Gokhale PC, Kasid UN (2009) Systemic delivery and pre-clinical evaluation of nanoparticles containing antisense oligonucleotides and siRNAs. Methods Mol Biol 480:65–83
116. Beisner J, Dong M, Taetz S, Piotrowska K, Kleideiter E, Friedel G, Schaefer U, Lehr CM, Klotz U, Murdter TE (2009) Efficient telomerase inhibition in human non-small cell lung cancer cells by liposomal delivery of 2′-O-methyl-RNA. J Pharm Sci 98:1765–1774
117. Dinh TD, Higuchi Y, Kawakami S, Yamashita F, Hashida M (2011) Evaluation of osteoclastogenesis via NFkappaB decoy/mannosylated cationic liposome-mediated inhibition of pro-inflammatory cytokine production from primary cultured macrophages. Pharm Res 28:742–751
118. Kohn DB, Hershfield MS, Carbonaro D, Shigeoka A, Brooks J, Smogorzewska EM, Barsky LW, Chan R, Burotto F, Annett G, Nolta JA, Crooks G, Kapoor N, Elder M, Wara D, Bowen T, Madsen E, Snyder FF, Bastian J, Muul L, Blaese RM, Weinberg K, Parkman R (1998) T lymphocytes with a normal ADA gene accumulate after transplantation of transduced autologous umbilical cord blood CD34+ cells in ADA-deficient SCID neonates. Nat Med 4:775–780
119. Baum BJ, Goldsmith CM, Kok MR, Lodde BM, van Mello NM, Voutetakis A, Wang J, Yamano S, Zheng C (2003) Advances in vector-mediated gene transfer. Immunol Lett 90:145–149
120. Raper SE, Yudkoff M, Chirmule N, Gao GP, Nunes F, Haskal ZJ, Furth EE, Propert KJ, Robinson MB, Magosin S, Simoes H, Speicher L, Hughes J, Tazelaar J, Wivel NA, Wilson JM, Batshaw ML (2002) A pilot study of in vivo liver-directed gene transfer with an adenoviral vector in partial ornithine transcarbamylase deficiency. Hum Gene Ther 13:163–175
121. Cavazzana-Calvo M, Hacein-Bey S, de Saint BG, Gross F, Yvon E, Nusbaum P, Selz F, Hue C, Certain S, Casanova JL, Bousso P, Deist FL, Fischer A (2000) Gene therapy of human severe combined immunodeficiency (SCID)-X1 disease. Science 288:669–672
122. Hacein-Bey-Abina S, Le Deist F, Carlier F, Bouneaud C, Hue C, De Villartay JP, Thrasher AJ, Wulffraat N, Sorensen R, Dupuis-Girod S, Fischer A, Davies EG, Kuis W, Leiva L, Cavazzana-Calvo M (2002) Sustained correction of X-linked severe combined immunodeficiency by ex vivo gene therapy. N Engl J Med 346:1185–1193

123. Hacein-Bey-Abina S, Von Kalle C, Schmidt M, McCormack MP, Wulffraat N, Leboulch P, Lim A, Osborne CS, Pawliuk R, Morillon E, Sorensen R, Forster A, Fraser P, Cohen JI, de Saint Basile G, Alexander I, Wintergerst U, Frebourg T, Aurias A, Stoppa-Lyonnet D, Romana S, Radford-Weiss I, Gross F, Valensi F, Delabesse E, Macintyre E, Sigaux F, Soulier J, Leiva LE, Wissler M, Prinz C, Rabbitts TH, Le Deist F, Fischer A, Cavazzana-Calvo M (2003) LMO2-associated clonal T cell proliferation in two patients after gene therapy for SCID-X1. Science 302:415–419
124. Edelstein ML, Abedi MR, Wixon J (2007) Gene therapy clinical trials worldwide to 2007 – an update. J Gene Med 9:833–842
125. Alton EW, Stern M, Farley R, Jaffe A, Chadwick SL, Phillips J, Davies J, Smith SN, Browning J, Davies MG, Hodson ME, Durham SR, Li D, Jeffery PK, Scallan M, Balfour R, Eastman SJ, Cheng SH, Smith AE, Meeker D, Geddes DM (1999) Cationic lipid-mediated CFTR gene transfer to the lungs and nose of patients with cystic fibrosis: a double-blind placebo-controlled trial. Lancet 353:947–954
126. Hyde SC, Southern KW, Gileadi U, Fitzjohn EM, Mofford KA, Waddell BE, Gooi HC, Goddard CA, Hannavy K, Smyth SE, Egan JJ, Sorgi FL, Huang L, Cuthbert AW, Evans MJ, Colledge WH, Higgins CF, Webb AK, Gill DR (2000) Repeat administration of DNA/liposomes to the nasal epithelium of patients with cystic fibrosis. Gene Ther 7:1156–1165
127. Yoshida J, Mizuno M, Fujii M, Kajita Y, Nakahara N, Hatano M, Saito R, Nobayashi M, Wakabayashi T (2004) Human gene therapy for malignant gliomas (glioblastoma multiforme and anaplastic astrocytoma) by in vivo transduction with human interferon beta gene using cationic liposomes. Hum Gene Ther 15:77–86
128. Bedikian AY, Richards J, Kharkevitch D, Atkins MB, Whitman E, Gonzalez R (2010) A phase 2 study of high-dose Allovectin-7 in patients with advanced metastatic melanoma. Melanoma Res 20:218–226
129. Hortobagyi GN, Ueno NT, Xia W, Zhang S, Wolf JK, Putnam JB, Weiden PL, Willey JS, Carey M, Branham DL, Payne JY, Tucker SD, Bartholomeusz C, Kilbourn RG, De Jager RL, Sneige N, Katz RL, Anklesaria P, Ibrahim NK, Murray JL, Theriault RL, Valero V, Gershenson DM, Bevers MW, Huang L, Lopez-Berestein G, Hung MC (2001) Cationic liposome-mediated E1A gene transfer to human breast and ovarian cancer cells and its biologic effects: a phase I clinical trial. J Clin Oncol 19:3422–3433
130. Voges J, Reszka R, Gossmann A, Dittmar C, Richter R, Garlip G, Kracht L, Coenen HH, Sturm V, Wienhard K, Heiss WD, Jacobs AH (2003) Imaging-guided convection-enhanced delivery and gene therapy of glioblastoma. Ann Neurol 54:479–487
131. Cavazzana-Calvo M, Thrasher A, Mavilio F (2004) The future of gene therapy. Nature 427:779–781
132. Hyde SC, Gill DR, Higgins CF, Trezise AE, MacVinish LJ, Cuthbert AW, Ratcliff R, Evans MJ, Colledge WH (1993) Correction of the ion transport defect in cystic fibrosis transgenic mice by gene therapy. Nature 362:250–255
133. McLachlan G, Ho LP, Davidson-Smith H, Samways J, Davidson H, Stevenson BJ, Carothers AD, Alton EW, Middleton PG, Smith SN, Kallmeyer G, Michaelis U, Seeber S, Naujoks K, Greening AP, Innes JA, Dorin JR, Porteous DJ (1996) Laboratory and clinical studies in support of cystic fibrosis gene therapy using pCMV-CFTR-DOTAP. Gene Ther 3:1113–1123
134. Griesenbach U, Ferrari S, Geddes DM, Alton EW (2002) Gene therapy progress and prospects: cystic fibrosis. Gene Ther 9:1344–1350
135. Hyde SC, Pringle IA, Abdullah S, Lawton AE, Davies LA, Varathalingam A, Nunez-Alonso G, Green AM, Bazzani RP, Sumner-Jones SG, Chan M, Li H, Yew NS, Cheng SH, Boyd AC, Davies JC, Griesenbach U, Porteous DJ, Sheppard DN, Munkonge FM, Alton EW, Gill DR (2008) CpG-free plasmids confer reduced inflammation and sustained pulmonary gene expression. Nat Biotechnol 26:549–551
136. Griesenbach U, Alton EW (2009) Gene transfer to the lung: lessons learned from more than 2 decades of CF gene therapy. Adv Drug Deliv Rev 61:128–139

137. Wakabayashi T, Natsume A, Hashizume Y, Fujii M, Mizuno M, Yoshida J (2008) A phase I clinical trial of interferon-beta gene therapy for high-grade glioma: novel findings from gene expression profiling and autopsy. J Gene Med 10:329–339
138. Hung MC, Chang JY, Xing X (1998) Pre-clinical and clinical study of HER-2/neu-targeting cancer gene therapy. Adv Drug Deliv Rev 30:219–227
139. Villaret D, Glisson B, Kenady D, Hanna E, Carey M, Gleich L, Yoo GH, Futran N, Hung MC, Anklesaria P, Heald AE (2002) A multicenter phase II study of tgDCC-E1A for the intratumoral treatment of patients with recurrent head and neck squamous cell carcinoma. Head Neck 24:661–669

Biomaterial-Based Vectors for Targeted Delivery of Nucleic Acids to the Nervous System

Ana Paula Pêgo, Hugo Oliveira, and Pedro Miguel Moreno

Abstract One of the challenges currently facing neuroscientists is the development of effective therapies based on the advances achieved on basic research. The use of genes as pro-drugs can be faced as an approach to reduce this gap. Furthermore, gene downregulation through the use of antisense strategies, including the recent introduction of RNA interference, is yet another tool with great therapeutic potential.

While viruses have proved to be the most efficient system to mediate the delivery of nucleic acids (mostly DNA), their use in a clinic scenario raises various safety concerns. Additionally, the regional specialization of the nervous system function dictates that a therapeutic intervention may be best achieved by the local and specific delivery of a therapeutic agent. In that sense a targeted delivery may avoid unwanted adverse effects that could result from a more ubiquitous delivery. In this chapter, the alternative strategies being explored to the delivery of nucleic acids (DNA, antisense and siRNA oligonucleotides) to the nervous system will be discussed, with particular focus on biomaterial-based systems.

Keywords Gene therapy • Nanomedicine • Neurodegenerative diseases • Neuropathies • Non-viral vectors • Nucleic acid based therapeutics • DNA • siRNA • Antisense oligonucleotides • Personalized medicine

---

A.P. Pêgo (✉) • P.M. Moreno
INEB – Instituto de Engenharia Biomédica, Universidade do Porto, Rua do Campo Alegre, 823, 4150-180 Porto, Portugal
e-mail: apego@ineb.up.pt

H. Oliveira
CNRS, Laboratoire de Chimie des Polymères Organiques (UMR5629), Université de Bordeaux/IPB, ENSCBP, 16 Avenue Pey Berland, 33607 Pessac Cedex, France

J. Coelho (ed.), *Drug Delivery Systems: Advanced Technologies Potentially Applicable in Personalised Treatment*, Advances in Predictive, Preventive and Personalised Medicine 4, DOI 10.1007/978-94-007-6010-3_7,

1 Introduction

The nervous system is constituted of a network of nerve cells that by transmitting signals between different body parts are responsible for the orchestration of the body actions. In humans the nervous system can be divided in the central nervous system (CNS) that includes the brain, the spinal cord and the retina, and the peripheral nervous system (PNS) constituted by sensory and motor neurons outside the CNS. Sensory nerves, originated from clusters of neurons called ganglia, conduct sensory signals from organs and skin to the CNS. Motor nerves, originated from motor neurons situated in the spinal cord, transmit stimulatory signals from the CNS towards muscles and glands.

Due to its crucial role, the impact of disease or a physical injury on the nervous system can have devastating consequences to the individual and very frequently significant societal and economical repercussions. Nervous system problems are common and encompass a large spectrum of traumatic injuries, diseases, tumors or iatrogenic lesions. Depending on the extent of the disorder it may result in a partial or total loss of motor, sensory and autonomic functions in the involved segments of the body.

Neurodegenerative diseases are hereditary and sporadic conditions characterized by a progressive nervous system dysfunction. Alzheimer's disease and other dementias, brain cancer, epilepsy, stroke, Parkinson's disease, multiple sclerosis, amyotrophic lateral sclerosis, Huntington's disease, among others, engross the list of neurodegenerative diseases that affect the CNS to which there is no cure [1].

Peripheral neuropathies may be caused either by diseases of or trauma to peripheral nerves or result as a consequence of a systemic illness [2]. One of the most common peripheral neuropathy is due to diabetes mellitus, with approximately 50% of the individuals who have had diabetes for more than 25 years suffering symptomatic peripheral neuropathy [3]. In the course of human immunodeficiency virus type 1 (HIV-1) infection, up to 50% of the patients suffer from neurological complications affecting either the central CNS or the PNS [4]. Genetic factors may also contribute to the onset of PNS pathologies [2].

Conventional treatments for the above mentioned diseases have primarily been palliative rather than curative [1, 5, 6]. Perhaps most importantly, they have often been ineffective. In recent years, enormous efforts have been made in the understanding of the biology of these disorders and the observed advances have suggested new therapeutic approaches that may arrest or reverse these conditions. Studies have demonstrated that a number of neurotrophic factors, including nerve growth factor (NGF), brain derived neurotrophic factor (BDNF), neurotrophin-3 (NT-3), insulin-like growth factor (IGF) and vascular endothelial growth factor (VEGF) can prevent the degeneration of peripheral sensory axons [5, 7], lead to an improvement of function in the scenario of certain neurodegenerative diseases [8, 9] or promote the regeneration of axons upon spinal cord or peripheral nerve injury [10]. The use of ribonucleic acid (RNA) inhibitors has also proved to be a therapeutic tool with potential application in fields that range from the treatment of malignant brain gliomas [11] to spinal cord lesions [12].

The lack of effective response presented by current therapeutics and the crescent demand on the design of effective treatments that translate the advances obtained in the basic science has brought new interest in the gene therapy approach. There are several key points that favor the use of therapeutic gene transfer or gene therapy in the treatment of neuropathologies. More unique RNA sequences are expressed in the nervous system and a large proportion of the identified genetic diseases display a neurologic component to the phenotype. In many cases, the regional specialization of nervous system function dictates that a therapeutic intervention may be best achieved by the local and specific expression of a transgene product, such as a neurotrophic or anti-apoptotic factor, or by down-regulation of a certain protein. In addition, the local delivery of a gene, antisense oligonucleotide or RNA, may be used to achieve the desired outcome avoiding unwanted adverse effects that could result from the systemic administration of therapeutic agents.

Extensive work has been developed in the recent years in order to optimize and accomplish the successful transfer of nucleic acids to mammalian cells. This effort made at the research level is mirrored by the growing number of gene therapy-based clinical trials, with over 1,786 studies completed, ongoing or pending worldwide [13]. Both non-viral and viral strategies have been attempted but some of the key drawbacks presented by each of the strategies were not yet solved, what has been hampering the translation of such systems from bench to the bedside.

2 Gene Therapy

The original concept of gene therapy was based on the simple idea that the transfer of DNA material into cells could serve as a therapeutic strategy in order to correct a certain genetic disorder. The field quickly saw the large potential of this tool and a broader definition emerged.

Gene therapy can thus be defined as a strategy that provides to somatic cells the genetic information required for producing specific therapeutic proteins, or increase their overall expression levels, in order to correct or modulate specific pathologies. In this sense, somatic gene therapy aims at overcoming some of the drawbacks presented by direct protein administration, including low bioavailability, immune response, in vivo instability, high hepatic and renal clearance (in the case of systemic administration), and high cost of production [14]. Therefore, providing a therapeutic gene as a "pro-drug" to a patient may indeed circumvent some of the limitations associated with the direct use of recombinant proteins. Beyond this, the gene therapy concept can be even further expanded to include specific regulation of gene expression by the use of short oligonucleotides targeting DNA or RNA, in what can be also defined as antisense gene therapy or oligonucleotide gene therapy. In general terms the classical antisense concept accommodated the notion of hybridization of an oligonucleotide with RNA in order to block its function through steric blockage of translation [15] or RNA cleavage mediated by RNase H [16]. It has since been further expanded to include new technologies such as RNA cleaving ribozymes [17] and DNAzymes [18], small-interfering RNAs (siRNA) [19, 20] or DNA binding oligos such as triplex-forming-oligonucleotides (TFO), with capacity to arrest transcription [21] or block promoter activation [22]. Yet another oligonucleotide-mediated gene therapy is represented by the splice-switching technology where, in a non-destructive fashion, an oligonucleotide binds to the pre-mRNA and redirects the splicing reaction [23, 24].

To achieve the full realization of the gene therapy concept, successful means to deliver the nucleic acid payload to cells must be developed. An effective delivery system should assure the protection of the nucleic acid from degradation in the extracellular environment, provide for a sufficient circulation time by avoiding premature clearance when the vector is administered intravenously (IV), allow the specific targeting of different cell populations, efficiently deliver the nucleic acid to a sufficient number of cells, and, in certain instances, result in expression of a gene at high enough concentration and with sufficient persistence to produce a therapeutic effect [25, 26].

The vectors that have successfully demonstrated the delivery of exogenous nucleic acids in vitro and in vivo, can be divided into two major groups: viral and non-viral. In the following paragraphs, an overview of the most relevant delivery systems from both groups will be presented, as well as a brief discussion of their main advantages and drawbacks.

3 Nucleic Acid Delivery

3.1 Viral Vectors

Viruses are the most popular vectors in laboratory studies and clinical trials, mainly due to their efficiency, with the spectrum of virus used in research still surpassing the few applied in clinical trials. The viruses most commonly used include: retrovirus, lentivirus (from the *Retroviridae* family), adenovirus, adeno-associated virus (AAV) and herpes simplex virus (HSV) [27]. An overview of the main features presented by the aforementioned viral vectors with the highest potential for application in the nervous system can be observed in Table 1.

Despite the promising and encouraging results obtained so far, the in vivo (pre)clinical application of viral vectors has demonstrated problems of toxicity, immunogenicity and inflammatory potential. And, while phenomena of insertional mutagenesis and oncogenesis are associated to integrating viruses, the immune and inflammatory response and the ability to recombine are transversal to all [65–68]. Indeed, the potential of viruses to adapt to new, or changing cellular environments, as well as ecological niches via genetic variation appears to be an important point to have in mind. And in spite of the fact that the major mechanism that drives adaptation is based on accumulation of point mutations, evolution of viruses can also occur through recombination, which takes place in both DNA and RNA viruses. Consequently, this process may result in the emergence of new virus variants. For

Table 1 Comparison of properties of various viral vector systems currently used in nervous system applications

Vector	Genome	Maximum insert size (kb)	Integration	Duration of expression in vivo	Pre-existing host immunity	Immunogenicity	Safety problems	Examples of application in the peripheral and central nervous system
Lentiviral	Single stranded RNA	$\approx$8 kb	Yes	Long	Unlikely	Low	Insertional mutagenesis[c]	[28–38] Note: Used in a clinical trial for treatment of Parkinson's disease [13].
Adenoviral	Linear double stranded DNA	8–36[a]	No	Short	Possible	High	Inflammatory response; toxicity	[39–46] Note: Has been used as an anti-cancer agent due to oncolytic potential [13]. Their use in the clinic to treat other neurological diseases has been limited primarily due to vector toxicity.
Adeno-associated virus	Single stranded DNA	$\approx$4.7 kb	No[b]	Potentially long	Possible	Low	Inflammatory response; toxicity; tumorigenesis[c]	[47–54] Note: Used extensively in clinical trials [13].
Herpes simplex virus	Double stranded DNA	30 kb	No	Short	Possible	Low	Inflammatory response; toxicity	[55–63] Note: Used in clinical trials for the treatment of intractable pain in patients with malignancies and painful diabetic neuropathy [13].

[a]Depending on the vector generation [64]
[b]Adeno-associated viruses are able to integrate with low frequency [64], so they are normally depicted as non-integrating
[c]Under investigation

recombination to take place, the co-infection of a cell by at least two different virus variants is required. Additionally, in some cases the recombination between viruses and cellular nucleic acid can lead to the capture of cellular coding sequences [69].

In order to develop an efficient and safe gene delivery system applicable in a clinical scenario, new approaches are awaited. Fundamental drawbacks presented by the generality of viral-vectors, including pathogenicity, insertional mutagenesis, toxicity, limitations in scaling up, cost and ethical concerns [70], have drawn interest towards non-viral vectors.

3.2 Non-viral Vectors in Gene Therapy

Over the past decades there have been many attempts to design a non-viral vector that could mediate the level of gene expression and specificity attained by viral vectors, allow a greater flexibility in terms of the size of the nucleic acid cargo that can be transported, bypass the immune system, and reduce safety concerns. Several lipid-, peptide-, polymer- and nanoparticle-based systems are currently being developed for gene therapy applications [27, 71–73]. The majority of these systems are cationic in nature, interacting with the negatively charged nucleic acids through electrostatic interactions. Classically, the complex total charge (zeta potential) should present a positive net value. This allows the carrier to efficiently interact with the negatively charged cell membrane and undergo internalization by the cell, mainly through endocytosis [74].

In the following paragraphs the different material and strategies applied in the design of non-viral gene delivery systems will be discussed.

3.2.1 Cationic Lipids

Since the introduction of (N-[1-(2,3,-dioleyloxy)propyl]-N,N,N-trimethylamonium chloride) (DOTMA) as gene carrier in 1987 [75], liposomes have become one of the most studied non-viral vectors. Since then, several other cationic lipids (Fig. 1a) have been synthesized and studied for gene, antisense and siRNA oligonucleotide delivery [76–87].

All cationic lipids possess a hydrophobic group, which may be comprised of one or two fatty acid or an alkyl moiety of 12–18 carbons in length or a cholesterol moiety. The hydrophobic tails assure that the cationic lipids assemble into bi-layer vesicles and disperse in aqueous media, by effectively exposing the amine head groups towards the aqueous medium while shielding the hydrophobic portion of the molecule. The amine group represents an absolute necessity for the transfection competence of these vectors, as this group interacts electrostatically with the negatively charged groups of nucleic acids, condensing the anionic molecule and forming small transportable units—lipoplexes [88].

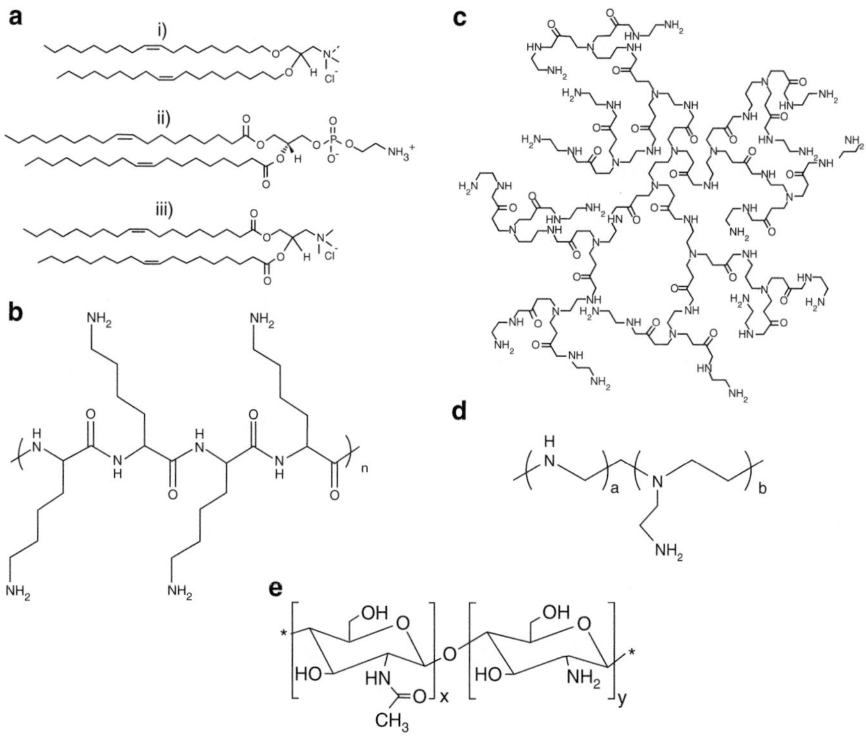

Fig. 1 Chemical formulas for different biomaterial based vectors. (**a**) Commonly used lipids for the formation of liposomes, (i) N-[1-(2,3,-dioleyloxy)propyl]-N,N,N-trimethylamonium chloride (DOTMA), (ii) dioleoylphosphatidyl ethanolamine (DOPE) and (iii) N-(1-(2,3-dioleoyloy)propyl)-N,N,N-trimethylammonium chloride (DOTAP); (**b**) Poly(L-lysine) (PLL); (**c**) Poly(ethylene imine) (PEI); (**d**) Chitosan (CH) and (**e**) Poly(amidoamine) (PAMAM) dendrimer

The transfection mechanism of cationic lipoplexes was initially associated with direct plasma membrane fusion [89, 90], but it is nowadays accepted that the cell entry occurs primarily via endocytosis [91, 92]. Many physical factors may influence the stability, formation and transfection efficiency of lipoplexes such as particle size, zeta potential, the nucleic acid:liposome ratio, ionic strength of the medium and presence of serum [93–95].

So far, in 6% of the gene therapy clinical trials, lipofection has been the selected method to mediate delivery [13]. Nevertheless, in the field of neurosciences, liposomes have been mainly applied in in vitro or, at most, preclinical studies. Indeed, despite the great success obtained in vitro, cationic lipoplexes often exhibit significant drawbacks for in vivo delivery. The interaction with plasma proteins leads to a rapid clearance from the circulation via the reticuloendothelial system [96]. Moreover, the in vivo assessment of lipid-base systems has revealed that, in the case of systemic administration, gene transfer was mainly observed in highly vascularized tissues (i.e., heart, lung, liver, spleen) [96–98]. To overcome these drawbacks, efforts have been made to allow the use of these systems in an in vivo scenario [99], however, further improvements regarding vector toxicity, efficiency and stability are still awaited.

3.2.2 Cationic Polymers

Polymers bearing groups that are protonated, or at least partially protonated at physiological pH, have been widely employed as gene carriers. The electrostatic interaction between the cationic amino groups on the polymer and the negatively charged phosphate groups in the nucleic acids results in a particulate complex, the polyplex, which is the transfecting unit [88]. When compared with lipoplexes, polyplexes seem to be more stable, less toxic, and have shown to improve the protection of the nucleic acids from endonuclease degradation. They show higher structural versatility to accommodate functionalization, including the possibility of covalent binding of targeting moieties for gene expression mediated through specific receptors [100, 101]. However, the efficiency of gene delivery mediated by these vectors is still modest when compared to viral-based systems [102]. This has motivated the research community to improve the efficacy of these vectors in order to render them useful in the clinical setting. The most widely studied cationic polymers for gene therapy will be further discussed in the following paragraphs.

Poly(L-lysine)

Poly(L-lysine) (PLL) is a linear polypeptide having the amino acid lysine (positively charged at physiological pH) as the repeating unit (Fig. 1b). The first polycation to be employed for gene delivery was PLL conjugated with asialoorosomucoid for hepatocellular gene targeted delivery [103]. The gene transfer activity of PLL polyplexes is poor, unless endosomolytic (e.g., imidazole) or lysosomotropic agents (e.g., chloroquine) are added [104, 105]. In an elegant study Akinc and Langer determined that after internalization, most of the polyplexes were retained in the lysosomal pathway instead of being released to the cytoplasm [106]. This is an important difference in the biological activity of the amphiphilic cationic lipids and the soluble polymer PLL [88]. In fact, PLL has poor transfection ability when applied alone or without modifications. However, it has been shown that DNA condensation and transfection efficiency increased with the increase of PLL molecular weight, though, also associated with a toxicity increase [107]. The creation of amphiphilic PLL, by linking both poly(ethylene glycol) (PEG) and palmitoyl groups to the polymer, reduced toxicity without compromising the gene delivery efficiency [107]. The conjugation of histidine residues to PLL resulted in a more efficient transfection than the one mediated by the PLL-chloroquine mixture [108, 109]. Poly(L-histidine) (PLH) has been shown to destabilize lipid bilayers when in a slightly acidic medium and induce fusion upon protonation of the imidazole groups by the increase of interaction between this polymeric cation and the membrane phospholipids [110]; explaining in part the effect of histidine on the increased transfection efficiency of PLL copolymers with PLH. Grafting PEG to the side chains of PLL was also found to augment the lifetime in blood circulation of complexes with nucleic acids [111], as when PLL polyplexes are administrated IV these are rapidly bound to plasma proteins and cleared from circulation [112, 113].

The chain length heterogeneity of the commercially available PLL is one of the major causes of variability in the preparation of reproducible and stable formulations. The extreme heterogeneity greatly complicates both the kinetics of DNA-PLL interaction and the thermodynamic stability of the final DNA complexes [107]. Although the peptide linkage is naturally found in proteins, the repeating unit of PLL, the L-lysine amino acid, is not found in nature. As a result, PLL has been found to be cytotoxic [114], possibly arising from its slow degradation in vivo.

Due to the above mentioned drawbacks the wide spread application of this polymer for nucleic acid delivery purposes has been limited.

Dendrimers

Poly(amidoamine) (PAMAM) dendrimers consist of a class of highly branched spherical polymers (Fig. 1c) that due to the ease of synthesis and commercial availability have also seen widespread use in gene therapy applications. The number of synthetic steps—generations—required for the preparation of any given PAMAM dendrimer determines its surface charge and diameter [115].

One of the major structural differences in PAMAM dendrimers are related to the core molecules, either ammonia (NH_3) or ethylenediamine (EDA), with which the stepwise polymerization process begins and which dictates the final overall shape, density and surface charge of the molecule. With each new generation, the molecular weight of the dendrimers more than doubles, and the number of surface amine groups exactly doubles [116]. Additional dendrimer chemistries have been successfully developed, including poly(propylene imine) (PPI) [117] and poly(L-Lysine) dendrimers [118], among others (see [119] for a review).

Dendrimers condense nucleic acids via electrostatic interactions of their terminal primary amines with the nucleic acids phosphate groups. The particle size, surface charge and transfection efficiency are influenced by the dendrimer concentration in the complexes [120, 121]. The dendriplexes are presumably internalized by endocytosis with several advantages associated with the star shape of the polymer, as the nucleic acids appear to interact with the surface primary amines only, leaving the internal tertiary amines available for the neutralization of the acidic pH within the endosomal/lysosomal compartment [122]. Moreover the increase in the amount of terminal amino groups (that is proportional to the generation of the dendrimer) was found to enhance the transfection efficiency up to a certain degree. It is known that the cytotoxicity of dendrimers is influenced by the characteristics of the surface functional groups. Dendrimers with free amine groups in the macromolecule periphery are reported to have a concentration- and generation number-dependent toxicity [123, 124] what, without doubt, has limited their application in a clinical scenario. Even so, several alterations to the basic PAMAM structure have been performed aiming at the reduction of these polymers cytotoxicity [125, 126]. Another drawback of the currently used dendrimers is their non-degradability under physiological conditions, leading to cytotoxicity, induced by the accumulation of non-degradable compounds inside cells. Being established that the degradation may solve the toxicity problems of these compounds with high charge densities and high molecular weights, recent interest has focused on the design of biodegradable dendrimers [127–129].

Poly(ethylene imine)

Poly(ethylene imine) (PEI) was introduced as a gene delivery vector by Behr in 1995 [130], and has become one of the gold standards among the non-viral based vectors. PEI can be synthesized in two forms, branched and linear. Branched PEI is a cationic polymer composed of primary, secondary and tertiary amines (Fig. 1d). It has been shown to effectively condense plasmid DNA, antisense and siRNA oligonucleotides into colloidal particles achieving transfection into a variety of cells, both in vitro and in vivo [130, 131].

PEI has a high charge density, as it contains a nitrogen atom for every three atoms of the backbone. In linear PEI all of those nitrogen atoms are protonable, whereas in branched PEI only two thirds of them can be charged [102]. As the overall protonation level of PEI increases from 20% at pH 7 to 40% at pH 5 [132], it is thought that the presence of PEI in the endosome can lead to an accumulation of protons brought in by the endosomal ATPase, coupled to an influx of chloride anion. PEI ability to accept the protons pumped to the endolysosome will oppose the pH decrease, inhibiting nucleases and unbalancing the endosome osmolarity by the increase of Cl^- concentration, with resulting osmotic swelling of the endosome. This effect is commonly referred as the "proton sponge effect" [130]. Therefore, PEI may enhance intracellular delivery, as by buffering the endosomal compartments it will induce lysosomal disruption, leading to endosomal release and subsequently protecting DNA from lysosomal degradation [130]. Many factors affect the efficiency/cytotoxicity profile of PEI polyplexes such as the polymer molecular weight and degree of branching, ionic strength of the solution, zeta potential, and particle size [133, 134]. The transfection efficiency of PEI has been tested for a wide range of polymer molecular weights and found to increase with the increase of the polymer molecular weight (in the range of 0.6–70 kDa) [135]. However, the toxicity associated to the transfection process mediated by PEI polymers was also shown to augment with the PEI molecular weight [136–138]. It is currently accepted that the optimal molecular weight exists between 11.9 and 70 kDa [88]. Moreover, besides molecular weight, the degree of PEI branching has also shown to play a role, with linear PEI being less efficient on complex condensation [139]. Complex stability was shown to be higher for PEI with increased primary amines, making, in that sense, branched PEI a more suitable polymer for gene delivery [140].

siRNA transfection was found to mainly rely on the overall PEI complexation, with higher outcomes when using 25 kDa branched PEI [141].

Interesting effects were also found regarding antisense oligonucleotide transfections, the chemistry of which having a determinant role. Hence, differences were seen when complexing phosphodiester (PO)- or phosphorothioate (PS)-based oligonucleotides with different molecular weight branched PEI. The PO achieving better transfections with an intermediate molecular weight PEI (25 kDa) and PS with PEI of higher molecular weight, such as 70 or 600 kDa [142].

Focus has been drawn on the improvement of PEI transfection efficiency and toxicity profile. One of the most common modifications is polymer PEGylation. In vivo studies have shown that IV injected PEI-PEG copolymer-based complexes had increased blood circulation and reduced toxicity. Still, no gene expression was detected what has been attributed to the lower stability of the copolymer based complexes, in relation to PEI complexes, leading to rapid clearance [143]. Nonetheless, the reduced charge of the copolymer-based complexes can also impair their interaction with the cell membrane and hinder intracellular gene transfer. Walker et al. have proposed a PEI-PEG copolymer in which the PEG is bound by a labile bond that upon a pH-induced trigger, within the endosome, led to the shield removal and proved advantageous both in vitro and in vivo [144]. Other molecules have been grafted to PEI with similar aims, but different outcomes. Namely, cholesterol moieties [145], cyclodextrin [146] and dexamethasone [147] and different amino acids [148, 149]. The synthesis of PEI compounds that could break down to less toxic lower molecular weight compounds was also attempted [150, 151].

Despite of the limitations described, PEI is still one of the most applied cationic polymer in nucleic acids delivery, mainly due to its high efficiency even in difficult to transfect cells, like neurons [152].

Chitosan

Chitosan (CH) is a linear polysaccharide composed of glucosamine and N-acetyl-D-glucosamine residues (Fig. 1e), obtained from the N-deacetylation of chitin [153]. CH is a biocompatible, biodegradable polymer that breaks down to harmless products (amino sugars that are completely absorbed by the human body). This polymer is a weak base with a pKa value of about 6.2–7.0 and is therefore insoluble at neutral and alkaline pHs. In acidic mediums, the amine groups of CH will be positively charged, conferring a high charge density to the polysaccharide. CH cationic characteristics provide strong electrostatic interactions with mucus, negatively charged mucosal surfaces and other macromolecules such as DNA and other nucleic acids [154]. Since the early application of CH for gene delivery, pioneered by Mumper et al. [155], CH has become one of the most prominent naturally derived non-viral vector. CH-based nucleic acid transfer depends greatly on the ionic and non-ionic interactions established between the carbohydrate backbone of CH and the surface proteins of cells [156].

CH particles can be prepared by different methods, including coacervation [157], gelation [158] and desolvation [159]. The use of the different methods together with other parameters like nucleic acid concentration, salt concentration, pH, charge ratio and temperature can influence the size of the formed complexes [160–163]. DNA binding with CH was also found to be dependent on both the molecular weight and the degree of acetylation (DA), CHs of lower molecular weight and higher DAs requiring higher polymer to DNA ratios to completely complex DNA. Based on these observations one can be lead to think that CH with lower DAs will increase the transfection efficacy. However, results can differ in vitro and in vivo. Kiang et al. tested CH with a starting molecular weight of 390 kDa with DAs that varied from 10 to 38% [153]. While the increase of DA resulted in a decrease of transfection activity in the cells tested in vitro, consequence of particle destabilization in the presence of serum proteins, the levels of transfection in vivo increased with the DA. This was associated with the DNA being more readily available to the surrounding tissue at the injection site (intramuscular), due to a more rapid release of DNA from the complexes in the presence of serum. Additionally, in vivo CH degradation has been shown to significantly increase for DAs above 30% [164]. However, CH-based delivery systems, as other non-viral gene delivery vectors, have shown transfection efficiencies that vary greatly with cell type [164, 165].

An advantage of CH over other systems is that its action is not hampered in the presence of serum [166], making it an interesting vector particularly for in vivo applications. Furthermore, the levels of cytotoxicity are residual by comparison. Despite of these arguments, the fact that the transfection efficiency of CH-based system is still very modest has triggered many research groups to improve this situation.

Lactosylated CH based vectors when tested in HeLa cells demonstrated to have transfection levels comparable to that of PEI [167]. PEI/CH co-polymerization has shown to lead to improvements in terms of transfection levels, with a reduction of cell toxicity being observed in relation to the PEI homopolymer [168]. To mimic the mechanism of endosomal escape of PEI, the functionalization of CH with buffering moieties like the imidazole group was also found to improve the vector transfection ability [169]. Trimethylated CH oligomers are also being explored in order to improve CH-based vectors, allowing increased CH solubility and polyplex stability in physiological conditions [170].

In the case of siRNA and antisense oligonucleotides, CH based vectors are still in their infancy. Regarding siRNA delivery perhaps the first example of CH/siRNA nanocomplexes is given by the work of Katas et al. In this work, the size of the siRNA/CH particles were shown to be proportional to the molecular weight of the CH, with smaller particles being formed when using lower molecular weight CH (110 kDa compared to 270 kDa) [171]. Moreover, the method of preparation was also playing a role since for the three methods of siRNA association to CH—simple complexation, adsorption and ionic gelation (siRNA entrapment)—it was the latter that gave the best results in terms of the target protein downregulation. The method of preparation of complexes seems to be not so relevant for plasmid

DNA (pDNA) [171]. It was also shown by Liu et al. that the use of low molecular weight (10 kDa) CH did not result in the formation of stable siRNA/CH complexes in contrast to higher molecular weight CH (64.8–170 kDa), which in conjugation with high N/P ratios (N/P 150; being the N/P ratio the theoretical charge ratio between amino groups of CH and phosphate groups of siRNA), showed good gene silencing efficiencies [172]. Interestingly, these abovementioned characteristics are in contrast to the use of CH for pDNA complexation. In fact, the molecular weight of CH seems to be less relevant in this case, since low molecular weight CH (4.7 kDa) has already been successfully used for preparation of stable pDNA/CH complexes [173]. Although still in its early days, siRNA/CH complexes have already been used for nerve regeneration studies showing its potential for nervous system applications [174].

With regard to antisense oligonucleotides (AONs) there are an even more limited number of reports [175–179]. There is also little information regarding the optimal properties of the CH and complexation conditions for creation of stable AON complexes. Although it can be speculated that it could follow similar rules as for siRNAs this might not be entirely true. It has been proposed that one of the difficulties could be the often huge difference in size between the AON (usually around 12–25 nt) and the polymer but also the possible high surface charge densities that could complex the AON too strongly, inhibiting their release once inside the cell [175]. In fact, the reports so far more specifically mentioning complexation characteristics have used relatively low molecular weight CHs (25 and 40 kDa) and relatively low N/P ratios (N/P around 5–6) when comparing to siRNA [175, 176].

Thus, favorable conditions for the formation of CH-based complexes seem to be quite dependent on the intrinsic characteristics of the nucleic acid, and this most likely holds true to most of the other polymer-based systems.

The qualities presented by CH have boosted the interest on this polymer regarding gene delivery applications. However, the modest transfection potential of CH has limited its success. Indeed new approaches are needed in order for the application of CH-based nanoparticles in neuronal cell transfection in vitro or in vivo to materialize.

3.2.3 Other Biomaterial-Based Non-viral Vectors

Despite liposomes, cationic polymer and dendrimer being the most studied non-viral gene delivery systems, the list of systems under investigation is much longer, including, but not confined to: nanoparticle systems, based on gold [180], silica [181] and carbon nanotubes [182, 183]; biodegradable polymers, including poly(4-hydroxy-L-proline ester) [184], poly[α-(4-aminobutyl)-L-glycolic acid] [185], poly(amino-ester) [186, 187], phosphorus containing degradable polymers and cationic synthetic peptides [72]. Although for certain applications some of these vectors have found some degree of success they come out of the scope of the topic of this chapter. For a more detailed review see [73].

4 What Are the Barriers for an Efficient Delivery?

To establish the requirements for an efficient vector mediated delivery of a nucleic acid to a particular cell population, is the first step towards the evaluation of the suitability of different delivery systems. As refereed in Sect. 3.1, safety is the main bottleneck for viral-mediated gene delivery. Consequently, the following paragraphs will focus the attention on non-viral based vectors as these offer the advantage of increased controllability and safety over viral vectors.

When targeting the nervous system, the vector of choice needs to overcome extra barriers. Neurons are considered one of the most difficult cell type to transfect [188, 189]. This can be due, at least in part, to the multiple cellular barriers encountered by the vectors. While most of the barriers are transversal to the majority of cell types, neurons have additional specifications. Morphologically neurons can be over 1 m long, which can impair the easy access of a vector to the cell nucleus, or perinuclear space in the case of antisense oligonucleotides and siRNA. As terminally differentiated neurons do not divide, one can consider the nuclear membrane as one of the main barriers regarding neuron cells. The understanding of the cellular obstacles conditioning the route of a non-viral nucleic acid delivery vector is of paramount importance in order to efficiently design and improve such a system. The main barriers encountered by a non-viral vector are summarized in Fig. 2, and will be discussed in further detail in the following paragraphs.

4.1 Vector Stability

The first step to consider in the journey of an efficient non-viral gene delivery vector is the carrier stability in the biological milieu. The existence of nucleases in the extra cellular environment imposes the necessity of genetic material protection at this step. The delivery systems have to display stability in physiological ionic strength and when in contact with existent endogenous anionic molecules (i.e., glycosaminoglycans (GAGs), serum albumin and other extra-cellular proteins) [190–193].

4.2 Cell Entrance and Intracellular Trafficking

The route followed by non-viral gene delivery vectors may differ depending whether the vector does or does not have targeting moieties. In the case of non-targeted vectors data suggests that the complexes first associate with cell membrane via electrostatic interactions with the anionic cell surface proteoglycans. It was shown that for PLL-DNA complexes, the inhibition of proteoglycan sulfation by the use of sodium chlorate, the removal of cell surface GAGs, or the addition of GAGs to the

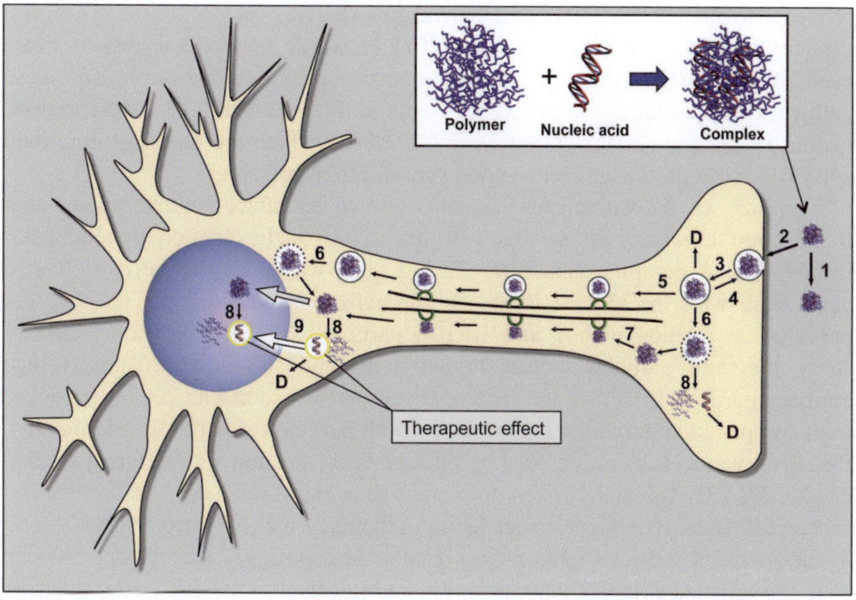

Fig. 2 Possible barriers encountered by a non-viral gene carrier. When bound to the cell surface a complex can be: (*1*) neutralized by polyanionic compounds or degraded when in the extracellular milieu or (*2*) endocytosed. Normally, the endocytosed material is trafficked in the early endosome compartment (*3*), which can be recycled to the cell surface and excreted (*4*), or the carrier can then undergo association with cell active transport (i.e., retrograde transport in the case of a neuronal cell) (*5*), or while the endosome maturates it can lead to endosomal escape (*6*) or to the fusion with lysosomes, leading to degradation (*D*). When free in the cytoplasm the complex can: associate with active transport mechanism and go through retrograde transport towards the cell body (*7*) or can disassemble (*8*) and the nucleic acid can be degraded (*D*) or enter the nucleus (in the case of DNA), free or associated with the carrier (*9*). The aimed therapeutic effect can, according to the nature of the nucleic acid, occur at the nucleus or at the perinuclear location

transfection media, dramatically decreases transfection levels [194]. A similar trend was observed for cationic liposome complexes, where lipoplexes where unable to transfect a cell line lacking proteoglycans [195]. In vivo studies also showed that cationic liposome mediated transfection was inhibited when mice where pre-treated with heparinases by IV administration [196].

Evidence exists indicating that cationic substances lead to adsorptive endocytosis and subsequent complex internalization [197, 198]. It has been reported the involvement of clathrin-coated pits in the internalization of the complexes of DNA with cationic lipids and polymers [199, 200]. Phagocytosis of cationic lipid or polymer/DNA complexes have also been described, even in cell lines that are not professional phagocytes [201–203]. The possibility that caveolae may be involved in the internalization process is supported by the increased internalization of folate (which resides in caveolae) targeted cationic lipid or polymer based complexes in tumor cells [204–206]. Another possible route of entry is macropinocytosis, proposed to mediate the uptake of non-viral gene delivery vectors due to the intrinsic capacity to internalize large structures [207]. However, macropinocytosis is a rare event in non-phagocytic cells. Further, one can postulate that several of these pathways can participate to different extents in different cell types. Nevertheless, further systematic studies are necessary to clarify which are the most important entry routes for particular vectors and/or for different cell types.

The size of the complexes has also shown to affect cellular uptake and transfection efficiency in several cell lines. Poly(lactic acid-co-glycolic acid) (50:50)—DNA complex uptake by Caco-2 cells was shown to be particle size dependent, with the highest levels of internalization observed for the 100 nm particles, in relation with 1 and 10 μm particles [208]. Moreover, in another study, the same particles showed higher transfection levels for 70 nm average diameter particles in COS-7 and HEK-293 cells as compared with 200 nm particles, with comparable internalization levels for both particle types [209]. Additionally, PEI/DNA nanogels were prepared by photo-Fenton reaction with different sizes—38, 75, 87, 121, 132 and 167 nm of average diameter—and were tested in several cancer cell lines. The highest transfection efficiency was obtained for the 75 and 87 nm particles [210]. In light of these results one can suggest that the optimal size of a non-targeted complex internalization and transfection-wise is in the range of 70–100 nm.

The incorporation of targeting moieties in the vector design is expected to modify the particle uptake, with the ligand-receptor interaction governing the internalization pathway and the subsequent intracellular route. A wide range of ligands has been found to improve the internalization and targeting of cationic nanoparticles. These include: asialoorosomucoid [211–213], mannose [214–216], lactose [217–219], transferrin (Tf) [220–222], and metabolites such as folate [223–226]. One of the major influences of ligand incorporation in a complex resides in the modulation of the cell uptake rate. Transferrin-PEI polyplexes were shown to enter K562 cells in a 1 h period, while PEI800 polyplexes required up to 4 h [227], with the targeted polyplexes promoting higher transfection efficiency (25-fold). In another study the grafting of CH with lactose (8%) has shown similar internalization levels in HepG2 cells as CH alone, however, the intracellular trafficking towards the nucleus was faster and promoted higher transfection efficiency [218]. The use of such ligands to access and/or target the nervous system will be discussed in further detail in the next section.

Despite the fact that receptor mediated gene delivery can augment selectivity and the rate of particle internalization, the intracellular fate seems to be more important to an efficient gene delivery. ATP-mediated proton accumulation provides the endosome and lysosome of cells with an acidic environment (pH 5.0–6.2) in contrast with the cytosol conditions (pH 7.4) [228]. Non-viral gene delivery systems that can use the acidic environment to escape the endosomal and lysosomal degradation pathway often mediate efficient transfection. Chloroquine is a well-defined lysosomotropic agent that raises the pH of the lysosomal environment, preventing enzymatic degradation [229]. Thus, one of the strategies used considers the incorporation of chloroquine in the non-viral gene delivery systems. In the same way, the incorporation of membrane destabilizing peptides, such as the synthetic N-terminal peptides of rhinovirus VP-1 or influenza virus HA-2, into cationic complexes can mediate endosomal release. Under acidic conditions, these peptides rearrange to form an amphipathic alpha-helical structure that can interact with the endosomal membrane and promote the endosomal escape [230]. Alternatively, several polymers that bear amine groups with low pKa have been shown to exhibit "proton sponge" potential, as discussed for PEI and CH. These compounds are able to buffer the acidification process inside the endosome vesicle, leading to swelling and finally lysis, allowing the release to the cytosol [231].

Following endosomal escape, the nucleic acid containing complexes must traverse the cytosol to either access the nucleus or encounter their final destination when the cargo is siRNA or antisense oligonucleotides. Diffusion of free DNA in the cytoplasm has been shown to be significantly lower than in a dilute solution [232, 233]. Indeed a plasmid DNA with over 2,000 base pairs in the cytosol has less than 1% of the diffusion coefficient than when in water. This difference may be explained by the presence of cytoskeletal elements in the cytoplasm that function as molecular sieves, preventing the free diffusion of large molecules [234]. In that sense, carriers that could compact genetic material in smaller particles could be beneficial. However, in the case of cationic lipid carriers the release of DNA seems to occur during endosomal escape. The injection of cationic lipid-based complexes in the cytoplasm has shown to be much less efficient in terms of transfection than injection of DNA alone [235]. Conversely, in cationic polymer based systems a partial association of DNA and the carrier seems to occur after endosomal release. A beneficial effect in transgene expression was observed when PEI and PLL based complexes were injected in the cytoplasm [236], indicating that an association of the complexes with cell transport processes could be occurring. However, this mechanism is still not fully understood. Enhanced diffusion due to the condensed nature of the complex and the enhanced protection of DNA could be partially involved. In the case of neurons, microtubule-based transport is fundamental for normal cellular function since it allows the rapid transport of molecules between the cell body and the axon. The main motor for retrograde transport, towards the cell nucleus, is associated with dynein [237], and it has been shown to be critical in the infectious process of neurotropic viruses like HSV and adenovirus [238]. As non-viral vectors are in a similar range of size of those of viruses it is expected that the association with active retrograde transport could be beneficial for the transfection process. However, until now, non-viral delivery systems that could specifically recruit the motor retrograde machinery have not yet been demonstrated.

4.3 *Nuclear Uptake*

In order to access the cell transcriptional machinery, a gene (double stranded DNA) must enter the nucleus. Except during mitosis, when the nuclear envelope disappears, the only way macromolecules can enter the nucleus is through the nuclear pore complex. Indeed, dividing cells often exhibit higher transfection levels than non-dividing cells, indicating that the DNA access to the nucleus is improved during the nuclear envelope disruption [239]. Given the post-mitotic nature of neurons, the nuclear import of plasmid DNA is *sine qua non* for the efficient gene expression.

The nuclear pore complex (NPC) is built of a diverse set of nucleoporins and associated nuclear and cytoplasmic filaments surrounding a central channel structure [240]. This structure allows passive diffusion of small molecules (up to 9 nm in diameter, or proteins up to 50 kDa) or active transport of larger molecules (up to 25 nm in diameter, or $\approx$1,000 kDa) [240]. The nuclear import mechanism of macromolecules like proteins, RNAs and ribonucleoproteins is known to be energy dependent and carrier-mediated [241]. Consequently, the import of an external gene or plasmid DNA is expected to be subject to a similar mechanism of entry [242].

As previously discussed, in the case of lipoplexes one believes that their cargo is released during the process of endosome destabilization. In the case of cationic polymer-based systems, while a number of intracellular trafficking studies have been conducted, the importance of complex disassembling on transfection is yet to be established [243]. Some authors suggested that DNA can be transcribed while complexed with PEI [244], whereas in other studies it was hypothesized that the release of DNA from CH is the limiting step for an efficient in vitro transfection [173]. In a study using an in vitro transcription/translation assay we were able to show that the production of a reporter protein is impaired when DNA is complexed with CH, clearly suggesting that disassembly is required for gene expression to occur. If this separation need to occur prior to the entrance in the nucleus or inside the nucleus is still a matter of discussion.

Targeting and transport of viral genomes to the nucleus have been shown to depend on nuclear localizing sequence (NLS) exposed in the surface of the capsid [245]. Karyophilic proteins bear one or more nuclear localization signal peptides designated NLS [246]. Depending on the imported protein, one or two molecules of the cytoplasmic transport factor karyopherin, associate with the NLS sequence to form a nuclear pore-targeting complex. The complex then translocates through the nuclear pore into the nucleus via direct interactions of the karyopherins with the NPC nucleoporins most likely through a series of dissociation/re-association steps [247]. Directionality of transport is assured by an energy dependent process that maintains high concentration levels of RanGTP in the nucleus, responsible for inducing the dissociation of the cargo-import complex [248]. Covalent attachment of NLS to plasmid DNA has shown to increase nuclear translocation and improve gene transfer [242, 249–252]. Nonetheless, conflicting studies have shown no improvement upon incorporation of NLS moieties in plasmid DNA [253, 254], indicating that additional variables can influence the nuclear import process. It has also been shown that an optimal degree of NLS exists, as too many NLS signals can lead to translocation inhibition due to the interaction with multiple nuclear pores [249]. Additionally, the grafting to the plasmid DNA can lead to the inhibition of the transgene by blocking the enzymatic access to the expression cassette. In order to circumvent this problem peptide nucleic acid sequences have been developed, which function as sequence-specific DNA binding domains, allowing the control of the number and the binding site of the NLS sequences [255].

5 How Can We Develop a Non-viral Gene Delivery Vector Targeted to the Nervous System?

As pointed out in the early sections of this chapter, the development of gene therapy and non-viral gene delivery protocols for the treatment of nervous system diseases/injury is warranted. Perhaps one of the most exciting, and daunting, prospects for nucleic acids delivery to the nervous system is the possibility to reach the CNS via systemic administration. This, however, does not extinguish the pressing question on how to specifically target the different neuronal cell populations. In the following paragraphs an overview of the strategies proposed so far to target such systems to neuronal cells will be presented. In Table 2 a summary of the functional entities and vector strategies for nervous system targeting is presented.

5.1 The "Closed" BBB Gateway

To fully realize the prospects of achieving a minimally invasive and safe therapeutical application of nucleic acids for the CNS, it is well recognized that among the routes of administration the systemic pathway would be the preferable one. This is, nonetheless, primarily hampered by the fact that access to the CNS is limited by an important obstacle, which is the blood-brain barrier (BBB) [283] or blood-spinal cord barrier (BSCB) [284]. The BBB is formed primarily by endothelial cells lining the cerebral vasculature and surrounding perivascular elements. The complex tight junction system that closely connects the endothelial cells is responsible for a high transendothelial electrical resistance, which severely restricts paracellular transport of water and solutes [285, 286].

In order to overcome this barrier some delivery strategies are trying to take advantage of receptor-mediated transcytosis (RMT). To achieve this, the surface of biomaterial-based vectors could be modified in such a way that they efficiently display a certain ligand or a number of ligands, that can be of different nature, such as antibodies, proteins and peptides or even DNA and RNA.

With regard to nucleic acid delivery, monoclonal antibodies (mAbs) against Tf and insulin have been used in the context of plasmid DNA delivery by liposome formulations, also designated immunoliposomes [256, 257]. Similarly, these mAbs have also been used for naked siRNA delivery through a biotin-avidin conjugation approach [287, 288] or together with other antisense molecules such as the nucleic acid analogue, peptide nucleic acid (PNA) [289]. In this context, there is one

Table 2 Blood-brain-barrier (BBB) and neuronal cell targeting ligands

Target	Functional moiety	Associated vectors	Reference
BBB	Monoclonal antibodies (α-Tf, α-insulin)	Liposomes	[256–260]
		PMLA	
		Chitosan	
	Transferrin	Liposomes	[221, 261, 262]
	Lactoferrin	PAMAM	
	Angiopep-2 (peptide)	PAMAM	[263]
	RVG29 (peptide)	PEI	[264–266]
		PAMAM	
		9R (arginines)	
	Pep-TGN (peptide)	PLGA	[267]
	Leptin30 (peptide)	PLL	[268]
	Glutathione	PLGA	[269]
Neuronal cells	Neurotensin (peptide)	PLL	[270, 271]
	NGF (loop regions 2 and 4)	PLL	[272, 273]
		10 K (lysines)	
		PEI	
	Monoclonal antibody (a-p^{75}NTR)	PLL	[274]
	Molossin (peptide)	PLL	[275]
	CTb (non-toxic fragment)	100 K (lysines)	[276]
	HC TeNT (non-toxic fragment)	PLL	[277–279]
		PEI	
		Chitosan	
	Tet1 (peptide)	PEI	[280, 281]
	DNA aptamer17	n.a.	[282]

PMLA poly(malic acid), *PAMAM* poly(amido amine) dendrimer, *PEI* poly(ethylene imine), *PLGA* poly(lactic-co-glycolic acid), *PLL* poly (L-Lysine), *R* arginine, *K* lysine
n.a. not applied yet

report concerning the joint use of mAbs and polymer-based vectors for nucleic acid delivery. Interestingly, the polymer chosen, poly(malic acid) (PMLA), was negatively charged. The nucleic acid used was a non-charged morpholino antisense oligonucleotide, which, due to non-available charge interactions in this concept, was coupled via disulfide linkages to the anionic polymer. The purpose was to direct the nanoconjugate, after IV injection, to brain tumor cells with the brain targeting function given by a rat anti-mouse Tf receptor (TfR) mAb [258]. Other positively charged polymers such as CH have also been used in conjugation with an OX26 mouse mAb to the rat Tf receptor, albeit for the delivery of a caspase inhibitor peptide to the brain [259, 260].

The Tf and lactoferrin (Lf) proteins, themselves, have been used as the targeting ligands for biomaterial-based vectors. Tf-lipoplexes (DOTAP: Cholesterol liposomes associated with Tf) have been used for the delivery of siRNA in vivo and in vitro [261]. In another example, Tf was conjugated to a PAMAM dendrimer via a bifunctional PEG, for plasmid DNA delivery to the brain after IV administration. This targeting strategy conferred a 2-fold increase in gene expression in the brain over the untargeted vector [221]. Transferrin was also compared to Lf as ligands for a PAMAM-based gene delivery vector. It was actually shown that after IV injection the use of Lf-modified PAMAM conferred a further 2.3-fold increase over the Tf-modified PAMAM. This increase efficiency could in part be attributed to a lower distribution of the Lf-particles in other organs such as liver, kidney, spleen and lung, in comparison to Tf-particles, which would render them more available in circulation to interact and be taken up by the brain capillary endothelial cells [262]. This example shows that a strategy using the naturally occurring ligands can suffer in part from competition effects. Moreover, the specific use of Tf has a potential disadvantage that is the abundant expression of the Tf receptor (TfR) in other tissues in the body, what lowers its specificity.

Small synthetic peptides are a good alternative to antibodies and more complex protein ligands precisely due to their simplicity and relative easy preparation (automated synthesis) and handling (they are normally quite stable). One such peptide, Angiopep-2 (19 amino acids long), has been conjugated to PAMAM, via bifunctional PEG linker, to originate a nanoparticle for the delivery of TRAIL-protein-encoding plasmid DNA, to glial tumors [263]. This peptide is derived from the Kunitz domains of aprotinin and has been shown to target the low-density lipoprotein receptor-related protein-1 (LRP-1), which is highly expressed in brain endothelial cells as well as brain tumor cells, neurons and astrocytes [260, 290]. It can however also be found in lung, ovary, uterus and liver.

Another example comes from a 29-amino acid peptide, derived from the rabies virus glycoprotein (RVG29) that has been found to bind to the acetylcholine receptor (AchR) [264]. This peptide was used in a chimeric form together with nona-arginine (RVG-9R) with the purpose to first bind siRNA through the electrostatic interactions of 9R while mediating its targeted delivery across the BBB and into the brain where gene silencing was achieved [264]. The RVG29 peptide has also been used together with other polymers such as PEI- and PAMAM-based nanoparticles for plasmid DNA delivery, although with moderate efficiency levels. In the case of PEI based nanoparticles, in vivo gene expression was shown to be enhanced 1.3-fold when comparing RVG modified to non-modified particles after IV administration [265]. For the PAMAM-based nanoparticles a first in vivo imaging analysis of whole body distribution showed a promising enhanced accumulation in the brain for those modified with the RVG29. The gene expression studies showed, however, a more modest 2-fold increase in the brain [266].

Recently, a novel 12-amino acid-peptide, Pep-TGN, selected through the use of phage display peptide library has been described. When conjugated to PEG-PLGA nanoparticles, it facilitated the targeted delivery across the BBB [267]. These nanoparticles were loaded with Coumarin-6 to show, as proof-of-principle, the accumulation of this marker in the brain. The peptide is yet to be used in the context of a nucleic acids delivery system.

In another study using dendrigraft PLL (DGL), a 30-amino acid peptide (leptin30) derived from an endogenic hormone-leptin was exploited as a targeting moiety for the delivery of DGL-PEG-leptin30/DNA nanoparticles to the brain. The peptide is thought to target the leptin-hormone receptor, which is expressed in the hypothalamus and other regions in the brain and is in parallel considered to participate in receptor mediated transport of leptin (a 16 kDa molecule) across the BBB. In vivo results were less preeminent when comparing to the obtained in vitro results, which has been attributed in part to some degree of competition with endogenous leptin. Still, around a 2-fold increase in brain expression was seen in comparison to non-targeted vector, after IV injections [268].

Also worth mentioning is the effect of surface modification of paclitaxel-loaded PLGA with glutathione, which was shown to enhance the drugs availability in the brain by improving BBB translocation [269].

5.2 Delivering It Right (into the Neuronal Cell)

As previously mentioned, the association of the vector with any cell membrane is the initial step of the vector internalization process. The majority of the non-viral gene delivery systems developed so far associates to cells via non-specific interactions. Conversely, targeted systems have primarily to interact with specific cell surface molecules. This strategy avoids widespread distribution of the vector and delivery of its payload to undesired cells, thus limiting possible toxicity effects.

Hence, receptor-mediated uptake into neuronal cell types has been explored, taking advantage of different classes of receptors expressed at their plasma membrane. This strategy holds true even if the vector has already overcome the BBB, since in the brain there might be specific types of neural cells that for therapeutic purposes need to be targeted.

The receptors that have been explored to achieve vector targeting can be divided in three major groups: neuropeptides, neurothrophins and neurotoxin receptors.

One of the first targeting moieties explored was neurotensin (NT), a 13 amino acid neuropeptide initially isolated from bovine hypothalami [291]. NT receptors have subsequently been identified in the brain of other species [292] and NT was found to undergo rapid internalization after receptor binding [293]. NT was conjugated to PLL in order to attain a vector that specifically transfected cells expressing the NT high affinity receptor (NTRH) [270]. The vector specificity was further proved by the absence of internalization observed in cell lines lacking NTRH (Cos-7 and L-929) [270]. Further work included the use of the hemagglutinin HA2 fusogenic peptide and the Vp1 nuclear localization signal of SV40 to the NT-polyplex. Both internalization in and transfection of dopamine neurons improved [271], enabling these vectors as promising for the treatment of Parkinson's disease. More recently this group has applied the same vector with a gene encoding for the human glial cell line-derived neurotrophic factor to hemiparkinsonism-induced rats, and showed biochemical, anatomical and functional recovery in this animal model of the disease [294].

Neurotrophin receptors have also been explored in order to design targeted gene delivery vectors to neurons, as their expression is specific for determined neuron cell populations. Henceforth, the NGF was used as a ligand capable of binding to the receptor protein tyrosine kinases, TrkA and the 75 kDa transmembrane glycoprotein, p^{75}NTR [295, 296]. A recombinant peptide was then derived from nerve growth factor loop regions 2 and 4 (regions shown to be important for TrkA binding) and utilized for creating, in combination with PLL, a targeted gene delivery vector [297]. In a subsequent study, a chimeric peptide composed of a targeting sequence derived from the NGF loop 4 and a DNA-binding moiety (ten lysine residues) was used to form complexes with plasmid DNA and shown to increase reporter gene expression in PC12 cells in a DNA dose dependent manner [272]. Moreover, the transfection levels were shown to be inhibited by an excess of NGF in the media but not by the use of a low-affinity NGF receptor p75 antibody, suggesting that the internalization was due to interaction with the high-affinity NGF receptor TrkA [272]. The electrostatic association of this peptide to low molecular weight PEI (600 Da) was able to increase the targeted transfection levels in cell lines, which express TrkA, as well as to the dorsal root ganglia after intrathecal injection in rats [273].

The targeting of up regulated receptors in injured neurons has also been demonstrated. It has been shown that upon injury the p75NTR neurotrophin receptor is up regulated in peripheral nerves [298]. By using a PLL based system coupled to a monoclonal antibody (MC192) to the p75NTR a previous study demonstrated transgene expression (glial derived neurotrophic factor) in targeted neurons leading to an almost complete neuronal rescue after axotomy [274].

Neurotoxins have also been exploited due to their natural ability to bypass the cellular barriers in order to access neurons. A 31 amino acids bi-functional peptide (PLL-molossin) has been developed for gene delivery and was tested in primary culture of rat cerebral cortex cells. With the best combination, more than 30% of cells strongly expressed the reporter gene, with no observed toxicity [275]. The peptide consisted of the 15 amino acid integrin target domain from the venom of the American pit viper *Crotalus molossus molossus* [299] RGDNP [300], which was found to bind strongly to the CNS of rats [299], and a 16-lysine chain for electrostatic binding of DNA [275].

The non-toxic pentameric cholera toxin b (CTb) chain of the *Vibrio cholerae bacterium* [301] has also been used due to its specific binding to the pentasaccharide moieties of the ganglioside GM1 [302]. The CTb fragment was grafted to a 100-lysine chain and the obtained complexes were shown to increase both transfection activity and efficiency of PC12 cells in relation to unmodified complexes [276].

Another non-viral gene delivery system targeted to the CNS was developed employing the non-toxic HC fragment of the tetanus toxin (TeNT). The TeNT is a large protein (150 kDa), which targets with specificity neuronal cells [303]. When applied intramuscularly the TeNT binds to pre-synaptic motor neuron terminals, and after internalization is trafficked to the cell body by retrograde axonal transport [303]. The toxin then undergoes trans-synaptic spread to other cells in the CNS, where it inhibits pre-synaptic neurotransmitter release by the cleavage of synaptobrevin [304]. TeNT contains three structural and functional domains. The L-chain is responsible for intracellular toxicity [304], while the HC and HN chains contain cell binding and putative endosome disruption activities, respectively [305]. HC has been chemically or genetically conjugated to several enzymes such as superoxide dismutase [306], horseradish peroxidase [307], and β-galactosidase [308]. These conjugates have been applied in a variety of experiments and showed to target neurons, both in vivo and in vitro. The intramuscular administration of these conjugates resulted in their delivery to the brain stem, motor neurons of the spinal cord and to a lesser extent the dorsal root ganglia [277, 306]. In a previous study using a vector based on the construct of PLL:HC, it was found to target neuronal cells, N18 RE 105 (neuroblastoma x glioma mouse/rat hybrid) and F98 (glioma) cell lines, and mediated transfection efficiency values ten times higher than without HC [277]. In contrast the transfection of epithelial cell lines (CaCo-2 cells, HeLa cells) showed only a 2-fold increase [277]. TeNT has also been applied to other polymers such as PEI and chitosan for the formation of plasmid DNA containing nanoparticles. These nanoparticles were shown to efficiently discriminate between a fibroblast cell line (NIH 3 T3) and a neuronal cell line (ND7/23) achieving increased transgene expression efficiency in the later ones. This demonstrated the improved specificity of the construct. Ultimately, they were also shown to successfully target and transfect dorsal root ganglia (DRG) primary neuron cultures [278, 279].

More recently, a 12 amino acid peptide (Tet1), which mimics the HC fragment receptor binding properties, has been identified by phage display [309]. Tet1 was grafted to 25 kDa branched PEI and showed to improve transfection in PC12 cells, in relation to PEI alone, and also to improve association with dissociated dorsal root ganglia primary culture [280]. In posterior work, the Tet1 peptide was grafted to 25 kDa branched PEI via a 5 kDa PEG and the obtained complexes were shown to improve gene expression in rat brain. Moreover, the Tet1-PEG-PEI based complexes were shown to exclusively transfect neural progenitor cells, whereas the PEG-PEI based complexes were found to transfect a heterogeneous cell population [281].

It is also worth mentioning a completely different category of possible ligands for neural cells, which is the case of nucleic acid-based aptamers. In this case, a specific DNA or RNA sequence gives rise to a determined tri-dimensional structure capable, much like peptide-ligands, to interact with specific receptors [310–312]. Although many of the aptamers have been developed so far against tumor cells [313, 314], it has already been described an aptamer, designated aptamer17, shown to be capable of distinguishing differentiated PC12 rat adrenal pheochromocytoma cell lines from undifferentiated ones [282]. Although this has not translated into any targeted delivery of specific cargoes, it bodes well for neural cell targeting applications of aptamers.

6 Future Prospects

New therapeutic approaches that may arrest or reverse the disease process underlying many types of neuropathies are awaited. In the search of such a solution, focus has been drawn to gene therapy. Due to unsolved hurdles presented by viral vectors crescent interest is currently being drawn to non-viral vectors. The main challenge ahead consists on the development of an effective non-viral nucleic acid delivery system aiming at this particular therapeutic clinical application. In face of the large number of barriers that such vector has to face in vivo, new designs are required. Furthermore, as the nervous system presents a number of specific characteristics that hinder the action of such a vehicle, it is conceivable that the non-viral vectors to be developed will have to include a number of features enabling it to sequentially address the different barriers posed. Being the crossing of the BBB and the targeting of the neuronal cell population of interest, among the more specific and challenging hurdles. Thus, although simplicity should be kept in mind when designing a delivery vector, it is likely that a multifunctional type of vector will be necessary. The challenge lies in the development of smart strategies that coordinate the biomaterial-based vector backbone with an efficient incorporation and display of functional units so that they exert their functions in a timely and successful manner. As one cannot neglect the efficiency and specificity attained by some virus, product of millions of years of evolution, to mimic some of their mechanisms of action can serve as the inspiration in finding the solution to the development of a not only safe but also efficient delivery system. Advances in the field of neurosciences and on the knowledge of the mechanisms of disease will also be key for the design of such systems.

Acknowledgements The authors would like to acknowledge the FEDER funds through the *Programa Operacional Factores de Competitividade*—COMPETE and the Portuguese funds through FCT—*Fundação para a Ciência e a Tecnologia* (contracts PTDC/CTM-NAN/115124/2009, HMSP-ICT/0020/2010 and PEst-C/SAU/LA0002/2011) that supported this work. PM is supported by a FCT post-doctoral fellowship (SFRH/BPD/76936/2011).

References

1. Lunn JS, Sakowski SA, Hur J, Feldman EL (2011) Stem cell technology for neurodegenerative diseases. Ann Neurol 70:353–361
2. Shy ME (2007) Peripheral neuropathies. In: Goldman L, Ausiello D (eds) Cecil medicine. Saunders Elsevier, Philadelphia
3. Pirart J (1978) Why don't we teach and treat diabetic patients better? Diabetes Care 1:139–140
4. von Giesen HJ, Koller H, Hefter H, Arendt G (2002) Central and peripheral nervous system functions are independently disturbed in HIV-1 infected patients. J Neurol 249:754–758
5. Apfel SC (1999) Neurotrophic factors in peripheral neuropathies: therapeutic implications. Brain Pathol 9:393–413
6. Hughes RA (2002) Peripheral neuropathy. BMJ 324:466–469
7. Calcutt NA, Jolivalt CG, Fernyhough P (2008) Growth factors as therapeutics for diabetic neuropathy. Curr Drug Targets 9:47–59
8. Schulte-Herbruggen O, Braun A, Rochlitzer S, Jockers-Scherubl MC, Hellweg R (2007) Neurotrophic factors – a tool for therapeutic strategies in neurological, neuropsychiatric and neuroimmunological diseases? Curr Med Chem 14:2318–2329
9. Zuccato C, Cattaneo E (2009) Brain-derived neurotrophic factor in neurodegenerative diseases. Nat Rev Neurol 5:311–322
10. Lykissas MG, Batistatou AK, Charalabopoulos KA, Beris AE (2007) The role of neurotrophins in axonal growth, guidance, and regeneration. Curr Neurovasc Res 4:143–151

11. Wurdinger T, Tannous BA (2009) Glioma angiogenesis: towards novel RNA therapeutics. Cell Adh Migr 3:230–235
12. Otsuka S, Adamson C, Sankar V, Gibbs KM, Kane-Goldsmith N, Ayer J, Babiarz J, Kalinski H, Ashush H, Alpert E, Lahav R, Feinstein E, Grumet M (2011) Delayed intrathecal delivery of RhoA siRNA to the contused spinal cord inhibits allodynia, preserves white matter, and increases serotonergic fiber growth. J Neurotrauma 28:1063–1076
13. Gene therapy clinical trials worldwide (February 2012) J Gen Med, Wiley. http://www.wiley.com/legacy/wileychi/genmed/clinical/. Accessed 7 Jan 2013
14. Ledley FD (1996) Pharmaceutical approach to somatic gene therapy. Pharm Res 13:1595–1614
15. Arzumanov A, Walsh AP, Rajwanshi VK, Kumar R, Wengel J, Gait MJ (2001) Inhibition of HIV-1 Tat-dependent trans activation by steric block chimeric 2′-O-methyl/LNA oligoribonucleotides. Biochemistry 40:14645–14654
16. Wu H, Lima WF, Zhang H, Fan A, Sun H, Crooke ST (2004) Determination of the role of the human RNase H1 in the pharmacology of DNA-like antisense drugs. J Biol Chem 279:17181–17189
17. Serganov A, Patel DJ (2007) Ribozymes, riboswitches and beyond: regulation of gene expression without proteins. Nat Rev Genet 8:776–790
18. Achenbach JC, Chiuman W, Cruz RP, Li Y (2004) DNAzymes: from creation in vitro to application in vivo. Curr Pharm Biotechnol 5:321–336
19. Fire A, Xu S, Montgomery MK, Kostas SA, Driver SE, Mello CC (1998) Potent and specific genetic interference by double-stranded RNA in Caenorhabditis elegans. Nature 391:806–811
20. Reynolds A, Leake D, Boese Q, Scaringe S, Marshall WS, Khvorova A (2004) Rational siRNA design for RNA interference. Nat Biotechnol 22:326–330
21. Carbone GM, McGuffie EM, Collier A, Catapano CV (2003) Selective inhibition of transcription of the Ets2 gene in prostate cancer cells by a triplex-forming oligonucleotide. Nucleic Acids Res 31:833–843
22. Hewett PW, Daft EL, Laughton CA, Ahmad S, Ahmed A, Murray JC (2006) Selective inhibition of the human tie-1 promoter with triplex-forming oligonucleotides targeted to Ets binding sites. Mol Med 12:8–16
23. Dominski Z, Kole R (1993) Restoration of correct splicing in thalassemic pre-mRNA by antisense oligonucleotides. Proc Natl Acad Sci USA 90:8673–8677
24. Bauman J, Jearawiriyapaisarn N, Kole R (2009) Therapeutic potential of splice-switching oligonucleotides. Oligonucleotides 19:1–13
25. Mahato RI, Smith LC, Rolland A (1999) Pharmaceutical perspectives of nonviral gene therapy. Adv Genet 41:95–156
26. Mahato RI, Takakura Y, Hashida M (1997) Nonviral vectors for in vivo gene delivery: physicochemical and pharmacokinetic considerations. Crit Rev Ther Drug Carrier Syst 14:133–172
27. Liu F, Huang L (2002) Development of non-viral vectors for systemic gene delivery. J Control Release 78:259–266
28. Palu G, Bonaguro R, Marcello A (1999) In pursuit of new developments for gene therapy of human diseases. J Biotechnol 68:1–13
29. Naldini L, Blomer U, Gage FH, Trono D, Verma IM (1996) Efficient transfer, integration, and sustained long-term expression of the transgene in adult rat brains injected with a lentiviral vector. Proc Natl Acad Sci USA 93:11382–11388
30. Blomer U, Naldini L, Kafri T, Trono D, Verma IM, Gage FH (1997) Highly efficient and sustained gene transfer in adult neurons with a lentivirus vector. J Virol 71:6641–6649
31. Dropulic B (2001) Lentivirus in the clinic. Mol Ther 4:511–512
32. Brooks AI, Stein CS, Hughes SM, Heth J, McCray PM Jr, Sauter SL, Johnston JC, Cory-Slechta DA, Federoff HJ, Davidson BL (2002) Functional correction of established central nervous system deficits in an animal model of lysosomal storage disease with feline immunodeficiency virus-based vectors. Proc Natl Acad Sci USA 99:6216–6221

33. Biffi A, De Palma M, Quattrini A, Del Carro U, Amadio S, Visigalli I, Sessa M, Fasano S, Brambilla R, Marchesini S, Bordignon C, Naldini L (2004) Correction of metachromatic leukodystrophy in the mouse model by transplantation of genetically modified hematopoietic stem cells. J Clin Invest 113:1118–1129
34. Abordo-Adesida E, Follenzi A, Barcia C, Sciascia S, Castro MG, Naldini L, Lowenstein PR (2005) Stability of lentiviral vector-mediated transgene expression in the brain in the presence of systemic antivector immune responses. Hum Gene Ther 16:741–751
35. Lundberg C, Bjorklund T, Carlsson T, Jakobsson J, Hantraye P, Déglon N, Kirik D (2008) Applications of lentiviral vectors for biology and gene therapy of neurological disorders. Curr Gene Ther 8:461–473
36. Tannemaat MR, Eggers R, Hendriks WT, de Ruiter GC, van Heerikhuize JJ, Pool CW, Malessy MJ, Boer GJ, Verhaagen J (2008) Differential effects of lentiviral vector-mediated overexpression of nerve growth factor and glial cell line-derived neurotrophic factor on regenerating sensory and motor axons in the transected peripheral nerve. Eur J Neurosci 28:1467–1479
37. Lattanzi A, Neri M, Maderna C, di Girolamo I, Martino S, Orlacchio A, Amendola M, Naldini L, Gritti A (2010) Widespread enzymatic correction of CNS tissues by a single intracerebral injection of therapeutic lentiviral vector in leukodystrophy mouse models. Hum Mol Genet 19:2208–2227
38. Tosi J, Sancho-Pelluz J, Davis RJ, Hsu CW, Wolpert KV, Sengillo JD, Lin CS, Tsang SH (2011) Lentivirus-mediated expression of cDNA and shRNA slows degeneration in retinitis pigmentosa. Exp Biol Med (Maywood) 236:1211–1217
39. Akli S, Caillaud C, Vigne E, Stratford-Perricaudet LD, Poenaru L, Perricaudet M, Kahn A, Peschanski MR (1993) Transfer of a foreign gene into the brain using adenovirus vectors. Nat Genet 3:224–228
40. Davidson BL, Allen ED, Kozarsky KF, Wilson JM, Roessler BJ (1993) A model system for in vivo gene transfer into the central nervous system using an adenoviral vector. Nat Genet 3:219–223
41. Boulis NM, Turner DE, Dice JA, Bhatia V, Feldman EL (1999) Characterization of adenoviral gene expression in spinal cord after remote vector delivery. Neurosurgery 45:131–137, discussion 137–138
42. Miwa H, Shibata M, Okado H, Hirano S (2001) Tracing axons in the peripheral nerve using lacZ gene recombinant adenovirus and its application to regeneration of the peripheral nerve. J Neuropathol Exp Neurol 60:671–675
43. Boulis NM, Turner DE, Imperiale MJ, Feldman EL (2002) Neuronal survival following remote adenovirus gene delivery. J Neurosurg 96:212–219
44. Millecamps S, Mallet J, Barkats M (2002) Adenoviral retrograde gene transfer in motoneurons is greatly enhanced by prior intramuscular inoculation with botulinum toxin. Hum Gene Ther 13:225–232
45. Sweigard JH, Cashman SM, Kumar-Singh R (2010) Adenovirus vectors targeting distinct cell types in the retina. Invest Ophthalmol Vis Sci 51:2219–2228
46. Kusano K, Enomoto M, Hirai T, Wakabayashi Y, Itoh S, Ichinose S, Okabe S, Shinomiya K, Okawa A (2011) Enhancement of sciatic nerve regeneration by adenovirus-mediated expression of dominant negative RhoA and Rac1. Neurosci Lett 492:64–69
47. Bartlett JS, Samulski RJ, McCown TJ (1998) Selective and rapid uptake of adeno-associated virus type 2 in brain. Hum Gene Ther 9:1181–1186
48. McCown TJ, Xiao X, Li J, Breese GR, Samulski RJ (1996) Differential and persistent expression patterns of CNS gene transfer by an adeno-associated virus (AAV) vector. Brain Res 713:99–107
49. Davidson BL, Stein CS, Heth JA, Martins I, Kotin RM, Derksen TA, Zabner J, Ghodsi A, Chiorini JA (2000) Recombinant adeno-associated virus type 2, 4, and 5 vectors: transduction of variant cell types and regions in the mammalian central nervous system. Proc Natl Acad Sci USA 97:3428–3432

50. Kaspar BK, Erickson D, Schaffer D, Hinh L, Gage FH, Peterson DA (2002) Targeted retrograde gene delivery for neuronal protection. Mol Ther 5:50–56
51. Boulis NM, Willmarth NE, Song DK, Feldman EL, Imperiale MJ (2003) Intraneural colchicine inhibition of adenoviral and adeno-associated viral vector remote spinal cord gene delivery. Neurosurgery 52:381–387, discussion 387
52. Lu YY, Wang LJ, Muramatsu S, Ikeguchi K, Fujimoto K, Okada T, Mizukami H, Matsushita T, Hanazono Y, Kume A, Nagatsu T, Ozawa K, Nakano I (2003) Intramuscular injection of AAV-GDNF results in sustained expression of transgenic GDNF, and its delivery to spinal motoneurons by retrograde transport. Neurosci Res 45:33–40
53. Ehlert EM, Eggers R, Niclou SP, Verhaagen J (2010) Cellular toxicity following application of adeno-associated viral vector-mediated RNA interference in the nervous system. BMC Neurosci 11:20
54. Snyder BR, Gray SJ, Quach ET, Huang JW, Leung CH, Samulski RJ, Boulis NM, Federici T (2011) Comparison of adeno-associated viral vector serotypes for spinal cord and motor neuron gene delivery. Hum Gene Ther 22:1129–1135
55. Geller AI, Freese A (1990) Infection of cultured central nervous system neurons with a defective herpes simplex virus 1 vector results in stable expression of Escherichia coli beta-galactosidase. Proc Natl Acad Sci USA 87:1149–1153
56. Fink DJ, Sternberg LR, Weber PC, Mata M, Goins WF, Glorioso JC (1992) In vivo expression of beta-galactosidase in hippocampal neurons by HSV-mediated gene transfer. Hum Gene Ther 3:11–19
57. Mata M, Zhang M, Hu X, Fink DJ (2001) HveC (nectin-1) is expressed at high levels in sensory neurons, but not in motor neurons, of the rat peripheral nervous system. J Neurovirol 7:476–480
58. Latchman DS (2001) Gene delivery and gene therapy with herpes simplex virus-based vectors. Gene 264:1–9
59. Lilley CE, Groutsi F, Han Z, Palmer JA, Anderson PN, Latchman DS, Coffin RS (2001) Multiple immediate-early gene-deficient herpes simplex virus vectors allowing efficient gene delivery to neurons in culture and widespread gene delivery to the central nervous system in vivo. J Virol 75:4343–4356
60. Perez MCP, Hunt SP, Coffin RS, Palmer JA (2004) Comparative analysis of genomic HSV vectors for gene delivery to motor neurons following peripheral inoculation in vivo. Gene Ther 11:1023–1032
61. Anesti AM, Peeters PJ, Royaux I, Coffin RS (2008) Efficient delivery of RNA interference to peripheral neurons in vivo using herpes simplex virus. Nucleic Acids Res 36:e86
62. Fink DJ, Wechuck J, Mata M, Glorioso JC, Goss J, Krisky D, Wolfe D (2011) Gene therapy for pain: results of a phase I clinical trial. Ann Neurol 70:207–212
63. Glorioso JC, Fink DJ (2009) Herpes vector-mediated gene transfer in the treatment of chronic pain. Mol Ther 17:13–18
64. Lentz TB, Gray SJ, Samulski RJ (2012) Viral vectors for gene delivery to the central nervous system. Neurobiol Dis 48:179–188
65. Verma IM, Somia N (1997) Gene therapy—promises, problems and prospects. Nature 389:239–242
66. Somia N, Verma IM (2000) Gene therapy: trials and tribulations. Nat Rev Genet 1:91–99
67. Dobbelstein M (2003) Viruses in therapy – royal road or dead end? Virus Res 92:219–221
68. Nayak S, Herzog RW (2010) Progress and prospects: immune responses to viral vectors. Gene Ther 17:295–304
69. Meyers G, Rumenapf T, Tautz N, Dubovi EJ, Thiel HJ (1991) Insertion of cellular sequences in the genome of bovine viral diarrhea virus. Arch Virol Suppl 3:133–142
70. Verma IM, Weitzman MD (2005) Gene therapy: twenty-first century medicine. Annu Rev Biochem 74:711–738
71. Dietz GP, Bähr M (2004) Delivery of bioactive molecules into the cell: the Trojan horse approach. Mol Cell Neurosci 27:85–131

72. Zhang S, Xu Y, Wang B, Qiao W, Liu D, Li Z (2004) Cationic compounds used in lipoplexes and polyplexes for gene delivery. J Control Release 100:165–180
73. Mintzer MA, Simanek EE (2009) Nonviral vectors for gene delivery. Chem Rev 109:259–302
74. Behr JP (1994) Gene transfer with synthetic cationic amphiphiles: prospects for gene therapy. Bioconjug Chem 5:382–389
75. Felgner PL, Gadek TR, Holm M, Roman R, Chan HW, Wenz M, Northrop JP, Ringold GM, Danielsen M (1987) Lipofection: a highly efficient, lipid-mediated DNA-transfection procedure. Proc Natl Acad Sci USA 84:7413–7417
76. Felgner JH, Kumar R, Sridhar CN, Wheeler CJ, Tsai YJ, Border R, Ramsey P, Martin M, Felgner PL (1994) Enhanced gene delivery and mechanism studies with a novel series of cationic lipid formulations. J Biol Chem 269:2550–2561
77. Remy JS, Sirlin C, Vierling P, Behr JP (1994) Gene transfer with a series of lipophilic DNA-binding molecules. Bioconjug Chem 5:647–654
78. Gao X, Huang L (1995) Cationic liposome-mediated gene transfer. Gene Ther 2:710–722
79. Balasubramaniam RP, Bennett MJ, Aberle AM, Malone JG, Nantz MH, Malone RW (1996) Structural and functional analysis of cationic transfection lipids: the hydrophobic domain. Gene Ther 3:163–172
80. Budker V, Gurevich V, Hagstrom JE, Bortzov F, Wolff JA (1996) pH-sensitive, cationic liposomes: a new synthetic virus-like vector. Nat Biotechnol 14:760–764
81. Stephan DJ, Yang ZY, San H, Simari RD, Wheeler CJ, Felgner PL, Gordon D, Nabel GJ, Nabel EG (1996) A new cationic liposome DNA complex enhances the efficiency of arterial gene transfer in vivo. Hum Gene Ther 7:1803–1812
82. Lee RJ, Huang L (1997) Lipidic vector systems for gene transfer. Crit Rev Ther Drug Carrier Syst 14:173–206
83. Rosenzweig HS, Rakhmanova VA, McIntosh TJ, MacDonald RC (2000) O-Alkyl dioleoylphosphatidylcholinium compounds: the effect of varying alkyl chain length on their physical properties and in vitro DNA transfection activity. Bioconjug Chem 11:306–313
84. Serikawa T, Suzuki N, Kikuchi H, Tanaka K, Kitagawa T (2000) A new cationic liposome for efficient gene delivery with serum into cultured human cells: a quantitative analysis using two independent fluorescent probes. Biochim Biophys Acta 1467:419–430
85. Rosenzweig HS, Rakhmanova VA, MacDonald RC (2001) Diquaternary ammonium compounds as transfection agents. Bioconjug Chem 12:258–263
86. Mendonca LS, Firmino F, Moreira JN, Pedroso de Lima MC, Simoes S (2010) Transferrin receptor-targeted liposomes encapsulating anti-BCR-ABL siRNA or asODN for chronic myeloid leukemia treatment. Bioconjug Chem 21:157–168
87. Chen CW, Lu DW, Yeh MK, Shiau CY, Chiang CH (2011) Novel RGD-lipid conjugate-modified liposomes for enhancing siRNA delivery in human retinal pigment epithelial cells. Int J Nanomedicine 6:2567–2580
88. Brown MD, Schatzlein AG, Uchegbu IF (2001) Gene delivery with synthetic (non viral) carriers. Int J Pharm 229:1–21
89. Felgner PL, Ringold GM (1989) Cationic liposome-mediated transfection. Nature 337:387–388
90. Bennett CF, Chiang MY, Chan H, Shoemaker JE, Mirabelli CK (1992) Cationic lipids enhance cellular uptake and activity of phosphorothioate antisense oligonucleotides. Mol Pharmacol 41:1023–1033
91. Legendre JY, Szoka FC Jr (1992) Delivery of plasmid DNA into mammalian cell lines using pH-sensitive liposomes: comparison with cationic liposomes. Pharm Res 9:1235–1242
92. Zhou X, Huang L (1994) DNA transfection mediated by cationic liposomes containing lipopolylysine: characterization and mechanism of action. Biochim Biophys Acta 1189:195–203
93. Ross PC, Hui SW (1999) Lipoplex size is a major determinant of in vitro lipofection efficiency. Gene Ther 6:651–659

94. Pedroso de Lima MC, Simoes S, Pires P, Faneca H, Duzgunes N (2001) Cationic lipid-DNA complexes in gene delivery: from biophysics to biological applications. Adv Drug Deliv Rev 47:277–294
95. Almofti MR, Harashima H, Shinohara Y, Almofti A, Baba Y, Kiwada H (2003) Cationic liposome-mediated gene delivery: biophysical study and mechanism of internalization. Arch Biochem Biophys 410:246–253
96. Tavitian B, Marzabal S, Boutet V, Kuhnast B, Terrazzino S, Moynier M, Dollé F, Deverre JR, Thierry AR (2002) Characterization of a synthetic anionic vector for oligonucleotide delivery using in vivo whole body dynamic imaging. Pharm Res 19:367–376
97. Thierry AR, Lunardi-Iskandar Y, Bryant JL, Rabinovich P, Gallo RC, Mahan LC (1995) Systemic gene therapy: biodistribution and long-term expression of a transgene in mice. Proc Natl Acad Sci USA 92:9742–9746
98. Li S, Rizzo MA, Bhattacharya S, Huang L (1998) Characterization of cationic lipid-protamine-DNA (LPD) complexes for intravenous gene delivery. Gene Ther 5:930–937
99. Dass CR, Choong PF (2006) Selective gene delivery for cancer therapy using cationic liposomes: in vivo proof of applicability. J Control Release 113:155–163
100. De Smedt SC, Demeester J, Hennink WE (2000) Cationic polymer based gene delivery systems. Pharm Res 17:113–126
101. El-Aneed A (2004) An overview of current delivery systems in cancer gene therapy. J Control Release 94:1–14
102. Garnett MC (1999) Gene-delivery systems using cationic polymers. Crit Rev Ther Drug Carrier Syst 16:147–207
103. Wu GY, Wu CH (1987) Receptor-mediated in vitro gene transformation by a soluble DNA carrier system. J Biol Chem 262:4429–4432
104. Wadhwa MS, Collard WT, Adami RC, McKenzie DL, Rice KG (1997) Peptide-mediated gene delivery: influence of peptide structure on gene expression. Bioconjug Chem 8:81–88
105. Pouton CW, Lucas P, Thomas BJ, Uduehi AN, Milroy DA, Moss SH (1998) Polycation-DNA complexes for gene delivery: a comparison of the biopharmaceutical properties of cationic polypeptides and cationic lipids. J Control Release 53:289–299
106. Akinc A, Langer R (2002) Measuring the pH environment of DNA delivered using nonviral vectors: implications for lysosomal trafficking. Biotechnol Bioeng 78:503–508
107. Brown MD, Schatzlein A, Brownlie A, Jack V, Wang W, Tetley L, Gray AI, Uchegbu IF (2000) Preliminary characterization of novel amino acid based polymeric vesicles as gene and drug delivery agents. Bioconjug Chem 11:880–891
108. Pichon C, Goncalves C, Midoux P (2001) Histidine-rich peptides and polymers for nucleic acids delivery. Adv Drug Deliv Rev 53:75–94
109. Midoux P, Monsigny M (1999) Efficient gene transfer by histidylated polylysine/pDNA complexes. Bioconjug Chem 10:406–411
110. Wang CY, Huang L (1984) Polyhistidine mediates an acid-dependent fusion of negatively charged liposomes. Biochemistry 23:4409–4416
111. Kano A, Moriyama K, Yamano T, Nakamura I, Shimada N, Shimada N, Maruyama A (2011) Grafting of poly(ethylene glycol) to poly-lysine augments its lifetime in blood circulation and accumulation in tumors without loss of the ability to associate with siRNA. J Control Release 149:2–7
112. Dash PR, Read ML, Barrett LB, Wolfert MA, Seymour LW (1999) Factors affecting blood clearance and in vivo distribution of polyelectrolyte complexes for gene delivery. Gene Ther 6:643–650
113. Ward CM, Read ML, Seymour LW (2001) Systemic circulation of poly(L-lysine)/DNA vectors is influenced by polycation molecular weight and type of DNA: differential circulation in mice and rats and the implications for human gene therapy. Blood 97:2221–2229
114. Choi YH, Liu F, Park JS, Kim SW (1998) Lactose-poly(ethylene glycol)-grafted poly-L-lysine as hepatoma cell-targeted gene carrier. Bioconjug Chem 9:708–718
115. Haensler J, Szoka FC Jr (1993) Polyamidoamine cascade polymers mediate efficient transfection of cells in culture. Bioconjug Chem 4:372–379

116. Tang MX, Redemann CT, Szoka FC Jr (1996) In vitro gene delivery by degraded polyamidoamine dendrimers. Bioconjug Chem 7:703–714
117. Zinselmeyer BH, Mackay SP, Schatzlein AG, Uchegbu IF (2002) The lower-generation polypropylenimine dendrimers are effective gene-transfer agents. Pharm Res 19:960–967
118. Shah DS, Sakthivel T, Toth I, Florence AT, Wilderspin AF (2000) DNA transfection and transfected cell viability using amphipathic asymmetric dendrimers. Int J Pharm 208: 41–48
119. Dufes C, Uchegbu IF, Schatzlein AG (2005) Dendrimers in gene delivery. Adv Drug Deliv Rev 57:2177–2202
120. Santos JL, Oramas E, Pego AP, Granja PL, Tomas H (2009) Osteogenic differentiation of mesenchymal stem cells using PAMAM dendrimers as gene delivery vectors. J Control Release 134:141–148
121. Kukowska-Latallo JF, Bielinska AU, Johnson J, Spindler R, Tomalia DA, Baker JR Jr (1996) Efficient transfer of genetic material into mammalian cells using Starburst polyamidoamine dendrimers. Proc Natl Acad Sci USA 93:4897–4902
122. Lee RJ, Wang S, Low PS (1996) Measurement of endosome pH following folate receptor-mediated endocytosis. Biochim Biophys Acta 1312:237–242
123. Malik N, Wiwattanapatapee R, Klopsch R, Lorenz K, Frey H, Weener JW, Meijer EW, Paulus W, Duncan R (2000) Dendrimers: relationship between structure and biocompatibility in vitro, and preliminary studies on the biodistribution of 125I-labelled polyamidoamine dendrimers in vivo. J Control Release 65:133–148
124. Roberts JC, Bhalgat MK, Zera RT (1996) Preliminary biological evaluation of polyamidoamine (PAMAM) Starburst dendrimers. J Biomed Mater Res 30:53–65
125. Lee JH, Lim YB, Choi JS, Lee Y, Kim TI, Kim HJ, Yoon JK, Kim K, Park JS (2003) Polyplexes assembled with internally quaternized PAMAM-OH dendrimer and plasmid DNA have a neutral surface and gene delivery potency. Bioconjug Chem 14:1214–1221
126. Kim TI, Seo HJ, Choi JS, Jang HS, Baek JU, Kim K, Park JS (2004) PAMAM-PEG-PAMAM: novel triblock copolymer as a biocompatible and efficient gene delivery carrier. Biomacromolecules 5:2487–2492
127. Gingras M, Raimundo JM, Chabre YM (2007) Cleavable dendrimers. Angew Chem Int Ed Engl 46:1010–1070
128. Medina SH, El-Sayed ME (2009) Dendrimers as carriers for delivery of chemotherapeutic agents. Chem Rev 109:3141–3157
129. Ma XP, Tang JB, Shen YQ, Fan MH, Tang HD, Radosz M (2009) Facile synthesis of polyester dendrimers from sequential click coupling of asymmetrical monomers. J Am Chem Soc 131:14795–14803
130. Boussif O, Lezoualc'h F, Zanta MA, Mergny MD, Scherman D, Demeneix B, Behr JP (1995) A versatile vector for gene and oligonucleotide transfer into cells in culture and in vivo: polyethylenimine. Proc Natl Acad Sci USA 92:7297–7301
131. Dheur S, Saison-Behmoaras TE (2000) Polyethyleneimine-mediated transfection to improve antisense activity of 3′-capped phosphodiester oligonucleotides. Methods Enzymol 313:56–73
132. Han S, Mahato RI, Sung YK, Kim SW (2000) Development of biomaterials for gene therapy. Mol Ther 2:302–317
133. Kunath K, von Harpe A, Fischer D, Petersen H, Bickel U, Voigt K, Kissel T (2003) Low-molecular-weight polyethylenimine as a non-viral vector for DNA delivery: comparison of physicochemical properties, transfection efficiency and in vivo distribution with high-molecular-weight polyethylenimine. J Control Release 89:113–125
134. Kircheis R, Schuller S, Brunner S, Ogris M, Heider KH, Zauner W, Wagner E (1999) Polycation-based DNA complexes for tumor-targeted gene delivery in vivo. J Gene Med 1:111–120
135. Godbey WT, Wu KK, Mikos AG (1999) Size matters: molecular weight affects the efficiency of poly(ethylenimine) as a gene delivery vehicle. J Biomed Mater Res 45:268–275

136. Shin JY, Suh D, Kim JM, Choi HG, Kim JA, Ko JJ, Lee YB, Kim JS, Oh YK (2005) Low molecular weight polyethylenimine for efficient transfection of human hematopoietic and umbilical cord blood-derived CD34+ cells. Biochim Biophys Acta 1725:377–384
137. Thomas M, Ge Q, Lu JJ, Chen J, Klibanov AM (2005) Cross-linked small polyethylenimines: while still nontoxic, deliver DNA efficiently to mammalian cells in vitro and in vivo. Pharm Res 22:373–380
138. Lampela P, Raisanen J, Mannisto PT, Yla-Herttuala S, Raasmaja A (2002) The use of low-molecular-weight PEIs as gene carriers in the monkey fibroblastoma and rabbit smooth muscle cell cultures. J Gene Med 4:205–214
139. Dunlap DD, Maggi A, Soria MR, Monaco L (1997) Nanoscopic structure of DNA condensed for gene delivery. Nucleic Acids Res 25:3095–3101
140. Reschel T, Konak C, Oupicky D, Seymour LW, Ulbrich K (2002) Physical properties and in vitro transfection efficiency of gene delivery vectors based on complexes of DNA with synthetic polycations. J Control Release 81:201–217
141. Grayson AC, Doody AM, Putnam D (2006) Biophysical and structural characterization of polyethylenimine-mediated siRNA delivery in vitro. Pharm Res 23:1868–1876
142. Sundaram S, Lee LK, Roth CM (2007) Interplay of polyethyleneimine molecular weight and oligonucleotide backbone chemistry in the dynamics of antisense activity. Nucleic Acids Res 35:4396–4408
143. Merdan T, Kunath K, Petersen H, Bakowsky U, Voigt KH, Kopecek J, Kissel T (2005) PEGylation of poly(ethylene imine) affects stability of complexes with plasmid DNA under in vivo conditions in a dose-dependent manner after intravenous injection into mice. Bioconjug Chem 16:785–792
144. Walker GF, Fella C, Pelisek J, Fahrmeir J, Boeckle S, Ogris M, Wagner E (2005) Toward synthetic viruses: endosomal pH-triggered deshielding of targeted polyplexes greatly enhances gene transfer in vitro and in vivo. Mol Ther 11:418–425
145. Wang DA, Narang AS, Kotb M, Gaber AO, Miller DD, Kim SW, Mahato RI (2002) Novel branched poly(ethylenimine)-cholesterol water-soluble lipopolymers for gene delivery. Biomacromolecules 3:1197–1207
146. Pun SH, Bellocq NC, Liu A, Jensen G, Machemer T, Quijano E, Schluep T, Wen S, Engler H, Heidel J, Davis ME (2004) Cyclodextrin-modified polyethylenimine polymers for gene delivery. Bioconjug Chem 15:831–840
147. Mi Bae Y, Choi H, Lee S, Ho Kang S, Tae Kim Y, Nam K, Sang Park J, Lee M, Sig Choi J (2007) Dexamethasone-conjugated low molecular weight polyethylenimine as a nucleus-targeting lipopolymer gene carrier. Bioconjug Chem 18:2029–2036
148. Creusat G, Zuber G (2008) Tyrosine-modified PEI: a novel and highly efficient vector for siRNA delivery in mammalian cells. Nucleic Acids Symp Ser (Oxf) 52:91–92
149. Zaghloul EM, Viola JR, Zuber G, Smith CI, Lundin KE (2010) Formulation and delivery of splice-correction antisense oligonucleotides by amino acid modified polyethylenimine. Mol Pharm 7:652–663
150. Gosselin MA, Guo W, Lee RJ (2001) Efficient gene transfer using reversibly cross-linked low molecular weight polyethylenimine. Bioconjug Chem 12:989–994
151. Lee Y, Mo H, Koo H, Park JY, Cho MY, Jin GW, Park JS (2007) Visualization of the degradation of a disulfide polymer, linear poly(ethylenimine sulfide), for gene delivery. Bioconjug Chem 18:13–18
152. Abdallah B, Hassan A, Benoist C, Goula D, Behr JP, Demeneix BA (1996) A powerful nonviral vector for in vivo gene transfer into the adult mammalian brain: polyethylenimine. Hum Gene Ther 7:1947–1954
153. Kiang T, Wen J, Lim HW, Leong KW (2004) The effect of the degree of chitosan deacetylation on the efficiency of gene transfection. Biomaterials 25:5293–5301
154. Barbosa MA, Pêgo AP, Amaral IF (2011) Chitosan. In: Ducheyne P, Hutmacher DW, Kirkpatrick J, Healy K (eds) Comprehensive biomaterials. Elsevier, London
155. Mumper RJ, Wang J, Clapell JM, Rolland AP (1995) Novel polymeric condensing carriers for gene delivery. Proc Intern Symp Control Release Bioact Mater 22:178–179

156. Venkatesh S, Smith TJ (1998) Chitosan-membrane interactions and their probable role in chitosan-mediated transfection. Biotechnol Appl Biochem 27(Pt 3):265–267
157. Mao HQ, Roy K, Troung-Le VL, Janes KA, Lin KY, Wang Y, August JT, Leong KW (2001) Chitosan-DNA nanoparticles as gene carriers: synthesis, characterization and transfection efficiency. J Control Release 70:399–421
158. Mooren FC, Berthold A, Domschke W, Kreuter J (1998) Influence of chitosan microspheres on the transport of prednisolone sodium phosphate across HT-29 cell monolayers. Pharm Res 15:58–65
159. Hamidi M, Azadi A, Rafiei P (2008) Hydrogel nanoparticles in drug delivery. Adv Drug Deliv Rev 60:1638–1649
160. Janes KA, Calvo P, Alonso MJ (2001) Polysaccharide colloidal particles as delivery systems for macromolecules. Adv Drug Deliv Rev 47:83–97
161. Borchard G (2001) Chitosans for gene delivery. Adv Drug Deliv Rev 52:145–150
162. Liu WG, De Yao K (2002) Chitosan and its derivatives – a promising non-viral vector for gene transfection. J Control Release 83:1–11
163. Mansouri S, Lavigne P, Corsi K, Benderdour M, Beaumont E, Fernandes JC (2004) Chitosan-DNA nanoparticles as non-viral vectors in gene therapy: strategies to improve transfection efficacy. Eur J Pharm Biopharm 57:1–8
164. Koping-Hoggard M, Tubulekas I, Guan H, Edwards K, Nilsson M, Vårum KM, Artursson P (2001) Chitosan as a nonviral gene delivery system. Structure-property relationships and characteristics compared with polyethylenimine in vitro and after lung administration in vivo. Gene Ther 8:1108–1121
165. Corsi K, Chellat F, Yahia L, Fernandes JC (2003) Mesenchymal stem cells, MG63 and HEK293 transfection using chitosan-DNA nanoparticles. Biomaterials 24:1255–1264
166. Sato T, Ishii T, Okahata Y (2001) In vitro gene delivery mediated by chitosan. Effect of pH, serum, and molecular mass of chitosan on the transfection efficiency. Biomaterials 22:2075–2080
167. Erbacher P, Zou SM, Bettinger T, Steffan AM, Remy JS (1998) Chitosan-based vector/DNA complexes for gene delivery: biophysical characteristics and transfection ability. Pharm Res 15:1332–1339
168. Wong K, Sun G, Zhang X, Dai H, Liu Y, He C, Leong KW (2006) PEI-g-chitosan, a novel gene delivery system with transfection efficiency comparable to polyethylenimine in vitro and after liver administration in vivo. Bioconjug Chem 17:152–158
169. Pires LR, Oliveira H, Barrias CC, Sampaio P, Pereira AJ, Maiato H, Simões S, Pêgo AP (2011) Imidazole-grafted chitosan-mediated gene delivery: in vitro study on transfection, intracellular trafficking and degradation. Nanomedicine (Lond) 6:1499–1512
170. Thanou M, Florea BI, Geldof M, Junginger HE, Borchard G (2002) Quaternized chitosan oligomers as novel gene delivery vectors in epithelial cell lines. Biomaterials 23:153–159
171. Katas H, Alpar HO (2006) Development and characterisation of chitosan nanoparticles for siRNA delivery. J Control Release 115:216–225
172. Liu X, Howard KA, Dong M, Andersen MO, Rahbek UL, Johnsen MG, Hansen OC, Besenbacher F, Kjems J (2007) The influence of polymeric properties on chitosan/siRNA nanoparticle formulation and gene silencing. Biomaterials 28:1280–1288
173. Koping-Hoggard M, Varum KM, Issa M, Danielsen S, Christensen BE, Stokke BT, Artursson P (2004) Improved chitosan-mediated gene delivery based on easily dissociated chitosan polyplexes of highly defined chitosan oligomers. Gene Ther 11:1441–1452
174. Mittnacht U, Hartmann H, Hein S, Oliveira H, Dong M, Pêgo AP, Kjems J, Howard KA, Schlosshauer B (2010) Chitosan/siRNA nanoparticles biofunctionalize nerve implants and enable neurite outgrowth. Nano Lett 10:3933–3939
175. Gao S, Chen J, Dong L, Ding Z, Yang YH, Zhang J (2005) Targeting delivery of oligonucleotide and plasmid DNA to hepatocyte via galactosylated chitosan vector. Eur J Pharm Biopharm 60:327–334
176. Kim ST, Kim CK (2007) Water-soluble chitosan-based antisense oligodeoxynucleotide of interleukin-5 for treatment of allergic rhinitis. Biomaterials 28:3360–3368

177. Springate CM, Jackson JK, Gleave ME, Burt HM (2008) Clusterin antisense complexed with chitosan for controlled intratumoral delivery. Int J Pharm 350:53–64
178. Ozbas-Turan S, Akbuga J, Sezer AD (2010) Topical application of antisense oligonucleotide-loaded chitosan nanoparticles to rats. Oligonucleotides 20:147–153
179. Dong L, Huang Z, Cai X, Xiang J, Zhu YA, Wang R, Chen J, Zhang J (2011) Localized delivery of antisense oligonucleotides by cationic hydrogel suppresses TNF-alpha expression and endotoxin-induced osteolysis. Pharm Res 28:1349–1356
180. Ghosh P, Han G, De M, Kim CK, Rotello VM (2008) Gold nanoparticles in delivery applications. Adv Drug Deliv Rev 60:1307–1315
181. Slowing II, Vivero-Escoto JL, Wu CW, Lin VS (2008) Mesoporous silica nanoparticles as controlled release drug delivery and gene transfection carriers. Adv Drug Deliv Rev 60:1278–1288
182. Liu Z, Tabakman S, Welsher K, Dai H (2009) Carbon nanotubes in biology and medicine: in vitro and in vivo detection, imaging and drug delivery. Nano Res 2:85–120
183. Al-Jamal KT, Gherardini L, Bardi G, Nunes A, Guo C, Bussy C, Herrero MA, Bianco A, Prato M, Kostarelos K, Pizzorusso T (2011) Functional motor recovery from brain ischemic insult by carbon nanotube-mediated siRNA silencing. Proc Natl Acad Sci USA 108:10952–10957
184. Putnam D, Langer R (1999) Poly(4-hydroxy-l-proline ester): low-temperature polycondensation and plasmid DNA complexation. Macromolecules 32:3658–3662
185. Lim YB, Han SO, Kong HU, Lee Y, Park JS, Jeong B, Kim SW (2000) Biodegradable polyester, poly[alpha-(4-aminobutyl)-L-glycolic acid], as a non-toxic gene carrier. Pharm Res 17:811–816
186. Lynn DM, Anderson DG, Putnam D, Langer R (2001) Accelerated discovery of synthetic transfection vectors: parallel synthesis and screening of a degradable polymer library. J Am Chem Soc 123:8155–8156
187. Tzeng SY, Guerrero-Cazares H, Martinez EE, Sunshine JC, Quinones-Hinojosa A, Green JJ (2011) Non-viral gene delivery nanoparticles based on poly(beta-amino esters) for treatment of glioblastoma. Biomaterials 32:5402–5410
188. Washbourne P, McAllister AK (2002) Techniques for gene transfer into neurons. Curr Opin Neurobiol 12:566–573
189. Lo EH, Singhal AB, Torchilin VP, Abbott NJ (2001) Drug delivery to damaged brain. Brain Res Brain Res Rev 38:140–148
190. Wiethoff CM, Smith JG, Koe GS, Middaugh CR (2001) The potential role of proteoglycans in cationic lipid-mediated gene delivery. Studies of the interaction of cationic lipid-DNA complexes with model glycosaminoglycans. J Biol Chem 276:32806–32813
191. Li S, Tseng WC, Stolz DB, Wu SP, Watkins SC, Huang L (1999) Dynamic changes in the characteristics of cationic lipidic vectors after exposure to mouse serum: implications for intravenous lipofection. Gene Ther 6:585–594
192. Oupicky D, Konak C, Dash PR, Seymour LW, Ulbrich K (1999) Effect of albumin and polyanion on the structure of DNA complexes with polycation containing hydrophilic nonionic block. Bioconjug Chem 10:764–772
193. Ruponen M, Yla-Herttuala S, Urtti A (1999) Interactions of polymeric and liposomal gene delivery systems with extracellular glycosaminoglycans: physicochemical and transfection studies. Biochim Biophys Acta 1415:331–341
194. Mislick KA, Baldeschwieler JD (1996) Evidence for the role of proteoglycans in cation-mediated gene transfer. Proc Natl Acad Sci USA 93:12349–12354
195. Mounkes LC, Zhong W, Cipres-Palacin G, Heath TD, Debs RJ (1998) Proteoglycans mediate cationic liposome-DNA complex-based gene delivery in vitro and in vivo. J Biol Chem 273:26164–26170
196. Ruponen M, Honkakoski P, Tammi M, Urtti A (2004) Cell-surface glycosaminoglycans inhibit cation-mediated gene transfer. J Gene Med 6:405–414
197. Pratten MK, Cable HC, Ringsdorf H, Lloyd JB (1982) Adsorptive pinocytosis of polycationic copolymers of vinylpyrrolidone with vinylamine by rat yolk sac and rat peritoneal macrophage. Biochim Biophys Acta 719:424–430

198. Ghinea N, Hasu M (1986) Charge effect on binding, uptake and transport of ferritin through fenestrated endothelium. J Submicrosc Cytol 18:647–659
199. Zuhorn IS, Kalicharan R, Hoekstra D (2002) Lipoplex-mediated transfection of mammalian cells occurs through the cholesterol-dependent clathrin-mediated pathway of endocytosis. J Biol Chem 277:18021–18028
200. Friend DS, Papahadjopoulos D, Debs RJ (1996) Endocytosis and intracellular processing accompanying transfection mediated by cationic liposomes. Biochim Biophys Acta 1278:41–50
201. Matsui H, Johnson LG, Randell SH, Boucher RC (1997) Loss of binding and entry of liposome-DNA complexes decreases transfection efficiency in differentiated airway epithelial cells. J Biol Chem 272:1117–11126
202. Labat-Moleur F, Steffan AM, Brisson C, Perron H, Feugeas O, Furstenberger P, Oberling F, Brambilla E, Behr JP (1996) An electron microscopy study into the mechanism of gene transfer with lipopolyamines. Gene Ther 3:1010–1017
203. Harbottle RP, Cooper RG, Hart SL, Ladhoff A, McKay T, Knight AM, Wagner E, Miller AD, Coutelle C (1998) An RGD-oligolysine peptide: a prototype construct for integrin-mediated gene delivery. Hum Gene Ther 9:1037–1047
204. Gottschalk S, Cristiano RJ, Smith LC, Woo SL (1994) Folate receptor mediated DNA delivery into tumor cells: potosomal disruption results in enhanced gene expression. Gene Ther 1:185–191
205. Hofland HE, Masson C, Iginla S, Osetinsky I, Reddy JA, Leamon CP, Scherman D, Bessodes M, Wils P (2002) Folate-targeted gene transfer in vivo. Mol Ther 5:739–744
206. Turk MJ, Reddy JA, Chmielewski JA, Low PS (2002) Characterization of a novel pH-sensitive peptide that enhances drug release from folate-targeted liposomes at endosomal pHs. Biochim Biophys Acta 1559:56–68
207. Francis CL, Ryan TA, Jones BD, Smith SJ, Falkow S (1993) Ruffles induced by Salmonella and other stimuli direct macropinocytosis of bacteria. Nature 364:639–642
208. Desai MP, Labhasetwar V, Walter E, Levy RJ, Amidon GL (1997) The mechanism of uptake of biodegradable microparticles in Caco-2 cells is size dependent. Pharm Res 14:1568–1573
209. Prabha S, Zhou WZ, Panyam J, Labhasetwar V (2002) Size-dependency of nanoparticle-mediated gene transfection: studies with fractionated nanoparticles. Int J Pharm 244:105–115
210. Xu DM, Yao SD, Liu YB, Sheng KL, Hong J, Gong PJ, Dong L (2007) Size-dependent properties of M-PEIs nanogels for gene delivery in cancer cells. Int J Pharm 338:291–296
211. Cho CW, Cho YS, Lee HK, Yeom YI, Park SN, Yoon DY (2000) Improvement of receptor-mediated gene delivery to HepG2 cells using an amphiphilic gelling agent. Biotechnol Appl Biochem 32(Pt 1):21–26
212. Stankovics J, Crane AM, Andrews E, Wu CH, Wu GY, Ledley FD (1994) Overexpression of human methylmalonyl CoA mutase in mice after in vivo gene transfer with asialoglycoprotein/polylysine/DNA complexes. Hum Gene Ther 5:1095–1104
213. Wu GY, Wu CH (1988) Evidence for targeted gene delivery to Hep G2 hepatoma cells in vitro. Biochemistry 27:887–892
214. Gour N, Purohit CS, Verma S, Puri R, Ganesh S (2009) Mannosylated self-assembled structures for molecular confinement and gene delivery applications. Biochem Biophys Res Commun 378:503–506
215. Park IY, Kim IY, Yoo MK, Choi YJ, Cho MH, Cho CS (2008) Mannosylated polyethylenimine coupled mesoporous silica nanoparticles for receptor-mediated gene delivery. Int J Pharm 359:280–287
216. Kim TH, Jiang HL, Nah JW, Cho MH, Akaike T, Cho CS (2007) Receptor-mediated gene delivery using chemically modified chitosan. Biomed Mater 2:S95–S100
217. Satoh T, Kakimoto S, Kano H, Nakatani M, Shinkai S, Nagasaki T (2007) In vitro gene delivery to HepG2 cells using galactosylated 6-amino-6-deoxychitosan as a DNA carrier. Carbohydr Res 342:1427–1433
218. Hashimoto M, Morimoto M, Saimoto H, Shigemasa Y, Sato T (2006) Lactosylated chitosan for DNA delivery into hepatocytes: the effect of lactosylation on the physicochemical properties and intracellular trafficking of pDNA/chitosan complexes. Bioconjug Chem 17:309–316

219. Cook SE, Park IK, Kim EM, Jeong HJ, Park TG, Choi YJ, Akaike T, Cho CS (2005) Galactosylated polyethylenimine-graft-poly(vinyl pyrrolidone) as a hepatocyte-targeting gene carrier. J Control Release 105:151–163
220. Penacho N, Filipe A, Simoes S, Pedroso de Lima MC (2008) Transferrin-associated lipoplexes as gene delivery systems: relevance of mode of preparation and biophysical properties. J Membr Biol 221:141–152
221. Huang RQ, Qu YH, Ke WL, Zhu JH, Pei YY, Jiang C (2007) Efficient gene delivery targeted to the brain using a transferrin-conjugated polyethyleneglycol-modified polyamidoamine dendrimer. FASEB J 21:1117–1125
222. Lee KM, Kim IS, Lee YB, Shin SC, Lee KC, Oh IJ (2005) Evaluation of transferrin-polyethylenimine conjugate for targeted gene delivery. Arch Pharm Res 28:722–729
223. Nie Y, Zhang Z, Li L, Luo K, Ding H, Gu Z (2009) Synthesis, characterization and transfection of a novel folate-targeted multipolymeric nanoparticles for gene delivery. J Mater Sci Mater Med 20:1849–1857
224. Cheng H, Zhu JL, Zeng X, Jing Y, Zhang XZ, Zhuo RX (2009) Targeted gene delivery mediated by folate-polyethylenimine-block-poly(ethylene glycol) with receptor selectivity. Bioconjug Chem 20:481–487
225. Liang B, He ML, Xiao ZP, Li Y, Chan CY, Kung HF, Shuai XT, Peng Y (2008) Synthesis and characterization of folate-PEG-grafted-hyperbranched-PEI for tumor-targeted gene delivery. Biochem Biophys Res Commun 367:874–880
226. Lee D, Lockey R, Mohapatra S (2006) Folate receptor-mediated cancer cell specific gene delivery using folic acid-conjugated oligochitosans. J Nanosci Nanotechnol 6:2860–2866
227. Ogris M, Steinlein P, Carotta S, Brunner S, Wagner E (2001) DNA/polyethylenimine transfection particles: influence of ligands, polymer size, and PEGylation on internalization and gene expression. AAPS PharmSci 3:E21
228. Ohkuma S, Poole B (1978) Fluorescence probe measurement of the intralysosomal pH in living cells and the perturbation of pH by various agents. Proc Natl Acad Sci USA 75:3327–3331
229. De Duve C, de Barsy T, Poole B, Trouet A, Tulkens P, Van Hoof F (1974) Commentary. Lysosomotropic agents. Biochem Pharmacol 23:2495–2531
230. Wagner E (1999) Application of membrane-active peptides for nonviral gene delivery. Adv Drug Deliv Rev 38:279–289
231. Behr JP (1997) The proton sponge: a trick to enter cells the viruses did not exploit. Chimia 51:34–36
232. Lukacs GL, Haggie P, Seksek O, Lechardeur D, Freedman N, Verkman AS (2000) Size-dependent DNA mobility in cytoplasm and nucleus. J Biol Chem 275:1625–1629
233. Dauty E, Verkman AS (2005) Actin cytoskeleton as the principal determinant of size-dependent DNA mobility in cytoplasm: a new barrier for non-viral gene delivery. J Biol Chem 280:7823–7828
234. Luby-Phelps K, Castle PE, Taylor DL, Lanni F (1987) Hindered diffusion of inert tracer particles in the cytoplasm of mouse 3 T3 cells. Proc Natl Acad Sci USA 84:4910–4913
235. Zabner J, Fasbender AJ, Moninger T, Poellinger KA, Welsh MJ (1995) Cellular and molecular barriers to gene transfer by a cationic lipid. J Biol Chem 270:18997–19007
236. Pollard H, Remy JS, Loussouarn G, Demolombe S, Behr JP, Escande D (1998) Polyethylenimine but not cationic lipids promotes transgene delivery to the nucleus in mammalian cells. J Biol Chem 273:7507–7511
237. Vallee RB, Williams JC, Varma D, Barnhart LE (2004) Dynein: an ancient motor protein involved in multiple modes of transport. J Neurobiol 58:189–200
238. Mabit H, Nakano MY, Prank U, Saam B, Dohner K, Sodeik B, Greber UF (2002) Intact microtubules support adenovirus and herpes simplex virus infections. J Virol 76:9962–9971
239. Wilke M, Fortunati E, van den Broek M, Hoogeveen AT, Scholte BJ (1996) Efficacy of a peptide-based gene delivery system depends on mitotic activity. Gene Ther 3:1133–1142
240. Ryan KJ, Wente SR (2000) The nuclear pore complex: a protein machine bridging the nucleus and cytoplasm. Curr Opin Cell Biol 12:361–371

241. Featherstone C, Darby MK, Gerace L (1988) A monoclonal antibody against the nuclear pore complex inhibits nucleocytoplasmic transport of protein and RNA in vivo. J Cell Biol 107:1289–1297
242. Ludtke JJ, Zhang G, Sebestyen MG, Wolff JA (1999) A nuclear localization signal can enhance both the nuclear transport and expression of 1 kb DNA. J Cell Sci 112(Pt 12):2033–2041
243. Won YY, Sharma R, Konieczny SF (2009) Missing pieces in understanding the intracellular trafficking of polycation/DNA complexes. J Control Release 139:88–93
244. Bieber T, Meissner W, Kostin S, Niemann A, Elsasser HP (2002) Intracellular route and transcriptional competence of polyethylenimine-DNA complexes. J Control Release 82:441–454
245. Whittaker GR, Helenius A (1998) Nuclear import and export of viruses and virus genomes. Virology 246:1–23
246. Imamoto N (2000) Diversity in nucleocytoplasmic transport pathways. Cell Struct Funct 25:207–216
247. Conti E, Izaurralde E (2001) Nucleocytoplasmic transport enters the atomic age. Curr Opin Cell Biol 13:310–319
248. Cook A, Bono F, Jinek M, Conti E (2007) Structural biology of nucleocytoplasmic transport. Annu Rev Biochem 76:647–671
249. Zanta MA, Belguise-Valladier P, Behr JP (1999) Gene delivery: a single nuclear localization signal peptide is sufficient to carry DNA to the cell nucleus. Proc Natl Acad Sci USA 96:91–96
250. Chan CK, Jans DA (1999) Enhancement of polylysine-mediated transferrinfection by nuclear localization sequences: polylysine does not function as a nuclear localization sequence. Hum Gene Ther 10:1695–1702
251. Subramanian A, Ranganathan P, Diamond SL (1999) Nuclear targeting peptide scaffolds for lipofection of nondividing mammalian cells. Nat Biotechnol 17:873–877
252. Collas P, Husebye H, Alestrom P (1996) The nuclear localization sequence of the SV40 T antigen promotes transgene uptake and expression in zebrafish embryo nuclei. Transgenic Res 5:451–458
253. Ciolina C, Byk G, Blanche F, Thuillier V, Scherman D, Wils P (1999) Coupling of nuclear localization signals to plasmid DNA and specific interaction of the conjugates with importin alpha. Bioconjug Chem 10:49–55
254. van der Aa MA, Koning GA, d'Oliveira C, Oosting RS, Wilschut KJ, Hennink WE, Crommelin DJ (2005) An NLS peptide covalently linked to linear DNA does not enhance transfection efficiency of cationic polymer based gene delivery systems. J Gene Med 7:208–217
255. Branden LJ, Mohamed AJ, Smith CI (1999) A peptide nucleic acid-nuclear localization signal fusion that mediates nuclear transport of DNA. Nat Biotechnol 17:784–787
256. Shi N, Boado RJ, Pardridge WM (2001) Receptor-mediated gene targeting to tissues in vivo following intravenous administration of pegylated immunoliposomes. Pharm Res 18:1091–1095
257. Zhang Y, Jeong Lee H, Boado RJ, Pardridge WM (2002) Receptor-mediated delivery of an antisense gene to human brain cancer cells. J Gene Med 4:183–194
258. Ding H, Inoue S, Ljubimov AV, Patil R, Portilla-Arias J, Hu J, Konda B, Wawrowsky KA, Fujita M, Karabalin N, Sasaki T, Black KL, Holler E, Ljubimova JY (2010) Inhibition of brain tumor growth by intravenous poly (beta-L-malic acid) nanobioconjugate with pH-dependent drug release [corrected]. Proc Natl Acad Sci USA 107:18143–18148
259. Aktas Y, Yemisci M, Andrieux K, Gursoy RN, Alonso MJ, Fernandez-Megia E, Novoa-Carballal R, Quiñoá E, Riguera R, Sargon MF, Celik HH, Demir AS, Hincal AA, Dalkara T, Capan Y, Couvreur P (2005) Development and brain delivery of chitosan-PEG nanoparticles functionalized with the monoclonal antibody OX26. Bioconjug Chem 16: 1503–1511

260. Demeule M, Currie JC, Bertrand Y, Che C, Nguyen T, Régina A, Gabathuler R, Castaigne JP, Béliveau R (2008) Involvement of the low-density lipoprotein receptor-related protein in the transcytosis of the brain delivery vector angiopep-2. J Neurochem 106:1534–1544
261. Cardoso AL, Simoes S, de Almeida LP, Plesnila N, Pedroso de Lima MC, Wagner E, Culmsee C (2008) Tf-lipoplexes for neuronal siRNA delivery: a promising system to mediate gene silencing in the CNS. J Control Release 132:113–123
262. Huang R, Ke W, Liu Y, Jiang C, Pei Y (2008) The use of lactoferrin as a ligand for targeting the polyamidoamine-based gene delivery system to the brain. Biomaterials 29:238–246
263. Huang S, Li J, Han L, Liu S, Ma H, Huang R, Jiang C (2011) Dual targeting effect of Angiopep-2-modified, DNA-loaded nanoparticles for glioma. Biomaterials 32:6832–6838
264. Kumar P, Wu H, McBride JL, Jung KE, Kim MH, Davidson BL, Lee SK, Shankar P, Manjunath N (2007) Transvascular delivery of small interfering RNA to the central nervous system. Nature 448:39–43
265. Son S, Hwang Do W, Singha K, Jeong JH, Park TG, Lee DS, Kim WJ (2011) RVG peptide tethered bioreducible polyethylenimine for gene delivery to brain. J Control Release 155:18–25
266. Liu Y, Huang R, Han L, Ke W, Shao K, Ye L, Lou J, Jiang C (2009) Brain-targeting gene delivery and cellular internalization mechanisms for modified rabies virus glycoprotein RVG29 nanoparticles. Biomaterials 30:4195–4202
267. Li J, Feng L, Fan L, Zha Y, Guo L, Zhang Q, Chen J, Pang Z, Wang Y, Jiang X, Yang VC, Wen L (2011) Targeting the brain with PEG-PLGA nanoparticles modified with phage-displayed peptides. Biomaterials 32:4943–4950
268. Liu Y, Li J, Shao K, Huang R, Ye L, Lou J, Jiang C (2010) A leptin derived 30-amino-acid peptide modified pegylated poly-L-lysine dendrigraft for brain targeted gene delivery. Biomaterials 31:5246–5257
269. Geldenhuys W, Mbimba T, Bui T, Harrison K, Sutariya V (2011) Brain-targeted delivery of paclitaxel using glutathione-coated nanoparticles for brain cancers. J Drug Target 19:837–845
270. Martinez-Fong D, Navarro-Quiroga I, Ochoa I, Alvarez-Maya I, Meraz MA, Luna J, Arias-Montaño JA (1999) Neurotensin-SPDP-poly-L-lysine conjugate: a nonviral vector for targeted gene delivery to neural cells. Brain Res Mol Brain Res 69:249–262
271. Navarro-Quiroga I, Gonzalez-Barrios JA, Barron-Moreno F, Gonzalez-Bernal V, Martinez-Arguelles DB, Martinez-Fong D (2002) Improved neurotensin-vector-mediated gene transfer by the coupling of hemagglutinin HA2 fusogenic peptide and Vp1 SV40 nuclear localization signal. Brain Res Mol Brain Res 105:86–97
272. Zeng J, Too HP, Ma Y, Luo ES, Wang S (2004) A synthetic peptide containing loop 4 of nerve growth factor for targeted gene delivery. J Gene Med 6:1247–1256
273. Zeng J, Wang X, Wang S (2007) Self-assembled ternary complexes of plasmid DNA, low molecular weight polyethylenimine and targeting peptide for nonviral gene delivery into neurons. Biomaterials 28:1443–1451
274. Barati S, Hurtado PR, Zhang SH, Tinsley R, Ferguson IA, Rush RA (2006) GDNF gene delivery via the p75(NTR) receptor rescues injured motor neurons. Exp Neurol 202:179–188
275. Collins L, Asuni AA, Anderton BH, Fabre JW (2003) Efficient gene delivery to primary neuron cultures using a synthetic peptide vector system. J Neurosci Methods 125:113–120
276. Barrett LB, Berry M, Ying WB, Hodgkin MN, Seymour LW, Gonzalez AM, Read ML, Baird A, Logan A (2004) CTb targeted non-viral cDNA delivery enhances transgene expression in neurons. J Gene Med 6:429–438
277. Knight A, Carvajal J, Schneider H, Coutelle C, Chamberlain S, Fairweather N (1999) Non-viral neuronal gene delivery mediated by the HC fragment of tetanus toxin. Eur J Biochem 259:762–769
278. Oliveira H, Fernandez R, Pires LR, Martins MCL, Simoes S, Barbosa MA, Pêgo AP (2010) Targeted gene delivery into peripheral sensorial neurons mediated by self-assembled vectors composed of poly(ethylene imine) and tetanus toxin fragment c. J Control Release 143:350–358

279. Oliveira H, Pires LR, Fernandez R, Martins MCL, Simoes S, Pêgo AP (2010) Chitosan-based gene delivery vectors targeted to the peripheral nervous system. J Biomed Mater Res A 95:801–810
280. Park IK, Lasiene J, Chou SH, Horner PJ, Pun SH (2007) Neuron-specific delivery of nucleic acids mediated by Tet1-modified poly(ethylenimine). J Gene Med 9:691–702
281. Kwon EJ, Lasiene J, Jacobson BE, Park IK, Horner PJ, Pun SH (2010) Targeted nonviral delivery vehicles to neural progenitor cells in the mouse subventricular zone. Biomaterials 31:2417–2424
282. Liu JJ, Wang CL, Xi Q, Xu J, Deng B, Ding HM, Chu B, Su DH (2011) Dissection of the functional structure of aptamer17, which specifically recognizes differentiated PC12 cells. Nucleic Acid Ther 21:225–229
283. Wohlfart S, Gelperina S, Kreuter J (2012) Transport of drugs across the blood-brain barrier by nanoparticles. J Control Release 161:264–273
284. Bartanusz V, Jezova D, Alajajian B, Digicaylioglu M (2011) The blood-spinal cord barrier: morphology and clinical implications. Ann Neurol 70:194–206
285. Wolburg H, Lippoldt A (2002) Tight junctions of the blood-brain barrier: development, composition and regulation. Vascul Pharmacol 38:323–337
286. Weksler BB, Subileau EA, Perriere N, Charneau P, Holloway K, Leveque M, Tricoire-Leignel H, Nicotra A, Bourdoulous S, Turowski P, Male DK, Roux F, Greenwood J, Romero IA, Couraud PO (2005) Blood-brain barrier-specific properties of a human adult brain endothelial cell line. FASEB J 19:1872–1874
287. Xia CF, Zhang Y, Boado RJ, Pardridge WM (2007) Intravenous siRNA of brain cancer with receptor targeting and avidin-biotin technology. Pharm Res 24:2309–2316
288. Xia CF, Boado RJ, Pardridge WM (2009) Antibody-mediated targeting of siRNA via the human insulin receptor using avidin-biotin technology. Mol Pharm 6:747–751
289. Suzuki T, Wu D, Schlachetzki F, Li JY, Boado RJ, Pardridge WM (2004) Imaging endogenous gene expression in brain cancer in vivo with 111In-peptide nucleic acid antisense radiopharmaceuticals and brain drug-targeting technology. J Nucl Med 45:1766–1775
290. Bertrand Y, Currie JC, Demeule M, Regina A, Che C, Abulrob A, Fatehi D, Sartelet H, Gabathuler R, Castaigne JP, Stanimirovic D, Béliveau R (2010) Transport characteristics of a novel peptide platform for CNS therapeutics. J Cell Mol Med 14:2827–2839
291. Carraway R, Leeman SE (1973) The isolation of a new hypotensive peptide, neurotensin, from bovine hypothalami. J Biol Chem 248:6854–6861
292. Kitabgi P, Carraway R, Van Rietschoten J, Granier C, Morgat JL, Menez A, Leeman S, Freychet P (1977) Neurotensin: specific binding to synaptic membranes from rat brain. Proc Natl Acad Sci USA 74:1846–1850
293. Castel MN, Malgouris C, Blanchard JC, Laduron PM (1990) Retrograde axonal transport of neurotensin in the dopaminergic nigrostriatal pathway in the rat. Neuroscience 36:425–430
294. Gonzalez-Barrios JA, Lindahl M, Bannon MJ, Anaya-Martinez V, Flores G, Navarro-Quiroga I, Trudeau LE, Aceves J, Martinez-Arguelles DB, Garcia-Villegas R, Jiménez I, Segovia J, Martinez-Fong D (2006) Neurotensin polyplex as an efficient carrier for delivering the human GDNF gene into nigral dopamine neurons of hemiparkinsonian rats. Mol Ther 14:857–865
295. Patapoutian A, Reichardt LF (2001) Trk receptors: mediators of neurotrophin action. Curr Opin Neurobiol 11:272–280
296. Roux PP, Barker PA (2002) Neurotrophin signaling through the p75 neurotrophin receptor. Prog Neurobiol 67:203–233
297. Ma N, Wu SS, Ma YX, Wang X, Zeng J, Tong G, Huang Y, Wang S (2004) Nerve growth factor receptor-mediated gene transfer. Mol Ther 9:270–281
298. Rende M, Giambanco I, Buratta M, Tonali P (1995) Axotomy induces a different modulation of both low-affinity nerve growth factor receptor and choline acetyltransferase between adult rat spinal and brainstem motoneurons. J Comp Neurol 363:249–263
299. Collins L, Gustafsson K, Fabre JW (2000) Tissue-binding properties of a synthetic peptide DNA vector targeted to cell membrane integrins: a possible universal nonviral vector for organ and tissue transplantation. Transplantation 69:1041–1050

300. Scarborough RM, Rose JW, Naughton MA, Phillips DR, Nannizzi L, Arfsten A, Campbell AM, Charo IF (1993) Characterization of the integrin specificities of disintegrins isolated from American pit viper venoms. J Biol Chem 268:1058–1065
301. Gill DM (1976) The arrangement of subunits in cholera toxin. Biochemistry 15:1242–1248
302. King CA, Van Heyningen WE (1973) Deactivation of cholera toxin by a sialidase-resistant monosialosylganglioside. J Infect Dis 127:639–647
303. Price DL, Griffin J, Young A, Peck K, Stocks A (1975) Tetanus toxin: direct evidence for retrograde intraaxonal transport. Science 188:945–947
304. Schiavo G, Benfenati F, Poulain B, Rossetto O, Polverino de Laureto P, DasGupta BR, Montecucco C (1992) Tetanus and botulinum-B neurotoxins block neurotransmitter release by proteolytic cleavage of synaptobrevin. Nature 359:832–835
305. Halpern JL, Neale EA (1995) Neurospecific binding, internalization, and retrograde axonal transport. Curr Top Microbiol Immunol 195:221–241
306. Figueiredo DM, Hallewell RA, Chen LL, Fairweather NF, Dougan G, Savitt JM, Parks DA, Fishman PS (1997) Delivery of recombinant tetanus-superoxide dismutase proteins to central nervous system neurons by retrograde axonal transport. Exp Neurol 145:546–554
307. Fishman PS, Savitt JM, Farrand DA (1990) Enhanced CNS uptake of systemically administered proteins through conjugation with tetanus C-fragment. J Neurol Sci 98:311–325
308. Coen L, Osta R, Maury M, Brulet P (1997) Construction of hybrid proteins that migrate retrogradely and transynaptically into the central nervous system. Proc Natl Acad Sci USA 94:9400–9405
309. Liu JK, Tenga QS, Garrity-Moses M, Federici T, Tanase D, Imperiale MJ, Boulis NM (2005) A novel peptide defined through phage display for therapeutic protein and vector neuronal targeting. Neurobiol Dis 19:407–418
310. Tan W, Wang H, Chen Y, Zhang X, Zhu H, Yang C, Yang R, Liu C (2011) Molecular aptamers for drug delivery. Trends Biotechnol 29:634–640
311. Ni X, Castanares M, Mukherjee A, Lupold SE (2011) Nucleic acid aptamers: clinical applications and promising new horizons. Curr Med Chem 18:4206–4214
312. Zhou J, Rossi JJ (2011) Cell-specific aptamer-mediated targeted drug delivery. Oligonucleotides 21:1–10
313. Dassie JP, Liu XY, Thomas GS, Whitaker RM, Thiel KW, Stockdale KR, Meyerholz DK, McCaffrey AP, McNamara JO II, Giangrande PH (2009) Systemic administration of optimized aptamer-siRNA chimeras promotes regression of PSMA-expressing tumors. Nat Biotechnol 27:839–849
314. Guo J, Gao X, Su L, Xia H, Gu G, Pang Z, Jiang X, Yao L, Chen J, Chen H (2011) Aptamer-functionalized PEG-PLGA nanoparticles for enhanced anti-glioma drug delivery. Biomaterials 32:8010–8020

Biofabrication of Hydrogel Constructs

Rúben F. Pereira, Henrique A. Almeida, and Paulo J. Bártolo

Abstract Hydrogels are three dimensional (3D) hydrophilic networks with the ability to absorb and retain large amounts of water without dissolution, as a result of the establishment of physical or chemical bonds between the polymeric chains. Hydrogels obtained from either natural or synthetic polymers are attractive materials for tissue engineering applications due to their excellent biocompatibility, biodegradability, elasticity and compositional similarities to the extracellular matrix. Several techniques have been explored to produce hydrogel meshes, films or 3D constructs for cell attachment, differentiation and proliferation or to release drugs and growth factors according to specific release profiles. This chapter describes the current state-of-the-art of biomanufacturing additive processes to produce hydrogel constructs for tissue engineering. Biomanufacturing processes are described in detail and the major advantages and limitations outlined.

Keywords Biofabrication • Tissue engineering • Hydrogels • Scaffolds • Stereolithography • Personalized medicine

1 Introduction

Tissue engineering is a multidisciplinary field that requires the combined effort of cell biologists, engineers, material scientists, mathematicians, geneticists, and clinicians towards the development of biological substitutes that restore, maintain,

R.F. Pereira • H.A. Almeida • P.J. Bártolo (✉)
Centre for Rapid and Sustainable Product Development, Polytechnic Institute of Leiria, Leiria, Portugal
e-mail: pbartolo@ipleiria.pt

J. Coelho (ed.), *Drug Delivery Systems: Advanced Technologies Potentially Applicable in Personalised Treatment*, Advances in Predictive, Preventive and Personalised Medicine 4, DOI 10.1007/978-94-007-6010-3_8,

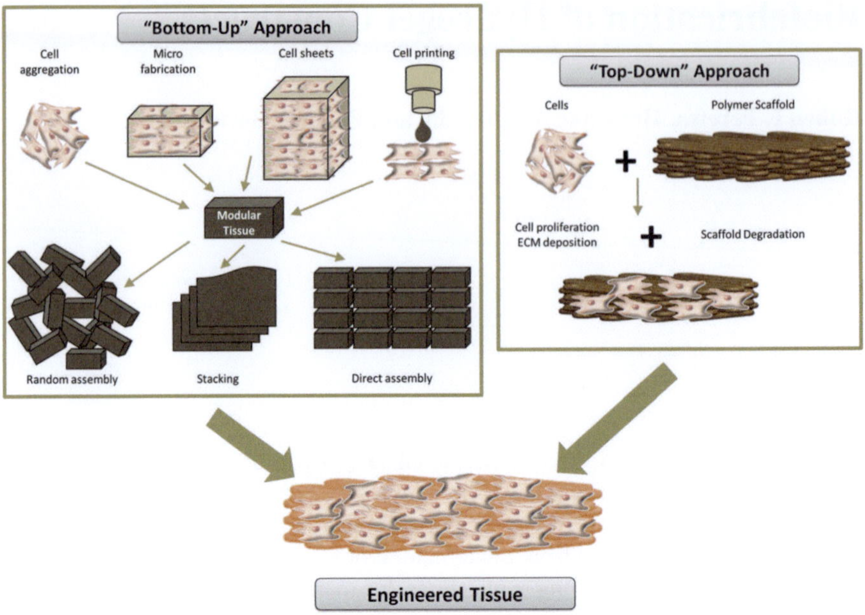

Fig. 1 Fundamental strategies for tissue engineering: bottom-up and top-down approaches [3]

or improve tissue function [1]. Two fundamental strategies can be considered [1–5] (Fig. 1):

1. Bottom–up approaches;
2. Top–down approaches.

The bottom-up approach employs different techniques for creating modular tissues, which are then assembled into engineered tissues with specific micro-architectural features [6]. Tissue modules can be created through self-assembled aggregation, microfabrication of cell-laden hydrogels, fabrication of cell sheets or direct printing. The ability of cell aggregates to fuse is based on the concept of tissue fluidity, according to which embryonic tissues can be considered as liquids [7]. The major drawback of this approach is that some cell types are unable to produce sufficient extracellular matrix (ECM), migrate or form cell-cell junctions [3].

The top-down or scaffold-based approach is based on the use of a temporary scaffold that provides a substrate for implanted cells and a physical support to organize the formation of the new tissue [1, 8]. In this approach, transplanted cells adhere to the scaffold, proliferate, secrete their own ECM and stimulate new tissue formation.

The top-down approach is the most commonly used strategy for tissue engineering. This strategy strongly depends on both materials and manufacturing processes. Generally, four classes of biomaterials have been used for engineering tissues: naturally derived polymeric materials (e.g. collagen, gelatine, starch, alginate, and chitosan), acellular tissue matrices (e.g. bladder submucosa and small intestinal submucosa), synthetic polymers (e.g. poly(glycolic acid) (PGA), poly(lactic acid) (PLA), poly(caprolactone) (PCL), poly(lactide-co-glycolide) (PLGA)) and ceramic materials (e.g. hydroxyapatite (HA) and β-tricalcium phosphate (TCP)) [9, 10].

Conventional techniques, such as solvent casting, freeze-drying, phase separation, gas foaming, particle-leaching and electrospinning techniques, have been widely applied for the fabrication of three-dimensional (3D) scaffolds [11, 12]. Despite being possible to produce porous polymer architectures, with tuneable micro- and macro-scale features, these techniques present several drawbacks, namely the lack of control over pore size, pore interconnectivity, porosity and spatial distribution of the pores. This can lead to pore pathways that are only partially connected and that follow contorted routes, which could impede the supply of nutrients and the ingrowth of tissue into the scaffold. Moreover, the incorporation of cells and proteins within the scaffolds becomes clearly limited by the use of toxic organic solvents. Therefore, biomanufacturing additive fabrication processes, such as stereolithography (SLA), extrusion-based processes, powder-based fusion techniques and inkjet printing processes, are considered as viable alternatives to fabricate scaffolds for tissue engineering as they offer better control and the ability to actively design the porosity and interconnectivity of scaffolds without the use of toxic organic solvents [1, 2, 8, 9].

This chapter describes the actual state-of-the-art of biomanufacturing additive processes to produce hydrogel constructs for tissue engineering.

2 Hydrogels

Hydrogels are crosslinked hydrophilic polymers exhibiting excellent biocompatibility and absorbing large amounts of water without dissolution [13–15]. They are soft and elastic materials, generally used above their glass transition temperature, presenting physiochemical characteristics similar to soft tissues and gelling under mild, cytocompatible conditions [16–18].

Both natural and synthetic polymers can be used for the production of hydrogels [19–21]. Synthetic hydrogels have been formed from poly(ethyleneglycol) (PEG), poly(vinyl alcohol) (PVA), propylene fumarate (PF), poly(butylene oxide) (PBO), PCL, poly(hydroxybuttyrate) (PHB), polyacylamide, poly(vinyl acetate) (PVac,), polyacrylonitrile (PAN), poly(ethylene oxide) (PEO), poly(propyleneoxide) (PPO), poly(hydroxypropyl methacrylamide) (PHPMA) and poly(2-hydroxyethyl methacrylate) (PHEMA). Natural hydrogels have been formed from hyaluronic acid, alginic acid, agarose, chitin, fibrin, collage, dextran, agarose, gelatine, pullulan and carrageenan.

Hydrogels are formed by physical (reversible hydrogels) or chemical (irreversible hydrogels) crosslinks of homopolymers or copolymers [20–22] (Fig. 2). The crosslink can be formed by covalent or non-covalent interactions.

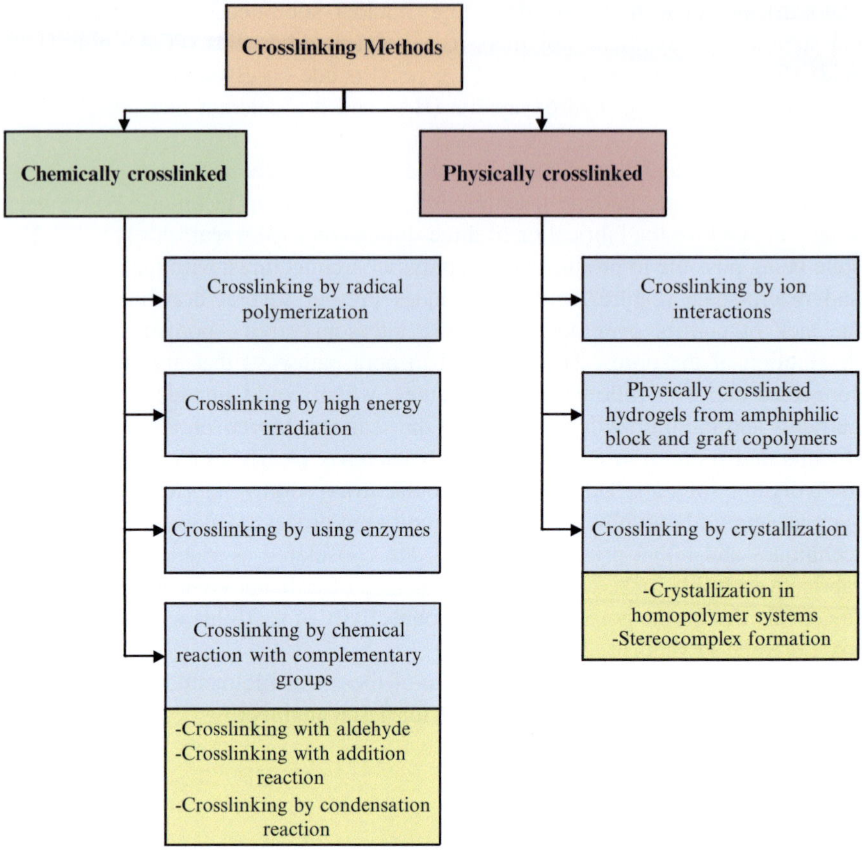

Fig. 2 Crosslinking methods used in hydrogels [23]

In the swollen state, hydrogels are usually highly stretched exhibiting limited mechanical properties [24]. Several strategies have been explored to circumvent this limitation including double network gels, interpenetrating network gels and gels reinforced with nanoparticles. Hydrogels also present high permeability for oxygen, nutrients, and other water-soluble metabolites.

Hydrogels are relevant materials for tissue engineering and drug delivery applications [20, 25]. They may be functionalized with proteins and peptides to promote cell proliferation, migration and adhesion. The pore size of the gel network can be manipulated to allow optimal diffusion and transport of biological materials. The hydrophilicity of some hydrogels prevents protein adsorption, minimizing the foreign body responses when implanted in vivo. However, natural polymers including collagen, gelatin and glycosaminoglycans often possess a high affinity for proteins present in serum [20, 26]. Hydrogels are also often used as wound dressings, absorbing the excess of wound exudates, protecting the wound from secondary infection and promoting wound healing by providing a moisturized wound-healing environment [27, 28].

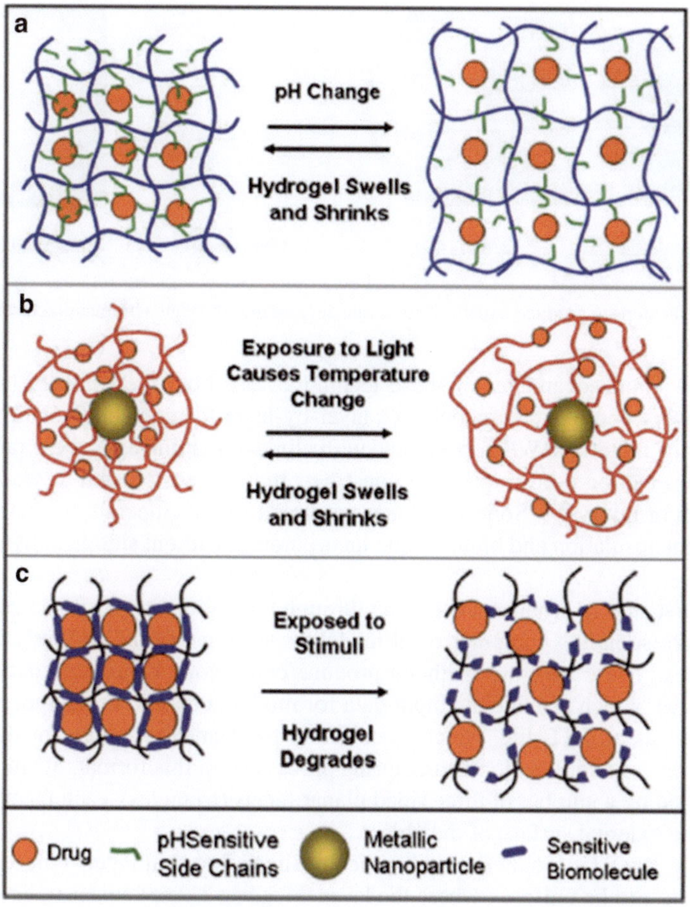

Fig. 3 Smart hydrogels: (**a**) pH-sensitive hydrogels. (**b**) Temperature sensitive hydrogels. (**c**) Biomolecule sensitive hydrogels [30]

Hydrogels can also be chemically tailored to respond to certain environmental stimuli, as the so-called temperature-responsive, potential-specific and pH-sensitive gels [29] (Fig. 3).

3 Biomanufacturing

Biomanufacturing was initially defined in 2005 during the Biomanufacturing Workshop hosted by Tsinghua University in China as "the use of additive technologies, biodegradable and biocompatible materials, cells, growth factors, etc., to produce biological structures for tissue engineering applications", representing a new group of non-conventional fabrication techniques recently introduced in the medical field

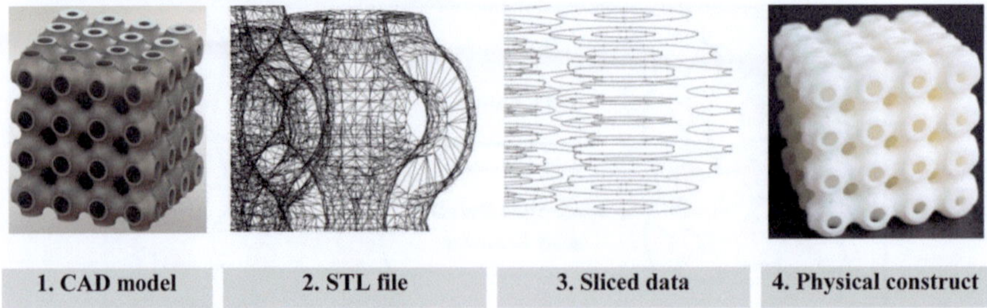

Fig. 4 Main steps to produce a scaffold for tissue engineering through a biomanufacturing process

[31]. The main advantages of these techniques are both the capacity to rapidly produce very complex 3D models in a layer-by-layer fashion and the ability to use various raw materials. When combined with clinical imaging data, these fabrication techniques can be used to produce constructs that are customised to the shape of the defect or injury [9]. Some processes operate at room temperature, thus allowing for cell encapsulation and biomolecule incorporation without significantly affecting viability.

The first step to produce a construct through biomanufacturing is the generation of the corresponding computer solid model through one of the currently available techniques. These imaging methods produce continuous volumetric data (voxel-based data), which provide the input data for model generation. The model is then tessellated as an STL file, which is currently the standard format for data input for all types of additive biomanufacturing processes. In this format, 3D models are represented by a number of three sided planar facets (triangles), each facet defining part of the external surface of the object.

Finally, the STL model is mathematically sliced into thin layers (sliced model). The slicing can be uniform, where the layer thickness is kept constant, or adaptive, where the layer thickness changes based on the surface geometry of the CAD model [32].

Finally, the sliced data is sent to a biomanufacturing device in order to produce the constructs containing or not cells and growth factors. Figure 4 illustrates the main steps to produce a scaffold for tissue engineering through a biomanufacturing process.

Biomanufacturing comprises a wide range of additive processes as indicated in Table 1. However, stereolithographic and inkjet printing processes are the most suitable techniques to process hydrogel materials.

3.1 *Stereolithography*

Stereolithographic processes produce 3D solid objects in a multi-layer procedure through the selective photo-initiated curing reaction of a photopolymeric material containing a low-molecular weight pre-polymer, additives and photo-initiators

Table 1 Biomanufacturing processes

Process	Description
Stereolithographic processes	
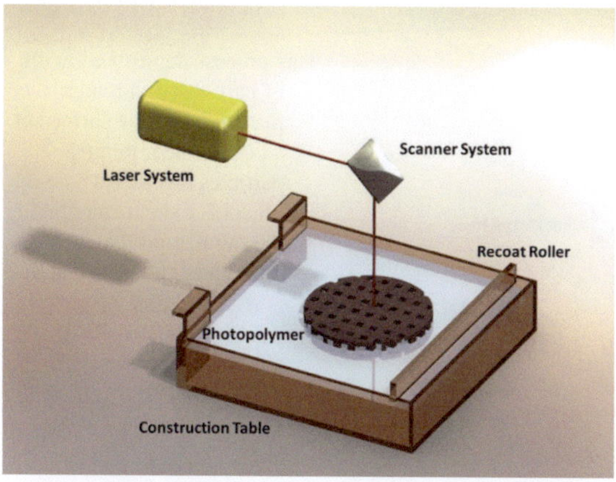	This technology involves the curing or solidification of a liquid photosensitive polymer through the use of an irradiation light source, which supplies the energy that is needed to induce a chemical reaction, bonding large numbers of small molecules and forming a highly cross-linked polymer.
Powder-bed fusion processes	
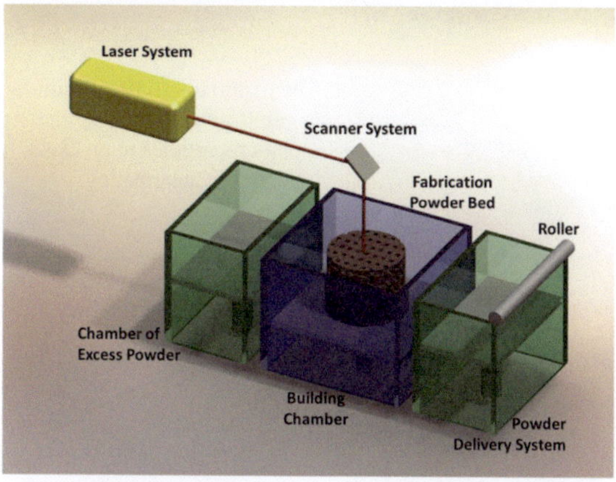	Powder-bed fusion processes comprise SLS (Selective Laser Sintering), SLM (Selective Laser Melting) and EBM (Electron Beam Melting) techniques, which use high-energy sources to consolidate powder material.
Extrusion-based processes	
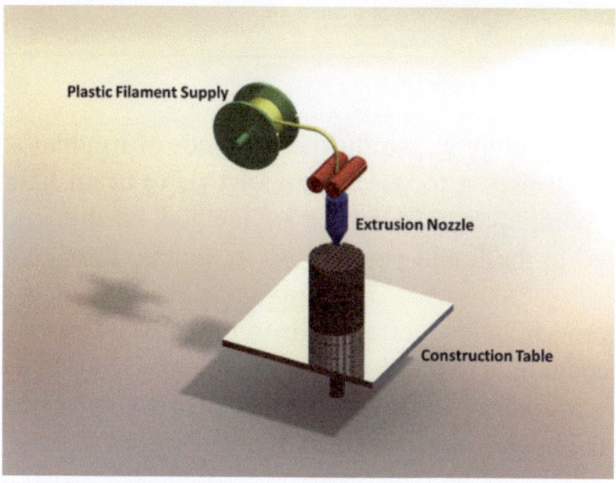	By this process, thin thermoplastic filaments are melted by heating and deposited by a numerical-controlled robotic device. The material leaves the extruder in a liquid form and hardens immediately.

(continued)

Table 1 (continued)

Process	Description
Inkjet printing processes	
(a) Binder jetting method	
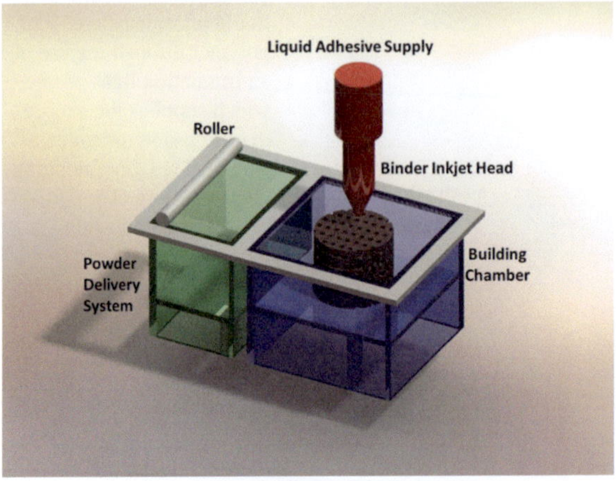	This process deposits a stream of particles of a binder material over the surface of a powder bed, joining particles together where the object is to be formed. After the fabrication the unbounded powder material is removed and the part is usually submitted to a sintering process.
(b) Material jetting method	
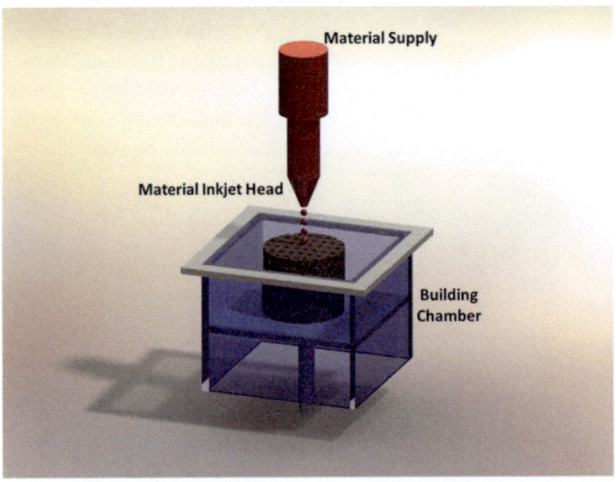	The material jetting techniques emit a stream of microparticles to an exact co-ordinate creating, layer-by-layer, the construct.

[1, 8, 33–36]. These processes usually employ two distinct methods of irradiation [37]. The first method, laser writing, uses a focused ultraviolet (UV) beam light to selectively solidify the liquid photopolymer. In the second method, mask-based irradiation, an image is transferred to the liquid polymer by irradiating through a patterned mask.

The direct or laser writing approach, which corresponds to the most frequently used method, consists of a vat containing a photosensitive material, a moveable platform on which the model is built, a laser to irradiate and cure the polymer, and a dynamic mirror system to direct the laser beam over the polymer surface "writing"

Table 2 Comparison between LCD and DMD technologies

	LCD	DMD
UV compatibility	No	Yes
Modulation efficiency	12.5%	88%
Light transmission	21%	71%
Optical fill factor	64%	85%
Pitch size	26 μm × 24 μm	14–17 μm
Pixel size	33 μm × 33 μm	13–16.2 μm
Contrast	100:1	350:1
Switching speed	20 ms	20 μs

each layer [37]. After drawing a layer, the platform dips into the polymer vat, leaving a thin film from which the next layer will be formed. The curing reaction can be induced using both single-photon polymerization and two-photon polymerization (2PP). The chemical principle of these two processes is similar, being the main difference the number of absorbed photons required to induce the polymerization process [37].

Two-photon-initiated polymerization processes usually employ a femtosecond infrared laser without photo-masks [38]. This method allows a submicron 3D resolution on top of enabling 3D fabrication at greater depth and an ultra-fast fabrication. However, 2PP systems are very expensive and operate with a single material type.

In the mask-based irradiation method, a dynamical pattern generator is used to shape a light beam from a source of irradiation, according to the image of the layer to be built [39]. This image is then focused over the resin surface to be solidified, through an appropriated optical device. This process has a high built velocity and absence of undesirable polymerizations, due to the low-density flux of light over the resin surface. LCD (Liquid Crystal Display) panels and DMDs (Digital Micromirror Devices) can be used as dynamic pattern generators [37]. The DMD's, developed by Texas Instruments, are micro-electro-mechanical systems with square mirrors in aluminum, capable of reflect the light in the proper direction. The DMD technology presents several advantages regarding the LCD as indicated in Table 2. Initially, these devices had a short lifetime when exposed to a radiation of wavelength less than 400 nm. Recent developments in this area allow Texas Instruments to produce optimized DMD's to work with wavelengths between 350–400 nm with expected lifetime of 2,000 h when exposed to a radiation with light intensities of 10 w/cm^2. Texas Instruments also offers DMD's operating in the near infrared spectrum (DMD-IR).

Some types of hydrogels can be photopolymerized using UV radiation and appropriate photo-initiators. The UV polymerization of hydrogels occurs at sufficiently mild conditions (low light intensity, short irradiation time, physiological temperature and low organic solvent levels), enabling the reaction to be carried out in the presence of cells [40]. Hydrogels used in stereolithographic applications comprise natural polymers, synthetic polymers or combination of both natural and synthetic polymers [40–44]. Natural polymeric hydrogels include polymers like hyaluronic acid, chitosan, collagen, gelatine, dextran, etc. Among synthetic polymers are those based on poly (propylene fumarate) (PPF), PEO, poly (HEMA), hyaluronic acid-based materials, PVA derivatives, etc. The UV polymerization of hydrogels for medical applications can be performed using two types of photo-initiators:

- Initiators that promote radical polymerization by photo-cleavage such as benzoin and acetophenone compounds;
- Initiators that promote radical polymerization by hydrogen abstraction such as aromatic ketones.

However, few photo-initiators have been identified as suitable for the fabrication of cell-based hydrogel constructs owing to their water solubility, crosslinking properties and non-toxic reaction conditions [45]. Cationic photo-initiators are not used as they generate protonic acids.

Bryant and Anseth used a photopolymerization process to encapsulate chondrocytes in PEO hydrogels structures with thickness varying from 2 to 8 mm [46]. The hydrogel structures were photocured at low light intensity ($\sim$10 mW/cm^2) during 10 min. The chondrocytes encapsulated in the hydrogel structures and cultured in vitro during 6 weeks remained viable and produced cartilaginous tissue. The results suggested that an increase in the hydrogel thickness from 2 to 8 mm do not change both cell viability and uniformity.

Lan et al. produced PPF scaffolds with highly interconnected porous structure and porosity of 65% [47]. The scaffolds were coated by applying accelerated biomimetic apatite and arginini-glycine-aspartic acid peptide coating to promote cell behavior. The coated scaffolds were seeded with MC3T3-E1 pre-osteoblasts and their biologic properties were evaluated using an MTS assay and histologic staining.

Stereolithography was also used to produce 3D scaffolds with embedded growth factor-delivering microspheres [48]. In this study, BMP-2-loaded PLGA microspheres (Fig. 5) were incorporated into a 3D scaffold (Fig. 6) that was photo-fabricated using a suspension of microspheres and a PPF/diethyl fumarate (DEF) photopolymer. The effects of BMP-2 release were assessed in vitro by observing cell differentiation using MC3T3-E1 pre-osteoblasts.

Seck et al. produced porous and non-porous biodegradable hydrogel structures using an aqueous photocurable resin based on methacrylate-functionalised poly (ethylene glycol)/poly (D,Llactide) macromers and Lucirin TPO-L as visible light photo-initiator [49] (Fig. 7). After photopolymerization, the obtained structures were extracted with distilled water to remove soluble compounds and dried at ambient conditions for 3 days. The structures showed good cell seeding characteristics, and human mesenchymal stem cells adhered and proliferated well on these structures.

Chan et al. modified a stereolithographic apparatus to produce poly (ethylene glycol) diacrylate (PEGDA) constructs with encapsulated/3 T3 cells [50]. Each layer of cell-containing photopolymer was manually added to prevent cells settling to the bottom of the stereolithographic vat due to gravity (Fig. 8). This strategy allows to use multiple hydrogel compositions and cell types and to control the

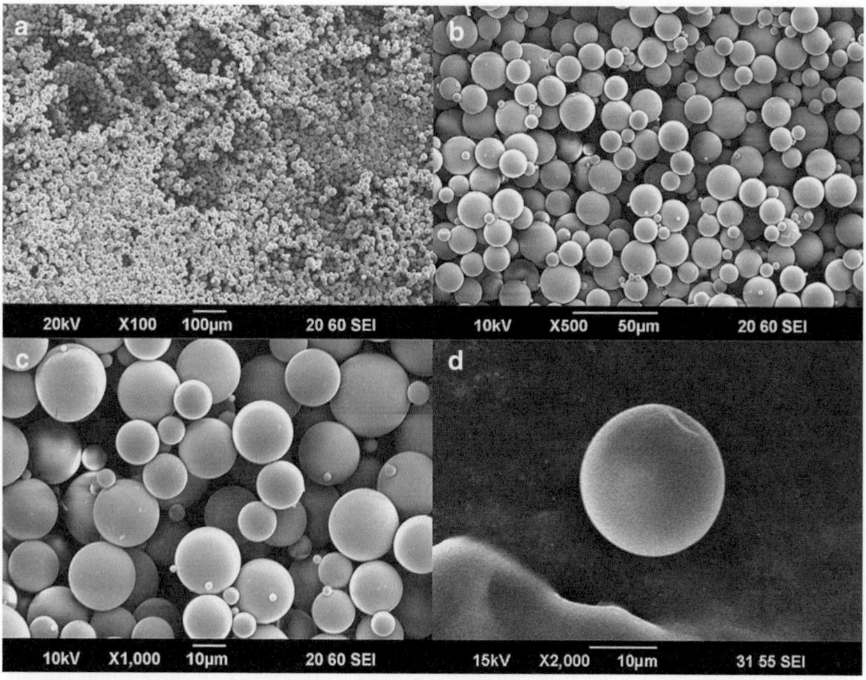

Fig. 5 SEM images of the BMP-2-loaded microspheres [48]

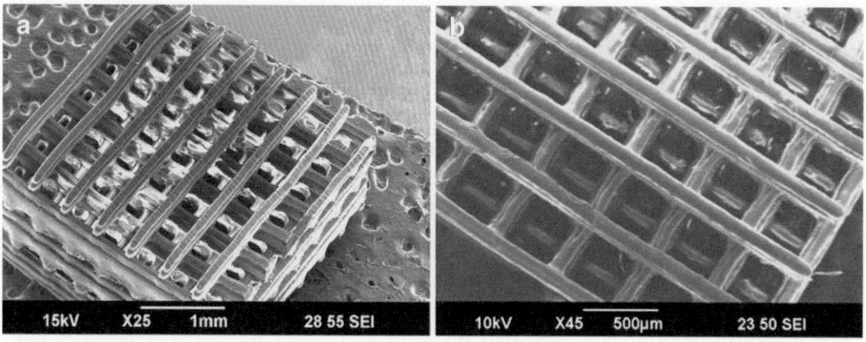

Fig. 6 SEM images of the BMP-2-loaded scaffolds [48]

spatial distribution of cells and bioactive molecules. A similar approach was used by Zorlutuna et al. to produce spatially organized 3D co-culture of multiple cell types to investigate cell-cell interaction and the microenvironments of complex tissues [51]. Two layer-constructs using different cell types and hydrogels were produced as described in Fig. 9. The first layer was produced by polymerizing poly (ethylene glycol) methyl ether methacrylate (PEGMA3400) containing adipose-derived stem cells. The second layer contains primary hippocampus neurons and skeletal muscle

Fig. 7 Hydrogel structures built by stereolithography using methacrylate-functionalized poly(ethylene glycol)poly(D-Lactide) [49]

myoblast cells encapsulated in OMA-PEGMA1100 (oxidized methacrylic alginate and poly (ethylene glycol) methyl ether methacrylate 1100).

Arcaute et al. explored stereolithography for fabricating multi-material spatially controlled bioactive scaffolds [41, 52, 53]. To accomplish multi-material fabrication, a mini-vat setup was designed allowing for self-aligning X–Y registration during fabrication. The mini-vat setup allowed the part to be easily removed and rinsed, and different photocrosslinkable solutions to be easily removed and added to the vat. Multi-material scaffolds were fabricated by including controlled concentrations of fluorescently labeled dextran, fluorescently labeled bioactive PEG or bioactive PEG in different regions of the scaffold (Fig. 10). Human dermal fibroblast cells were seeded on top of the fabricated scaffolds. Spatial control was successfully demonstrated in features down to 500 μm.

2PP has been explored to produce scaffolds using the photosensitive organic-inorganic hybrid sol-gel ORMOCER and the negative photoresist SU8 [54, 55]. Recently, Ovsianikov et al. demonstrated the potential of porous gelatin-based hydrogels produced using two-photon polymerization to be applied as carriers for mesenchymal stem cell [56]. Additionally, upon applying osteogenic stimulation, the seeded cells differentiated into the anticipated lineage.

Mask-based writing systems can be used to pattern hydrogel structures with high resolution. Liu and Bhatia reported a method, where multiple steps of micropatterned photopolymerization processes can be coupled to produce 3D cell matrix structures with micro-scale resolution [57] (Fig. 11).

Figure 12 illustrates a mask-based stereolithographic system. This system, called Stereo-thermal-lithography (STLG) uses a mercury lamp of 350 W as a light source. Appropriated filters split the radiation into two different wavelengths: UV radiation and near-infrared (near-IR) radiation. Optical fibres, projection and focal lenses irradiated the UV-DMD and the IR-DMD. A dichroic mirror captures the images projected on both DMDs (1,024 × 768 pixels, 14 mm in size) combining them into a single image that is transferred to the liquid polymer. The equipment also includes

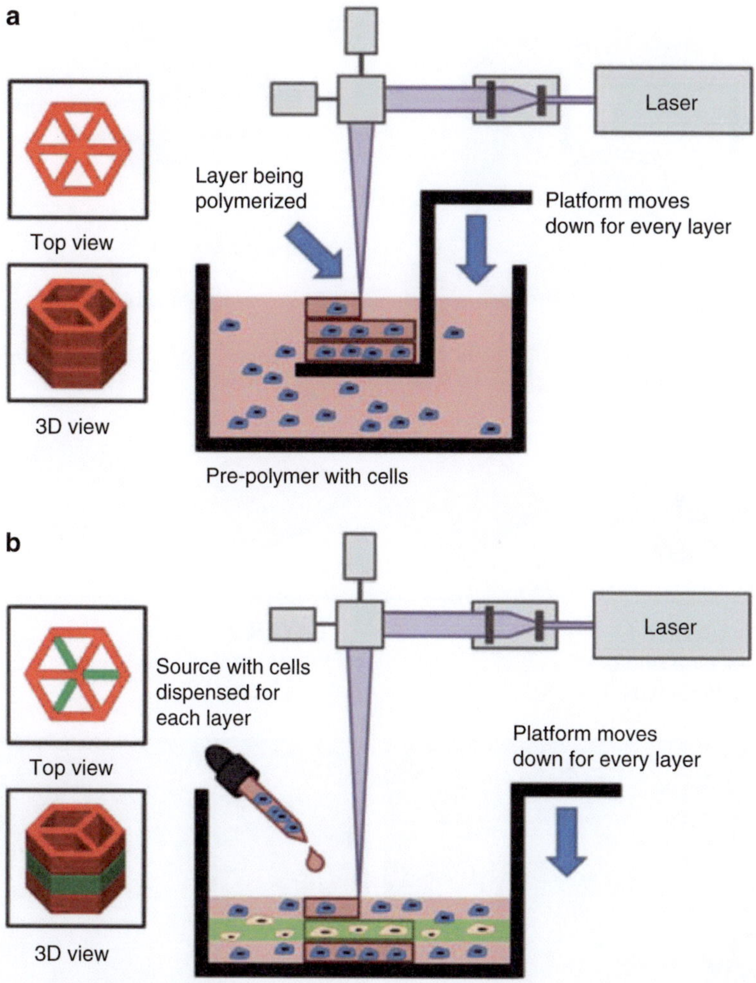

Fig. 8 (**a**) Common stereolithgraphic approach to produce hydrogel constructs encapsulating cells. (**b**) Manually deposition of each layer of cell-containing photopolymer [50]

a multi-vat system enabling the fabrication of multi-material constructs. The vertical displacement of the platform is secured by a positioner uniaxial MYCOSIS® Translation Stage VT-80. This positioner allows vertical increments of 1 μm at a speed ranging between 0.001 and 20 mm/s.

STLG intends to solve some of the major limitations of stereolithography, such as efficiency, speed and accuracy. Its main advantages over the other stereolithographic processes are:

- The generation of radicals is more efficient;
- Small concentrations of the two types of initiator (UV and near-IR initiators) are used, enabling the radiation to penetrate deeper into the polymer;

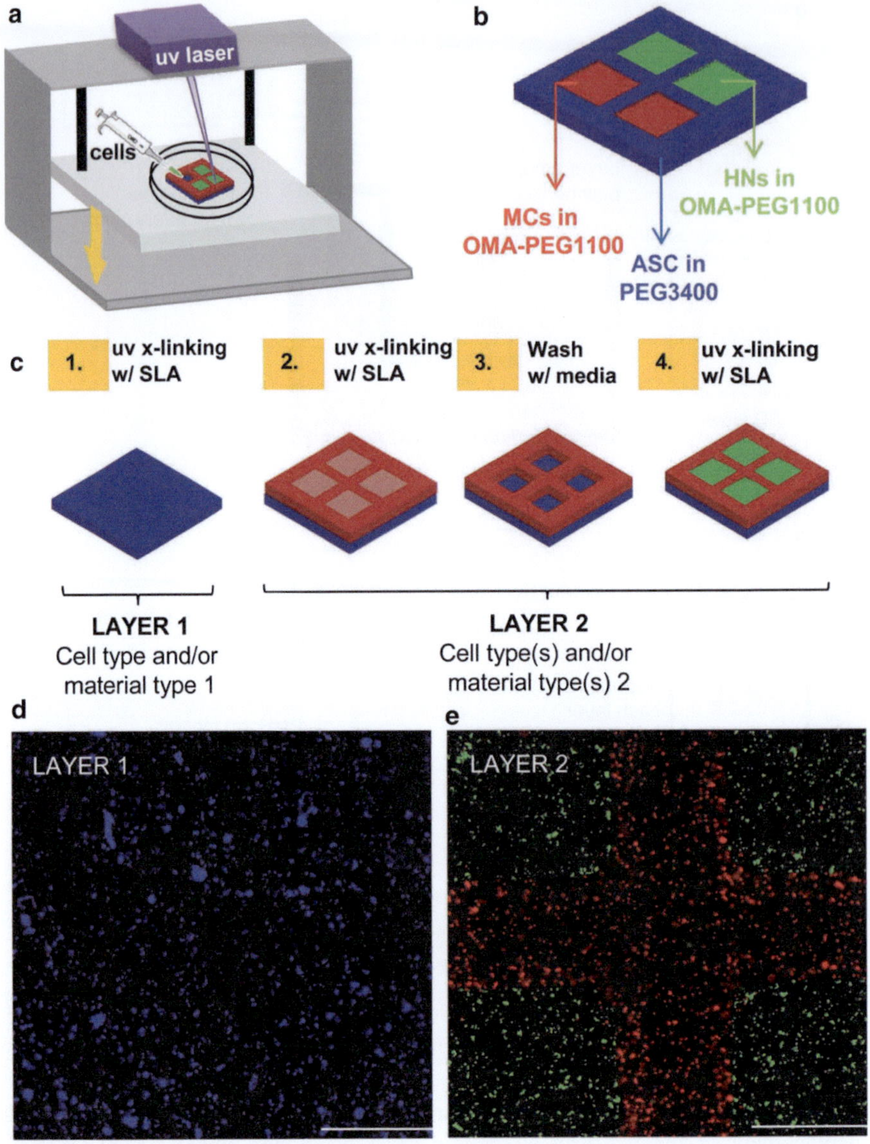

Fig. 9 (**a**) Fabrication procedure. (**b**) CAD model. (**c**) Fabrication sequence. (**d**) Fluorescence microscopy image of the adipose-derived stem cells (*blue*) in the first layer (the scale bar is 1 mm). (**e**) Fluorescence microscopy image of the primary hippocampus neurons (*green*) and the skeletal muscle myoblast cells (*red*) in the second layer (the scale bar is 1 mm) [51]

- The curing reaction is more localized, improving the accuracy of the produced models;
- The system has more tunability. Three subsystems can be considered. ***Subsystem A:*** uses UV radiation to solidify a photopolymer that contains a certain

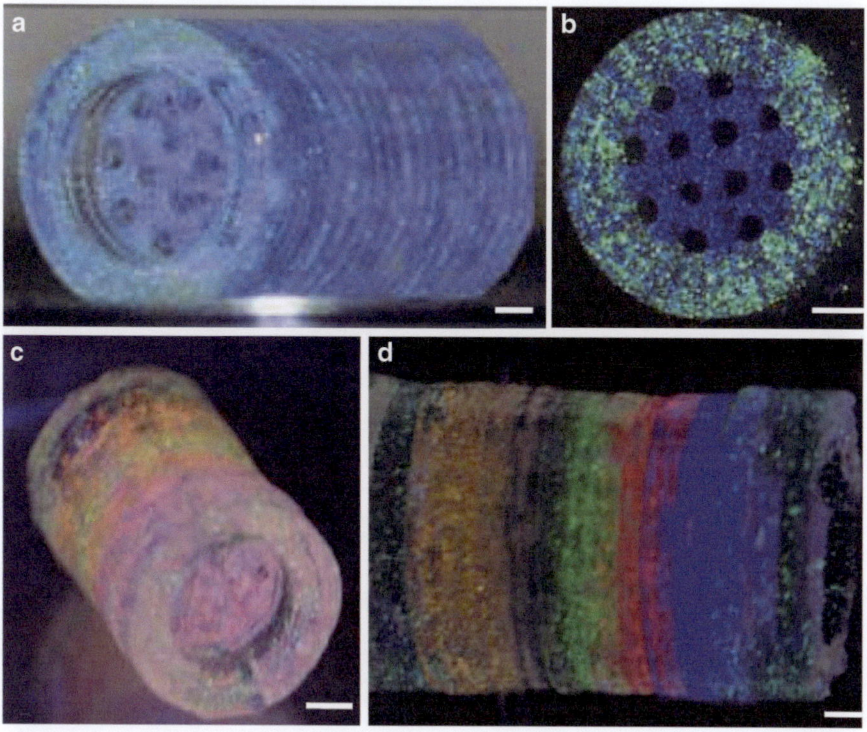

Fig. 10 Nerve guidance conduit (NGC) produced using different colors of fluorescent particles in the PEG-based solution (the scale bar is 1 mm) [53]

amount of UV initiator. ***Subsystem B:*** uses IR radiation to solidify a photopolymer that contains a certain amount of IR initiator. ***Subsystem C:*** uses both UV and IR radiation to solidify a photopolymer that contains a certain amount of UV and IR initiators.

The STLG process has been used to produce PVA and polyHEMA scaffolds [58] (Fig. 13).

Lu et al. used a dynamic masking system to produce hydrogel scaffolds with different pore geometries (hexagons, triangles, honeycombs with triangles, and squares) and pore sizes (165–650 μm), through the polymerization of a polymeric system consisting of PEGDA dissolved in phosphate buffered saline and 0.1 wt.% of photo-initiator (Irgacure 2959) [59]. Murine OP-9 marrow stromal cells were also encapsulated within the PEGDA hydrogel. An encapsulation efficiency of 73% and cell viability after 24 h of incubation was obtained. To achieve an efficient cell seeding, the scaffold surfaces were covalently functionalized with fibronectin. The modified scaffolds were seeded with murine mesenchymal stem cells, showing the ability to promote cell attachment and the osteogenic differentiation of cells through the addition of osteogenic culture medium. Using different fluorescently-labeled

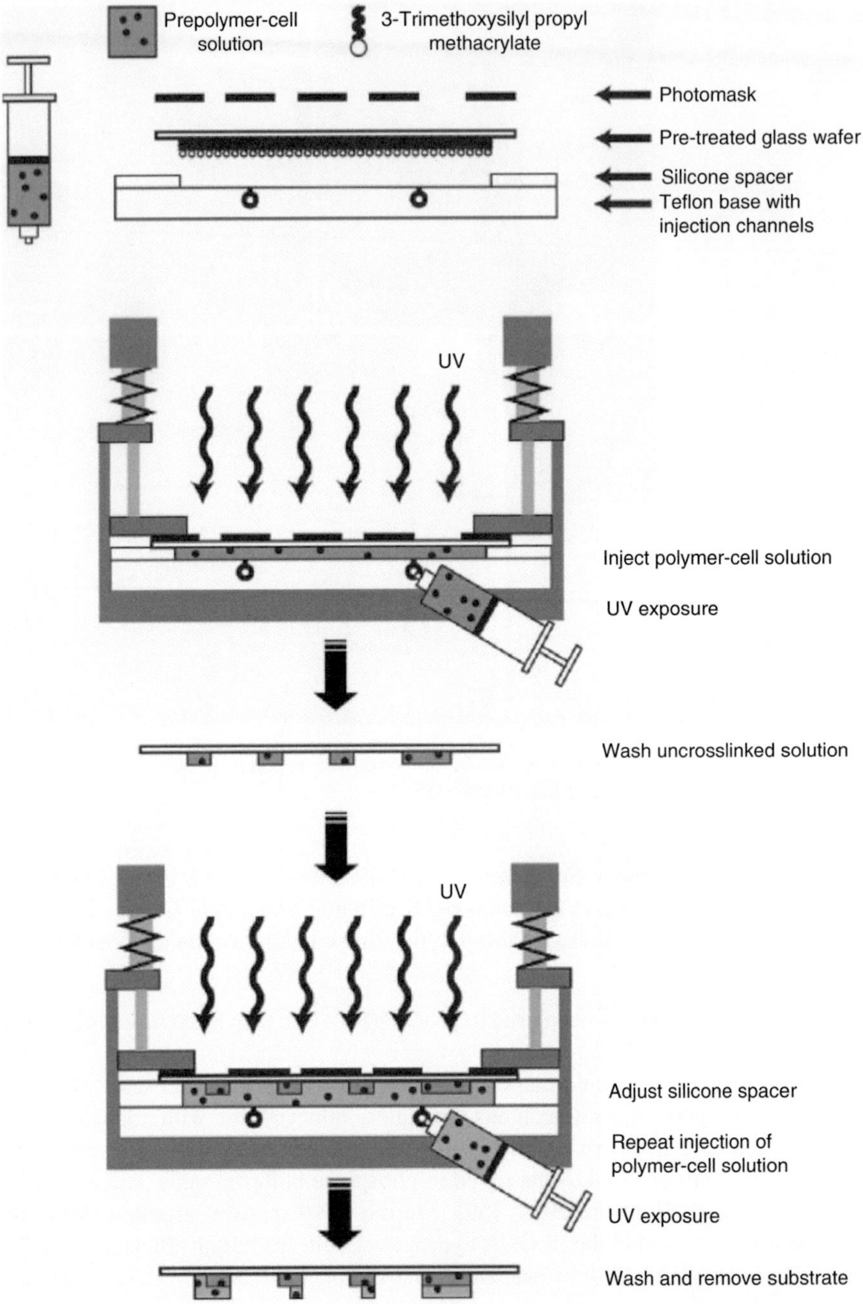

Fig. 11 Process for the formation of hydrogel microstructures containing living cells [57]

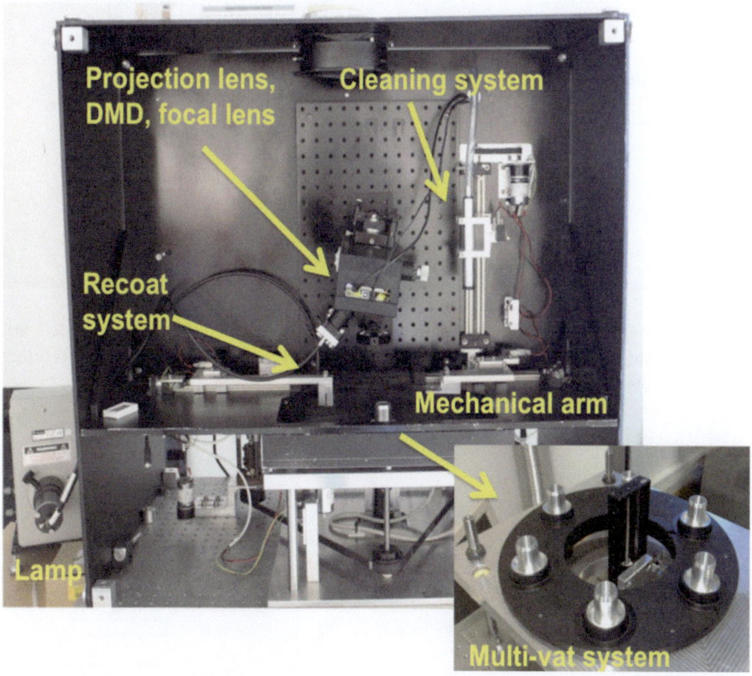

Fig. 12 Stereo-thermal-lithography system

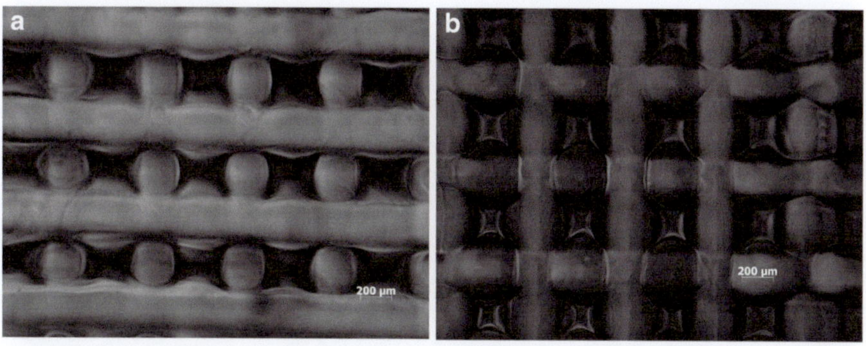

Fig. 13 (**a**) Poly(HEMA) scaffolds containing 2 wt.% initiator. (**b**) Poly(HEMA) scaffolds containing 1 wt.% initiator

polystyrene microparticles, the authors also demonstrated the feasibility of the system to produce scaffolds with entrapped multiple biochemical factors with precise pre-designed and spatially-patterned layers (Fig. 14).

Du et al. used a mask-based irradiation approach to direct the assembly of cell-laden microgels to generate tissue constructs with tunable microarchitecture and complexity [4] (Fig. 15). Microgel units were synthesized by photopolymerization, transferred into a dish containing mineral oil, subjected to mechanical agitation

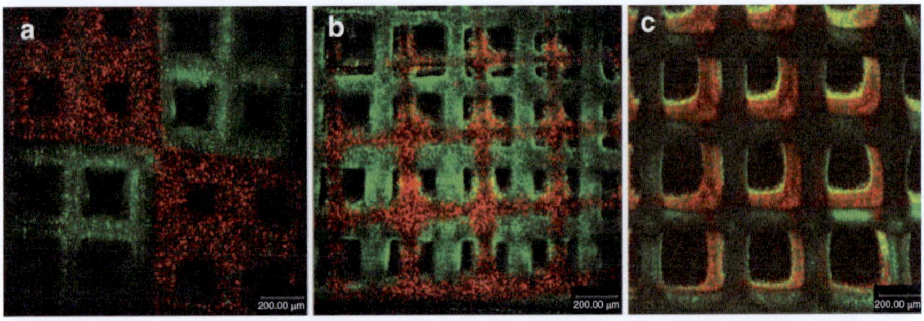

Fig. 14 Fluorescence confocal microscopy of PEGDA hydrogel scaffolds produced with pre-defined spatial-pattern in a single layer (**a**) and multi-layered (**b**, **c**) scaffolds carried either FITC or Cy5-labeled polystyrene particles [59]

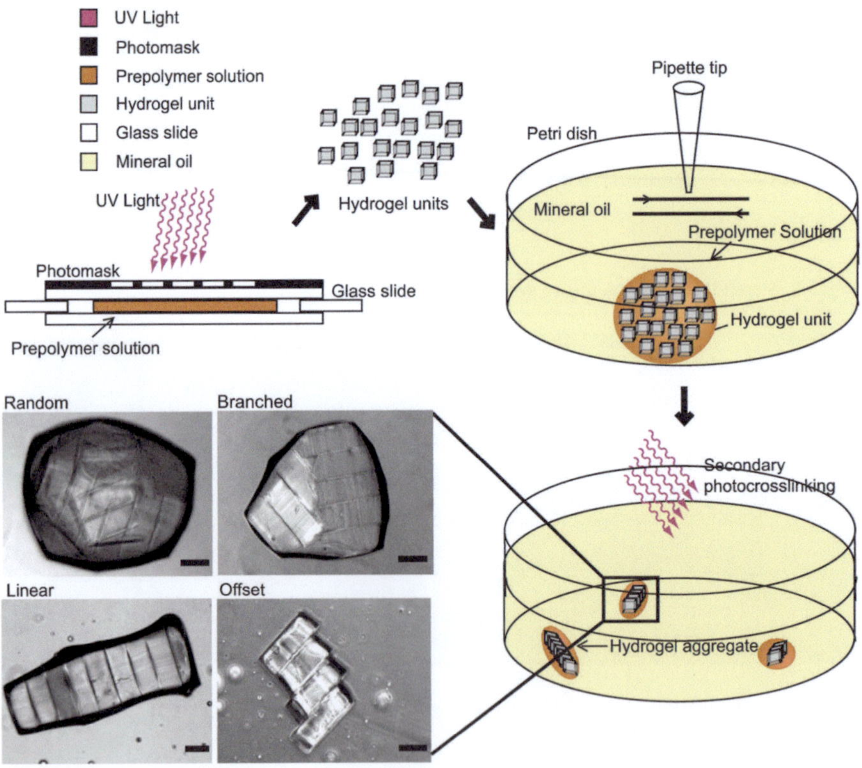

Fig. 15 Microgel assembly process using a mask-based irradiation procedure [4]

applied by manually manipulating a pipette-tip. Four structural types of microgel assemblies were observed: linear, branched, random, and offset. The assembly process was driven by the tendency of multiphase liquid-liquid systems to minimize the surface area and the resulting surface free energy between the phases. Microgels were exposed to a secondary irradiation mechanism for further crosslinking.

3.2 *Inkjet Printing Processes*

Inkjet printing systems have been used to print both aqueous solutions onto supports and cells within a scaffold [60, 61]. During the droplet formation process the liquid material experiences shear rates close to $10^4 s^{-1}$ and similar strains occurs during the impact [60]. Therefore, cells in suspension are subjected to large stresses and deformation. Nevertheless, inkjet printing has been reported to be a viable method for cell deposition and patterning [60].

Inkjet printing technology comprises two main configurations:

1. Binder jetting processes or bonding method;
2. Material jetting processes or build-up method.

The binder jetting process was developed at MIT (USA) under the name of thee-dimensional printing (3DP). The process deposits a stream of particles of a binder material over the surface of a powder bed, joining particles together where the object is to be formed [9]. A piston lowers the powder bed so that a new layer of powder can be spread over the surface of the previous layer and then selectively joined it. Examples of the direct use of binder jetting systems together with hydrogels are rather limited. In some cases, the use of an indirect approach have been reported.

Lam et al. developed a blend of starch-based powder containing cornstarch (50%), dextran (30%) and gelatine (20%), bounded by printing distilled water [62]. Cylindrical scaffolds were produced measuring 12.5 mm (diameter) by 12.5 mm (height) and infiltrated with different amounts of a copolymer solution consisting of 75% L-PLA and 25% PCL in dichloromethane to improve their mechanical properties.

Sachlos et al. used an indirect approach to produce collagen scaffolds with complex internal morphology and macroscopic shape by using a 3DP sacrificial mould [63]. A dispersion of collagen was cast into the mould and frozen. The mould was then dissolved with ethanol and the collagen scaffold was critical point dried with liquid carbon dioxide. Other research works, like the ones of Limpanuphap and Derby and Park et al., have also exploited the capabilities of 3DP for tissue engineering [64, 65].

Boland et al. used an inkjet printing system to print 0.25 M calcium chloride ($CaCl_2$) onto liquid alginate/gelatin solutions [66]. Similarly, Cohen et al. developed a robotic system to deposit alginate hydrogels seeded with articular chondrocytes [67]. The alginate-cell suspension at a concentration of 33×10^6 cells/mL was crosslinked with calcium sulfate. After printing, $94 \pm 5\%$ of cell viability was observed. An analogous process, called alginate-based rapid prototyping, has been developed at the Polytechnic Institute of Leiria (Portugal). This process produces alginate solid structures, by extruding a solution of sodium alginate, mixed with a solution of $CaCl_2$, providing a temporary support for the seeded cells in culture [68–70]. In the same way, Lee et al. produced porous scaffolds of calcium phosphate cement (CPC) combined with alginate, by printing a solution of α-tricalcium phosphate-based powder and sodium alginate in a calcium chloride bath [71]

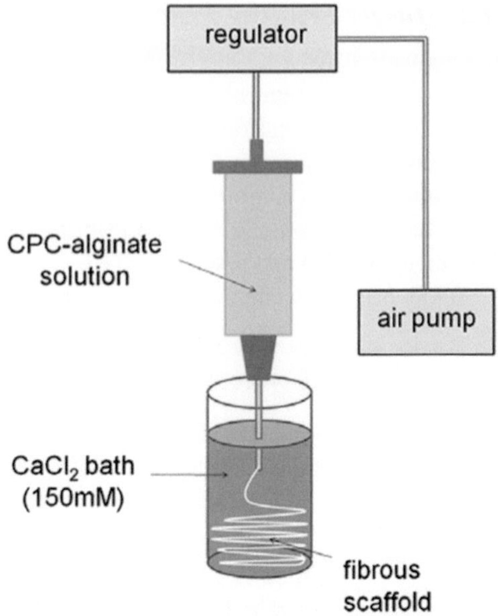

Fig. 16 Production of calcium phosphate cement/alginate composite fibrous network [71]

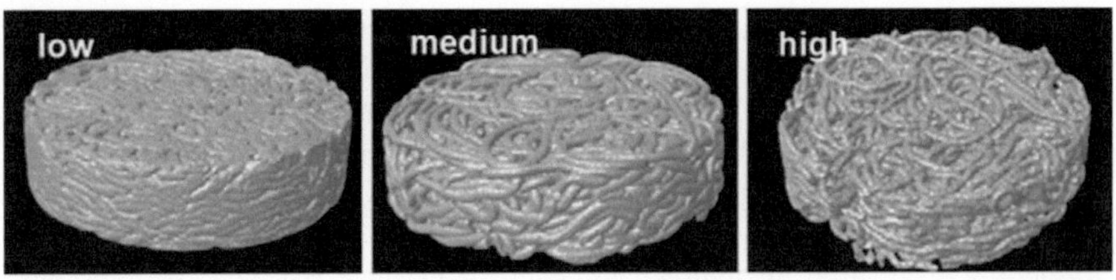

Fig. 17 Micro-CT images of calcium phosphate cement/alginate scaffolds with low (13.6 ± 2.9%), medium (34.0 ± 3.1%) and high (53.7 ± 2.6%) porosity [71]

(Fig. 16). A rapid hardening reaction of the alginate with Ca^{2+} shaped the composite into a fibrous form with diameters of hundreds of micrometers. Subsequently, structures were pressed in a mould allowing the formation of porous scaffolds. Scaffolds with different porosities were produced by changing the needle diameter and composition of the suspension (Fig. 17). Mesenchymal stem cells were able to grow and populate the scaffolds.

Material jetting processes emit a stream of hydrogel microparticles to an exact co-ordinate offering a possible solution for organ printing, which has been defined as the engineering of 3D living structures supported by the self-assembly and self-organizing capabilities of cells [40, 72]. This process (Fig. 18) is based on the deposition of bio-ink particles in well-defined topological patterns into bio-paper sheets of biocompatible gels.

After the deposition is finished, the construct is transferred to a bioreactor to fuse the bio-ink particles. Three dimensional myocardial patches were formed using this approach [74].

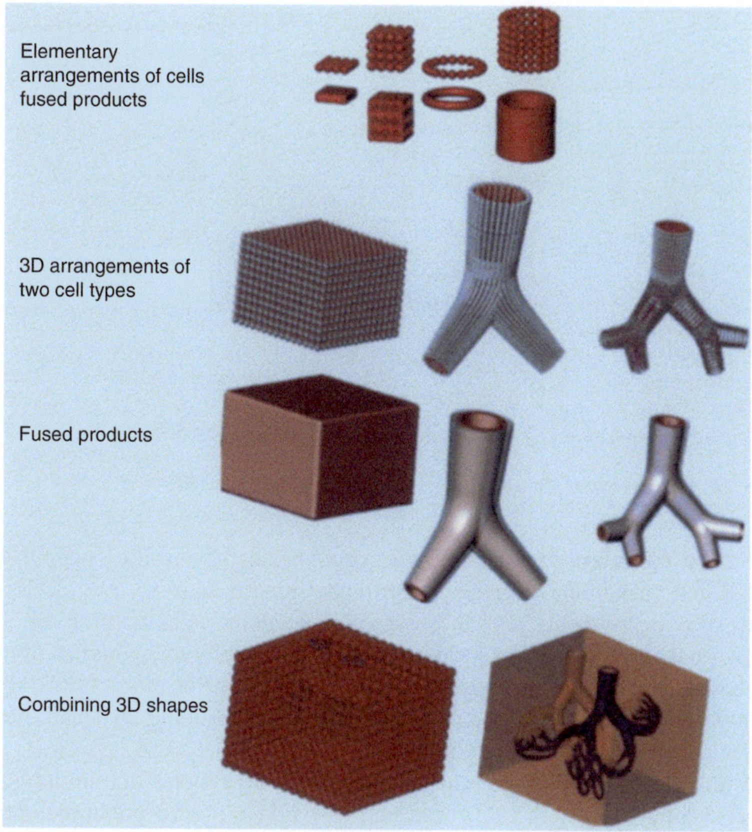

Fig. 18 Organ printing process [73]

The bio-paper (agarose, collagen, fibrin, fibrinogen, polyethylene glycol tetra-acrylates), which provides a temporary support for the deposited bio-inks, can be removed after construction if required [75]. In this technique the hydrogels must preserve the printed shape after the deposition and maintain cell viability and cell function [40]. The bio-ink building blocks (natural and synthetic physical hydrogels, concentrated cell pellets and collagen solutions), typically have a spherical or cylindrical shape and consist of single or multiple cell types.

Several researchers modified commercial inkjet printing systems to print living cells [76–79]. Two types of inkjet systems have been used (Fig. 19): thermal inkjet printers and piezoelectric inkjet printers [77, 80–84].

Cui and Boland reported the fabrication of functional human microvasculature networks by printing fibrin gel and human microvascular endothelial cells, using a modified thermal inkjet printer [75]. It was possible to observe that cells aligned inside the fibrin channels forming a tubular structure. The produced microvasculature presented good integrity after 21 days of culture. The same research group

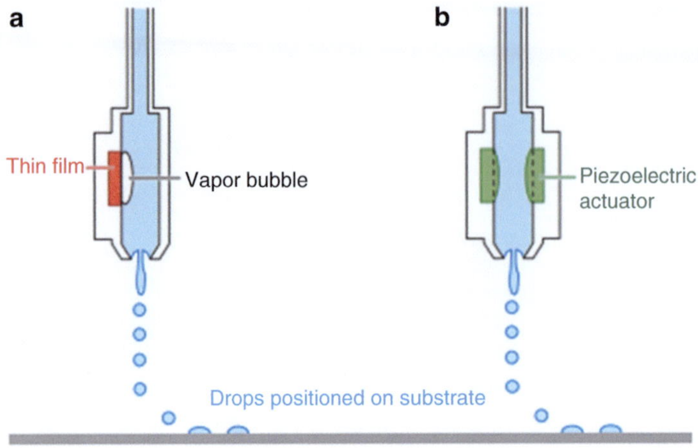

Fig. 19 (**a**) Thermal inkjet printers, (**b**) piezoelectric inkjet printers [80]

investigated the effect of thermal inkjet printing on the viability, apoptosis and possible cell membrane damages of printed Chinese hamster ovary cells [78]. Results show cell viability of 89% and 3.5% of apoptotic cell after printing.

Piezoelectric inkjet printers have been also used to print living human fibroblast cells [85] and bovine vascular endothelial cells [86]. However, piezoelectric systems are characterized by small diameter jetting heads, resulting in small material droplets and lower cell concentrations (10^4–10^6 cells/ml).

Several researchers used pressure-operated printing systems to print living cells and cell aggregates [67, 87–89]. Yan and co-workers used a pressure-controlled syringe to print hepatocyte cells with gelatin/chitosan [90]. In order to improve the vascularization, the constructs were designed with cylindrical channels (diameters in the range of 100–300 μm). The produced 3D constructs were submitted to a crosslinked process to stabilize the structures and then cultured. After 2 months of culture, it was possible to observe that cells were well distributed maintaining their viability and function. Lee et al. used a similar approach to print collagen, fibroblasts and keratinocytes [91]. For multi-material printing, Khalil and Sun developed the concept of multinozzle deposition [89] (Fig. 20). Recently, Schuurman et al. used a multi-head printer to produce viable hybrid construct through the deposition of thermoplastic PCL fibers and cell-laden hydrogels composed by alginate and chondrocyte cells [87] (Fig. 21).

Laser-assisted cell-printing techniques have also been developed, allowing the deposition of cell solutions with high concentrations and velocity ($\geq$10 m/s) [83]. These techniques comprise the so-called laser guidance direct write (LG DW), laser-induced forward transfer (LIFT) and modified laser-induced forward transfer (modified-LIFT) processes.

The LG DW process employs radiation pressure from the scattering of energetic photons in a laser beam, to guide and deposit cells with micrometer resolution [84]. This process has been used to deposit multiple cell types, including embryonic

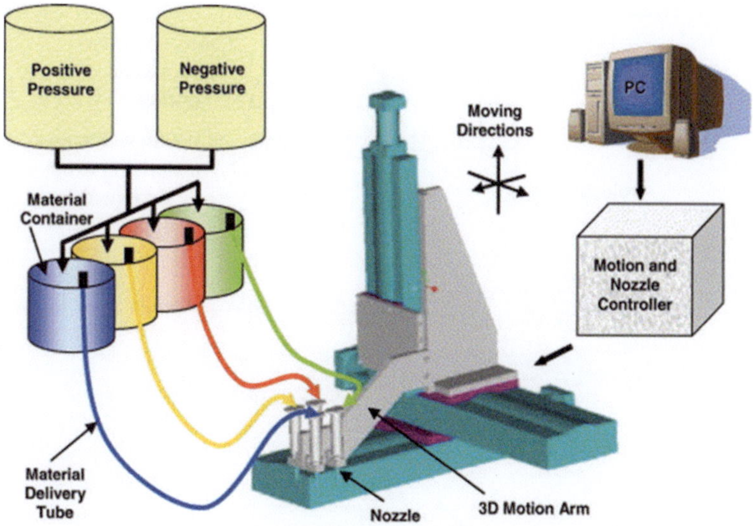

Fig. 20 The multinozzle biopolymer deposition system [89]

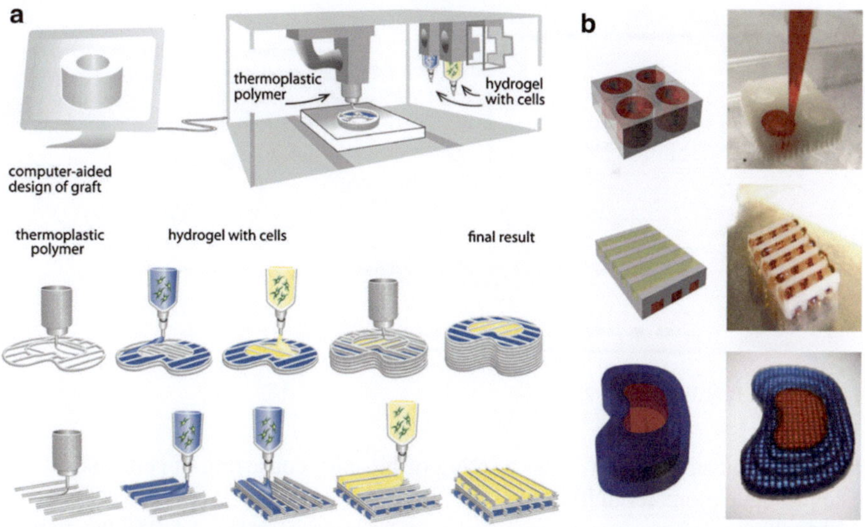

Fig. 21 (**a**) Generic representation of the multi-head printer system to produce hybrid constructs. (**b**) Hybrid constructs produced through the layer-by-layer deposition of thermoplastic PCL filaments and cell-laden hydrogels [87]

chick spinal cord cells [92], human umbilical vein endothelial cells [93] and mouse embryonic stem cells [94]. The modified-LIFT technique uses a pulsed focused beam (UV light) to print cells with micron-scale resolution from a transfer layer or reservoir [83, 95–97]. The modified-LIFT uses two main strategies for cell deposition: (i) the deposition of cells onto an ECM-like layer with variable

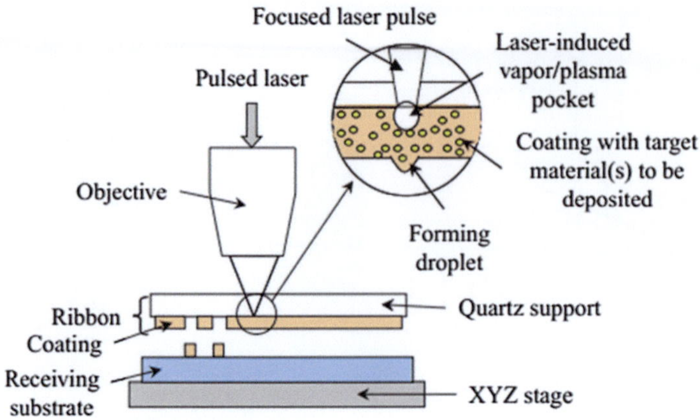

Fig. 22 The MAPLE DW system [98]

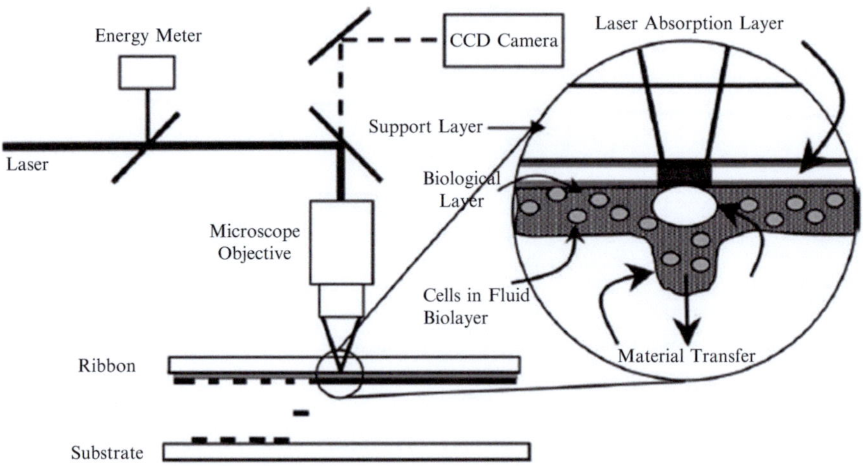

Fig. 23 The BioLP system [97]

thickness (substrate) and (ii) the printing of bio-inks consisting of cells and ECM-like printable biomaterial [95].

Modified-LIFT processes comprise two main techniques [83, 84, 95]: the matrix-assisted pulsed laser evaporation direct write or matrix assisted pulsed laser direct write (MAPLE DW) (Fig. 22) and the biological laser printing (BioLP) or absorbing film-assisted LIFT (Fig. 23).

All variants of the LIFT process include a pulsed laser source, a target from which the biomaterial is printed and a CAD/CAM controlled substrate, placed few hundred micrometers of distance from the target, that collects the cells [84, 95]. LIFT-based processes have been used to print viable mammalian cells, including human osteoprogenitors cells [99], human osteosarcoma cells [100], human endothelial cells [95, 101], rabbit carcinoma cells and human umbilical vein endothelial cells [96].

Table 3 Characteristics of the most commonly used additive biomanufacturing processes to process hydrogels [38, 83, 84]

Process	Advantages	Limitations	Resolution/spot size (μm)
SLA	High accuracy and precision; fabrication of complex geometries with high resolution; easy to achieve small features	Limited number of available photocrosslinkable materials; requires pos-processing and support structures	70–250
3DP	Wide range of processable materials; fast processing; supports are not required; possible use of non-organic binders; low heat effect on the raw powder; induce microporosity	Material in powder form; low mechanical strength; surface finish; entrapped powder; might require pos-processing; use of toxic solvents	200
LGDW	High resolution in cell deposition	Use of capillary to print the cells; low cell throughput; risk of clogging; limited cell concentration	10–30
MAPLE DW	Capillary free; no risk of clogging; more concentrated cell solutions; ability to process viscous fluids and print cells from multiple adjacent reservoirs	Direct interaction between the laser light and the cells; low reproducibility and resolution	30–100
BioLP	Capillary free; no interaction between the laser light and biological material; more concentrated cell solutions; possible to use viscous fluids; high reproducibility; ability to print cells from multiple adjacent reservoirs; no risk of clogging	Limited size of the constructs; difficult to produce constructs in the micrometric range; can only construct small arrays of cells with few hundred cells; work with small volumes	30–100
Ink jet printing	Non-contact printing technology; ability to use multiple printer heads and print multiple types of "inks" in parallel; high print speeds and reduced processing time	Use of nozzles to print the cells; risk of clogging; limited cell concentration, resulting in low overall cell throughput; temperature and shear forces can damage the biological materials	>300

3.3 Summary

Table 3 summarizes the actual state of the art of hydrogel processing with appropriate stereolithographic and inkjet printing processes.

4 Future Challenges

Relevant challenges to be addressed in the future include:

- Achieving a fundamental understanding of manufacturing processes and convergence of techniques for best and affordable healthcare. The combination of multiple fabrication methods to produce a single construct will enable to control its nano, micro and macro characteristics and to produce functional graded structures;
- Automation and industrialization. Traditional laboratory processes are often characterized by a high degree of manual handling operations;
- Establishing a directory of materials and related processes, and correlations between processing conditions and the mechanical and biological behavior of the constructs;
- Development of in situ manufacturing strategies;
- Enhancing multidisciplinarity, linking clinicians and engineers to facilitate further developments and the clinical translation of the devices being developed;
- Scaling up the biomanufacturing processes towards clinical application.

Acknowledgements This work was supported by the Portuguese Foundation for Science and Technology through the projects PTDC/EME-PME/098037/2008 and Pest-OE/EME/UI4044/2011.

References

1. Bártolo PJ, Domingos M, Patrício T, Cometa S, Mironov V (2011) Biofabrication strategies for tissue engineering. In: Fernandes PR, Bártolo PJ (eds) Advances on modeling in tissue engineering. Springer, Dordrecht
2. Liao H-T, Chang K-H, Jiang Y, Chen J-P, Lee M-Y (2011) Fabrication of tissue engineered PCL scaffold by selective laser-sintered machine for osteogenesis of adipose-derived stem cells. Virtual Phys Prototyp 6:57–60
3. Nichol JW, Khademhosseini A (2009) Modular tissue engineering: engineering biological tissues from the bottom up. Soft Matter 5:1312
4. Du Y, Lo E, Ali S, Khademhosseini A (2008) Directed assembly of cell-laden microgels for fabrication of 3D tissue constructs. Proc Natl Acad Sci U S A 105:9522–9527
5. Khademhosseini A, Langer R, Borenstein J, Vacanti JP (2006) Microscale technologies for tissue engineering and biology. Proc Natl Acad Sci U S A 103:2480–2487
6. Rivron NC, Rouwkema J, Truckenmüller R, Karperien M, De Boer J, Van Blitterswijk CA (2009) Tissue assembly and organization: developmental mechanisms in microfabricated tissues. Biomaterials 30:4851–4858
7. Jakab K, Neagu A, Mironov V, Markwald RR, Forgacs G (2004) Engineering biological structures of prescribed shape using self-assembling multicellular systems. Proc Natl Acad Sci U S A 101:2864–2869
8. Bártolo PJ, Chua CK, Almeida HA, Chou SM, Lim ASC (2009) Biomanufacturing for tissue engineering: present and future trends. Virtual Phys Prototyp 4:203–216
9. Bartolo P, Kruth J-P, Silva J, Levy G, Malshe A, Rajurkar K, Mitsuishi M, Ciurana J, Leu M (2012) Biomedical production of implants by additive electro-chemical and physical processes. CIRP Ann Manuf Technol 61:635–655

10. Melchels FPW, Domingos MAN, Klein TJ, Malda J, Bartolo PJ, Hutmacher DW (2012) Additive manufacturing of tissues and organs. Prog Polym Sci 37:1079–1104
11. Meinel AJ, Germershaus O, Luhmann T, Merkle HP, Meinel L (2012) Electrospun matrices for localized drug delivery: current technologies and selected biomedical applications. Eur J Pharm Biopharm 81:1–13
12. Mitchell GR, Ahn K, Davis FJ (2011) The potential of electrospinning in rapid manufacturing processes. Virtual Phys Prototyp 6:63–77
13. Tomatsu I, Peng K, Kros A (2011) Photoresponsive hydrogels for biomedical applications. Adv Drug Deliv Rev 63:1257–1266
14. Deligkaris K, Tadele TS, Olthuis W, van den Berg A (2010) Hydrogel-based devices for biomedical applications. Sensors Actuat B-Chem 147:765–774
15. Lee KY, Mooney DJ (2001) Hydrogels for tissue engineering. Chem Rev 101:1869–1879
16. Fedorovich NE, Alblas J, Hennink WE, Oner FC, Dhert WJA (2011) Organ printing: the future of bone regeneration? Trends Biotechnol 29:601–606
17. Hoare TR, Kohane DS (2008) Hydrogels in drug delivery: progress and challenges. Polymer 49:1993–2007
18. Yang X, Liu Q, Chen X, Yu F, Zhu Z (2008) Investigation of PVA/ws-chitosan hydrogels prepared by combined γ-irradiation and freeze-thawing. Carbohydr Polym 73:401–408
19. Lee KY, Mooney DJ (2012) Alginate: properties and biomedical applications. Prog Polym Sci 37:106–126
20. Van Vlierberghe S, Dubruel P, Schacht E (2011) Biopolymer-based hydrogels as scaffolds for tissue engineering applications: a review. Biomacromolecules 12:1387–1408
21. Van Tomme SR, Storm G, Hennink WE (2008) In situ gelling hydrogels for pharmaceutical and biomedical applications. Int J Pharm 355:1–18
22. Pescosolido L, Vermonden T, Malda J, Censi R, Dhert WJA, Alhaique F, Hennink WE, Matricardi P (2011) In situ forming IPN hydrogels of calcium alginate and dextran-HEMA for biomedical applications. Acta Biomater 7:1627–1633
23. Hamidi M, Azadi A, Rafiei P (2008) Hydrogel nanoparticles in drug delivery. Adv Drug Deliv Rev 60:1638–1649
24. Guvendiren M, Heiney PA, Yang S (2009) Precipitated calcium carbonate hybrid hydrogels: structural and mechanical properties. Macromolecules 42:6606–6613
25. Biondi M, Ungaro F, Quaglia F, Netti PA (2008) Controlled drug delivery in tissue engineering. Adv Drug Deliv Rev 60:229–242
26. Van Vlierberghe S, Vanderleyden E, Dubruel P, De Vos F, Schacht E (2009) Affinity study of novel gelatin cell carriers for fibronectin. Macromol Biosci 9:1105–1115
27. Pereira R, Tojeira A, Vaz DC, Mendes A, Bártolo P (2011) Preparation and characterization of films based on alginate and aloe Vera. Int J Polym Anal Ch 16:449–464
28. Jayakumar R, Prabaharan M, Sudheesh Kumar PT, Nair SV, Tamura H (2011) Biomaterials based on chitin and chitosan in wound dressing applications. Biotechnol Adv 29:322–337
29. Sato K, Yoshida K, Takahashi S, Anzai J (2011) pH-and sugar-sensitive layer-by-layer films and microcapsules for drug delivery. Adv Drug Deliv Rev 63:809–821
30. Caldorera-Moore M, Peppas NA (2009) Micro-and nanotechnologies for intelligent and responsive biomaterial-based medical systems. Adv Drug Deliv Rev 61:1391–1401
31. Bártolo P, Chua CK (2008) Editorial: celebrating the 70th anniversary of professor Yongnian Yan: a life dedicated to science and technology. Virtual Phys Prototyp 3:189–191
32. Danjou S, Köhler P (2010) Improving part quality and process efficiency in layered manufacturing by adaptive slicing. Virtual Phys Prototyp 5:183–188
33. Matias JM, Bartolo PJ, Pontes AV (2009) Modeling and simulation of photofabrication processes using unsaturated polyester resins. J Appl Polym Sci 114:3673–3685
34. Bartolo PJ, Gaspar J (2008) Metal filled resin for stereolithography metal part. CIRP Ann Manuf Technol 57:235–238
35. da Silva BP (2007) Photo-curing modelling: direct irradiation. Int J Adv Manuf Technol 32:480–491

36. Bartolo PJ, Mitchell G (2003) Stereo-thermal-lithography: a new principle for rapid prototyping. Rapid Prototyping J 9:150–156
37. Bartolo PJ (2011) Stereolithographic processes. In: Bártolo PJ (ed) Stereolithography: materials, processes and applications. Springer, New York
38. Peltola SM, Melchels FPW, Grijpma DW, Kellomäki M (2008) A review of rapid prototyping techniques for tissue engineering purposes. Ann Med 40:268–280
39. Bertsch A, Zissi S, Jézéquel JY, Corbel S, André JC (1997) Microstereophotolithography using a liquid crystal display as dynamic mask-generator. Microsyst Technol 3:42–47
40. Fedorovich NE, Oudshoorn MH, van Geemen D, Hennink WE, Alblas J, Dhert WJA (2009) The effect of photopolymerization on stem cells embedded in hydrogels. Biomaterials 30:344–353
41. Arcaute K, Mann B, Wicker R (2010) Stereolithography of spatially controlled multi-material bioactive poly(ethylene glycol) scaffolds. Acta Biomater 6:1047–1054
42. Melchels FPW, Feijen J, Grijpma DW (2010) A review on stereolithography and its applications in biomedical engineering. Biomaterials 31:6121–6130
43. Schuster M, Turecek C, Weigel G, Saf R, Stampfl J, Varga F, Liska R (2009) Gelatin-based photopolymers for bone replacement materials. J Polym Sci A Polym Chem 47:7078–7089
44. Lee K-W, Wang S, Fox BC, Ritman EL, Yaszemski MJ, Lu L (2007) Poly(propylene fumarate) bone tissue engineering scaffold fabrication using stereolithography: effects of resin formulations and laser parameters. Biomacromolecules 8:1077–1084
45. Mironi-Harpaz I, Wang DY, Venkatraman S, Seliktar D (2012) Photopolymerization of cell-encapsulating hydrogels: crosslinking efficiency versus cytotoxicity. Acta Biomater 8:1838–1848
46. Bryant SJ, Anseth KS (2001) The effects of scaffold thickness on tissue engineered cartilage in photocrosslinked poly(ethylene oxide) hydrogels. Biomaterials 22:619–626
47. Lan PX, Lee JW, Seol Y-J, Cho D-W (2009) Development of 3D PPF/DEF scaffolds using micro-stereolithography and surface modification. J Mater Sci Mater Med 20:271–279
48. Lee JW, Kang KS, Lee SH, Kim J-Y, Lee B-K, Cho D-W (2011) Bone regeneration using a microstereolithography-produced customized poly(propylene fumarate)/diethyl fumarate photopolymer 3D scaffold incorporating BMP-2 loaded PLGA microspheres. Biomaterials 32:744–752
49. Seck TM, Melchels FPW, Feijen J, Grijpma DW (2010) Designed biodegradable hydrogel structures prepared by stereolithography using poly(ethylene glycol)/poly(D, L-lactide)-based resins. J Control Release 148:34–41
50. Chan V, Zorlutuna P, Jeong JH, Kong H, Bashir R (2010) Three-dimensional photopatterning of hydrogels using stereolithography for long-term cell encapsulation. Lab Chip 10:2062–2070
51. Zorlutuna P, Jeong JH, Kong H, Bashir R (2011) Stereolithography-based hydrogel microenvironments to examine cellular interactions. Adv Funct Mater 21:3642–3651
52. Arcaute K, Mann BK, Wicker RB (2006) Stereolithography of three-dimensional bioactive poly(ethylene glycol) constructs with encapsulated cells. Ann Biomed Eng 34:1429–1441
53. Arcaute K, Mann BK, Wicker RB (2011) Practical use of hydrogels in stereolithography for tissue engineering applications. In: Bártolo PJ (ed) Stereolithography: materials, processes and applications. Springer, New York
54. Doraiswamy A, Jin C, Narayan RJ, Mageswaran P, Mente P, Modi R, Auyeung R, Chrisey DB, Ovsianikov A, Chichkov B (2006) Two photon induced polymerization of organic-inorganic hybrid biomaterials for microstructured medical devices. Acta Biomater 2:267–275
55. Narayan RJ, Jin C, Doraiswamy A, Mihailescu IN, Jelinek M, Ovsianikov A, Chichkov B, Chrisey DB (2005) Laser processing of advanced bioceramics. Adv Eng Mater 7:1083–1098
56. Ovsianikov A, Deiwick A, Van Vlierberghe S, Dubruel P, Möller L, Dräger G, Chichkov B (2011) Laser fabrication of three-dimensional CAD scaffolds from photosensitive gelatin for applications in tissue engineering. Biomacromolecules 12:851–858
57. Liu VA, Bhatia SN (2002) Three-dimensional photopatterning of hydrogels containing living cells. Biomed Microdevices 4:257–266

58. Domingos MA, Amalvy JI, Oliveira LM, Pinto EM, Almeida HA, Bártolo PJ (2011) Biofabrication of poly(HEMA) scaffolds through Stereolithography. In: 2nd International Conference on Tissue Engineering An ECCOMAS Thematic Conference (ICTE2011), Lisboa, Portugal
59. Lu Y, Mapili G, Suhali G, Chen S, Roy K (2006) A digital micro-mirror device-based system for the microfabrication of complex, spatially patterned tissue engineering scaffolds. J Biomed Mater Res A 77:396–405
60. Saunders RD (2004) Ink-jet printing of human cells. In: Materials Research Society Symposium Proceedings. https://www.escholar.manchester.ac.uk/uk-ac-man-scw:2a158. Accessed 9 Oct 2012
61. Pardo L, Wilson WC, Boland T (2003) Characterization of patterned self-assembled monolayers and protein arrays generated by the Ink-Jet method. Langmuir 19:1462–1466
62. Lam CXF, Mo XM, Teoh SH, Hutmacher DW (2002) Scaffold development using 3D printing with a starch-based polymer. Mater Sci Eng C 20:49–56
63. Sachlos E, Reis N, Ainsley C, Derby B, Czernuszka JT (2003) Novel collagen scaffolds with predefined internal morphology made by solid freeform fabrication. Biomaterials 24:1487–1497
64. Limpanuphap S, Derby B (2002) Manufacture of biomaterials by a novel printing process. J Mater Sci Mater Med 13:1163–1166
65. Park A, Wu B, Griffith LG (1998) Integration of surface modification and 3D fabrication techniques to prepare patterned poly (L-lactide) substrates allowing regionally selective cell adhesion. J Biomater Sci Polym Ed 9:89–110
66. Boland T, Tao X, Damon BJ, Manley B, Kesari P, Jalota S, Bhaduri S (2007) Drop-on-demand printing of cells and materials for designer tissue constructs. Mater Sci Eng C 27:372–376
67. Cohen DL, Malone E, Lipson H, Bonassar LJ (2006) Direct freeform fabrication of seeded hydrogels in arbitrary geometries. Tissue Eng 12:1325–1335
68. Rezende RA, Bártolo PJ, Mendes A, Filho RM (2009) Rheological behavior of alginate solutions for biomanufacturing. J Appl Polym Sci 113:3866–3871
69. Bártolo PJS (2006) State of the art of solid freeform fabrication for soft and hard tissue engineering. WIT Press, Southampton
70. Bartolo P, Mendes A, Jardini A (2004) Bio-prototyping. In: Collins M, Brebbia J (eds) Design and nature II. WIT Press, Southampton
71. Lee G-S, Park J-H, Shin US, Kim H-W (2011) Direct deposited porous scaffolds of calcium phosphate cement with alginate for drug delivery and bone tissue engineering. Acta Biomater 7:3178–3186
72. Mironov V, Kasyanov V, Drake C, Markwald RR (2008) Organ printing: promises and challenges. Regen Med 3:93–103
73. Varghese D, Deshpande M, Xu T, Kesari P, Ohri S, Boland T (2005) Advances in tissue engineering: cell printing. J Thorac Cardiovasc Surg 129:470–472
74. Marga F, Neagu A, Kosztin I, Forgacs G (2007) Developmental biology and tissue engineering. Birth Defects Res C Embryo Today 81:320–328
75. Cui X, Boland T (2009) Human microvasculature fabrication using thermal inkjet printing technology. Biomaterials 30:6221–6227
76. Tirella A, Vozzi F, De Maria C, Vozzi G, Sandri T, Sassano D, Cognolato L, Ahluwalia A (2011) Substrate stiffness influences high resolution printing of living cells with an ink-jet system. J Biosci Bioeng 112:79–85
77. Zheng Q, Lu J, Chen H, Huang L, Cai J, Xu Z (2011) Application of inkjet printing technique for biological material delivery and antimicrobial assays. Anal Biochem 410:171–176
78. Cui X, Dean D, Ruggeri ZM, Boland T (2010) Cell damage evaluation of thermal inkjet Printed Chinese hamster ovary cells. Biotechnol Bioeng 106:963–969
79. Wilson WC Jr, Boland T (2003) Cell and organ printing 1: protein and cell printers. Anat Rec A Discov Mol Cell Evol Biol 272:491–496
80. Derby B (2010) Inkjet printing of functional and structural materials: fluid property requirements, feature stability, and resolution. Ann Rev Mater Res 40:395–414

81. Poncelet D, de Vos P, Suter N, Jayasinghe SN (2012) Bio-electrospraying and cell electrospinning: progress and opportunities for basic biology and clinical sciences. Adv Healthc Mater 1:27–34
82. Scoutaris N, Alexander MR, Gellert PR, Roberts CJ (2011) Inkjet printing as a novel medicine formulation technique. J Control Release 156:179–185
83. Ringeisen BR, Othon CM, Barron JA, Wu PK, Spargo BJ (2009) The evolution of cell printing. In: Meyer U, Meyer T, Handschel J, Wiesmann HP (eds) Fundamentals of tissue engineering and regenerative medicine fundamentals of tissue engineering and regenerative medicine. Springer, Berlin
84. Ringeisen BR, Othon CM, Barron JA, Young D, Spargo BJ (2006) Jet-based methods to print living cells. Biotechnol J 1:930–948
85. Saunders RE, Gough JE, Derby B (2008) Delivery of human fibroblast cells by piezoelectric drop-on-demand inkjet printing. Biomaterials 29:193–203
86. Nakamura M, Kobayashi A, Takagi F, Watanabe A, Hiruma Y, Ohuchi K, Iwasaki Y, Horie M, Morita I, Takatani S (2005) Biocompatible inkjet printing technique for designed seeding of individual living cells. Tissue Eng 11:1658–1666
87. Schuurman W, Khristov V, Pot MW, van Weeren PR, Dhert WJA, Malda J (2011) Bioprinting of hybrid tissue constructs with tailorable mechanical properties. Biofabrication 3:021001
88. Moon S, Hasan SK, Song YS, Xu F, Keles HO, Manzur F, Mikkilineni S, Hong JW, Nagatomi J, Haeggstrom E, Khademhosseini A, Demirci U (2010) Layer by layer three-dimensional tissue epitaxy by cell-laden hydrogel droplets. Tissue Eng Part C Methods 16:157–166
89. Khalil S, Sun W (2007) Biopolymer deposition for freeform fabrication of hydrogel tissue constructs. Mater Sci Eng C 27:469–478
90. Yan Y, Wang X, Pan Y, Liu H, Cheng J, Xiong Z, Lin F, Wu R, Zhang R, Lu Q (2005) Fabrication of viable tissue-engineered constructs with 3D cell-assembly technique. Biomaterials 26:5864–5871
91. Lee W, Debasitis JC, Lee VK, Lee J-H, Fischer K, Edminster K, Park J-K, Yoo S-S (2009) Multi-layered culture of human skin fibroblasts and keratinocytes through three-dimensional freeform fabrication. Biomaterials 30:1587–1595
92. Odde DJ, Renn MJ (2000) Laser-guided direct writing of living cells. Biotechnol Bioeng 67:312–318
93. Nahmias Y, Schwartz RE, Verfaillie CM, Odde DJ (2005) Laser-guided direct writing for three-dimensional tissue engineering. Biotechnol Bioeng 92:129–136
94. Raof NA, Schiele NR, Xie Y, Chrisey DB, Corr DT (2011) The maintenance of pluripotency following laser direct-write of mouse embryonic stem cells. Biomaterials 32:1802–1808
95. Catros S, Guillotin B, Bačáková M, Fricain J-C, Guillemot F (2011) Effect of laser energy, substrate film thickness and bioink viscosity on viability of endothelial cells printed by laser-assisted bioprinting. Appl Surf Sci 257:5142–5147
96. Guillotin B, Souquet A, Catros S, Duocastella M, Pippenger B, Bellance S, Bareille R, Rémy M, Bordenave L, Amédée J, Guillemot F (2010) Laser assisted bioprinting of engineered tissue with high cell density and microscale organization. Biomaterials 31:7250–7256
97. Barron JA, Wu P, Ladouceur HD, Ringeisen BR (2004) Biological laser printing: a novel technique for creating heterogeneous 3-dimensional cell patterns. Biomed Microdevices 6:139–147
98. Lin Y, Huang G, Huang Y, Tzeng T-RJ, Chrisey D (2010) Effect of laser fluence in laser-assisted direct writing of human colon cancer cell. Rapid Prototyp J 16:202–208
99. Catros S, Fricain J-C, Guillotin B, Pippenger B, Bareille R, Remy M, Lebraud E, Desbat B, Amédée J, Guillemot F (2011) Laser-assisted bioprinting for creating on-demand patterns of human osteoprogenitor cells and nano-hydroxyapatite. Biofabrication 3:025001
100. Barron JA, Krizman DB, Ringeisen BR (2005) Laser printing of single cells: statistical analysis, cell viability, and stress. Ann Biomed Eng 33:121–130
101. Guillemot F, Souquet A, Catros S, Guillotin B, Lopez J, Faucon M, Pippenger B, Bareille R, Rémy M, Bellance S, Chabassier P, Fricain JC, Amédée J (2010) High-throughput laser printing of cells and biomaterials for tissue engineering. Acta Biomater 6:2494–2500

The Warburg Effect Is a Guide to Multipurpose Cancer Therapy Including Trace Element Delivery

Anatoliy V. Popov and Leonid G. Menchikov

Abstract The Nobel Prize Winner (1931) Dr. Otto H. Warburg had established that the main energy source of the cancer cell is aerobic glycolysis (the Warburg effect). He also postulated the hypothesis about "the prime cause of cancer", which is a matter of debate till nowadays. On the contrary to the hypothesis, his discovery was completely recognized. However, the discovery had been almost vanished in the heat of battle about the hypothesis. The prime cause of cancer is important for the prevention and diagnosis, and yet the effects that influence on tumor growth are more important for cancer treatment. As a consequence of the Warburg effect, the recovery of the oxygen respiration of cancer cells can inhibit the tumor growth and lead to a remission. Here, we review the current knowledge of the inhibition of abnormal glycolysis and normalization of oxygen respiration of malignant tumor and a potential role of ultra trace element germanium in this process. Efficiency of different Ge-delivery molecules is also discussed.

Keywords Warburg effect • Glycolysis • Cancer cells • Ge-delivery • Positron emission tomography • Treatment tailored to the person

---

A.V. Popov (✉)
Department of Radiology, University of Pennsylvania, 231 South 34th Street, Philadelphia, PA 19104, USA
e-mail: avpopov@mail.med.upenn.edu

L.G. Menchikov (✉)
N.D. Zelinsky Institute of Organic Chemistry, Russian Academy of Sciences, 47 Leninsky prosp., 119991 Moscow, Russian Federation
e-mail: mlg@ioc.ac.ru

J. Coelho (ed.), *Drug Delivery Systems: Advanced Technologies Potentially Applicable in Personalised Treatment*, Advances in Predictive, Preventive and Personalised Medicine 4, DOI 10.1007/978-94-007-6010-3_9,

Otto Heinrich Warburg
(1883–1970)

1 Introduction

The Nobel Prize Winner (1931) Dr. Otto H. Warburg had established [1–7] that the main energy source of the cancer cell is aerobic glycolysis (the Warburg effect). In his opinion, the normalization of the oxygen respiration can slow the growth of cancer cells and, possibly, lead to a full recovery. Alas, no one in the 1920's knew how to accomplish that (i.e. how to stop glycolysis and normalize the respiration of oxygen).

On the basis of his works Warburg formulated the hypothesis, that "the prime cause of cancer is the replacement of the respiration of oxygen in normal body cells by a fermentation of sugar" [8]. A comprehensive review of the works of Warburg was done by his student, H. Krebs [9], Nobel prize-winner (1953) (who thought that the Warburg effect just might be a symptom of an underlying root cause, that could be found at the gene level). For many years, Warburg's theory raised controversy in scientific circles, but the generally accepted theory became that of genetic carcinogenesis [10–12]. It is now assumed that Warburg's discovery only demonstrates a side effect, but is not the prime cause of cancer. Moreover, some tumor cells do not show the Warburg effect. Despite all this, Warburg's hypothesis to this day attracts the attention of scientists, finds an increasing amount of scientific evidence and interpretation [13–30].

On the contrary to the hypothesis, his discovery was completely recognized. Moreover, the positron emission tomography (18-FDG PET) [31] widely used in clinical diagnosis of cancer is based on his discovery (on the fact that cancer cells are highly glycolytic and actively uptake glucose). However, the discovery had been almost vanished in the heat of battle about the hypothesis. The prime cause of cancer is important for the prevention and diagnosis, and yet the effects that influence on tumor growth are more important for cancer treatment. As a consequence of the Warburg effect, the recovery of the oxygen respiration of cancer cells can inhibit the tumor growth and lead to a remission.

Here, we review the current knowledge of the inhibition of abnormal glycolysis and normalization of oxygen respiration of malignant tumor.

2 Towards Warburg Paradigms for Cancer Therapy

Independent of the true causes of carcinogenesis, a number of significant changes in the biochemical parameters in a cancer patient organism can be observed. What is more, the biochemical changes, differ from the genetic ones, are observed not only on the cancer cellular level, but significant changes happen on the level of the whole organism. In particular, lactic acidolysis and enhanced activity of lactate dehydrogenase (LDH) are due to the Warburg effect i.e. alternations in energetic exchange [32]. And even if Warburg's discovery is not the prime cause of cancer, then the normalization of biochemical parameters can, in addition to traditional treatments, improve one's situation. Moreover, the methods of treatment, corresponding to the Warburg's theory complement the traditional molecular genetic hypotheses about the prime cause of cancer, rather than contradict them!

At the moment, the root cause of most diseases is unknown, yet they are effectively treated, saving and prolonging lives. Whenever someone shows deviations in physiological (or biochemical) parameters, modern medicine does its best to normalize them. When temperature is elevated, aspirin is prescribed; for heartburn—antacids; when hemoglobin is low—iron-containing medicines, for diabetes—insulin etc. In addition to that, a diet is prescribed to treat minor deviations in blood parameters (bilirubin, sugar and so on). While the primary weapon of choice for oncology is the knife (as well as X-ray irradiation and cytostatics), no one is concerned with normalizing biochemical parameters during cancer treatment (i.e. the treatment of the disease, not the patient). If they were to be normalized then the tumor would be deprived of vital importance to its components (foremost, nutrition, mechanism of grow, defense from the immune system, etc.), as the result of which, the effectiveness of traditional treatment (now we have synergy) and the rate of survival would increase, which would reduce metastases. And, to the contrary, when tissue experiences a lack of oxygen, it undergoes malignant metamorphosis [33]. Enhanced lactic acid content initiates angiogenesis and intense tumor growth [34].

> And if tomorrow some other kind of therapy were invented—radiation, chemical or herbal, or even something worked by light, color or telepathy—which could save her patients without the knife and would mean that surgery would completely vanish from human practice ... [the surgeon] would not have defended her craft even for a day. (A. Solzhenitsyn, Cancer Ward) [35]

In order to implement this approach we will focus on biochemical changes in cells and organisms that are caused by the Warburg effect. In the time since the discovery of the Warburg effect, many things about its mechanism came to light.

3 Origin of the Warburg Effect, the Modern Ideas on Its Mechanism

Tumor cells have high rate of aerobic glycolysis and defects in oxidative phosphorylation. In a normal cell the sequence of aerobic glycolysis includes a conversion of glucose via pyruvate into acetyl CoA, followed by its complete oxidation into CO_2 and H_2O in the tricarboxylic acid cycle (TCA cycle, Krebs cycle, taken place in mitochondria) and oxidative phosphorylation. As the result, 38 molecules of ATP are generated per one glucose molecule [36]. In contrast to the above process in the cancer tumor cells the main pathway of the aerobic glycolysis includes the conversion of glucose into pyruvate in cytosol (forming only two ATP molecules per one glucose molecule) and onwards into lactic acid, which is emerged from the cell (Fig. 1).

In order to normalize biochemical parameters caused by the Warburg effect we need to first, inhibit glycolysis and neutralize its metabolites (lactic acid and toxins), and second, to activate oxygen-based cleavage of glucose (Krebs cycle). Moreover, it is vital to act on all these manifestations simultaneously.

4 Aerobic Glycolysis

The main final product of a cancer cell's glycolysis is lactic acid, which is formed from the pyruvate under the action of the enzyme lactate dehydrogenase (LDH). The reduction of pyruvate to lactate by means of NADH oxidation is the only source of NAD+ under anaerobic conditions.

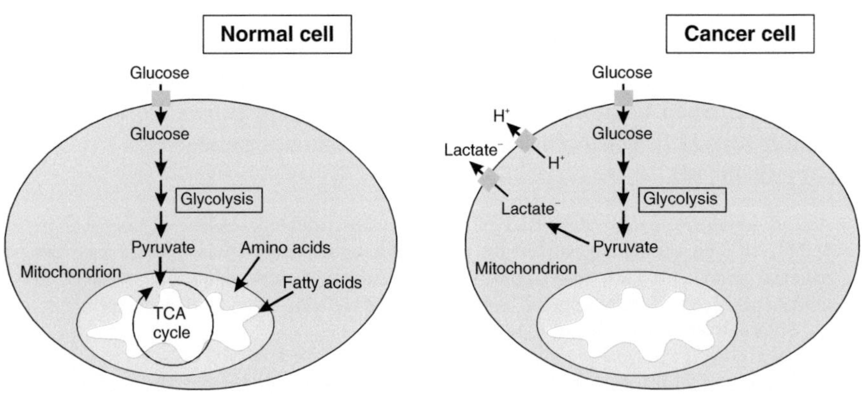

Fig. 1 Main pathways of glucose transformations in normal and cancer cells

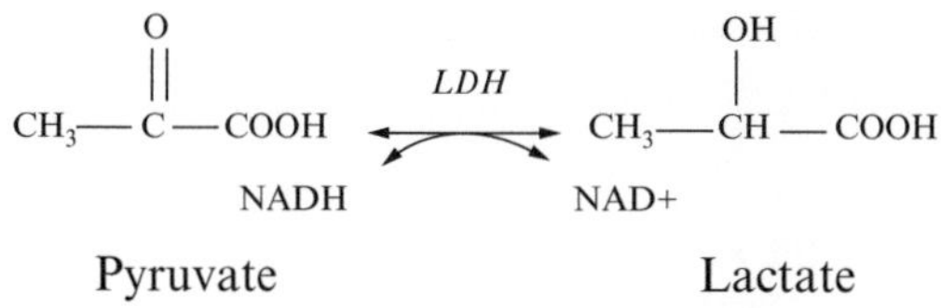

Otto Warburg found [2, 3, 6] that the rate of glycolysis in cancer cells is many times higher than in normal cells. A decrease in glycolysis [37] leads to the loss of nutrition in a cancerous growth (i.e. energy). Because of this, glycolysis inhibitors suppress the growth of cancer cells and can be used, in combination with traditional methods, to successfully treat cancer. The reviews [38–44] summarize various mechanics of glycolysis in cancer cells and other related metabolic processes, in addition to various methods that can be used to suppress glycolysis in every stage (Fig. 2). Currently a large number of tumor cell targets and glycolysis-suppressing compounds [38, 40, 45–48] are known which inhibit glycolysis at various stages and are being intensively researched for their antitumor properties.

Another consequence of the Warburg effect is that not only do metabolites of cancer cells provoke the intoxication of an organism with paraneoplastic syndrome, but also trigger mechanisms that help the tumor successfully defend itself against the immune system, block apoptosis, inhibit the respiration of oxygen in the cell (the Warburg effect, anti-Pasteur), initiate angiogenesis, and ensure the growth of the tumor. (Apart from this there is observed intoxication by tumor decay products under a common therapy with cytostatic medicines). It is therefore important to suppress the aerobic glycolysis, as well as to neutralize byproducts and remove its from the organism.

Currently, several things are offered for the multipurpose treatment of cancer separately, such as inhibitors of angiogenesis [49], inducers of apoptosis [50, 51] etc. But in order for their actions during the treatment of cancer to be effective it is imperative to neutralize existing metabolites of glycolysis. Otherwise even a direct impact on glycolysis could not give the desired results [52]. Lactic acid is the one of the major metabolites of glycolysis, which characterizes the Warburg effect. It is one of the strongest monocarboxylic acids (Table 1) [53].

That is why an excessive formation of lactic acid leads inevitably to the excretion of a number of minerals and trace elements from the organism, and alteration of their ratio in body. Such changes in element composition in cancer are well known (e.g. see [54–60]).

It is minerals and trace elements together with enzymes and other organic compounds participate in the processes of detoxification and neutralization of excess of the acid and, as a consequence, they are excreted from the body. Thus, for successful disease treatment it is necessarily to ensure adequate intake of minerals and trace elements as well as to inhibit glycolysis in order to avoid their loss on the organism.

In general, minerals and trace elements play an important role in a living cell. Thank to them there happens delivery of different nutrients (vitamins, enzymes, etc)

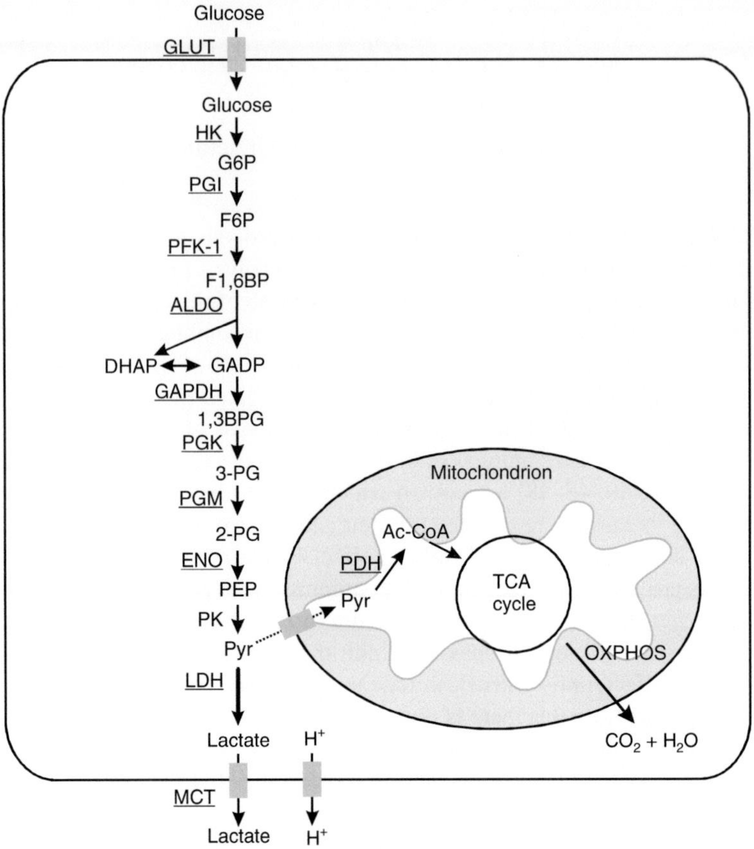

Fig. 2 Glucose metabolism. *G6P* glucose 6-phosphate; *F6P* fructose 6-phosphate; *F1,6BP* fructose 1,6-bisphosphate; *DHAP* dihydroxyacetone phosphate; *GADP* glyceraldehyde 3-phosphate; *1,3BPG* 1,3-bisphosphoglycerate; *3-PG* 3-phosphoglycerate; *2-PG* 2-phosphoglycerate; *PEP* phosphoenolpyruvate; *Pyr* pyruvate; *GLUT* glucose transporter; *MCT* monocarboxylic acid transporter; *OXPHOS* oxidative phosphorylation. Enzymes: *HK* hexokinase; *PGI* phosphoglucose isomerase; *PFK-1* phosphofructokinase; *ALDO* fructose bisphosphate aldolase; *TPI* triosephosphate isomerase; *GAPDH* glyceraldehyde phosphate dehydrogenase; *PGK* phosphoglycerate kinase; *PGM* phosphoglycerate mutase; *ENO* enolase; *PK* pyruvate kinase; *LDH* lactate dehydrogenase; *PDH* pyruvate dehydrogenase

Table 1 Acid dissociation constants of monocarboxylic acids [53]

Formula	Name	pK_a
HCOOH	Formic acid	3.75
$CH_3CH(OH)COOH$	**Lactic acid**	**3.86**
CH_3COOH	Acetic acid	4.76
C_2H_5COOH	Propionic acid	4.87
$C_{17}H_{33}COOH$	Oleic acid	4.86

to the right sites of the body (trace elements act as a delivery system); and trace element composition is changed during almost any disease, even with burns and bruises.

Minerals and trace elements play crucial roles in the active sites of enzymes; they are part of the key fragments proteins (e.g. zinc finger of p53); they are needed to perform the vast majority of bodily functions (i.e. global biochemical reactions). Interaction at the level of genes and proteins are just one of the links in a long chain of biochemical reactions in the cell and organism as a whole. But at the elementary level, the end result is provided by the classical biochemical reactions. And here is the presence or absence of the micronutrient is critical for the entire chain, and if you do not have the necessary trace element, then any manipulation at the genetic level will be fruitless. For example, if zinc deficiency is observed, then the p53 tumor suppressor can not perform its functions and fail to run apoptosis with its help [61]. Thus, the normalization of the trace element composition is a prerequisite for the normalization of biochemistry of the organism as a whole.

5 About Lactic Acid

The "enemy" of minerals and trace elements, lactic acid, can also cause symptoms of acidosis [62] and run the mechanisms for tumor growth [63, 64].

To neutralize the excess lactic acid and maintain a normal pH in the most obvious is a simple method of baking soda ($NaHCO_3$) or other alkali metal salts. This hypothesis has received theoretical basis and practical confirmation [62, 65–68] (although for some time the effectiveness of treatment with baking soda was questionable [69]). Thus, it was shown that oral administration of $NaHCO_3$ selectively increases the pH of tumors and reduces the formation of spontaneous metastases [66]. Inhibition of tumor acidity by bicarbonate will reduce the incidence of in vivo metastases [66].

There are also other ways to deal with acidity [70], for example, proton pump inhibitors [71, 72], which prevent the release of lactic acid from the tumor cell. For the detoxification of the organism from other metabolites of cancerous tumors, in clinical practice it is possible to use, for instance, hemoperfusion [73, 74], and hemodialysis [75]. For the detoxification of separate metabolites plenty of different methods and medications are suggested.

6 Oxygen Respiration

To the contrast of the inhibition of the glycolysis of cancer cells, mechanisms that are sufficiently researched, [38, 39] the normalization of the oxygen (aerobic) respiration cell presents itself as a more difficult problem. Currently, an intensive search for drugs that can effectively impact the respiration of cancer cells is underway.

One of the classes of substances normalizing aerobic respiration of cells are germanium compounds [76–78]. Their stimulating effects on oxidizing enzymes, such as aldose reductase [79] are also proven. The well-known natural source of germanium, Ginseng, is also a stimulator of aerobic respiration in cells [80]. Germanium is one of the key ultra trace elements in organisms, the lack of which causes several diseases, including those that are oncogenic [81].

Among the plant sources of germanium, the ones with highest concentration are Shelf fungus (*Ganoderma lucidum, Chaga*), Ginseng, garlic, aloe, and Echinacea [78, 82–87]. All of the above fungi, and plants, are used (and successfully) not only in traditional, but also in modern medicine, including in the treatment of cancer [88–90].

The antitumor properties of germanium compounds (in the form of inorganic compounds) were established in 1928 [91] and have been confirmed multiple times since then [92–95]. But widespread implementations of germanium compounds in medical practice only came about with the advent of water-soluble organic compounds of germanium.

Water-soluble organic compounds of germanium were first discovered in 1965 during the hydrolysis of adducts of $HGeCl_3$ to 1,3-enones (methyl methacrylate, as well as aldol condensation product of cyclohexanone) [96, 97]. Later, other alkene compounds were involved into this reaction.

$$HGeCl_3 + CH_2 = CHR \rightarrow Cl_3GeCH_2CH_2R \rightarrow (O_{1.5}GeCH_2CH_2R)_2$$

By far the most well-known one is bis-2-carboxyethyl germanium sesquioxide (Ge-132, CEGS) [98]. It started the active research of the biological activity of organic compounds of germanium and its widespread use in medical treatments, especially in complex treatments of cancer [78]. Several successful clinical confirmed cases of its use in the treatment of cancer are known, for instance, the complete remission of pulmonary spindle cell carcinoma after treatment with oral germanium sesquioxide [99]. In multipurpose cancer therapy, germanium sesquioxide can significantly reduce the dose and toxicity of conventional cytotoxic agents [100]. Its antitumor activity and toxicity is very well known at this time [101, 102]. The toxicity of germanium-based organic compounds is no more than that of the table salt, NaCl [102, 103]; Mice: GeO_2 – LD_{50}: Oral: 5,400 mg/kg. Germartranol—LD_{50}: Oral: 8,400 mg/kg, Intravenous: LD50: 300 mg/kg; for Ge-132 – LD50 is >6.300 mg/kg orally for mice, >10,000 mg/kg orally for rats, and >1,000 mg/kg intravenously for rats. Discussion of the supposedly high toxicity of germanium compounds—that lasted for a long time—was explained away as simply a typo [104, 105] in a paper that contained erroneous values. Even though the mistake was quickly noticed in the same journal and fixed in an errata, many works cited and specifically discussed the erroneous data.

The observed clinical toxicity of Ge-132 [106–110] (in particular, nephrotoxicity and neurotoxicity in humans) is due to the current practice of applying cytostatic drugs in doses that are borderline toxic, aggravated by the extremely low solubility of Ge-132. The toxicity of Ge-132 is due to the fact that, in massive doses, its use in a live organism leads to its hydrolysis into GeO_2, which precipitates and, in particular, blocks the renal pelvis. One needs to be a *kamikaze* to take per oral 328 g (!!!) **ultra trace** element germanium and thus determine its lethal dose in Humans. Using a readily soluble organogermanium it is easily possible to solve this problem and to take medication in dosages that are too far from toxic. From 1965, the moment when germanium sesquioxide was discovered, many new soluble organogermanium compounds were synthesized (germanium sesquioxide derivatives, germatranes, etc.) and their properties and physiological activity were investigated [111–117]. Biological activity of readily soluble compounds of germanium was found to be orders of magnitude higher than that of Ge-132 [114, 118–121] (i.e., water solubility determines physiological activity of Ge-containing compounds). This allows them to be effectively used in trace doses.

In reality, the spectrum of the biological activity of germanium compounds is quite large [76], and their known properties have several special features [122]. It is known that Ge-132 does not show direct cytotoxicity, and that its antitumor activity is purely secondary (in particular, the antitumor effect is the result of activation of macrophase, an increase of NK activity and increase in humoral immunity) [123–126]. During the research of the antitumor activity of germanium compounds, only traditional tests were conducted to determine direct cytotoxicity, which, according to Warburg, are a mistake. The things are, that even after the restoration of normal respiration of oxygen, a cell's full recovery can take an extended amount of time, and traditional cytotoxicity tests would be negative (or show poor performance) even in the best potential anti-cancer resources according to Warburg. As a result of this, in agreement with the research of Warburg, during the search of new drugs, one needs to first consider the degree of the normalization of the respiration of oxygen of a cell (reduction of glycolysis and an increase in oxygen consumption or indirect cell tests, for example, the activity of respiratory enzymes, or possibly even just the basic content of trace elements, which are directly responsible for carrying oxygen to the enzyme). The definition of such activity would also have to be modified and new methods developed in order to more effectively determine it. This can help in the development of new anti-tumor drugs that would be aimed to defects correcting in tumor cells without harming normal ones.

7 Conclusion

In this manner, at the current time a large number of different drugs are undergoing clinical trials that are targeted at separate stages of the Warburg effect, foremost those that inhibit glycolysis of cancer cells [42].

Of course, these drugs give way to positive results in several cases, e.g. the use of aspirin as an LDH inhibitor [65, 66], detoxication—hemoperfusion [73, 74]. As for germanium, the case of complete remission of pulmonary spindle cell carcinoma after treatment with oral germanium sesquioxide was described in the medicinal literature [99].

In accordance with the principles of Warburg's only their simultaneous use in order to normalize biochemical parameters may lead to success. But, unfortunately, these drugs are not being used all together in modern medical practice.

At the same time, Warburg's principles are successfully being used in traditional medicine (for example, in Chinese and Russian,) where several different drugs are being used based on the time of day (i.e. one taken in the morning, the second—in the afternoon, the third—in the evening).

The first group of drugs in traditional medicine consist of glycolysis inhibitors and means for the neutralization of its metabolites. The main component of the glycolysis inhibitor group is raw materials, rich in salicylic or oxalic acid (sorrel, horsetail, willow bark, meadowsweet, etc.) as well as urea and urinotherapy. It is known that oxalic acid [127, 128] and salicylic acid [129], urea [127] and urine peptides [130] are inhibitors of LDH, i.e. inhibit anaerobic glycolysis. Further, baking soda is used (or ash of hardwood trees, which is enriched with potash), to neutralize excess lactic acid (acidosis) and normalize pH. Finally, burdock root, *Bidens tripartita*, and other herbs are widely known to possess blood-cleansing properties. For example, burdock is used [131] for detoxification even during a fever, poisoning by salts of heavy metals, as well as insect and poisonous snake bites.

The next main group of drugs in traditional medicine are known normalizes of oxygen respiration in cells— saprophytes fungi (Bracket fungi, aka Shelf fungi), in particular, Chaga (*Inonotus obliquus*, Chaga [132], in Russian medicine) and tinder varnished (*Ganoderma lucidum* [133, 134] in Chinese medicine). In addition, this group of drugs includes ginseng [135], aloe [136, 137] and garlic [138–140]. It is necessary to note that all just mentioned plants contain Ge in enhanced level.

> ... Mightn't it be that the same *chaga* that had cured the Russian peasants of cancer for centuries without their even knowing it?
>
> You mean they used it as a prophylactic? (A. Solzhenitsyn, Cancer Ward) [35]

The Nobel Prize Winner (1972) and writer Alexander Solzhenitsyn (1918–2008) cured his own cancer with the help of traditional medicine, in particular, with the use of Chaga, a birch fungus. In 1952 Alexander Solzhenitsyn found out that he had seminoma cancer and was given a terminal prognosis. His experiences there became the basis of his novel "Cancer Ward" (1967).

In contrast to traditional medicine, almost all currently known methods and drugs in modern medicine are targeted towards the elimination of only one of the manifestations of the Warburg effect. That is why they are so ineffective! The time has come to unite all these parts together.

In order to accomplish this, it is necessary to—in additional to traditional methods—normalize biochemical markers caused by the Warburg effect, in particular, first, inhibit glycolysis and neutralize its metabolites (lactic acid and toxics), and second, activate oxygen cleavage of glucose. Moreover, it is necessary to treat these manifestations simultaneously.

> But how long prevention [of cancer] will be avoided depends on how long the prophets of agnosticism will succeed in inhibiting the application of scientific knowledge in the cancer field (Otto Warburg) [8].

Acknowledgements Supported in part (AVP) by Grant #IRG-78-002-31 from the American Cancer Society.

References

1. Warburg O, Posener K, Negelein E (1924) Metabolism of carcinoma cells. Biochem Z 152:309–344
2. Warburg O (1930) On metabolism of tumors. Constable, London
3. Warburg O (1931) The metabolism of tumours. Richard R Smith Inc., New York
4. Warburg OH (1931) Nobel prize in medicine. http://nobelprize.org/nobel_prizes/medicine/laureates/1931/warburg-bio.html. Accessed 10 Jan 2013
5. Warburg O (1956) On the origin of cancer cells. Science 123:309–314
6. Warburg O, Wind F, Negelein E (1927) The metabolism of tumors in the body. J Gen Physiol 8:519–530
7. Warburg O (2010) The classic: the chemical constitution of respiration ferment. Clin Orthop Relat Res 468:2833–2839
8. Warburg OH (1969) The prime cause and prevention of cancer. http://healingtools.tripod.com/primecause1.html/. Accessed 10 Jan 2013
9. Krebs HA (1972) Otto Heinrich Warburg, 1883–1970. Biogr Mem Fellows R Soc 18:629–699
10. Bertram JS (2000) The molecular biology of cancer. Mol Asp Med 21:167–223
11. Grandér D (1998) How do mutated oncogenes and tumor suppressor genes cause cancer? Med Oncol 15:20–26
12. Croce CM (2008) Oncogenes and cancer. N Engl J Med 358:502–511
13. Vazquez A, Liu J, Zhou Y, Oltvai ZN (2010) Catabolic efficiency of aerobic glycolysis: the Warburg effect revisited. BMC Syst Biol 4:1–9
14. Schulz TJ, Thierbach R, Voigt A, Drewes G, Mietzner B, Steinberg P, Pfeiffer AF, Ristow M (2006) Induction of oxidative metabolism by mitochondrial frataxin inhibits cancer growth: Otto Warburg revisited. J Biol Chem 281:977–981
15. Pedersen P (2007) Warburg, Me and Hexokinase 2: multiple discoveries of key molecular events underlying one of cancers' most common phenotypes, the "Warburg Effect", i.e., elevated glycolysis in the presence of oxygen. J Bioenerg Biomembr 39:211–222
16. Michelakis ED (2008) Mitochondrial medicine: a new era in medicine opens new windows and brings new challenges. Circulation 117:2431–2434
17. Kiebish MA, Han X, Cheng H, Chuang JH, Seyfried TN (2008) Cardiolipin and electron transport chain abnormalities in mouse brain tumor mitochondria: lipidomic evidence supporting the Warburg theory of cancer. J Lipid Res 49:2545–2556
18. Miles KA, Williams RE (2008) Warburg revisited: imaging tumour blood flow and metabolism. Cancer Imaging 8:81–86
19. Garber K (2004) Energy boost: the Warburg effect returns in a new theory of cancer. J Natl Cancer Inst 96:1805–1806
20. Kritikou E (2008) Cancer biology: Warburg effect revisited. Nat Rev Mol Cell Biol 9:264
21. Racker E, Spector M (1981) Warburg effect revisited: merger of biochemistry and molecular biology. Science 213:303–307
22. Racker E (1981) Warburg effect revisited. Science 213:1313
23. Liu X, Wang X, Zhang J, Lam EK, Shin VY, Cheng AS, Yu J, Chan FK, Sung JJ, Jin HC (2010) Warburg effect revisited: an epigenetic link between glycolysis and gastric carcinogenesis. Oncogene 29:442–450
24. Langbein S, Zerilli M, Zur Hausen A, Staiger W, Rensch-Boschert K, Lukan N, Popa J, Ternullo MP, Steidler A, Weiss C, Grobholz R, Willeke F, Alken P, Stassi G, Schubert P, Coy JF (2006) Expression of transketolase TKTL1 predicts colon and urothelial cancer patient survival: Warburg effect reinterpreted. Br J Cancer 94:578–585

25. Ristow M, Cuezva JM (2009) Oxidative phosphorylation and cancer: the ongoing Warburg hypothesis. In: Apte SP, Sarangarajan R (eds) Cellular respiration and carcinogenesis. Humana Press, New York
26. Czarnecka AM, Czarnecki JS, Kukwa W, Cappello F, Scińska A, Kukwa A (2010) Molecular oncology focus—is carcinogenesis a 'mitochondriopathy'? J Biomed Sci 17:31
27. Shanmugam M, McBrayer SK, Rosen ST (2009) Targeting the Warburg effect in hematological malignancies: from PET to therapy. Curr Opin Oncol 21:531–536
28. Singh K, Costello L (2009) Mitochondria and cancer. Springer Science, New York
29. Chance (2005) Was Warburg right? Or was it that simple? Cancer Biol Ther 4:125–126
30. Scatena R, Bottoni P, Pontoglio A, Giardina B (2010) Revisiting the Warburg effect in cancer cells with proteomics. The emergence of new approaches to diagnosis, prognosis and therapy. Proteomics Clin Appl 4:143–158
31. Pauwels EK, Sturm EJ, Bombardieri E, Cleton FJ, Stokkel MP (2000) Positron-emission tomography with [18F]fluorodeoxyglucose. J Cancer Res Clin Oncol 126:549–559
32. Weljie AM, Jirik FR (2011) Hypoxia-induced metabolic shifts in cancer cells: moving beyond the Warburg effect. Int J Biochem Cell Biol 43:981–989
33. Goldblatt H, Cameron G (1953) Induced malignancy in cells from rat myocardium subjected to intermittent anaerobiosis during long propagation in vitro. J Exp Med 97:525–552
34. Dhup S, Dadhich RK, Porporato PE, Sonveaux P (2012) Multiple biological activities of lactic acid in cancer: influences on tumor growth, angiogenesis and metastasis. Curr Pharm 18:1319–1330
35. Solzhenitsyn A (1969) Cancer ward. Farrar, Straus and Giroux, New York
36. Voet D, Voet JG (2004) Biochemistry, 3rd edn. Wiley, Hoboken
37. Kim JW, Dang CV (2006) Cancer's molecular sweet tooth and the Warburg effect. Cancer Res 66:8927–8930
38. Pelicano H, Martin DS, Xu RH, Huang P (2006) Glycolysis inhibition for anticancer treatment. Oncogene 25:4633–4646
39. Marín-Hernández A, Gallardo-Pérez JC, Rodríguez-Enríquez S, Encalada R, Moreno-Sánchez R, Saavedra E (2011) Modeling cancer glycolysis. Biochim Biophys Acta 1807: 755–767
40. Kroemer G, Pouyssegur J (2008) Tumor cell metabolism: cancer's Achilles' heel. Cancer Cell 13:472–482
41. Cairns RA, Harris IS, Mak TW (2011) Regulation of cancer cell metabolism. Nat Rev Cancer 11:85–95
42. Tennant DA, Duran RV, Gottlieb E (2010) Targeting metabolic transformation for cancer therapy. Nat Rev Cancer 10:267–277
43. Porporato PE, Dadhich RK, Dhup S, Copetti T, Sonveaux P (2011) Anticancer targets in the glycolytic metabolism of tumors: a comprehensive review. Front Pharmacol 2:49
44. Lu W, Huang P (2010) Glycolytic pathway as a target for tumor inhibition. In: Bagley RG (ed) The tumor microenvironment. Springer, New York
45. Scatena R, Bottoni P, Pontoglio A, Mastrototaro L, Giardina B (2008) Glycolytic enzyme inhibitors in cancer treatment. Expert Opin Investig Drugs 17:1533–1545
46. Israel M, Schwartz L (2011) On the metabolic origin of cancer: substances that target tumor metabolism. Biomed Res 22:132–166
47. Israel M, Schwartz L (2011) The metabolic advantage of tumor cells. Mol Cancer 10:70
48. Vander Heiden MG (2011) Targeting cancer metabolism: a therapeutic window opens. Nat Rev Drug Discov 10:671–684
49. Miller T, Isenberg J, Roberts D (2009) Molecular regulation of tumor angiogenesis and perfusion via redox signaling. Chem Rev 109:3099–3124
50. Senderowicz AM (2003) Cell cycle modulators for the treatment of lung malignancies. Clin Lung Cancer 5:158–168

51. Zerbini LF, Czibere A, Wang Y, Correa RG, Otu H, Joseph M, Takayasu Y, Silver M, Gu X, Ruchusatsawat K, Li L, Sarkar D, Zhou JR, Fisher PB, Libermann TA (2006) A novel pathway involving melanoma differentiation associated gene-7/interleukin-24 mediates nonsteroidal anti-inflammatory drug-induced apoptosis and growth arrest of cancer cells. Cancer Res 66:11922–11931
52. Sattler UG, Hirschhaeuser F, Mueller-Klieser WF (2010) Manipulation of glycolysis in malignant tumors: fantasy or therapy? Curr Med Chem 17:96–108
53. Dawson RMC, Elliot DC, Elliot WH, Jones KM (1986) Data for biochemical research. Oxford Science Publications, Oxford & New York
54. Schwartz MK (1975) Role of trace elements in cancer. Cancer Res 35:3481–3487
55. Gregoriadis GC, Apostolidis NS, Romanos AN, Paradellis TP (1983) A comparative study of trace elements in normal and cancerous colorectal tissues. Cancer 52:508–519
56. Drake EN II, Sky-Peck HH (1989) Discriminant analysis of trace element distribution in normal and malignant human tissues. Cancer Res 49:4210–4215
57. Zaichick V, Tsyb A, Vtyurin B (1995) Trace elements and thyroid cancer. Analyst 120: 817–821
58. Carvalho ML, Magalhães T, Becker M, von Bohlen A (2007) Trace elements in human cancerous and healthy tissues: a comparative study by EDXRF, TXRF, synchrotron radiation and PIXE. Spectrochim Acta B 62:1004–1011
59. Piacenti da Silva M, Zucchi OLAD, Ribeiro-Silva A, Poletti ME (2009) Discriminant analysis of trace elements in normal, benign and malignant breast tissues measured by total reflection X-ray fluorescence. Spectrochim Acta B 64:587–592
60. Cobanoglu U, Demir H, Sayir F, Duran M, Mergan D (2010) Some mineral, trace element and heavy metal concentrations in lung cancer. Asian Pac J Cancer Prev 11:1383–1388
61. Orlova M, Orlov A (2011) Role of zinc in an organism and its influence on processes leading to apoptosis. Br J Med Med Res 1:239–305
62. Fields ALA, Wolman SL, Halperin ML (1981) Chronic lactic acidosis in a patient with cancer: therapy and metabolic consequences. Cancer 47:2026–2029
63. Chen JL, Merl D, Peterson CW, Wu J, Liu PY, Yin H, Muoio DM, Ayer DE, West M, Chi JT (2010) Lactic acidosis triggers starvation response with paradoxical induction of TXNIP through MondoA. PLoS Genet 6:e1001093
64. Parks SK, Chiche J, Pouyssegur J (2011) pH control mechanisms of tumor survival and growth. J Cell Physiol 226:299–308
65. Silva AS, Yunes JA, Gillies RJ, Gatenby RA (2009) The potential role of systemic buffers in reducing intratumoral extracellular pH and acid-mediated invasion. Cancer Res 69: 2677–2684
66. Robey IF, Baggett BK, Kirkpatrick ND, Roe DJ, Dosescu J, Sloane BF, Hashim AI, Morse DL, Raghunand N, Gatenby RA, Gillies RJ (2009) Bicarbonate increases tumor pH and inhibits spontaneous metastases. Cancer Res 69:2260–2268
67. Forsythe SM, Schmidt GA (2000) Sodium bicarbonate for the treatment of lactic acidosis. Chest 117:260–267
68. Takigawa S, Sugano N, Ochiai K, Arai N, Ota N, Ito K (2008) Effects of sodium bicarbonate on butyric acid-induced epithelial cell damage in vitro. J Oral Sci 50:413–417
69. Cuhaci B, Lee J, Ahmed Z (2000) Sodium bicarbonate controversy in lactic acidosis. Chest 118:882–884
70. McCarty MF, Whitaker J (2010) Manipulating tumor acidification as a cancer treatment strategy. Altern Med Rev 15:264–272
71. Fais S (2010) Proton pump inhibitor-induced tumour cell death by inhibition of a detoxification mechanism. J Intern Med 267:515–525
72. Spugnini EP, Citro G, Fais S (2010) Proton pump inhibitors as anti vacuolar-ATPases drugs: a novel anticancer strategy. J Exp Clin Cancer Res 29:1–5
73. Tani T, Hanasawa K, Endo Y, Yoshioka T, Kodama M, Kaneko M, Uchiyama Y, Akizawa T, Takahasi K, Sugai K (1998) Therapeutic apheresis for septic patients with organ dysfunction: hemoperfusion using a polymyxin B immobilized column. Artif Organs 22:1038–1044

74. Nikolaev VG, Sakhno LA, Snezhkova EA, Sarnatskaya VV, Yushko LA (2011) Carbon adsorbents in oncology: achievements and perspectives. Exp Oncol 33:2–8
75. Chan FH, Carl D, Lyckholm LJ (2009) Severe lactic acidosis in a patient with B-cell lymphoma: a case report and review of the literature. Case Report Med 2009:534561
76. Goodman S (1988) Therapeutic effects of organic germanium. Med Hypotheses 26:207–215
77. Levine SA (1987) Organic germanium. A novel dramatic immunostimulant. J Orthomol Med 2:83–87
78. Asai K (1980) The miracle cure: organic germanium. Japan Publications, New York
79. Unakar NJ, Tsui J, Johnson M (1997) Effect of pretreatment of germanium-132 on Na(+)-K(+)-ATPase and galactose cataracts. Curr Eye Res 16:832–837
80. Chen XC, Zhu YG, Zhu LA, Huang C, Chen Y, Chen LM, Fang F, Zhou YC, Zhao CH (2003) Ginsenoside Rg1 attenuates dopamine-induced apoptosis in PC12 cells by suppressing oxidative stress. Eur J Pharmacol 473:1–7
81. Marczynski B (1988) Carcinogenesis as the result of the deficiency of some essential trace elements. Med Hypotheses 26:239–249
82. McMahon M, Regan F, Hughes H (2006) The determination of total germanium in real food samples including Chinese herbal remedies using graphite furnace atomic absorption spectroscopy. Food Chem 97:411–417
83. Ohri LK, Vicari SM, Malone PM (1993) Germanium use and associated adverse effects: a review. J Pharm Technol 9:237–241
84. Zhen Y, Gu X, He S, Wang L (1993) Determination of Germanium in lucid ganoderma and ginseng by GFASS. Chin J Modern Appl Pharm 10:11–12
85. Avula B, Wang YH, Smillie TJ, Duzgoren-Aydin NS, Khan IA (2010) Quantitative determination of multiple elements in botanicals and dietary supplements using ICP-MS. J Agric Food Chem 58:8887–8894
86. Rosenberg E (2009) Germanium: environmental occurrence, importance and speciation. Rev Environ Sci Bio/Technol 8:29–57
87. Yang LL, Zhang DQ (2002) Direct determination of germanium in botanical samples by graphite furnace atomic absorption spectrometry with palladium–zirconium as chemical modifier. Talanta 56:1123–1129
88. Jamison JR (2003) Ginseng leutherococcus senticosus and panax ginseng. In: Jamison JR (ed) Clinical guide to nutrition & dietary supplements in disease management. Churchill Livingstone, London
89. Pham PTK, Primack A (2003) Cancer. In: Spencer JW, Jacobs JJ (eds) Complementary and alternative medicine: an evidence-based approach, 2nd edn. Mosby, Saint Louis
90. Najm WI (2009) Cáncer de pulmón. In: Rakel D (ed) Medicina integrativa. Elsevier, Espana
91. Ishiwara F (1928) The influence of different types of metals on mouse carcinoma. Ber Gesamte Physiol Exp Pharmakol 49:615
92. Rothermundt M, Burschkies K (1936) Germanium in chemotherapy. Z Immunitätsforsch Exp Ther 87:445–448
93. Kanisawa M, Schroeder HA (1967) Life-term studies on the effects of arsenic, germanium, tin, and vanadium on spontaneous tumors in mice. Cancer Res 27A:1192–1195
94. Atassi G (1985) Antitumor and toxic effects of silicon, germanium, tin and lead compounds. Rev Silicon, Germanium, Tin Lead Compd 8:219–235
95. Ward SG, Taylor RC (1988) Antitumor activity of the main-group metallic elements: aluminum, gallium, indium, thallium, germanium, antimony, and bismuth. In: Gielen M (ed) Metal-based antitumor drugs, vol 1. Freund Publ. House Ltd., London
96. Kolesnikov SP (1966) Research in chemistry of trihalogermans and germanium analogues of dihalocarbenes. Thesis. ND Zelinsky Institute of Organic Chemistry, Academy of Sciences of USSR (in Russian)
97. Kolesnikov SP, Nefedov OM (1967) About interaction trichlorogermane with ketones (for the posts, TK Gar and VF Mironov). Zh Obshch Khim 37:746, English transl
98. Mironov VF, Berliner EM, Gar TK (1967) Reactions of trichlorogermane with acrylic acid and its derivatives. Zh Obshch Khim 37:911–912, English transl

99. Mainwaring MG, Poor C, Zander DS, Harman E (2000) Complete remission of pulmonary spindle cell carcinoma after treatment with oral germanium sesquioxide. Chest 117:591–593
100. Ishikawa M, Ozaki M, Sekimoto R, Takayanagi Y, Sasaki K (1993) Germanium sesquioxide, a biological response modifier, counteracts toxic but not antineoplastic effects of cis-diamminedichloroplatinum(II) in mice. Oyo Yakuri 46:1–8
101. Chase TA, Cupp MJ, Tracy TS (2003) Germanium. In: Cupp MJ, Tracy TS (eds) Dietary supplements. Toxicology and clinical pharmacology. Humana Press, Totowa
102. Baselt RC (2008) Disposition of toxic drugs and chemicals in man, 8th edn. Biomedical Publications, Foster City
103. Gerber GB, Leonard A (1997) Mutagenicity, carcinogenicity and teratogenicity of germanium compounds. Mutat Res 387:141–146
104. Kaplan BJ, Parish WW, Andrus GM, Simpson JSA, Field CJ (2004) Germane facts about germanium sesquioxide: I. Chemistry and anticancer properties. J Altern Complement Med 10:337–344
105. Kaplan BJ, Andrus GM, Parish WW (2004) Germane facts about germanium sesquioxide: II. Scientific error and misrepresentation. J Altern Complement Med 10:345–348
106. Schauss A (1991) Nephrotoxicity and neurotoxicity in humans from organogermanium compounds and germanium dioxide. Biol Trace Elem Res 29:267–280
107. Tao SH, Bolger PM (1997) Hazard assessment of germanium supplements. Regul Toxicol Pharmacol 25:211–219
108. Krapf R, Schaffner T, Iten PX (1992) Abuse of germanium associated with fatal lactic acidosis. Nephron 62:351–356
109. Raisin J, Hess B, Blatter M, Zimmermann A, Descoeudres C, Horber FF, Jaeger P (1992) Toxicity of an organic germanium compound: deleterious consequences of a "natural remedy". Schweiz Med Wochenschr 8:11–13
110. Takeuchi A, Yoshizawa N, Oshima S, Kubota T, Oshikawa Y, Akashi Y, Oda T, Niwa H, Imazeki N, Seno A, Fuse Y (1992) Nephrotoxicity of germanium compounds: report of a case and review of the literature. Nephron 60:436–442
111. Patai S (1995) The chemistry of organic germanium, tin, and lead compounds, vol 1. Wiley, Chichester
112. Thayer JS (1987) Germapharmaca: some recent studies on biologically active organogermanium compounds. Appl Organomet Chem 1:227–234
113. Ignatenko MA (1987) Antitumor activity of organosilicon and organogermanium compounds (a review). Chim Farmacol Zh 21:402–408
114. Lukevics E, Gar TK, Ignatovich LM, Mironov VF (1990) Biological activity of Germanium compounds (in Russian). Zinatne, Riga
115. Rappoport Z (2002) The chemistry of organic germanium, tin and lead compounds, vol 2. Wiley, Chichester
116. Lukevics E, Ignatovich L (2005) Biological activity of organogermanium compounds. In: Gielen M, Tiekink ER (eds) Metallotherapeutic drugs and metal-based diagnostic agents. The use of metals in medicine. Wiley, Chichester
117. Menchikov LG, Ignatenko MA (2012) Biological activity of organogermanium compounds (a review). Chim Farmacol Zh 46(11):3–7 (in Russian)
118. Lukevics EA, Germane SK, Zidemane AA, Dauvarte AK, Kravchenko IM, Trushule MA, Mironov VF, Gar TK, Khromanova NY, Viktorov NA, Shiryaev VI (1984) Synthesis and neurotropic and antineoplastic activity of a number of germatranes and germasesquioxanes and their organotin analogs. Pharm Chem J 18:89–94
119. Mironov VF, Chernyshev AE, Malochkin VV, Martynov AI, Kulikov GA (1998) Compound stimulating during experiment synthesis of immunoglobulins. RU patent 2108096
120. Shcherbinin VV, Chernyshev AE (1997) Method of decrease of toxicity of medicinal agents, metals, organic and inorganic compounds and other toxins. RU patent 2104033
121. Shcherbinin VV, Chernyshev AE (1997) Method of potentiation of medicinal agent curative effect. RU patent 2104032

122. Kluska M (2008) Some aspects of the analysis of biologically active organogermanium substances. Crit Rev Anal Chem 38:84–92
123. Akiba M, Kakimoto N (1994) Synthesis and properties of poly[3,3′-(1,3-dioxo-1,3-digermoxanediyl)bispropanoic acid](Ge-132) and related compounds as bioactive organogermanium compound. Nippon Kagaku Kaishi 3:286–300
124. Kuwabara M, Ohba S, Yukawa M (2002) Effect of germanium, poly-trans-[2-carboxyethyl] germasesquioxane on natural killer (NK) activity in dogs. J Vet Med Sci 64:719–721
125. Suzuki F (1985) Antitumor activity of Ge-132, a new organogermanium compound, in mice is expressed through the functions of macrophages and T lymphocytes. Gan To Kagaku Ryoho 12:1445–1452
126. Suzuki F, Brutkiewicz RR, Pollard RB (1985) Ability of sera from mice treated with Ge-132, an inorganic germanium compound, to inhibit experimental murine ascites tumor. Br J Cancer 52:757–763
127. Emerson PM, Wilkinson JH (1965) Urea and oxalate inhibition of the serum lactate dehydrogenase. J Clin Pathol 18:803–807
128. Yamamoto S (1983) H4-isozyme of lactate dehydrogenase in the solution of sodium chloride-4. Inhibition by oxalate and oxamate. Int J Biochem 15:355–360
129. Cheshire RM, Park MV (1971) Inhibition by salicylate of porcine lactate dehydrogenase isoenzyme 5. Biochem J 125:45P
130. Schoenenberger GA, Wacker WE (1966) Peptide inhibitors of lactic dehydrogenase (LDH). II. Isolation and characterization of peptides I and II. Biochemistry 5:1375–1379
131. Chan YS, Cheng LN, Wu JH, Chan E, Kwan YW, Lee SM, Leung GP, Yu PH, Chan SW (2011) A review of the pharmacological effects of Arctium lappa (burdock). Inflammopharmacology 19:245–254
132. Zheng W, Miao K, Liu Y, Zhao Y, Zhang M, Pan S, Dai Y (2010) Chemical diversity of biologically active metabolites in the sclerotia of Inonotus obliquus and submerged culture strategies for up-regulating their production. Appl Microbiol Biotechnol 87:1237–1254
133. Sudheesh N, Ajith T, Janardhanan K (2009) Ganoderma lucidum (Fr.) P. Karst enhances activities of heart mitochondrial enzymes and respiratory chain complexes in the aged rat. Biogerontology 10:627–636
134. Sheena N, Lakshmi B, Janardhanan KK (2005) Therapeutic potential of Ganoderma lucidum (Fr.) P. Karst. Nat Prod Rad 4:382–386
135. Shin HR, Kim JY, Yun TK, Morgan G, Vainio H (2000) The cancer-preventive potential of Panax ginseng: a review of human and experimental evidence. Cancer Causes Control 11:565–576
136. Surjushe A, Vasani R, Saple DG (2008) Aloe vera: a short review. Indian J Dermatol 53: 163–166
137. Vogler B, Ernst E (1999) Aloe vera: a systematic review of its clinical effectiveness. Br J Gen Pract 49:823–828
138. Abdullah TH, Kandil O, Elkadi A, Carter J (1988) Garlic revisited: therapeutic for the major diseases of our times? J Natl Med Assoc 80:439–445
139. Milner JA (2001) A historical perspective on garlic and cancer. J Nutr 131:1027S–1031S
140. Kim JY, Kwon O (2009) Garlic intake and cancer risk: an analysis using the Food and Drug Administration's evidence-based review system for the scientific evaluation of health claims. Am J Clin Nutr 89:257–264

In Silico Research in Drug Delivery Systems

André Filipe Ferreira, Rodrigo José Lopes, and Pedro Nuno Simões

Abstract Like in many other research fields, scientific simulation has been established as a crucial element in the design technology of drug delivery systems. Modern multi-scale modeling and simulation techniques, supported by advanced and high-performance computational resources, form a cost-effective complement and/or alternative to the experimentally based trial-and-error approach traditionally used in the development of new drugs. This chapter gives a short overview of the application of modern modeling and simulation techniques within the context of drug delivery systems. Different approaches will be considered depending on the quality and the scale of organization of matter, ranging from picometers to nanoscale and beyond. Molecular modeling and simulation tools will be put in the perspective of their important role in the development of new drugs and in the simulation of their behavior. Such approach enables the engineering of tailored carriers for a specific drug, the optimization of its effectiveness, as well as the understanding, at an atomistic level, of how they interact with the surroundings. The application of computational flow models to drug delivery systems will be systematically addressed for hydrophobic and hydrophilic molecules. The current development of drug transport modeling by applying state-of-art computational fluid dynamics will also be described based on the drug release mechanism for diffusion, swelling and erosion-controlled systems. Finally, a brief prospective view on the high-performance scientific techniques underlying the advanced scientific simulation methods will be given.

A.F. Ferreira (✉) • R.J. Lopes • P.N. Simões
Department of Chemical Engineering, University of Coimbra, Pólo II, Pinhal de Marrocos, Coimbra 3030-790, Portugal
e-mail: aff@eq.uc.pt; rodrigo@eq.uc.pt; pnsim@eq.uc.pt

J. Coelho (ed.), *Drug Delivery Systems: Advanced Technologies Potentially Applicable in Personalised Treatment*, Advances in Predictive, Preventive and Personalised Medicine 4, DOI 10.1007/978-94-007-6010-3_10,

Keywords Molecular modeling • Simulation • Computation flow models • Computational fluid dynamics • High performance computing • Personalized medicine

1 Introduction

In the article by Rosen and Abribat [1] on drug release systems technology, a look towards the future of the field in the light of its past has been proposed. The epigraph for the article was the well-known maxim attributed to Alan Kay: "The best way to predict the future is to invent it". Alan Kay is a computer scientist, noted for his contributions in personal computing and the invention of the concept of the laptop computer. This makes the choice of that quote rather curious, especially when considering that the mentioned article of Rosen and Abribat covers the topic essentially within the framework of the technological application, whereas the impact of computational issues in the research and development (R&D) of delivery systems is only marginally alluded.

The experimental approach has been, and will always be, a crucial element in product design and development. However, it is now clear that the computational approach is a key component of the overall R&D of new materials and processes *lato sensu*. The computer simulation has long been used in treating problems with industrial interest, mostly within the scope of the continuous domain and related methods, such as finite differences, finite elements, boundary element methods, etc. The relevance of the continuum based simulation is well recognized in engineering science, in which materials design and computational fluid dynamics are prominent examples. However, the need to understand and describe matter at finer scales has become increasingly important as other scientific and technological advances have either allowed or demanded a deep understanding of material structures and properties, starting at and below nanoscale. Nanotechnology and high (or super) resolution microscopy are paradigmatic examples of how the manipulation and the study of materials atom-by-atom became a reality.

Computational chemistry is a noticeable case of new computational technologies which emerged over the last years as a decisive practice in R&D, being increasingly used by modern chemical industries aiming at reach more and more ambitious goals. This topic appeared as one of the most promising paradigms leading to better R&D performance with great technological impact, not only in the chemistry field, but also in other related disciplines, such as physics, materials and biology. The role of computational chemistry has changed from a merely useful tool in research to a critical instrument with significant technological impact. Concomitantly, a unified experimental-theoretical-computational approach to scientific research and product project development has been advocated as a cost-effective and crucial strategy in modern R&D programs (Fig. 1), making the development of new products/processes not only more efficient, but also an invaluable source of fundamental scientific knowledge.

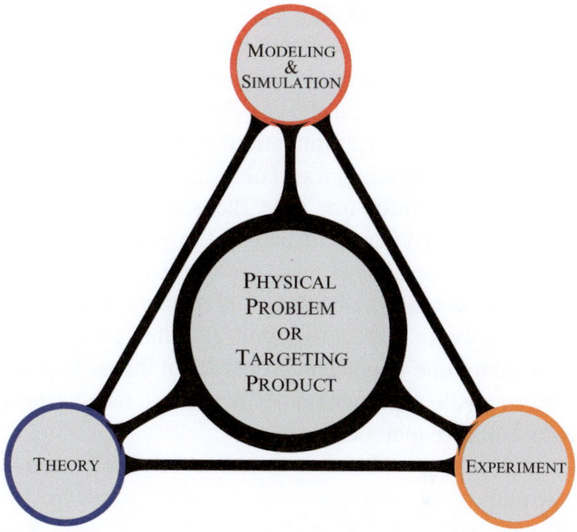

Fig. 1 Illustrative scheme of modern approach to scientific research and product project/development (adapted from Landau [2])

The pharmaceutical industry is a noticeable example in which computational technologies are an invaluable element. The "long, arduous, and expensive process" [3] of drug discovery has been increasingly accomplished by methods such as high-throughput screening, combinatorial chemistry, and computer-aided drug design as a cost-saving standard. A comprehensive review of these methods is beyond the scope of this chapter. The subject is quite vast and has deserved the interest of many authors, which has been manifested in recent years by a remarkable number of technical and scientific publications in the form of book chapters and articles in thematic collections (*e.g.* [4–9]), and, obviously, journal articles. Here we will focus on the application of modeling and simulation methods to the field of drug delivery systems (DDS). In doing so, some representative topics within the area will be considered with the purpose of showing how modern high-performance computing is a complementary alternative to the traditional experimental practice. In short, the authors claim nothing but a brief overview, avowedly subjective, on the topic.

2 Molecular Modeling and Simulation

2.1 Scope of the Application

As in many classification schemes, it is not easy to strictly bound the meaning of *molecular modeling and simulation*. Here we will take it in the somewhat broad sense, however precise enough, provided by IUPAC [10], according to which molecular modeling deals with the investigation of molecular structures and properties, and with the simulation of the molecular behavior in time and/or space, from the use of adequate mathematical frameworks and graphical visualization techniques. Simulation presupposes modeling. Hence, for the sake of simplicity, the phrase “molecular simulation” will hereafter be used.

Generally speaking, molecular simulation can be grouped into two classes: static and dynamic. In the first case, the geometry of the molecular system is optimized towards identifying local and/or the global minima, i.e. the most energetically favorable three-dimensional arrangements of the atoms. From these equilibrium structures several molecular properties are then calculated from the derivatives of the energy with respect to atomic coordinates. The dynamic simulations are usually targeted to time-dependent properties of the molecular system. Molecular dynamics (deterministic approach) and Monte Carlo (stochastic approach) simulations are the representative methods.

Molecular simulation is nowadays routinely used in modern research protocols. The purpose of such approach is twofold: to predict the properties and the behavior of materials; and to link theory and experiment. With this scope in mind, there has been a remarkable progress in the computational methods capable of treating problems at different scales. Thanks to the dramatic evolution of high-performance scientific techniques and computational resources thereof [11], broader range of length and time scales are now addressed, as schematically represented in Fig. 2. As a natural consequence of this development, the multiscale molecular simulation became the new issue, in the sense that the major challenge is now the integration of different scales towards a unified approach. This challenge is all the more difficult as the continuum domain is involved in the overall multiscale framework. Bridging the gaps between the diverse methods characterizing the different scales is a fascinating field of research which is still in infancy stage.

The field of DDS is plenty of examples of scientific and technological activities that call for a thorough support by advanced molecular simulation tools, the nanotechnology being a prominent case. Most of the commercial applications of nanoparticles in medicine are directed to drug delivery, for which several solutions have been proposed, namely liposomal and lipid-based colloidal nanoDDS, nanoparticulate polymeric micelles (as drug carriers) and polymer based nanoparticulate DDS [13–15]. Molecularly imprinted polymers [16, 17] is another field which includes applications in drug delivery. Drug-protein systems constitute on more typical case and possibly the most representative example of classes of ligand-protein complexes being actively investigated [18].

2.2 Mathematical Background

In what follows a very short introduction to common molecular simulation methods applicable within the framework of DDS will be given. It must be stressed that the

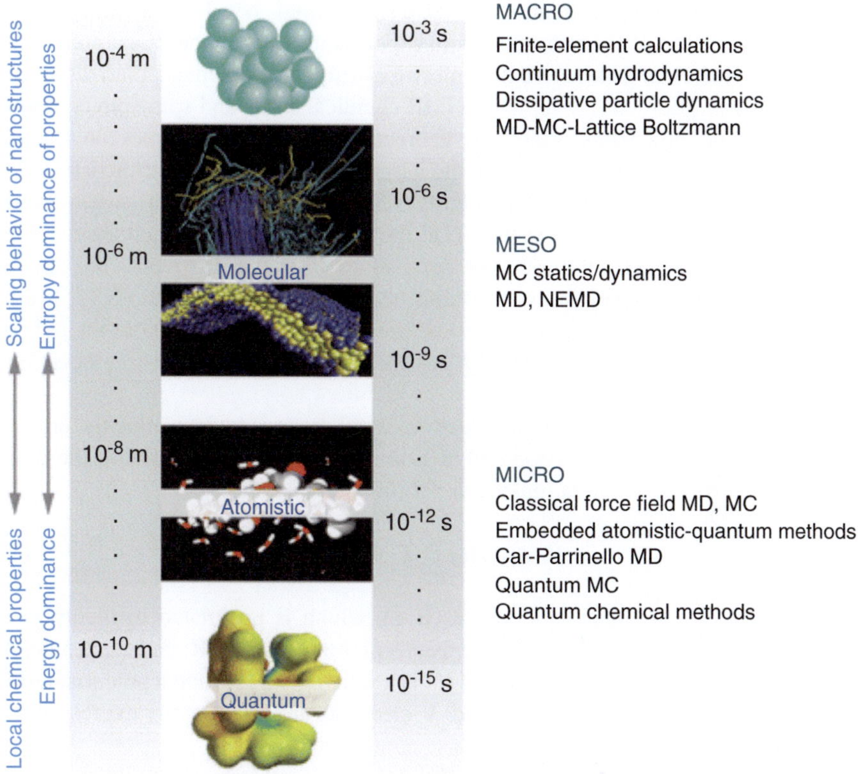

Fig. 2 Schematic representation of temporal and spatial scales accessible by simulation techniques and some representative methods thereof (from Praprotnik et al. [12])

topic is presented in a fairly simple manner, hence glossing over most details of the remarkably complex mathematical framework underlying it. For those readers interested in getting a deeper insight into both scientific and technical aspects related to the computational chemistry methods and its implementation, other more proper sources are recommended here [19–21].

2.2.1 Quantum Mechanics Based Approach

The preferential domain of the quantum mechanics based methods, such as *ab initio* and density functional theory (DFT) methods [18, 22, 23], falls within the approximate range of few picometers (1 pm $= 10^{-12}$ m) to few nanometers (1 nm $= 10^{-9}$ m). These electronic structure methods allow accurate theoretical studies to be performed on single molecules in different phases, molecular clusters, and, to a certain extent, macromolecules (*eg.* synthetic polymers and proteins), as well as condensed matter (liquids and solids).

Quantum mechanics calculations are also an essential complement or alternative in the interpretation of the outcomes from experiments by the theoretical prediction of various molecular characteristics: electrical (dipole moments, polarizabilities, hyperpolarizabilities, etc.); magnetic (NMR chemical shifts and spin-spin coupling constants); and properties related to geometrical derivatives (forces on nuclei, vibrational spectra within and beyond the harmonic-oscillator approximation) [24]. Such methods might be computationally very demanding, which poses severe restrictions as the size of the system (i.e. the number of atoms) increases. These cases can be addressed by alternative approaches whenever the electronic effects need be accounted. Apart the well established semi-empirical methods [25], different strategies have been proposed to address larger systems, notably the hybrid QM and molecular mechanics (QM/MM) methods [26, 27] or the fragmentation methods [28].

The main purpose of electronic structure calculations is to determine the energy (E_t) and subsequently the properties of a molecular system by solving the non-relativistic, time independent molecular Schrödinger equation,

$$\hat{H}_t\, \Psi_t(\mathbf{r}, \mathbf{R}) = E_t\, \Psi_t(\mathbf{r}, \mathbf{R}). \tag{1}$$

The total (molecular) wavefunction, $\Psi_t(\mathbf{R}, \mathbf{r})$, which is postulated to incorporate all the information about the system, depends on both nuclear, $\mathbf{R}$, and electronic, $\mathbf{r}$, positions. The molecular Hamiltonian, $\hat{H}_t$, is a sum of a kinetic and a potential term, which for a system with M nuclei and N electrons are respectively expressed (in atomic units) as

$$\hat{T} = \underbrace{-\frac{1}{2}\sum_{A}^{M} \nabla_A^2}_{\hat{T}_N(\mathbf{R})} \underbrace{-\frac{1}{2}\sum_{i}^{N} \nabla_i^2}_{\hat{T}_e(\mathbf{r})} \tag{2}$$

and

$$\hat{V} = \underbrace{-\sum_{A}^{M}\sum_{i}^{N} \frac{Z_A}{|\mathbf{R}_A - \mathbf{r}_i|}}_{\hat{V}_{eN}(\mathbf{r},\mathbf{R})} + \underbrace{\sum_{A}^{M}\sum_{B>A}^{M} \frac{Z_A Z_B}{|\mathbf{R}_A - \mathbf{R}_B|}}_{\hat{V}_{NN}(\mathbf{R})} + \underbrace{\sum_{i}^{N}\sum_{j>i}^{N} \frac{1}{|\mathbf{r}_i - \mathbf{r}_j|}}_{\hat{V}_{ee}(\mathbf{r})}. \tag{3}$$

The large difference between the electron and nuclei masses allows a separate treatment of the respective motions. The nuclei is then considered fixed relatively to electron motion, thus allowing a separation of variable and to consider the total wavefunction as the product of the electronic, $\Psi(\mathbf{r})$, and the nuclear, $\Psi_N(\mathbf{R})$, wavefunction, i.e. $\Psi_t(\mathbf{r}, \mathbf{R}) = \Psi(\mathbf{r})\, \Psi_N(\mathbf{R})$. This approach is known as the Born-Openheimmer approximation and considerably simplifies the complexity of the original problem, which is converted into the problem of solving parametrically the electronic Schrödinger equation (for each fixed nuclear position the nuclear kinetic energy vanishes and the nuclear-nuclear repulsion is constant),

$$\underbrace{\left[\hat{T}_{\mathrm{e}}(\mathbf{r}) + \hat{V}_{\mathrm{eN}}(\mathbf{r}, \mathbf{R}) + \hat{V}_{\mathrm{ee}}(\mathbf{r})\right]}_{\hat{H}} \Psi(\mathbf{r}, \mathbf{R}) = E\Psi(\mathbf{r}, \mathbf{R}) \tag{4a}$$

$$\hat{H}\,\Psi = E\Psi. \tag{4b}$$

For each nuclear position a ground state electron configuration energy can in principle be found. Then, the many different possible nuclei positions and the corresponding energies can be expressed as a function, the so-called potential energy surface (PES).

The Born-Openheimmer approximation is far from being sufficient to overcome the complexity underlying the electronic Schrödinger equation in virtue of the electronic-electronic interactions. The next approximation step relies on the independent particle approximation, a central approach in the Hartree-Fock (HF) theory, but also one of the most severe approximations of the *ab initio* HF based methods. The independent particle approximation assumes that each electron moves within its respective orbital independently from each others. That is, each electron experiences a mean field generated all the other electrons, so that the overall N-electron wavefunction is described as a product of single-electron wavefunctions. Thus, including also the concept of spinorbitals (apart the three spatial coordinates, it is necessary to account for electron spin to completely describe an electron),

$$\chi(\mathbf{x}) = \begin{cases} \psi(\mathbf{r})\alpha(\omega), \\ \psi(\mathbf{r})\beta(\omega) \end{cases} \quad \text{where } \mathbf{x} = \{\mathbf{r}, \omega\}, \tag{5}$$

as well as the Pauli antisymmetry principle (the N-electron wavefunction must be antisymmetric regarding the electron interchange), it comes that the exact N-electron wavefunction is expressed in an approximate form, usually known as the Slater determinant

$$\Psi_0(\mathbf{x}_1, \mathbf{x}_2, \cdots, \mathbf{x}_N) = \frac{1}{\sqrt{N!}} \begin{vmatrix} \chi_1(\mathbf{x}_1) & \cdots & \chi_N(\mathbf{x}_1) \\ \vdots & \ddots & \vdots \\ \chi_1(\mathbf{x}_N) & \cdots & \chi_N(\mathbf{x}_N) \end{vmatrix}. \tag{6}$$

The variational principle states that the energy calculated from an approximate wavefunction is always greater than the true energy. This gives a prescription to proceed with the calculation, provided that a trial wavefunction is available. This is usually constructed by expanding the molecular orbitals (MO) as a linear combination of predefined basis functions, usually called atomic orbitals (LCAO),

$$\psi_i(\mathbf{r}) = \sum_{\mu} C_{\mu i}\, \phi_{\mu}(\mathbf{r}), \tag{7}$$

where the set of functions $\{\phi_\mu(\mathbf{r})\}$ is called basis set, which, together with the method (*e.g.* HF, MP2, CI, DFT, etc.; see below) form the so-called model chemistry for the calculation. Hence, by expressing $\psi_i(\mathbf{r})$ as in Eq. 7, the problem is converted into an optimization problem consisting in minimizing the energy with respect the coefficients $C_{\mu i}$. Since the number of basis functions is necessarily finite, an additional approximation should be recognized here.

From a practical computational point of view, the implementation of the HF method relies on a matrix algebraic approach, compactly expressed as

$$\mathbf{FC} = \mathbf{SC}\boldsymbol{\varepsilon}, \tag{8}$$

known as Roothaan-Hall (RH) equation. Here **C** corresponds to the matrix of the coefficients, **S** is the overlap matrix, $\boldsymbol{\varepsilon}$ the eigenvector of the respective MOs energies, and **F** is the Fock operator. The Roothaan-Hall equation corresponds to a nonlinear eigenvalue problem. In fact, once the basis set is chosen, **S** is determined but **F** depends on **C**, meaning that the problem must be solved iteratively, i.e. by the so-called self-consistent field (SCF) method.

The HF method (and the RH computational scheme thereof) is the simplest *ab initio* electronic structure method. Beyond its own relevance in a wide range of applications, in which offers valuable qualitative insights on molecular systems and chemical processes, the HF framework has another important attribute. It is also the starting point of two other approaches, which diverge in complexity and in scope. In spite of being the simplest *ab initio* method, the HF scheme is computationally demanding enough to preclude its application in large molecular systems. If these are to be studied by including electronic effects, semiempirical methods might be the more practical solution (Fig. 3). These methods correspond to a step down in computational complexity because the HF-RH approach is simplified in different ways (minimal basis sets, some complex and computationally expensive integrals neglected and/or replaced by parametrized functions, etc.), thus determining the various types of semiempirical methods available, the Modified Neglect of Diatomic Overlap (MNDO), Austin Model 1 (AM1), and the Parametric Method Number $n = 3, 6$ (PMn) being examples. On the other hand, if the chemical accuracy is the issue and/or excited states are to be accurately described, the HF method fails in most of the cases. The lacking of an adequate treatment of the electronic correlations is its main weakness, which can be overcome by deriving the so-called post-HF or electron correlation methods (Fig. 3), *e.g.* Møller-Plesset perturbation theory (MPn; order $n = 2, \ldots, 5$), configuration interaction (CI), coupled cluster (CC) methods, etc.. These imply a noteworthy step up in computational complexity, meaning that their applicability is restricted to relatively small molecular systems.

It follows from the above that the complexity and the huge computational cost underlying the wavefunction based methods in tackling a many body problem prevents a generalization of pure *ab initio* methods in large molecular systems, and, with greater reason, in condensed matter. Density functional theory (DFT) has been the most prominent quantum mechanics based approach in dealing with

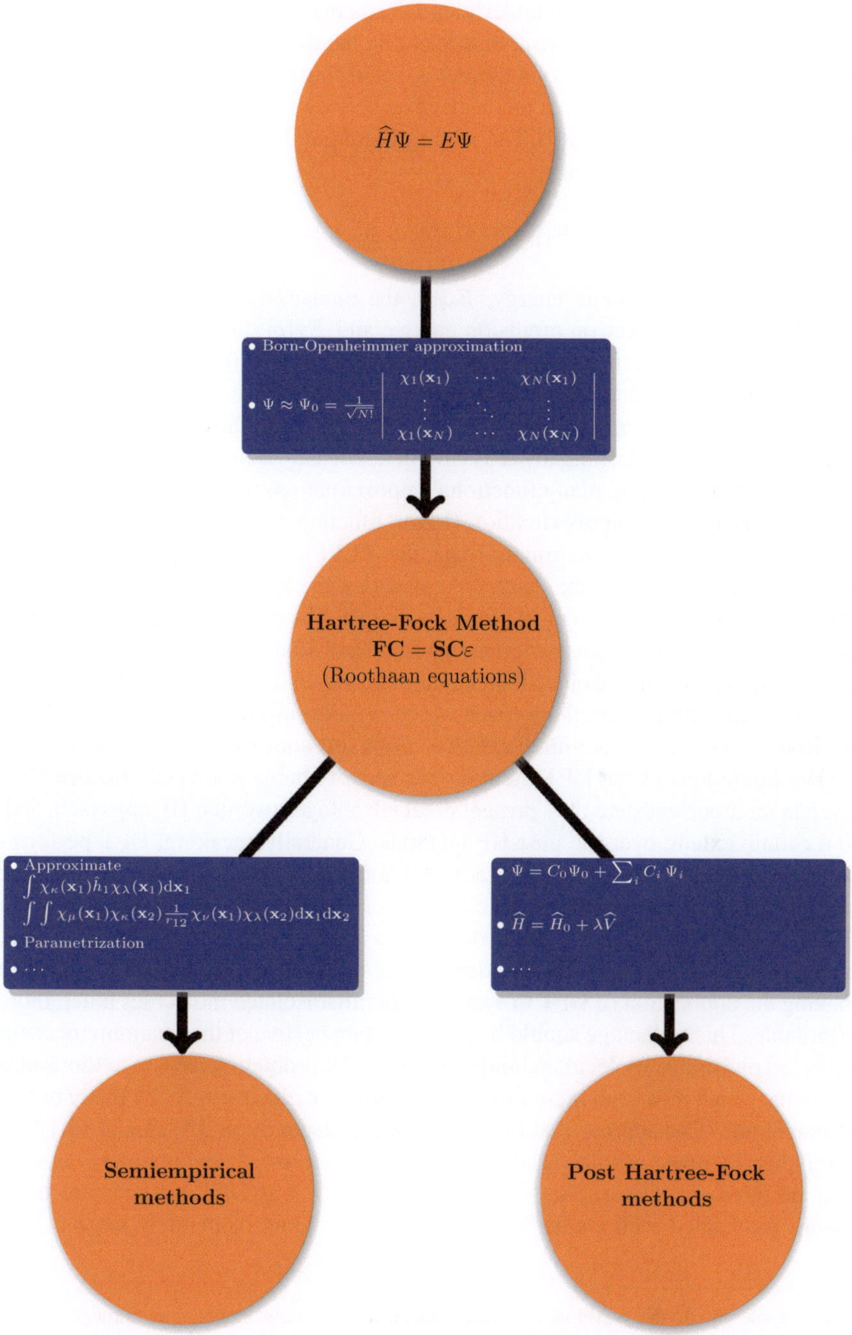

Fig. 3 Schematic overview of grouping of methods derived from the Hartree-Fock framework, either more approximated (semi-empirical) or more accurate (correlated methods)

such cases. The basic premise underlying DFT is that the electron density provides all the information on the system, thus avoiding the wavefunction calculation. DFT is rooted in the Hohenberg-Kohn theorems, according to which the exact energy $\mathbb{E}$, of a molecular system depends on its electron density, the latter being a function of the electronic coordinates, $\rho(\mathbf{r})$. The total (ground state) energy is then expressed as the sum of several functionals,

$$\mathbb{E}[\rho] = E_{\mathrm{T}}[\rho] + E_{\mathrm{ne}}[\rho] + E_{\mathrm{ee}}[\rho] + E_{\mathrm{xc}}[\rho]\,. \tag{9}$$

where $E_{\mathrm{T}}[\rho]$ is the kinetic energy, $E_{\mathrm{ne}}[\rho]$ the nucleus-electron potential energy, $E_{\mathrm{ee}}[\rho]$ the electron-electron repulsion energy, and $E_{\mathrm{X}}[\rho]$ the exchange-correlation functional. While the first three functionals (from the left) have explicit forms, the latter has not. In fact, "it is the approximate nature of the exchange-correction functional that is the reason both for the success and the failure of DFT applications" [22], and the searching towards new, better density functionals is still the great challenge. Nevertheless, many functional approximations are available, such as the simplest local density approximation (LDA) functionals, the generalized gradient approximation (GGA) functionals (e.g., the BLYP), and the hybrid functionals (that include a Hartree-Fock character), of which the B3LYP is probably the best performer, and certainly the most notable example. The practical implementation of the DFT approaches was firstly derived by Kohn and Sham. They established a set of equations (the so-called Kohn-Sham equations) and an iterative method for solving them by means of an overall approach which is quite similar to that used in solving the Roothaan equations within the Hartree-Fock self consistent field calculations.

The limitations of the DFT methods are actively being matter of concern [22]. Even in their current state they present great advantages over the HF approach, and, to a certain extent, over the post-HF methods. Generally speaking, DFT performs better than the HF method in accuracy at a comparable computational effort. The improved accuracy arises from the electron correlation that DFT accounts for. On the other hand, DFT methods are certainly less accurate than post-HF methods but, compared to these, are much less demanding on computational resources, hence making the application of DFT to systems other that isolated molecules much more affordable. This advantage should be put in the perspective of the quantum mechanics based multiscale molecular simulation. Here, the problem scales up to thousands of atoms, which means an increasing in the number of electrons, N, by many orders of magnitude. Considering that the wavefunction depends on $3N$ spatial variables (excluding by now the spin variable for each of N electrons) and that $\rho(\mathbf{r})$ depends only on three spatial variables[1], it is understandable why DFT methods are those usually associated with studies on systems which, being moderately large in size

[1]The complexity of the wavefunction based methods is usually emphasized through an example like the following. For a single oxygen atom, having eight electrons, its wavefunction $\psi(\mathbf{r}_1, \mathbf{r}_2, \cdots, \mathbf{r}_8)$ is a function of 24 coordinates. Considering that just the modest number of 10 ψ values are to be stored for each coordinate, a total of 10^{24} values need to be stored. Could this be

and complexity, claim for a quantum mechanical handling of electronic effects. The same can be said about dynamic simulations, i.e. those which include time as a fundamental variable. Most of the so-called *ab initio* molecular dynamics computer codes are based on the DFT framework. The calculations are typically accomplish using spatially periodic functions (instead of the spatially localized basis functions mentioned above). This approach is particularly adequate for modeling extended (infinite) systems, such as condensed phases, under periodic boundary conditions, i.e. by considering a simulation box (usually, but not necessarily, cubic) which is replicated throughout space (see below). It should be mentioned, however, that for very large systems and/or large simulation times, the DFT based *ab initio* molecular dynamics constitute a formidable task, if not an impossible one from a practical point of view. Such cases are better tackled by lighter molecular simulation methods, such as the classical molecular dynamics approach briefly described in what follows.

2.2.2 Classical Molecular Dynamics

Molecular mechanics has been the prevailing molecular simulation approach in drug design. The development of new materials, such as polymers, to be used as drug carriers is now one of the most active areas of biomedical research, both in academic laboratories and in the pharmaceutical industry. Polymeric nanoparticles have received attention as targeted delivery systems of anti-tumor drugs due to their capability to migrate to the tumor sites more efficiently than the free drug [29]. It is known that the ability to maintain drug stability is closely related to the physicochemical properties of the microenvironment in the delivery system, and to the interactions between the drug and the polymer [30, 31]. Thus, the prediction of phenomena such as drug permeability across biomembranes and the optimization of DDS carriers to improve their entrap drug capabilities would be highly desirable [32]. During the development of new drugs it is important to predict its lipid bilayer permeability and use the results for achieving the drug characteristics and release rates. Due to the experimental limitations to analyze the nanoscale [33], a tool capable to provide detailed information at this level has become a fundamental task. Thus, the resort to computational methods to model and simulate DDS presented itself as a great complementary technique.

Molecular dynamics has the potential to play a significant role in achieving better design and administration of controlled release systems and in designing patient-specific treatment regimens. It is widely used in drug design [34] to study the three-dimensional properties and interaction energies of molecules. Therefore, predictive

done on DVDs with a capacity of 5 GB, more than 10^{14} DVDs would be necessary. With a weight of 10 g per DVD, this would correspond to more than 10^{9} ton DVDs!

computational models based on macro-scale transport properties of the drug in a fluid, can be used to accelerate the selection process of drug compounds, from large polymer libraries, prior to the synthesis and biological characterization [35].

Such type of approach can save considerable time during the selection of the best polymer for a particular drug, to improve its effectiveness. Moreover, the simulations can be easily extended to incorporate different solvent mixtures, subjected to a wide range of temperatures and pressures and easily manipulated by changing the particles coordinates. Still, some key questions must be answered before beginning the simulations [36]. How large a system does one need to access the phenomenon of interest? Even more important, how long should the MD simulation run?

Classical molecular dynamics is the name used to refer to a MD method that resorts to Newton's classical mechanics. In its basic framework, MD allows to model and to simulate larger molecular systems and timescales. This enables the study of structure and key properties such as stability, diffusion, binding between molecules, as well as the of simulation of self-assembly and other macromolecular phenomena [33, 37] involving systems formed by large molecules like polymers [38]. Due to the mathematical formalism of MD, it is necessary to describe/parametrize the system in study, more specifically, each molecule structural and electrostatic properties. These parameters can be obtained both experimental and computationally. However, there is no established methodology to obtain these parameters, which makes the parameterization process a very difficult task. Therefore, high-quality parameters are mandatory to obtain accurate and reliable results.

As mentioned previously, MD resorts to classical molecular mechanics to solve the system, and to do so it is common to resort to the Hamiltonian formalism. The Hamiltonian (H) that describes a system is formed by two terms, the kinetic energy K, and potential energy, V [39, 40]:

$$H(\mathbf{p}, \mathbf{r}, m, s) = K(\mathbf{p}; m) + V(\mathbf{r}; s), \tag{10}$$

were $\mathbf{p}$ stands for momentum, $\mathbf{r}$ for the particle coordinates, m for mass and s are parameters.

The kinetic energy is simply the sum of the kinetic energy of all the system particles,

$$K(\mathbf{p}, m) = \sum_{i=1}^{N} \frac{\mathbf{p}_i^2}{2m_i} = \sum_{i=1}^{N} \frac{1}{2} m_i \mathbf{v}_i^2, \tag{11}$$

where v stands for velocity.

The determination of the potential energy is more complex. The classical approach considers that the potential energy can be defined by

$$V_{\text{total}} = \underbrace{V_{\text{stretch}} + V_{\text{angular}} + V_{\text{torsion}} + V_{\text{improper}}}_{\text{bonded}} + \underbrace{V_{\text{van der Waals}} + V_{\text{electrostatic}}}_{\text{non-bonded}}, \tag{12}$$

which basically results from the sum of the bonded and non-bonded interactions. The Eq. 12 can be expanded to:

$$V(\mathbf{r},s) = \sum_{\text{stretch}} \frac{k_b}{2}(r_{ij} - r_{ij,0})^2 + \sum_{\text{angular}} \frac{k_\theta}{2}(\theta_{ij} - \theta_{ij,0})^2$$
$$+ \sum_{\text{torsion}} \frac{k_\phi}{2}[1 + \cos(n\phi_{ijkl} - \gamma)] + \sum_{\text{improper}} \frac{k_\xi}{2}(\xi_{ijkl} - \xi_{ijkl,0})^2$$
$$+ \sum_{i,j} 4\varepsilon_{ij}\left[\left(\frac{\sigma_{ij}}{r_{ij}}\right)^{12} - \left(\frac{\sigma_{ij}}{r_{ij}}\right)^{6}\right] + \sum_{i,j} \frac{q_i q_j}{4\pi\varepsilon_0 r_{ij}}, \tag{13}$$

The set of parameters s are referred as force field (see below).

Force Field

A force field involves hundreds to thousands of parameters that need to be selected to accurately represent the Hamiltonian of the system. Currently there are four popular force fields, GROMOS [39, 41–44], AMBER [45, 46], CHARMM [47, 48] and OPLS [49, 50]. OPLS force field is more oriented to model lipid based systems. In contrast, GROMOS, AMBER and CHARMM force fields are more specific for proteins, amino acids and nucleic acids. This set of force fields partially reflects different parametrization philosophies.

The simplest way to develop a force field for a new molecule, based on one of the major biomolecular force fields, is by analogy with molecules that have equal or similar functional groups that are already described. Still, this "process is tedious, time-consuming, and error-prone" [51] for large molecules such as polymers. Moreover, to create a proper parameterization for a new molecule, a deep knowledge of the force field background is required, since some parameterizations are tuned with some empiricism. To overcome these problems, several research groups have developed tools to generate the parameterization of new molecules requiring only as input the atoms coordinates. Some of the most popular tools to generate new parameterizations are the PRODRG server [52], the Automated Topology Builder (ATB) [51] and SwissParam [53].

Some force fields allow to create simpler molecular models according to three types of molecule description: all-atoms (AA), when all atoms are described; united-atoms (UA), when the aliphatic carbons are treated as a single particle; and coarse grain (CG), when functional groups are agglutinated in a single particle. In Fig. 4 it is presented the referred representations for a 1,2-dimyristoyl-*sn*-glycero-3-phosphocholine (DMPC) lipid.

Simulation methods in which each atom is explicitly represented are well established but are problematic in addressing many cooperative effects of experimental and theoretical interest. In fact, there is a large gap between the timescale and spatial

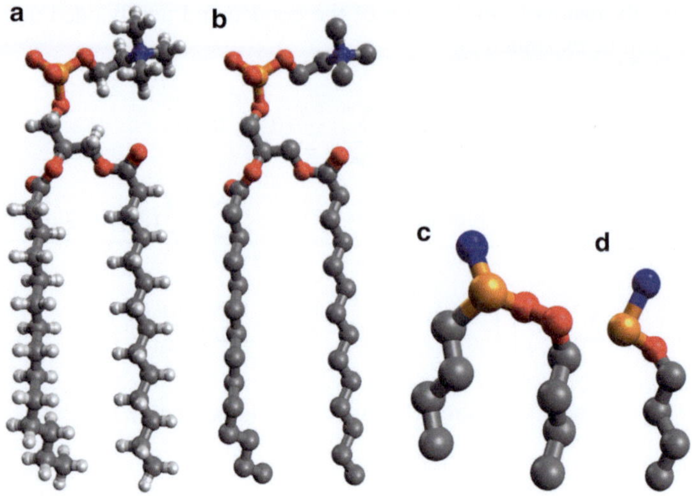

Fig. 4 Representation of a DMPC lipid: (**a**) *all-atoms*; (**b**) *united-atoms*; (**c**) *coarse grain*; (**d**) more simplistic *coarse-grain* model

scale that govern typical intramolecular events and those which are relevant for collective motions. CG models have been developed allowing MD to simulate at the mesoscale with system sizes approaching the micrometer range and phenomena taking place on time scales approaching milliseconds and beyond. Due to the considerable reduction in the number of particles, it can provide more than one order of magnitude gain over atomistic models, enabling the study large-scale phenomena such as lipid vesicles formation [54]. There are several coarse-grained models of lipid bilayers in the literature [55–57], but the MARTINI model [57–59] is the most widely used.

Regarding membrane simulations, recent published MD simulations that addressed membrane binding and transport-related problems have been limited to system sizes of 128 or 256 lipid molecules [60–62] (UA description), along with a sufficient number of water molecules to provide nearly full hydration. Artifacts are reduced by the use of periodic boundary conditions. Currently, typical time scales for UA simulations that are computationally feasible are in the range of 100–200 ns. These time scales are sufficient to adequately sample relatively fast motions of drug molecules or drug fragments over short distances within the bilayer, but in some cases not enough to detect larger lipid displacements and self-assembly [32].

2.3 Molecular Modeling and Simulation Case-Studies

As mentioned before, there is a broad range of instances that can be invoked to demonstrate the important role of computational chemistry in the overall field of DDS. For the sake of space this discussion will be confined just to few representative cases.

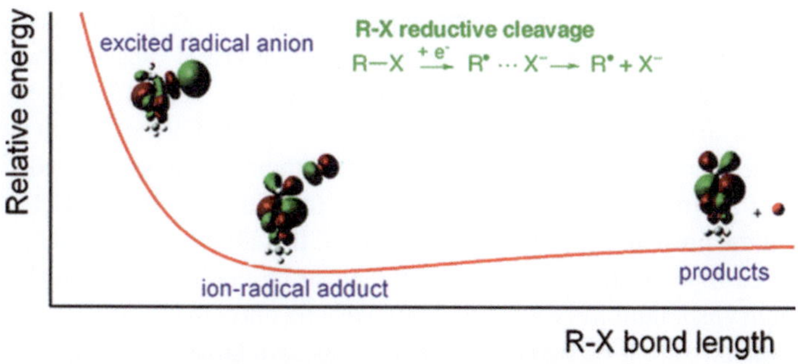

Fig. 5 Schematic reaction profile of a mechanistic step in living radical polymerization (LRP) studied by quantum mechanics based methods (from Isse et al. [65])

One can take lipid and/or polymer based systems within the domain from and below the nanoscale to exemplify how molecular simulation can accomplish several issues at different scales. For instance, the notable development of smart stimuli-responsive polymers [63] has demanded new synthetic routes to be investigated and developed, aiming at give a stringent control over the products, namely by tuning the polymer microstructure in terms of size, shape, topology, morphology, and functionalization. A fundamental knowledge of the polymerization reaction is critical, and there are several examples, notably the works of Coote et al. [64, 65], in which high level molecular simulation approaches are thoroughly used to clarify mechanistic aspects of modern controlled/living radical polymerization techniques. This type of studies recommend high accurate results in calculating energies (*e.g.* binding energies, heats of reaction, and activation energies) towards the ultimate (rather difficult) goal of reaching the *chemical accuracy* level (*ca.* 1 $\text{kcal} \cdot \text{mol}^{-1}$). Quantum mechanics based molecular simulation can be decisive in the fundamental study of mechanisms, kinetics, and thermodynamics of elementary steps characterizing chemical reactions of interest. The rate constants can be assessed within the transition state theory framework, which implies the evaluation of molecular partition functions related to relevant species, such as the reactants and transition state entities (Fig. 5). This requires knowledge on the respective molecular properties, such as geometries, accurate energies and vibrational frequencies.

For instance, calculated structural and spectroscopic data can be quite useful in characterizing the molecular system of interest, as well as related molecular systems. The compound N_α-benzoyl-*L*-argininate ethyl ester chloride ($BAEEH^+{\cdot}Cl^-$, Fig. 6a) can be referred as a representative example within this context. $BAEEH^+$ is an amino acid derivative in which the negative carboxylate moiety ($-COO^-$) of arginine is replaced by the neutral hydrophobic carboxylic ethyl ester group [$-C(=O)OC_2H_5$], and one of the hydrogen atoms of the R-amino group ($-NH_2$) is replaced by the hydrophobic and voluminous benzoyl moiety [$-C(=O)C_6H_6$]. $BAEEH^+{\cdot}Cl^-$ has been used as a model drug in the study of the potential of the polymeric system to delivery the active compound in the intestinal tract [66, 67]. Furthermore, $BAEEH^+{\cdot}Cl^-$ was used as a model substrate in the evaluation of the inhibitory capacity of proteolytic enzymes, namely trypsin. Studies on the evaluation of the attack mechanism of trypsin on $BAEEH^{+}{\cdot}Cl^{-}$ in the presence of a polymeric delivery system have been described [66, 68, 69]. Such exploratory investigations aimed to determine the best polymeric system to be used in the protein or peptide delivery in order to avoid or minimize their enzymatic degradation [69]. Dorkoosh et al. [66] showed that the action of trypsin towards $BAEEH^{+}{\cdot}Cl^{-}$ was inhibited in the presence of polymeric hydrogels. In another study, Vaidya et al. [69] demonstrated that the interaction polymer-substrate could be responsible for the decrease in trypsin inhibitory activity. Only very recently the structural and spectroscopic characterization of $BAEEH^{+}{\cdot}Cl^{-}$ became available. Fonseca et al. [70, 71] investigated $BAEEH^{+}{\cdot}Cl^{-}$ as model compound for poly(ester amide) (PEA) precursors. PEAs, which contain both ester and amide linkages in their structure, have been considered an important group of polymers in biomedical/pharmaceutical field. However, PEAs are lacking in structural and spectroscopic information, and with the goal of filling this gap a collection of studies were performed on $BAEEH^{+}{\cdot}Cl^{-}$. This compound was investigated in gas phase and in solution ($BAEEH^{+}$ cation) [71], and in the solid phase ($BAEEH^{+}{\cdot}Cl^{-}$), both experimentally and computationally (mostly at the DFT level of approximation, using the B3LYP functional, and the 6-31++G(d,p) Pople-type basis set). H-bond interactions are an issue in PEAs and in $BAEEH^{+}{\cdot}Cl^{-}$ as well. Such interactions, as other environmental effects (*e.g.* solvent effects) should be accounted as much as possible in the models used in quantum chemical calculations. For instance, solvent effects can be taken into account in the computations by using the polarized continuum model (PCM) (Fig. 6b), in which the solvent is modeled as a polarizable continuum, instead by considering explicit molecules. However, the PCM approach does not treat the effect of explicit H-bonding in condensed phase, and a possible alternative, although computationally more demanding, consists in considering a supermolecular model. This can be based on a $BAEEH^{+}(H_2O)_n$ complex to simulate specific intermolecular H-bonds (Fig. 6c) that are expected to actually be formed in solution and in the solid state. Figure 7 summarizes the effect of the choice of the model in the prediction of the ^{1}H chemical shifts for $BAEEH^{+}$. Despite the reasonable overall agreement between experimental and calculated data for isolated $BAEEH^{+}$, a considerable discrepancy is observed for the protons of the guanidinium group, probably due to possible interactions established between $BAEEH^{+}$ and the solvent, which are not accounted in the theoretical calculations for the isolated molecule. The inclusion of an implicit solvent in the theoretical model appeared not to solve the problem in the system under study as suggested by the rather modest results from the PCM calculations. The model based on the $BAEEH^{+}(H_2O)_5$ complex, in turn, led to fairly improved results, hence clearly supporting the experimental data, according to which the guanidinium group of $BAEEH^{+}$ is prone to establish interactions in solution, probably through intermolecular H-bonds [71].

The analysis of the solid phase data poses additional challenges which are difficult to accomplished by a simple model based on an isolated $BAEEH^{+}$ cation in vacuo, or even by the mentioned enhanced variants (PCM or supramolecular

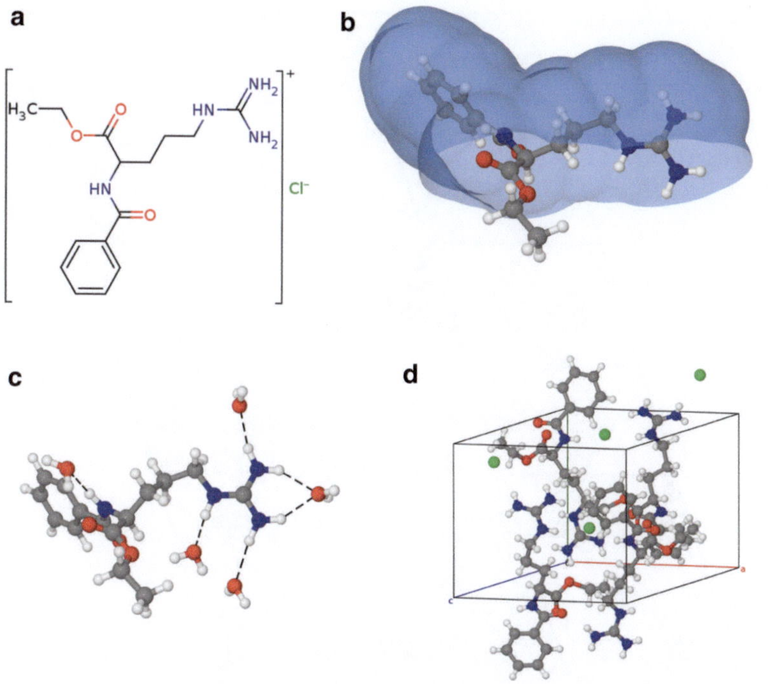

Fig. 6 N_α-benzoyl-*L*-argininate ethyl ester chloride ($BAEEH^+{\cdot}Cl^-$). (**a**) 2-D representation; (**b**) schematic representation of the polarized continuum model (PCM); (**c**) $BAEEH^+(H_2O)_5$ supermolecular model; (**d**) X-ray unit cell

Fig. 7 Comparison between experimental and calculated ^1H chemical shifts (δ/ppm) for $BAEEH^+{\cdot}Cl^-$ in solution. The models used in the calculations were: isolated molecule in vacuo; isolated molecule in DMF solution (PCM); and $BAEEH^+(H_2O)_5$ supermolecular model in vacuo

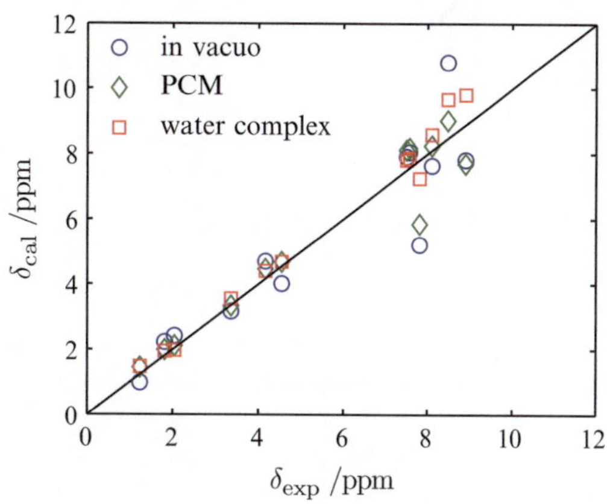

approaches). The effects of the crystalline force field is not duly accounted by those models. The same applies to specific kind of interactions within the crystal, the intermolecular H-bonding interactions being a notable example. These cases are better undertaken in the framework solid-state model under periodic boundary conditions (PBC) (Fig. 8).

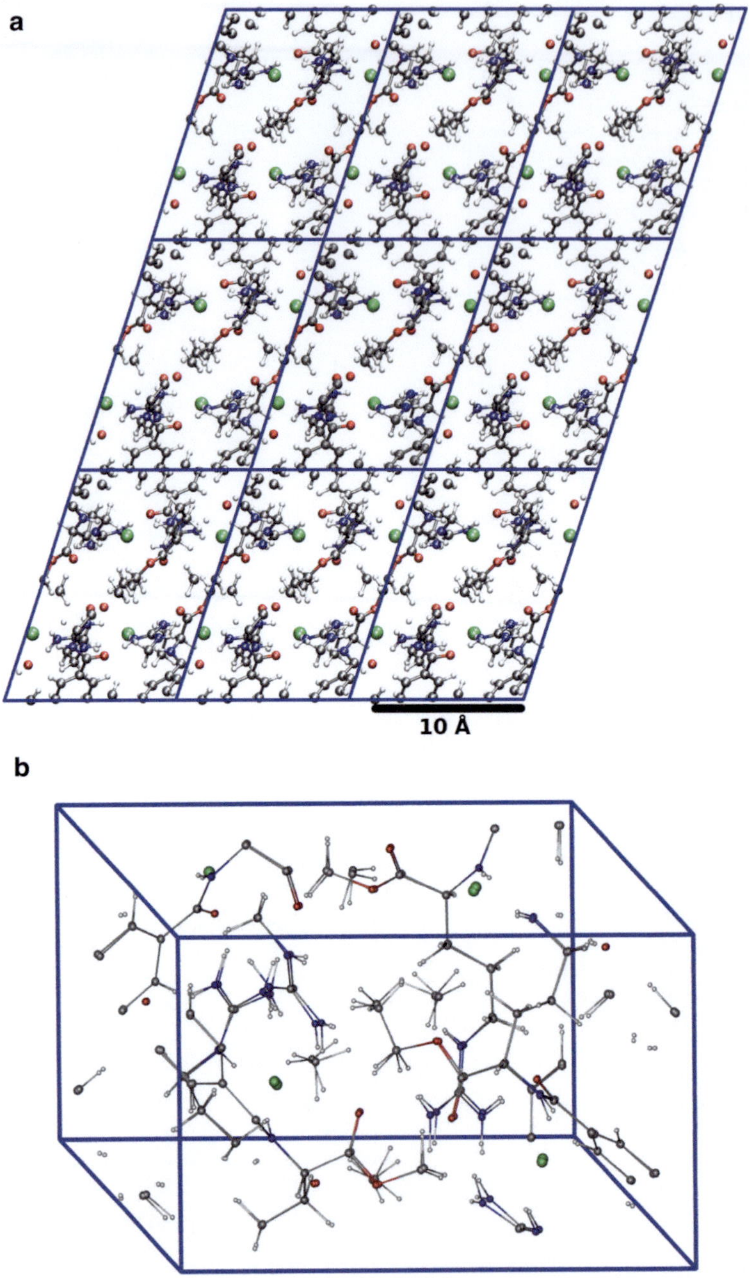

Fig. 8 BAEEH$^+$·Cl$^-$ studied under periodic boundary conditions. (**a**) Replicated unit cell of crystalline BAEEH$^+$·Cl$^-$ based on X-ray data (see Fig. 6d). (**b**) Comparison between experimental and calculated structural data for crystalline BAEEH$^+$·Cl$^-$; the figure shows the initial and the final snapshot of an *ab initio* optimization under periodic boundary conditions

In order to pursue a reliable spectroscopic and/or structural study when dealing with solid state data, there are some issues that should be considered. For instance, taking the $N = 46$ atoms in each $BAEEH^{+}{\cdot}Cl^{-}$ entity and Z molecules present in the unit cell, a total of $138 \times Z - 3$ normal vibrational modes [i.e. $(3 \times 46 - 6) \times Z = 528$ internal plus $6 \times Z - 3 = 21$ external modes] are expected. This should prompt us to the question of knowing whether the calculated vibrational spectrum for the isolated $BAEEH^{+}{\cdot}Cl^{-}$ molecule in vacuo is sufficient to the assignment of the normal modes. Actually, the observed frequencies are perturbed by a crystalline force field, which is unaccounted in the isolated-molecule calculations (apart the effects of neglecting anharmonicity). The same can be said about specific kind of interactions within the crystal, such as the prominent case of intermolecular H-bonding interactions. Issues related to structural data are also important within this context. For example, the uncertainty in locating the hydrogen atom positions from X-ray diffraction analysis and/or the difficulty in solving crystal structures exhibiting local disorders are well known problems. Again, the resort to quantum mechanics calculations is a possible strategy towards the necessary corrections, whose reliability, however, depends on the quality of the theoretical model. As in the case of the spectroscopic analysis, the results of calculations performed on an isolated molecule in vacuo must be cautiously taken if important environmental interactions actually exist in the crystal, such as H-bonding, torsions and/or other crystal packing effects. Moreover, because of the different environments experienced by a molecule in vacuo and in the crystal, their possible conformations, and hence their stablest geometries, can contrast each other to a greater or lesser extent. These issues recommend that the theoretical approach include calculations based on a solid-state model under PBC (Fig. 8).

Figure 8b compares the structural data obtained experimentally and computationally. The main differences between the two approaches were found in the CH and NH bond lengths (with the calculated values substantially longer than the experimental), as well as in some torsion angles. Such discrepancies can be mostly explained by the well-known uncertainty underlying the X-ray diffraction technique in locating the hydrogen atom positions and/or the difficulty in solving crystal structures exhibiting local disorders. Quantum mechanics calculations appear to be quite relevant in contributing to the necessary corrections.

The simulation of vibrational spectra including the calculation of intensities usually implies long *ab initio* molecular dynamics simulation runs, thus being both demanding computationally and time consuming. However, the calculation of the frequencies of normal vibrational modes is reasonably tractable, and when combined with data from cheaper theoretical approaches, such as the $BAEEH^{+}(H_2O)_5$ supermolecular model (see Fig. 9), provide a valuable support to the interpretation of the experimental findings.

The described example on the model compound $BAEEH^{+}{\cdot}Cl^{-}$ shows how an interplay between experiment (*e.g.* X-ray analysis, vibrational and NMR spectroscopy) and theory (quantum mechanics calculations) provides a wealthy of data. Great insights on $BAEEH^{+}{\cdot}Cl^{-}$ were made possible through a comprehensive

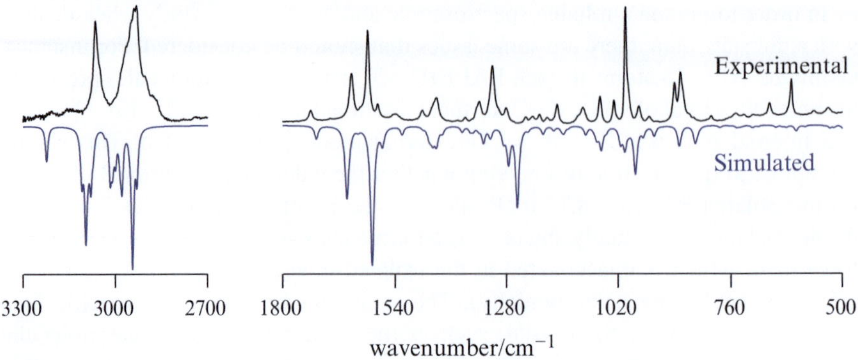

Fig. 9 Comparison between experimental and calculated vibrational spectroscopic data (Raman) for solid BAEEH^{+}·Cl^{-}. The calculated results are referred to the BAEEH^{+}(H$_2$O)$_5$ complex model

structural and spectroscopic characterization, which might be quite relevant in the study of more complex systems with similar structural moieties, such as PEAs [70, 71].

On going from small molecular systems to larger ones, and depending on the objective of the computational study, a considerable step might be necessary in terms of simulation methods and computational resources as well. As mentioned, for high level quantum mechanics methods the computational cost becomes prohibitive as the dimension of the targeted molecular system increases. Different approaches can be used to circumvent this practical limitation without giving up accurate methods, and macromolecular systems are again a good source of examples. One possible approach consists in systematically study by reasonably accurate methods small to medium size constitutive unites (*e.g.* from monomeric to oligomeric entities) of a polymer towards the prediction of some characteristics of the polymer itself. For instance, it is well known that the mechanical and thermal properties of a crystalline polymer depend in a great extent on its morphology and crystal structure. On the other hand, the chains conformations are crucial in determining the application of polymer materials. Moreover, the crystalline structure in many materials is largely determined by intramolecular interactions, whose study in the isolated polymer (oligomers) appears to be of interest. This sort of reasons suggest that the investigation of the preferential conformations adopted by an isolated polymeric chain can be very useful in assessing the crystalline structure of polymers. This strategy was applied in a recent work carried out on poly(L-lactic acid) homopolymer (PLLA), H–[OCH(CH$_3$)C=O]$_n$–OH [72], a chiral polyester analogue to poly(L-alanine). PLLA has deserved increasing attention in recent years due to its distinctive attributes among the family of environmentally friendly biodegradable polymers. Different applications are known, such as a biocompatible polymer in implant materials, surgical sutures, and controlled drug delivery systems. Apart from the spectroscopic and structural properties L-LA oligomers depicted in Fig. 10, PLLA has been comprehensively investigated to derive the low-energy

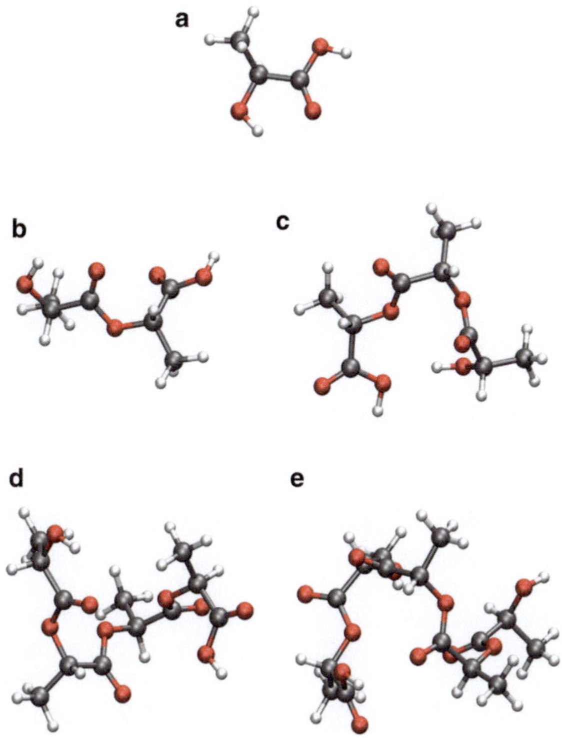

Fig. 10 (**a**) L-lactic acid, and correspondent (**b**) dimer, (**c**) trimer, (**d**) tetramer, and (**e**) pentamer

and stable conformations. This theoretical study was accompanied by experimental spectroscopic techniques as IR/Raman and 1H/13C NMR. Both computed results and experimental data were confronted to gather new findings for different L-LA oligomers. Accordingly, the optimized geometries adopted by these oligomers assisted the final design of PLLA polymer by reducing drastically the numerical simulation time and encompassed a deep understanding of experimental data [72].

The accurate modeling of proteins and their interactions with nucleic acids has been challenging the field of computational science regarding the investigation of molecules substrates and pharmaceutical drugs. Specifically, the interaction between proteins or nucleic acid macromolecules is described by a ligand agent, and this entity interrelates with the molecule's substrate such as the protein or nucleic acid. Indeed, the overall behavior can be affected by small molecules as drug molecular systems [6]. Often, ligand-protein complexes have been investigated as the most representative drug-protein systems. The mechanism of drug interaction is characteristically divided in a binding stage of drug compound to an enzyme active site avoiding the substrate from competing with the ligand on the desired reaction catalysis. The binding phenomenon is primarily described by hydrogen bonds, charge-charge interactions, and other strong electrostatic forces, and secondly by van der Waals forces which assume particular importance on these drug-protein interactions [6].

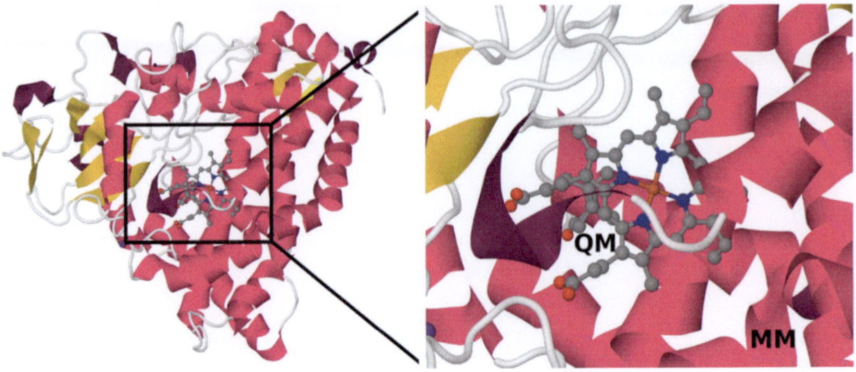

Fig. 11 Example of a QM/MM modeled system (center molecule described by QM and the surrounding protein by MM) [73]

By defining a target polymeric region, multiscale modeling techniques have been alleviating the overwhelming constraint on the accurate computation modeling for such molecular systems. This class of methods is labeled as quantum mechanical/molecular mechanical (QM/MM) framework so the core region is evaluated by QM, while MM describes the remainder of the case-study (Fig. 11). An intrinsic advantage relies on the arbitrary hierarchy between QM and MM as in the ONIOM (Our own N-layered Integrated molecular Orbital and molecular Mechanics) approach. However, the computational expense might become excessively high as the core region grows in spatial and temporal dimensions. This inherent drawback has opened a route to AM1/MM3 hybrid approaches aimed to simulate in a timely fashion the polymeric matrix system [8]. Still, there are other works that resort to different QM/MM combinations. Odoh et al. [74] applied B3LYP-6-311++G*/AMBER95 to analyze uranyl fluorides in the gas and aqueous phases and in the hydrophobic cavities of tetrabrachion (the inorganic tetrabrachion complex presents large cavities which can hold or bind to molecules, making them ideal drug-delivery vehicles). Li et al. [75] studied the conjugation of organic-metallic hybrid polymers with section of DNA, deploying a combined BLYP-def2-SV(P)/AMBER96 approach.

A typical example of these hierarchical models comprises the application of lumped parameters from shorter spatial and temporal scales to probe the pertinent phenomena for larger scale domains. An averaged number of polymeric entities is used to represent single beads with modifiable interactions, and further applied to calculate physicochemical properties to be subsequently used in continuum flow models for the macroscopic description of the selected materials [8].

Alternatively, coarse graining has been envisaged as an extreme multiscale technique from the conventional QM/MM framework. While the intended core region is thoroughly detailed either by accurate predictions or experimental evidence, it is succeedingly coalesced into a single system unit where the interactions are modeled by means of time-efficient methods. Both synthetic and natural polymers have been simulated by the coarse graining methodology to produce expedite and quite accurate findings for those case studies. Nonetheless particular attention should be paid to the scale limitations and the development and application of trustworthy force fields [8].

Regarding classical MD, Stepniewski et al. [76] carried out studies on a polymer-liposome complex using poly(ethylene glycol) (PEG), which has been broadly applied as a half-life-increasing coating for nanoparticles and as a conjugate for proteins and other drugs. The lipids head group were modified to incorporate the PEG, this complex is called PEGylated. It was analyzed the interaction between salt ions in the blood plasma an the PEGylated surface. They addressed the problem using classical MD to analyze the effect of the PEGylation on the bilayer properties, the interaction of PEG-lipid, the structuring of the PEGylated layer, the interaction between the ions and the PEG chains and the effect of PEGylation on the ion-headgroup interactions. Their study, combined with some experimental results, showed that the PEG surface layer can not be treated as a generic hydrophilic polymer layer, as was done previously.

There is ongoing some research regarding DDS in our laboratory, where computational techniques are being extensively applied. Our aim is to develop polymer-liposome complexes to improve the liposome characteristics as DDS. On of the studied polymers is the poly(acrylic acid) (PAA). With the input obtained from the QM calculations, it was possible to generate and tune a force field for PAA, essential to classical MD. It was simulated the interaction of bigger PAA oligomers with water and biomembranes (Fig. 12).

With these simulations, it was possible to understand the intramolecular behavior of the PAA oligomers in aqueous environment, and guess possible H-bonds (Fig. 12a). Regarding the intermolecular interactions, it was simulated a dispersion of PAA oligomers on water (Fig. 12b) to predict possible formation of clusters/micelles of PAA by itself. Finally, the simulation of the interaction of a PAA with a biomembrane was performed (Fig. 12c) to study the effect of a new drug, or a drug fragment, on the membrane.

There are several parameters that can be extracted from a classical MD simulation. The area per lipid (A_{lip}), Fig. 13a, and the deuterium order parameter (S_{CD}), Fig. 13b are two examples. A_{lip} provides information on the organization and stability of the bilayer, whereas S_{CD} is related to the flexibility of the alkyl tails of the lipid (higher S_{CD} means lower flexibility). These two parameters give important data on the stiffness and the arrangement of the lipid molecules in the bilayer. As an example, it is common to analyze the variation of these parameter with the temperature. In this way, it is possible to understand the impact of a drug in a bilayer surface within a range of temperatures, which is crucial in the case of thermal sensitive polymers. Other impressive, but more qualitative tool, is the ability to obtain density profiles of the system (Fig. 13c), thus making possible to determine the relative position of a molecule or a group of atoms in the system. In Fig. 13c it is possible to observe that Drug 2 (full red line) has the tendency to migrate to the membrane center (dashed black line).

Finally, there are several examples of MD using CG models for DDS. Figure 14 shows a schematic representation of a coarse grained PAA oligomer. The reduction in the number of particles is dramatic, from 47 to 11, which combined with a proper

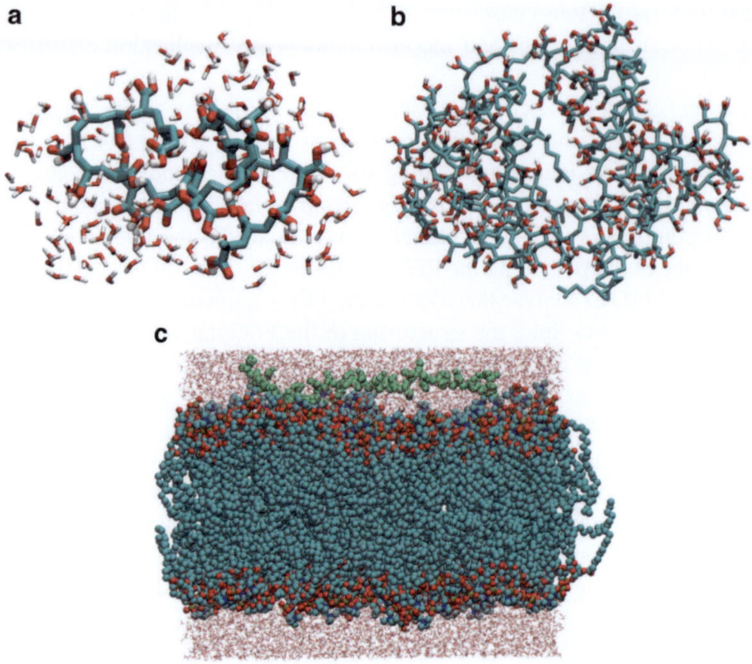

Fig. 12 Snapshots of some MD calculations. (**a**) Interaction of PAA with water; (**b**) self-assembled PAA oligomers; (**c**) interaction of a PAA oligomer with a biomembrane

parameterization can provide relevant results in the mesoscale. Loverde et al. [77] applied this approach to analyze how the shape of a nanoparticle can improve drug delivery. In this study, it was analyzed how the nanoparticle shape improves the delivery, which, due to the nanoparticles dimension, implies the use of a CG approach. Winter et al. [78] used CG model to study the permeability enhancement in DPPC bilayers by incorporation of lysolipid (with a base system formed by 512 DPPC lipids and 8,000 coarse-grained water molecules!).

3 Computational Fluid Dynamics

3.1 Scope of the Application of Computational Fluid Dynamics to DDS

Taking into account that computational flow models can assist the design and optimization of new and existing DDS, reduce economic and financial resources, increasing productivity and profit margins and, namely, improving wellness and

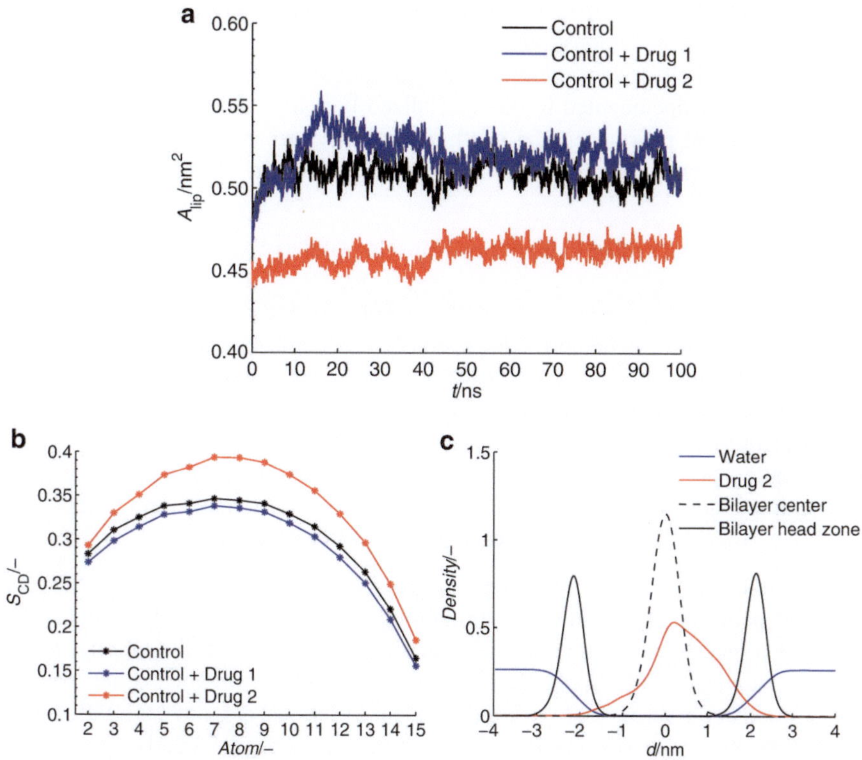

Fig. 13 Extracted information from a MD simulation of a bilayer. (**a**) Area per lipid (A_{lip}); (**b**) deuterium order parameter (S_{CD}); (**c**) density profile

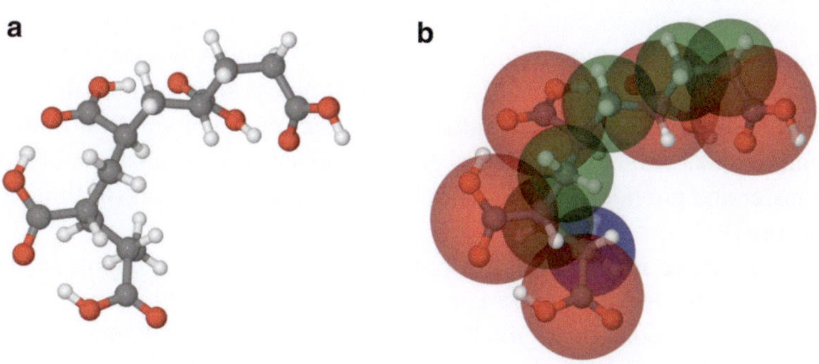

Fig. 14 Illustration of the coarse grain approach. (**a**) PAA oligomer; (**b**) coarse grain scheme of PAA oligomer

human health, the present review is driven by the great advantage that modern CFD models already demonstrated in chemical and pharmaceutical industries. The simulation of drug release profiles for hydrophobic and hydrophilic molecules are increasingly documented in the specialized literature and a broad segment of the chemical and pharmaceutical synthesis industry are anxiously waiting for its appliance. Numerous requirements have been identified on CFD tools so that they should be:

- Experimentally verified – simulations must be validated with experimental tissue results.
- Computationally efficient – adaptable to multiprocessor workstations and clusters of processors, which provide expediently valuable results.
- Fundamentally based – containing more and better physics coupled with drug release profile and tissue elimination kinetics, rather than non-mechanistic correlations that are not scalable.
- Versatile – portable to various temporal and spatial drug transport configurations, usable by generalist rather than specialists, scalable, reliable, and compatible with current chemical and pharmaceutical industry databases.

The realization of maximum benefit from CFD modeling for drug delivery design depends on how well CFD simulations can predict adequately an industrial unit either because of numerical inaccuracies or because of limitations in the physical models [79]. Sophisticated numerical methods enable better control on discretization errors but grid-independent solutions are not straightforward even with the actual advancement in computing resources. Underlying physical models should be calibrated with high-quality experimental data aided by new tools based on modern visualization techniques. Electrical, radiation, optical and acoustic methods have already been developed to provide noninvasive, spatially, and temporally resolved measurements of drug release profiles [80]. The major drawback rely on the technical knowledge needed to examine experimental data on fluid-structure interactions, viscous flow characterization, regime transition, and other important dynamic characteristics of anisotropic porous medium flows. High-performance computing such as parallel machines/cluster computing allow fine-grid computations that will bring insightful interpretations on the mechanism of momentum transfer and provide insights into mass transfer and diffusion/swelling/erosion phenomena [81].

From the above mentioned forewords, the application of computational flow modeling into the design of DDS is an actual fertile and wealthy area of research in which there is room for significant improvement. The development of active, cost effective, robust delivery systems for use in many different tissues could lead to significant breakthrough step for pharmaceutical and chemical actuators in the academic field and industry arena. With the present contribution, our main purpose is to present the integration and couple state of the art CFD codes with computational flow modeling applications in the design and optimization of drug delivery systems.

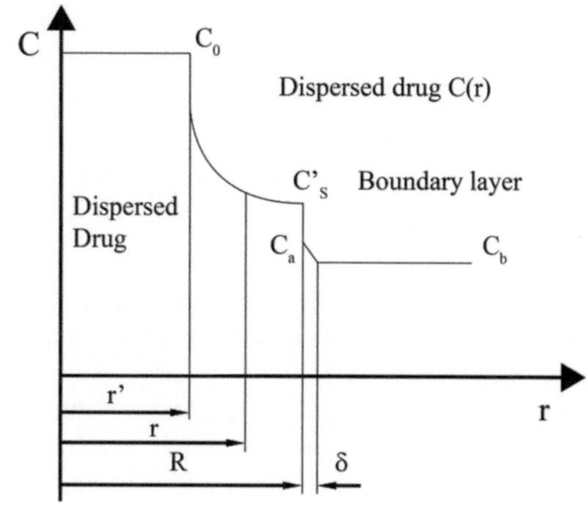

Fig. 15 Representative concentration profile in a drug delivery system accounting for the boundary layer in a perfect sink medium characterized by the drug concentrations in the bulk medium (C_b) and at the matrix surface (C_s)

3.2 Mathematical Background

The CFD code calculates the spatial and temporal evolution of drug concentration within the delivery system. The unsteady diffusion-advection equation is used to model the dissolution process and the simulation domain focuses on the boundary layer close to the solid-liquid interface. The numerical framework is solved using a moving-boundary finite element formulation. Generally, two distinct procedures are identified in this methodology. In a first stage, the dissolution process evolves at the solid-liquid interface and then it is considered the mixing of the drug concentration in the matrix system. A representative diagram for the concentration profile in a drug delivery system accounting for the boundary layer is illustrated in Fig. 15, which identifies the drug concentrations in the bulk medium (C_b) and at the matrix surface (C_s).

Whereas the second phase is not widely validated for polymer-based systems, the computed results handled agreeably better with experimental data during the first stage. However, CFD can overcome these limitations as long as it can be parametrically optimized in dealing with cumbersome designs. The realistic drug release profiles are attained by means of finite volume and finite element discretization methods. Both approaches confer reliable and numerically stable algorithms. In the finite volume method, the simulation advances by solving the governing equations over discrete control volumes. The partial differential Navier-Stokes equations are recasted in a conservative form followed by the discretization of the new equation. This numerical procedure guarantees the conservation of fluxes through a particular control volume. Notwithstanding the finite element method is weaker to ensure a conservative solution it brings an additional stability than the finite volume formulation, thereby requiring more memory for the same case study. Predominantly, the finite element code applies a pre-conditioned conjugate gradient solver and encompasses extra modules for moving mesh and mesh adaptation around the moving boundary.

First, a mathematical model of diffusion through the fluid structure is developed to account for the drug dissolution and release profiles. The diffusion process is governed by the time dependent diffusion-advection equation:

$$\frac{\partial c}{\partial t} = -(\mathbf{v}.\nabla)\, c + D\Delta c \tag{14}$$

where c [mg/cm^3] is the concentration of dissolving substance in the fluid, $\mathbf{v}$ [cm/min] is the velocity flow field of the fluid, and D [cm^2/min], the diffusion coefficient of the solute in the solvent which is a physical property of the substance. The phase transformation from the solid system into the surrounding fluid is captured by the boundary condition at the solid-liquid interface in where the concentration is at saturation conditions. In what follows, we have expressed the saturation concentration by C_s and the concentration in the solid by C_0. Hence, by Fick's law of diffusion, we have the following equation for each computational cell of the mathematical domain mimicking the delivery system:

$$(C_0 - C_s)\,(X(t) - X(t + \mathrm{d}t)) = -D \left(\left. \frac{\partial c}{\partial x} \right|_{x=X(t)} \right) \mathrm{d}t \tag{15}$$

The mathematical model described in Eq. 14 has been solved using a standard Galerkin finite element method with linear triangular shape functions, and a moving grid method has been also employed for polymer matrix systems. The mesh coordinates are functions of time, $x(t)$, so that the concentration is discretised as follows:

$$c\,(\mathbf{x}, t) = \sum_{j=1}^{N} c_j(t)\phi_j(\mathbf{x}, t) \tag{16}$$

where ϕ_j is the shape function, and the temporal dependence of the coordinates accounts for the time correction term that is needed to be incorporated in the temporal discretization schemes.

$$\frac{\partial c}{\partial t} = \sum_{j=1}^{N} \frac{\mathrm{d}c_j(t)}{\mathrm{d}t}\phi_j + \sum_{j=1}^{N} c_j(t)\frac{\mathrm{d}\mathbf{x}}{\mathrm{d}t}\nabla\phi_j \tag{17}$$

Traditionally, specific mathematical idealizations and assumptions have been made for the purpose of CFD simulations. The first and prominent one was related with the system arrangement that has been mimicked by a symmetric cylindrical shape. This fact was carried out by a two-dimensional cylindrical co-ordinate formulation which allows performing a comparative analysis of orders of magnitude for different delivery systems. The advection term is exceedingly dominant over the diffusion one, meaning that as soon as the drug concentration leaves the boundary layer region, it rapidly mixes in the bulk fluid system. Therefore, the CFD simulation is directed to the diffusion phenomena occurring within the boundary layer region. The numerical methodology for the moving boundary is calculated according to the mass solubilized of the drug in time which is described by:

$$m = -\Delta t D \int_{B} \frac{\partial c}{\partial n} \mathrm{d}\Gamma \tag{18}$$

where B stands for the solid-liquid interface boundary. Bearing in mind that the solid-liquid interface boundary moves outwards as the drug is advected within the delivery system, a new location of B after a certain time is calculated by determining the remaining drug concentration in the solid system. In case of two layered designs, two separate diffusion equations are computed being one for each drug component in the delivering system. Numerically, the resulting constitutive flow equations are solved autonomously and the diffusion processes are tackled interactively through the shape modification of the computational domain.

3.3 Probing CFD Tools for Drug Delivery Systems

Computational flow modeling and microscale flow simulation are two major areas that can bring important advances for the pharmaceutical and chemical industries. The development of new drugs has to encompass the proper design of delivery process that ensure the delivery of materials and energy at the right places and at the right times in the realm of tissue engineering [82]. The manipulation and optimization of underlying fluid dynamics is a top priority on the minimization of trial-and-error approach reducing the capital and costs and leading to competitive solutions.

Bearing in mind the physical or chemical characteristics of the polymer backbone, drug release mechanism from a polymer matrix can be categorized in accordance to three main process mechanisms: drug diffusion from the non-degraded polymer (diffusion-controlled system), enhanced drug diffusion due to polymer swelling (swelling-controlled system), and drug release due to polymer degradation and erosion (erosion-controlled system). In this regard, CFD is one of the essential technologies that enable the drug designer to predict, manipulate, and realize the desired fluid dynamics in DDS. CFD serves as a reengineering activity and provide careful analysis to improve their overall effectiveness and to confer new insights into the application of new hydrophobic and hydrophilic molecules. CFD embraces a body of knowledge and techniques to solve mathematical models of fluid dynamics on digital computers in comparison to traditional techniques, which are based overly in empirical correlations. Notable reviews identified CFD as one of the potential technologies to move forward the current development of drug transport modeling in tissues [81].

The specification of operating protocols and system configuration in the design of DDS provide suitable background to complete the desired process without compromising safety, since the patient validation for tissue modeling is still not fully attainable. CFD modeling plays a role in suggesting the appropriate release profile in order to optimize the therapeutic efficiency of an agent. The CFD simulation results can also be used to provide recommendation on the required and optimized release profile at the drug device fabrication stage. The application of CFD has been envisaged as a prevailing tool to gain more control in the delivery of drugs since any chemical or physical drug transformation process requires the addition or removal of different materials by flow transport mechanisms. Therefore, the distribution of chemical as well as the release profiles within the tissues has been improved noticeably if one has the capability to predict and control fluid dynamics by means of CFD. The overall methodology can be divided in four main stages: identification of system requirements to categorize the desired fluid dynamic characteristics; evaluation of possible drug release configurations and operating protocols to achieve the desired fluid dynamic characteristics; development of quantitative relationships between the drug delivery configuration and performance; and optimization and fine-tuning of the final design [83].

Although the importance of fluid dynamics in the whole process has been understandable in CFD methodology, until the development of modern CFD codes attended by sophisticated computer architectures, pharmaceutical and chemical engineers had to manipulate fluid dynamics either by harnessing diffusion, swelling, or erosion mechanisms in tissue applications. CFD has been used to avoid intensive laboratory and clinical tests and mass transfer allowed the development of several empirical correlations, which account for different characteristics of films/boundary layers and transport rates over different shapes. Instead of extensive and costly experimentation, CFD models allowed the optimization of drug delivery formulations to maximize health benefit following rigorous clinical regulations. CFD enabled the establishment of quantitative relationships between the throughput and operating flow regime avoiding a massive number of empirical parameters. Transport parameters such as mass transfer coefficients and several hydrodynamic parameters including film structures and thicknesses, boundary layer geometries, drug holdup, release area or even the residence time distribution were investigated and optimized for multiple system configurations. Consequently, CFD aided in the identification of most appropriate process configurations from the large selection of alternatives.

Among the three mechanisms of drug release from a polymer matrix, diffusion is always involved. In case of non-biodegradable polymer matrix, drug release is dictated by the concentration gradient by either diffusion or matrix swelling, mostly well-known by enhanced diffusion. For biodegradable polymer matrix, release is generally governed by the hydrolytic cleavage of polymer chains that lead to matrix erosion, even though diffusion may be still dominant when the erosion is slow. Based on the prevailing mechanisms for the DDS, the literature has followed the classification of mathematical models into those three different categories [81]. The computational flow modeling of drug release was then confined to different approaches for each case study and provided ultimately insightful meanings on the drug release profiles for reservoir and matrix based systems. Whereas the reservoir system has been consisted of a drug reservoir surrounded by the polymer matrix shell, alternatively, in matrix system, the drug has been accounted for in the polymer matrix in either dissolved or dispersed condition.

The optimization of new formulations and devices have been guided concerning the mass transport limitations and chemical processes involved in drug delivery system as well as the effect of design parameters, such as the device geometry and drug loading, on drug release mechanism. Accordingly, the device design of DDS was optimized for a required drug release profile and can be predicted using a systematic approach with a minimum number of material resources and experimental studies [84, 85]. During the last decade, in vitro release profile of drug from controlled release devices has been combined with state-of-the-art computational fluid dynamics simulation to predict the spatial and temporal variation of drug transport in the living tissues. A wide assortment of system configurations has been investigated accounting for different parameters including the solute transport mechanisms in the tissue, given by convection of fluid, diffusion, elimination in the extracellular space, receptor- and non-receptor-based cells internalization, and intracellular elimination [86].

3.4 Synopsis of Computational Flow Models

CFD models have been divided in two main classes: design models that attempt to provide a quantitative relationship between the drug delivery system and release profile, and learning models, which provide a basic understanding of different underlying processes. Drug design can thereby steered to different process arrangements and project concepts, which overcome usually the lack of experimental, time and even funding resources. From a concise literature survey, the current status of CFD on flow simulation was summarized into two distinct categories: single-phase and multiphase flows.

Laminar flow and mixing of Newtonian fluids have been simulated in complex matrix systems accounting for the composition dependent on physical properties. The quantification of viscous drag, form drag, and losses due to directional changes in the overall drug holdup provide the efficient design of pharmaceutical products. Recent developments on free surface flow allow a deep mathematical understanding of surface characteristics and surface phenomena (surface tension, contact angle, surface adhesion) involving contact of different phases. During the last decade, the boost and enlargement of computational resources allowed the flow simulation of rheologically complex fluids with viscoelastic characteristics and the quantitative prediction of the flow regime characteristics and release profiles [81].

The predictive capabilities of CFD codes have been always a discussion center on the viscous modeling. Viscous flow models and the three main categories: direct numerical simulations (DNS), large eddy simulations (LES), and Reynolds-averaged

Navier-Stokes equations (RANS) are the dominant flow frameworks to compute the drug release profile. If one moves from DNS to RANS, the flow of specific drug are estimated and, therefore, less computational resources are required. DNS models simulate all of the dynamically important scales of viscous flows directly being the specification of the initial and boundary conditions one of the imperative step. DNS provides valuable information about the interaction of small-scale and large-scale motions requiring huge computational resources. Therefore, applications of DNS for DDS design are not likely to be cost-effective for the near future unless one has a straightforward flow configuration with very low number of computational cells. In LES, the large-scale motions are resolved while the small-scale motions are modelled using the subgrid scale (SGS) models. Until recently, few theoretical studies have been accomplished with SGS models that exhibited low satisfactory levels. LES models are expected to be reformulated in the forthcoming years with practical applications.

Alternatively, RANS-based turbulence models are a ubiquitous approach that enables a good compromise between the width of application and computational economy. The quality of RANS simulations are strongly affected by the numerical issues, *e.g.* grid quality and/or density, discretization schemes, wall functions and unsteady flow features [87]. It should be pointed out that RANS-based models contain more adjustable constants that need to be determined by fitting the experimental data in comparison with DNS and LES models. Hence, careful verification and validation should be performed through a meticulous error analysis to avoid dangerous extrapolations. Release profile and flow simulations on linking the drug release profile and tissue elimination kinetics to predict temporal and spatial drug transport are one of the most intricate case-study since it requires the resolution of much finer scales and computational demands may increase significantly even with simple viscous flow models. Adequate closure models for the estimation of drug transport characteristics that embody the interactions of drug diffusion, swelling and erosion with viscous properties are needed to confer an accurate and secure design of advanced drug delivery technologies potentially applicable in personalized treatments.

Despite the advancement of CFD models in viscous flow modeling, quantitative predictions of drug release profile is obtained by solving Fick's second law of diffusion subject to appropriate boundary conditions. In addition to diffusion, erosion-controlled systems play a major role for bioerodible matrixes. These polymeric systems are versatile materials for a variety of biomedical applications, as long as their chemistry and surfaces can be tailored to stabilize macromolecular agents and enhance the tissue site-targeting. For such case of DDS, the erosion kinetics was tailored by careful selection of polymer and a variety of techniques of encapsulation to control the drug release profile. The methodology was based on the principle that the erosion kinetics can be promoted by modifying copolymer composition or the degree of crystallinity as crystalline and amorphous polymers erode at different rates [88]. Here, CFD codes provide valuable insights on release profiles, yet partially, in the improvement of drug design for a certain tissue. DNS models are foreseen as a breakthrough point to contribute with useful proficiency on several key issues from a singular investigation of drug transport and characteristics of generated dispersed-phase particles to the extrapolation of computed data to similar tissues and even the entrainment of the continuous phase due to diffusion, swelling and erosion-leaked surface. There was also a lack of appropriate closure models for Eulerian-Eulerian frameworks and whenever DNS fails to simulate most likely denser systems, Eulerian-Lagrangian approach is a common practice. Notwithstanding the widespread modeling efforts that have been accomplished on the development of reliable closure models in Eulerian-Eulerian, reliable mathematical approaches for dense and dispersed-phase drug delivery systems are scarce [89]. Mainly due to the limited availability of high-quality computed data to obtain different model parameters, the simulation of complex drug delivery systems have been retained from further progress and, therefore, state-of-art multiphase CFD models are envisaged to provide a feasible and quantitative prediction of intra-phase and inter-phase transport rates.

3.5 CFD Methodology for DDS

The increasing concern in modern DDS has been intensified in line with the advancement in the innovation, conception, development and synthesis of novel therapeutic macromolecules for targeted applications. This methodology has been dictated by maximizing the atom utilization from the raw materials to the final end products and following efficient routes via the so-called atom economy framework. Computational fluid dynamics as a micro- and nano-scale design tool allows designers of drugs and pharmaceutical products to evaluate different designs promptly and cost effectively.

During the last decade, the intrinsic advantages of CFD design tools within the pharmaceutical industry are ascribed to their applications on new macromolecules presenting considerable benefits in terms of yielding lower costs, a faster time to market, new insights of existing or novel innovations, and the awareness to make knowledgeable decisions to steer and shape the course of future research efforts. Indeed, the above-mentioned benefits assume a major role at the time of the investigation for novel effective DDS which confer relevant outcomes in view of the fact that traditional needle and syringe-based methodologies pose additional development issues.

CFD can assist the pharmaceutical and drug developer to overcome the recursive animal experimentation in the evaluation of DDS, allowing the analysis of various drug delivery designs timely and economically. CFD rely in a three-dimensional model of human physiology making insightful comparisons avoiding the practical restrictions on the level of testing unknown release profile for innovative macromolecules. CFD deals with numerical methods to solve the equations that govern fluid flow. First, a specific computational domain is generated and discretized into several thousands or millions of small three-dimensional cells. After the optimization of this computational grid, the constitutive and fundamental Navier-

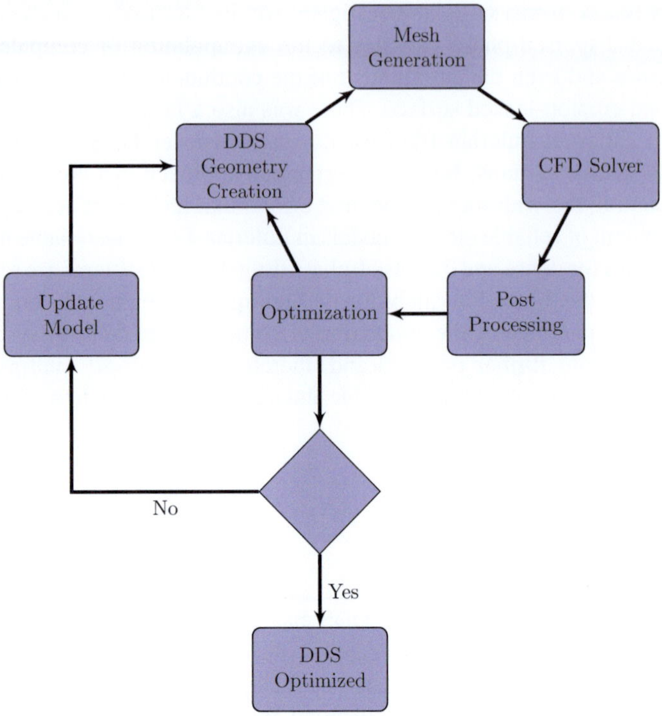

Fig. 16 Development stages of CFD frameworks for drug delivery systems

Stokes equations are solved numerically for each cell within the mesh to attain an overall and comprehensive solution. Following an integrated framework, the development stages of CFD methodologies for DDS are shown in Fig. 16.

Within this methodology, the drug developer can engineer several virtual experiments and computer-generated analyses reducing expressively the amount of physical experimentation required to search for an optimal pharmaceutical product formulation. Additionally, a major leap encompassed by CFD framework is attributed to the conception of multiple designs, dissimilarities of similar systems and numerous process parameters that are thoroughly evaluated to achieve maximum efficiency and consistency of a specified drug release profile. Moreover, the multifaceted investigation of an unrestricted number of system configurations can be accomplished concurrently, without being constrained by the number of time-points or regions of analysis. CFD virtualizes unique drug release patterns that are more cost-effective than expensive practical experimental methods. Correspondingly, particular phenomena or case scenarios can be straightforwardly isolated and subjected to further intensification and refinement. This window of opportunity authorizes a plain identification of potential final designs for further validation and entails fewer experimental prototypes to be evaluated accordingly.

Noticeably, the research and design activities provided by CFD techniques serve as an immensurable assessment of prospective designs for new drug delivery systems. CFD allows reducing the man-hours spent on evaluating novel pharmaceutical products and makes affordable a faster development of these feasible products in a timely fashion by greatly reducing the capital investment and operating stage periods. CFD generates a wide assortment of parametric data including drug release kinetics, dose concentration and spatial distributions. A post-processing phase involves the three-dimensional mapping of interstitial results and represents in a three-dimensional visualization different variables including temperature, pressure and fluid flow depicted in color maps.

In order to accomplish such a task, there is a variety of open-source and commercial CFD codes being OpenFOAM and ANSYS the most representative, respectively. The OpenFOAM (Open Field Operation and Manipulation) CFD Toolbox is a free, open source CFD software package produced by OpenCFD Ltd. OpenFOAM has an extensive range of features to solve complex fluid flows involving chemical reactions, turbulence and heat transfer, to solid dynamics and electromagnetics. It includes tools for meshing for complex CAD geometries, and for pre- and post-processing. Most importantly, different meshing, and pre- and post-processing tools run in parallel as standard, enabling users to take full advantage of computer hardware at their disposal. Having being developed under the GNU general public licence (GPL), OpenFOAM offers users complete freedom to customize and extend its existing functionality by implementing a highly modular code design in which collections of functionality such as numerical methods, meshing, physical models which are each compiled into their own shared library. Indeed, OpenFOAM includes over 80 solver applications that simulate specific problems in mechanics and fluid flow engineering and over 170 utility applications that perform pre- and post-processing tasks for meshing and data visualization activities.

3.6 CFD Case-Studies

Ubiquitous CFD case-studies for DDS embrace the investigation of brain, nasal and lung tissues. The drug simulation in brain tissue poses great challenges to provide more meaningful results for biomedical applications [82]. This fact is partly ascribed to the development of the accurate methodology of geometric reconstruction and transport parameters by fitting to the model firstly with the normal case. While the automatic mesh reconstruction of normal tissue and tumor geometry can be generated from medical images, i.e. magnetic resonance imaging and computed tomography scan, the intrinsic limitation to obtain transport parameters experimentally can be overcome by fitting these variables to the model with actual modeling grid owing to accurate boundary conditions of physiological variables [90].

The development and applications of in silico models of aerosol delivery to the respiratory tract have been reviewed by Longest [80]. The literature survey covered transport and deposition within the inhaler device, the extrathoracic (oral and nasal)

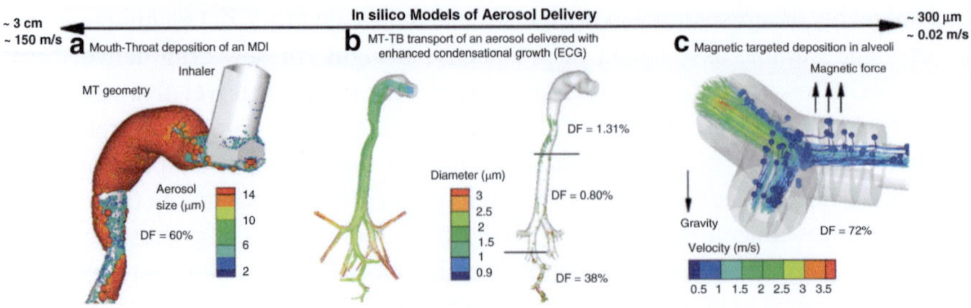

Fig. 17 Actuation scales of *in silico* models for aerosol delivery to the respiratory tract (from Longest and Holbrook [80])

region, conducting airways, and alveolar space. Three-dimensional transport and deposition simulations of inhaled pharmaceutical aerosols have been highlighted to confer realistic drug profiles, as schematically illustrated in Fig. 17.

Further recommendations on the need for refined model validations, and development of a stochastic individual path model of aerosol transport and deposition throughout the respiratory tract have been presented accordingly. Comprehensive numerical simulations of the USP drug delivery problem using CFD have been investigated by Crane [79]. The dissolution of layered compacts was simulated by adapting a moving boundary finite element approach to mimic the number of layers of different acids with different dissolution rates. The simulation results agreed quite accurately with experimental measurements apart from the idealizations that were adopted in the numerical model. Kleven [91] have used CFD to improve the bi-directional nasal drug delivery concept. The simulations were used to evaluate the prominent features of the bi-directional technique and how its design and optimization could be further accomplished. CFD computations were found to increase the adeptness of device design and reduce the need for expensive and time consuming laboratory experiments. Critical assertions were made to perform successful CFD calculations on the nasal tissue so the generation of a proper computational grid of the nasal cavity was a key issue. In silico modeling of aerosol deposition in lungs has been reported by Annapragada [92]. A wide range of interesting fields were presented from inhalation toxicology to infectious disease proliferation on drug delivery. Key phenomena were identified such as the complexities of the airway geometry, intermittent turbulence, the highly non-constant nature of the airflow involving inhalation-exhalation cycles, and the shedding of turbulent vortices into otherwise non-turbulent flows. These constraints resulted in complicated moving boundaries in the deep lung so the underlying physics has to be fully characterized for the successful implementation and mathematical representation of particle transport into sophisticated CFD codes. Three-dimensional simulations and experimental investigations of intrathecal drug delivery in the spinal canal and the brain were addressed by Hsu [93]. The CFD model hypothesized representative fluid pulsations inside the spinal canal and the results have demonstrated clinically relevant applications as the physiological cerebrospinal fluid pulsations were replicated in an anatomically consistent computational framework. Additional recommendations and guidelines were pointed out to achieve a quantified and validated drug dispersion profile inside the central nervous system with the computational flow model.

4 High-Performance Computing

Modern multiscale modeling and simulation techniques emerged over the last years as a decisive practice in science and engineering, allowing for a deep understanding of material structures, properties and interactions, starting at and below nanoscale. The theoretical study of crystalline and amorphous solids, nanostructures, supramolecular systems, liquids and solutions is now possible with acceptable simulation times thanks to the dramatic advances in high-performance computing (HPC) architectures and molecular modeling and simulation codes thereof. As the processing capabilities of recent machines have been dramatically increased, the bottleneck in HPC has moved to the network data transfer rate, thus motivating the search for new interconnect solutions.

The use of computational techniques to help scientists and engineers in solving their problems started since the creation of the first computers. Although it was difficult to predict that such approach would evolve to a mature science, it was already clear the need to understand molecular systems, and related phenomena, with enhanced detail [11, 94]. In those early days two of the most important tools to analyze chemical systems emerged, *viz.* computational quantum chemistry and molecular dynamics (MD). Computational quantum chemistry resorts to *ab initio* and Density Functional Theory (DFT) frameworks to describe the molecule/system under study, allowing for good accuracy levels in the results, although at expense of high computational resources. Classical MD requires lower computational power, which enables large timescale simulations of bigger systems to be tackled. Still, those codes demand tremendous computing power, and many companies and code developers turned their emphasis to the development of massively parallel high performance computing applications.

During the last decade, continuing advances in microprocessor design and semiconductor manufacturing provided an exponential performance growth. Still, beginning around year 2002 some practical limitations on power consumption and heat dissipation, resulting from the elevated clock rates, caused a stabilization on the performance growth of a single-core microprocessor. This left the community with no alternative but to begin development of parallel and multi-core versions of applications that require performance levels exceeding the capabilities of single-core graphics processing units (GPUs) [95]. In the search of higher processing power, GPUs presented itself as a great alternative.

Over the past 8 years, there has been a remarkable increase in the performance and capabilities of GPUs. Modern GPUs have evolved into fully programmable, massively parallel co-processors that can now be exploited to accelerate many scientific computations, typically providing speedups of one order of magnitude, sometimes tow orders of magnitude, over CPU code [95, 96]. This speedup is possible because GPUs contain high-bandwidth memory systems as a result of the needs of graphics, capable of intra-GPU transfer rates of over 100 GB/s. GPUs implement hardware multi threading and machine instructions for many of the mathematical functions used by molecular simulation applications. In combination, these attributes enable GPUs to outperform traditional CPU cores on highly data-parallel workloads by factors ranging from ten to twenty times faster in the majority of cases, up to as much as one hundred times faster in a few ideal cases [95,97–99].

References

1. Rosen H, Abribat T (2005) The rise and rise of drug delivery. Nat Rev Drug Discov 4(5): 381–385
2. Landau DP (2005) The future of simulation in materials science. In: Yip S (ed) Handbook of materials modeling. Springer, Dordrecht
3. Han C, Wang B (2005) Factors that impact the developability of drug candidates: an overview. Wiley, Hoboken, chap 1, pp 1–14. doi:10.1002/0471475734.ch1
4. Wheeler RA (ed) (2011) Annual reports in computational chemistry, vol 7. Elsevier, Amsterdam/London
5. DaCosta H (ed) (2011) Rate constant calculation for thermal reactions. Wiley, New York
6. Lipkowitz KB, Cundari TR (eds) (2010) Reviews in computational chemistry, vol 27. Wiley, Hoboken
7. Engquist B, Ltstedt P, Runborg O (eds) (2009) Multiscale modeling and simulation in science. Springer, Berlin/Heidelberg
8. Ross R, Mohanty S (eds) (2008) Multiscale simulation methods for nanomaterials. Wiley, Hoboken
9. Yip S (ed) (2005) Handbook of materials modeling. Springer, Dordrecht
10. IUPAC (1997) Compendium of chemical terminology, 2nd edn. (the "Gold Book"). Compiled by A. D. McNaught and A. Wilkinson. Blackwell Scientific Publications, Oxford. URL http://goldbook.iupac.org. Accessed 10 Dec 2012
11. de Jong WA, Bylaska E, Govind N, Janssen CL, Kowalski K, Muller T, Nielsen IMB, van Dam HJJ, Veryazov V, Lindh R (2010) Utilizing high performance computing for chemistry: parallel computational chemistry. Phys Chem Chem Phys 12(26):6896–6920
12. Praprotnik M, Site LD, Kremer K (2008) Multiscale simulation of soft matter: from scale bridging to adaptive resolution. Annu Rev Phys Chem 59(1):545–571. doi:10.1146/annurev.physchem.59.032607.093707
13. Timko BP, Whitehead K, Gao W, Kohane DS, Farokhzad O, Anderson D, Langer R (2011) Advances in drug delivery. Annu Rev Mater Res 41(1):1–20. doi:10.1146/annurev-matsci-062910-100359
14. Shi J, Votruba AR, Farokhzad OC, Langer R (2010) Nanotechnology in drug delivery and tissue engineering: from discovery to applications. Nano Lett 10(9):3223–3230. doi:10.1021/nl102184c
15. Thassu D, Pathak Y, Deleers M (2007) Nanoparticulate drug-delivery systems: an overview. In: Thassu D, Pathak Y, Deleers M (eds) Nanoparticulate drug delivery systems. Informa Healthcare, New York

16. Vasapollo G, Del Sole R, Mergola L, Lazzoi MR, Scardino A, Scorrano S, Mele G (2011) Molecularly imprinted polymers: present and future prospective. Int J Mol Sci 12(9): 5908–5945. doi:10.3390/ijms12095908
17. Puoci F, Iemma F, Picci N (2008) Stimuli-responsive molecularly imprinted polymers for drug delivery: a review. Curr Drug Del 5(2):85–96. doi:10.2174/156720108783954888
18. Utkov H, Livengood M, Cafiero M (2010) Using density functional theory methods for modeling induction and dispersion interactions in ligand-protein complexes. In: Wheeler RA (ed) Annual reports in computational chemistry. Elsevier, Amsterdam/Boston
19. McQuarrie DA (2007) Quantum chemistry, 2nd edn. University Science Books, Sausalito
20. Szabo A, Ostlund NS (1989) Modern quantum chemistry. McGraw-Hill, Inc., New York
21. Koch W, Holthausen MC (2001) A chemist's guide to density functional theory. Wiley-VCH Verlag GmbH, Weinheim/New York
22. Cohen AJ, Mori-Sánchez P, Yang W (2012) Challenges for density functional theory. Chem Rev 112(1):289–320. doi:10.1021/cr200107z
23. Friesner RA (2005) *Ab initio* quantum chemistry: methodology and applications. PNAS 102(19):6648–6653. doi:10.1073/pnas.0408036102
24. Gauss J (2000) Molecular properties. In: Grotendorst J (ed) NIC symposium 2000, Graphische Betriebe, Forschungszentrum Jülich, NIC series, vol 3, pp 541–592
25. Stewart JJP (2007) Semiempirical molecular orbital methods. In: Lipkowitz KB, Boyd DB (eds) Reviews in computational chemistry. Wiley, Hoboken, pp 45–81. ISBN: 9780470125786, URL: http://dx.doi.org/10.1002/9780470125786.ch2, doi:10.1002/9780470125786.ch2
26. Acevedo O, Jorgensen WL (2010) Advances in quantum and molecular mechanical (QM/MM) simulations for organic and enzymatic reactions. Acc Chem Res 43(1):142–151. doi:10.1021/ar900171c
27. Lin H, Truhlar D (2007) Qm/mm: what have we learned, where are we, and where do we go from here? Theor Chem Accounts Theor Comput Model Theor Chim Acta 117:185–199. doi:10.1007/s00214-006-0143-z
28. Gordon MS, Fedorov DG, Pruitt SR, Slipchenko LV (2012) Fragmentation methods: a route to accurate calculations on large systems. Chem Rev 112(1):632–672. doi:10.1021/cr200093j
29. Arcamone F, Animati F, Capranico G, Lombardi P, Pratesi G, Manzini S, Supino R, Zunino F (1997) New developments in antitumor anthracyclines. Pharmacol Ther 76(1–3):117–124. doi:10.1016/S0163-7258(97)00096-X
30. Yang L, Alexandridis P (2000) Physicochemical aspects of drug delivery and release from polymer-based colloids. Curr Opin Colloid Interface Sci 5(1–2):132–143. doi:10.1016/S1359-0294(00)00046-7
31. Subashini M, Devarajan PV, Sonavane GS, Doble M (2011) Molecular dynamics simulation of drug uptake by polymer. J Mol Model 17(5):1141–1147. doi:10.1007/s00894-010-0811-8
32. Xiang TX, Anderson BD (2006) Liposomal drug transport: a molecular perspective from molecular dynamics simulations in lipid bilayers. Adv Drug Deliver Rev 58(12–13): 1357–1378. doi:10.1016/j.addr.2006.09.002
33. Lu D, Aksimentiev A, Shih AY, Cruz-Chu E, Freddolino PL, Arkhipov A, Schulten K (2006) The role of molecular modeling in bionanotechnology. Phys Biol 3(1):S40–S53. doi:10.1088/1478-3975/3/1/S05
34. Xinhuai Z (2006) Three leading molecular dynamics simulation packages. SVU/Academic Computing, Computer Centre, National University of Singapore, Singapore
35. Haddish-Berhane N, Nyquist C, Haghighi K, Corvalan C, Keshavarzian A, Campanella O, Rickus J, Farhadi A (2006) A multi-scale stochastic drug release model for polymer-coated targeted drug delivery systems. J Control Release 110(2):314–322. doi:10.1016/j.jconrel.2005.09.046
36. Klein ML, Shinoda W (2008) Large-scale molecular dynamics simulations of self-assembling systems. Science 321(5890):798–800. doi:10.1126/science.1157834
37. Truhlar DG (2008) Molecular modeling of complex chemical systems. J Am Chem Soc 130(50):16824–16827. doi:10.1021/ja808927h

38. Wang Q, Keffer DJ, Petrovan S, Thomas JB (2010) Molecular dynamics simulation of poly(ethylene terephthalate) oligomers. J Phys Chem B 114(2):786–795. doi:10.1021/jp909762j
39. Oostenbrink C, Villa A, Mark AE, van Gunsteren WF (2004) A biomolecular force field based on the free enthalpy of hydration and solvation: the GROMOS force-field parameter sets 53A5 and 53A6. J Comput Chem 25(13):1656–1676. doi:10.1002/jcc.20090
40. Barth TJ, Griebel M, Keyes DE, Nieminen RM, Roose D, Schlick T, Knapek S, Zumbusch G (2007) Numerical simulation in molecular dynamics. Springer, Berlin
41. Soares Ta, Daura X, Oostenbrink C, Smith LJ, van Gunsteren WF (2004) Validation of the GROMOS force-field parameter set 45A3 against nuclear magnetic resonance data of hen egg lysozyme. J Biomol NMR 30(4):407–422. doi:10.1007/s10858-004-5430-1
42. Lins RD, Hünenberger PH (2005) A new GROMOS force field for hexopyranose-based carbohydrates. J Comput Chem 26(13):1400–1412. doi:10.1002/jcc.20275
43. Oostenbrink C, Soares TA, van der Vegt NFA, van Gunsteren WF (2005) Validation of the 53A6 GROMOS force field. Eur Biophys J 34(4):273–284. doi:10.1007/s00249-004-0448-6
44. Soares Ta, Hünenberger PH, Kastenholz Ma, Kräutler V, Lenz T, Lins RD, Oostenbrink C, van Gunsteren WF (2005) An improved nucleic acid parameter set for the GROMOS force field. J Comput Chem 26(7):725–737. doi:10.1002/jcc.20193
45. Jayaram B, Sprous D, Beveridge DL (1998) Solvation free energy of biomacromolecules: parameters for a modified generalized Born model consistent with the AMBER force field. J Phys Chem B 102(47):9571–9576. doi:10.1021/jp982007x
46. Wang J, Wolf RM, Caldwell JW, Kollman Pa, Case Da (2004) Development and testing of a general amber force field. J Comput Chem 25(9):1157–1174. doi:10.1002/jcc.20035
47. Patel S, Brooks CL (2004) CHARMM fluctuating charge force field for proteins: I parameterization and application to bulk organic liquid simulations. J Comput Chem 25(1):1–15. doi:10.1002/jcc.10355
48. Patel S, Mackerell AD, Brooks CL (2004) CHARMM fluctuating charge force field for proteins: II protein/solvent properties from molecular dynamics simulations using a nonadditive electrostatic model. J Comput Chem 25(12):1504–1514. doi:10.1002/jcc.20077
49. Jorgensen WL, Tirado-Rives J (1988) The OPLS [optimized potentials for liquid simulations] potential functions for proteins, energy minimizations for crystals of cyclic peptides and crambin. J Am Chem Soc 110(6):1657–1666. doi:10.1021/ja00214a001
50. Jorgensen WL, Maxwell DS, Tirado-Rives J (1996) Development and testing of the OPLS all-atom force field on conformational energetics and properties of organic liquids. J Am Chem Soc 118(45):11225–11236. doi:10.1021/ja9621760
51. Malde AK, Zuo L, Breeze M, Stroet M, Poger D, Nair PC, Oostenbrink C, Mark AE (2011) An automated force field topology builder (ATB) and repository: version 1.0. J Chem Theory Comput 7(12):4026–4037. doi:10.1021/ct200196m
52. Schüttelkopf AW, van Aalten DMF (2004) PRODRG: a tool for high-throughput crystallography of protein-ligand complexes. Acta Crystallogr D 60(Pt 8):1355–1363. doi:10.1107/S0907444904011679
53. Zoete V, Cuendet MA, Grosdidier A, Michielin O (2011) SwissParam: a fast force field generation tool for small organic molecules. J Comput Chem 32(11):2359–2368. doi:10.1002/jcc.21816
54. Gurtovenko AA, Anwar J, Vattulainen I (2010) Defect-mediated trafficking across cell membranes: insights from in silico modeling. Chem Rev 110(10):6077–6103. doi:10.1021/cr1000783
55. Shelley JC, Shelley MY, Reeder RC, Bandyopadhyay S, Klein ML (2001) A coarse grain model for phospholipid simulations. J Phys Chem B 105(19):4464–4470. doi:10.1021/jp010238p
56. Izvekov S, Voth Ga (2005) A multiscale coarse-graining method for biomolecular systems. J Phys Chem B 109(7):2469–2473. doi:10.1021/jp044629q
57. Marrink SJ, de Vries AH, Mark AE (2004) Coarse grained model for semiquantitative lipid simulations. J Phys Chem B 108(2):750–760. doi:10.1021/jp036508g

58. Marrink SJ, Risselada HJ, Yefimov S, Tieleman DP, de Vries AH (2007) The MARTINI force field: coarse grained model for biomolecular simulations. J Phys Chem B 111(27):7812–7824. doi:10.1021/jp071097f
59. Monticelli L, Kandasamy SK, Periole X, Larson RG, Tieleman DP, Marrink SJ (2008) The MARTINI coarse-grained force field: extension to proteins. J Chem Theory Comput 4(5):819–834. doi:10.1021/ct700324x
60. Kukol A (2009) Lipid models for united-atom molecular dynamics simulations of proteins. J Chem Theory Comput 5(3):615–626. doi:10.1021/ct8003468
61. Poger D, Mark AE (2010) On the validation of molecular dynamics simulations of saturated and cis -monounsaturated phosphatidylcholine lipid bilayers: a comparison with experiment. J Chem Theory Comput 6(1):325–336. doi:10.1021/ct900487a
62. Alwarawrah M, Dai J, Huang J (2010) A molecular view of the cholesterol condensing effect in DOPC lipid bilayers. J Phys Chem B 114(22):7516–23. doi:10.1021/jp101415g
63. Hu J, Liu S (2010) Responsive polymers for detection and sensing applications: current status and future developments. Macromolecules 43(20):8315–8330. doi:10.1021/ma1005815
64. Coote ML (2011) *Ab initio* kinetic modeling of free-radical polymerization. In: DaCosta H (ed) Rate constant calculation for thermal reactions: methods and applications. Wiley, New York
65. Isse AA, Gennaro A, Lin CY, Hodgson JL, Coote ML, Guliashvili T (2011) Mechanism of carbon halogen bond reductive cleavage in activated alkyl halide initiators relevant to living radical polymerization: theoretical and experimental study. J Am Chem Soc 133(16):6254–6264. doi:10.1021/ja110538b
66. Dorkoosh FA, Verhoef JC, Borchard G, Rafiee-Tehrani M, Junginger HE (2001) Development and characterization of a novel peroral peptide drug delivery system. J Control Release 71(3):307–318. doi:10.1016/S0168-3659(01)00232-2
67. Dorkoosh FA, Borchard G, Rafiee-Tehrani M, Verhoef JC, Junginger HE (2002) Evaluation of superporous hydrogel (SPH) and SPH composite in porcine intestine ex-vivo: assessment of drug transport, morphology effect, and mechanical fixation to intestinal wall. Eur J Pharm Biopharm 53(2):161–166
68. Sajeesh S, Sharma CP (2006) Cyclodextrin-insulin complex encapsulated polymethacrylic acid based nanoparticles for oral insulin delivery. Int J Pharm 325(1–2):147–154. doi:10.1016/j.ijpharm.2006.06.019
69. Vaidya A, Wigent R, Moore J, Schwartz J (2007) Protective effect of Carbopol on enzymatic degradation of a peptide-like substrate. I: effect of various concentrations and grades of Carbopol and other reaction variables on trypsin activity. Pharm Dev Technol 12(1):89–96. doi:10.1080/10837450601168656
70. Fonseca AC, Jarmelo S, Silva MR, Beja AMM, Fausto R, Gil MH, Simões PN (2011) Study of N_α-benzoyl-L-argininate ethyl ester chloride, a model compound for poly(ester amide) precursors: x-ray diffraction, infrared and Raman spectroscopies, and quantum chemistry calculations. J Chem Phys 134(12):124505. doi:10.1063/1.3565966
71. Fonseca AC, Jarmelo S, Carvalho RA, Fausto R, Gil MH, Simões PN (2010) H-1 NMR spectroscopic and quantum chemical studies on a poly(ester amide) model compound: N_α-benzoyl-L-argininate ethyl ester chloride. Structural preferences for the isolated molecule and in solution. J Phys Chem B 114(18):6156–6164. doi:10.1021/jp9114749
72. Jarmelo S, Marques DAS, Simões PN, Carvalho RA, Batista CMSG, Araujo-Andrade C, Gil MH, Fausto R (2012) Experimental (IR/Raman and 1H/13C NMR) and theoretical (DFT) studies of the preferential conformations adopted by l-lactic acid oligomers and poly(l-lactic acid) homopolymer. J Phys Chem B 116(1):9–21. doi:10.1021/jp205033c
73. Schlichting I, Berendzen J, Chu K, Stock AM, Maves SA, Benson DE, Sweet RM, Ringe D, Petsko GA, Sligar SG (2000) The catalytic pathway of cytochrome p450cam at atomic resolution. Science 287(5458):1615–1622. doi:10.1126/science.287.5458.1615
74. Odoh SO, Walker SM, Meier M, Stetefeld J, Schreckenbach G (2011) QM and QM/MM studies of uranyl fluorides in the gas and aqueous phases and in the hydrophobic cavities of tetrabrachion. Inorg Chem 50(7):3141–3152. doi:10.1021/ic2001706

75. Li J, Futera Z, Li H, Tateyama Y, Higuchi M (2011) Conjugation of organic-metallic hybrid polymers and calf-thymus DNA. Phys Chem Chem Phys 13(11):4839–4841. doi:10.1039/c0cp02037k
76. Stepniewski M, Pasenkiewicz-Gierula M, Róg T, Danne R, Orlowski A, Karttunen M, Urtti A, Yliperttula M, Vuorimaa E, Bunker A (2011) Study of PEGylated lipid layers as a model for PEGylated liposome surfaces: molecular dynamics simulation and Langmuir monolayer studies. Langmuir 27(12):7788–7798. doi:10.1021/la200003n
77. Loverde SM, Klein ML, Discher DE (2012) Nanoparticle shape improves delivery: rational coarse grain molecular dynamics (rCG-MD) of taxol in worm-like PEG-PCL micelles. Adv Mater 24(28):3823–3830. doi:10.1002/adma.201103192
78. Winter ND, Schatz GC (2010) Coarse-grained molecular dynamics study of permeability enhancement in DPPC bilayers by incorporation of lysolipid. J Phys Chem B 114(15): 5053–5060. doi:10.1021/jp911309s
79. Crane M, Hurley N, Crane L, Healy A, Corrigan O, Gallagher K, McCarthy L (2004) Simulation of the USP drug delivery problem using CFD: experimental, numerical and mathematical aspects. Simul Model Pract Th 12(2):147–158. doi:10.1016/S1569-190X(03)00089-3
80. Longest PW, Holbrook LT (2012) In silico models of aerosol delivery to the respiratory tract – development and applications. Adv Drug Deliver Rev 64(4):296–311. doi:10.1016/j.addr.2011.05.009
81. Arifin DY, Lee LY, Wang CH (2006) Mathematical modeling and simulation of drug release from microspheres: implications to drug delivery systems. Adv Drug Deliver Rev 58(12–13):1274–325. doi:10.1016/j.addr.2006.09.007
82. Teo CS, Hor Keong Tan W, Lee T, Wang CH (2005) Transient interstitial fluid flow in brain tumors: effect on drug delivery. Chem Eng Sci 60(17):4803–4821. doi:10.1016/j.ces.2005.04.008
83. Prabhakarpandian B, Shen MC, Pant K, Kiani MF (2011) Microfluidic devices for modeling cell-cell and particle-cell interactions in the microvasculature. Microvasc Res 82(3):210–220. doi:10.1016/j.mvr.2011.06.013
84. Fredenberg S, Wahlgren M, Reslow M, Axelsson A (2011) The mechanisms of drug release in poly(lactic-co-glycolic acid)-based drug delivery systems-A review. Int J Pharm 415(1–2):34–52. doi:10.1016/j.ijpharm.2011.05.049
85. Lao LL, Peppas Na, Boey FYC, Venkatraman SS (2010) Modeling of drug release from bulk-degrading polymers. Int J Pharm 418(1):28–41. doi:10.1016/j.ijpharm.2010.12.020
86. Longest PW, Hindle M (2010) CFD simulations of enhanced condensational growth (ECG) applied to respiratory drug delivery with comparisons to in vitro data. J Aerosol Sci 41(8):805–820. doi:10.1016/j.jaerosci.2010.04.006
87. Mortier STFC, De Beer T, Gernaey KV, Remon JP, Vervaet C, Nopens I (2011) Mechanistic modelling of fluidized bed drying processes of wet porous granules: a review. Eur J Pharm Biopharm 79:205–225. doi:10.1016/j.ejpb.2011.05.013
88. Paun IA, Moldovan A, Luculescu CR, Dinescu M (2011) Biocompatible polymeric implants for controlled drug delivery produced by MAPLE. Appl Surf Sci 257(24):10780–10788. doi:10.1016/j.apsusc.2011.07.097
89. Dudukovic M (2002) Opaque multiphase flows: experiments and modeling. Exp Therm Fluid Sci 26(6–7):747–761. doi:10.1016/S0894-1777(02)00185-1
90. Linninger Aa (2011) Biomedical systems research? New perspectives opened by quantitative medical imaging. Comput Chem Eng 36:1–9. doi:10.1016/j.compchemeng.2011.07.010
91. Kleven M, Melaaen M, Reimers M, Røtnes J, Aurdal L, Djupesland P (2005) Using computational fluid dynamics (CFD) to improve the bi-directional nasal drug delivery concept. Food Bioprod Process 83(2):107–117. doi:10.1205/fbp.04403
92. Annapragada A, Mishchiy N (2007) In silico modeling of aerosol deposition in lungs. Drug Disc Today 4(3):155–161. doi:10.1016/j.ddmod.2007.11.004
93. Hsu Y, Harris TJ, Hettiarachchi H, Penn R, Linninger AA (2011) Three dimensional simulation and experimental investigation of intrathecal drug delivery in the spinal canal and the brain. In: EN Pistikopoulos MG, Kokossis A (eds) 21st European symposium on computer aided process engineering. Elsevier, Amsterdam/Boston

94. Maginn EJ, Elliott JR (2010) Historical perspective and current outlook for molecular dynamics as a chemical engineering tool. Ind Eng Chem Res 49(7):3059–3078. doi:10.1021/ie901898k
95. Stone JE, Hardy DJ, Ufimtsev IS, Schulten K (2010) GPU-accelerated molecular modeling coming of age. J Mol Graphics Model 29(2):116–125. doi:10.1016/j.jmgm.2010.06.010
96. Owens J, Houston M, Luebke D, Green S, Stone J, Phillips J (2008) GPU computing. Proc IEEE 96(5):879–899. doi:10.1109/JPROC.2008.917757
97. Friedrichs MS, Eastman P, Vaidyanathan V, Houston M, Legrand S, Beberg AL, Ensign DL, Bruns CM, Pande VS (2009) Accelerating molecular dynamic simulation on graphics processing units. J Comput Chem 30(6):864–872. doi:10.1002/jcc.21209
98. Garland M, Le Grand S, Nickolls J, Anderson J, Hardwick J, Morton S, Phillips E, Zhang Y, Volkov V (2008) Parallel computing experiences with CUDA. IEEE Micro 28(4):13–27. doi:10.1109/MM.2008.57
99. Stone JE, Saam J, Hardy DJ, Vandivort KL, Hwu W-MW, Schulten K (2009) High performance computation and interactive display of molecular orbitals on GPUs and multi-core CPUs. Proceedings of 2nd workshop on general purpose processing on graphics processing units, GPGPU-2, Washington, DC. ACM, New York, pp 9–18. doi:10.1145/1513895.1513897, ISBN: 978-1-60558-517-8, URL: http://doi.acm.org/10.1145/1513895.1513897

The Importance of Controlled/Living Radical Polymerization Techniques in the Design of Tailor Made Nanoparticles for Drug Delivery Systems

Nuno Rocha, Patrícia Mendonça, Joana R. Góis, Rosemeyre Cordeiro, Ana Fonseca, Paula Ferreira, Tamaz Guliashvili, Krzysztof Matyjaszewski, Arménio Serra, and Jorge Coelho

Abstract Recent developments in controlled/living radical polymerization methods (CLRP) have created the opportunity to prepare polymeric based systems with site specific functionality that has significantly expanded the range of physical and chemical properties that can be generated in materials prepared by these systems. For example, CLRP prepared block copolymers can self-assemble into nanoparticles that can be used in drug delivery applications. The development of synthetic procedures for preparation of materials targeting new and more efficient drug delivery systems (DDS) is of great interest since ultimately they can mimic most of the properties of biological systems.

This chapter will initially discuss the key aspects of the development of nanotechnology for drug delivery. The cell internalization process will be described and related with the relevant properties required for the "nanocarrier systems". Afterwards, a summary of the polymeric systems that can be used for DDS will be provided and the importance of CLRP methods in the preparation of polymer-based

N. Rocha • P. Mendonça • J.R. Góis • R. Cordeiro • A. Fonseca • P. Ferreira • J. Coelho (✉)
Department of Chemical Engineering, University of Coimbra, Polo II,
Pinhal de Marrocos 3030-790, Coimbra, Portugal
e-mail: jcoelho@eq.uc.pt

T. Guliashvili
GE Power and Water, Water and Process Technologies, 4636 Somerton Road,
Trevose, PA 19053, USA

K. Matyjaszewski
Department of Chemistry, Carnegie Mellon University, 4400 Fifth Avenue,
Pittsburgh, PA 15213, USA

A. Serra
Chemistry Department, University of Coimbra, 3004-535 Coimbra, Portugal

J. Coelho (ed.), *Drug Delivery Systems: Advanced Technologies Potentially Applicable in Personalised Treatment*, Advances in Predictive, Preventive and Personalised Medicine 4, DOI 10.1007/978-94-007-6010-3_11,

systems will be discussed. Finally, mechanisms of block copolymers self-assembly will be discussed and supported with some examples of CLRP-based self-assembly systems for drug delivery applications.

Keywords Controlled/living radical polymerization • Self-assembly • Drug delivery systems • Targeting • Nanocarriers • Treatment tailored to the person

1 Nanotechnology for DDS

Over the last decades, the use of nanotechnology to develop materials for drug delivery, namely nanocarrier systems, has been extensively studied in many areas of medicine [1]. This research has been driven by the problems associated with the direct use of drugs with no protection, in the body it can raise problems mainly related to the systemic side effects, premature degradation, and bioavailability [2]. In this context, the use of well designed drug delivery systems (DDS) allows the drug to be delivered in a controlled manner and targeted to the diseased tissue or cell while preventing damage to the adjacent healthy tissues or cells. Polymers, either biopolymers or synthetic polymers, provide an ideal opportunity for the development of an effective DDS due to the combination of their processability and the level of control over the physical and chemical characteristics that can be accomplished during the polymerization procedure [3].

Physico-chemical properties of the DDS, such as size, shape, mechanical properties and surface chemistry (see Fig. 1) need to be carefully considered during the design and development in order to improve the efficiency of the DDS for a specific biological response [4, 5].

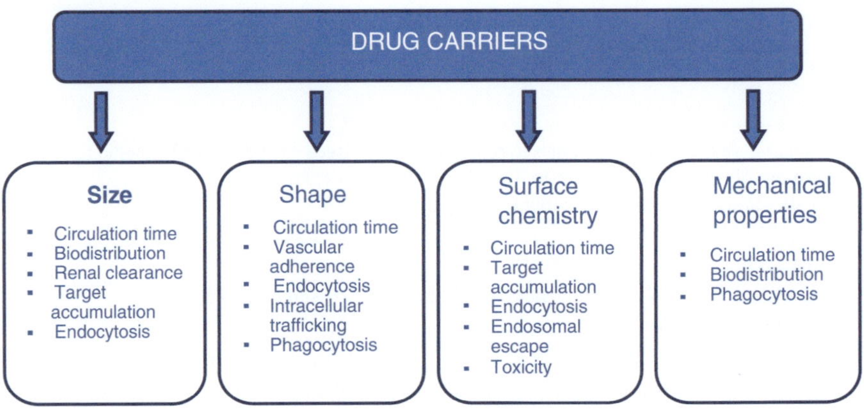

Fig. 1 Key properties in the design of nanoparticulate DDS (adapted from [4])

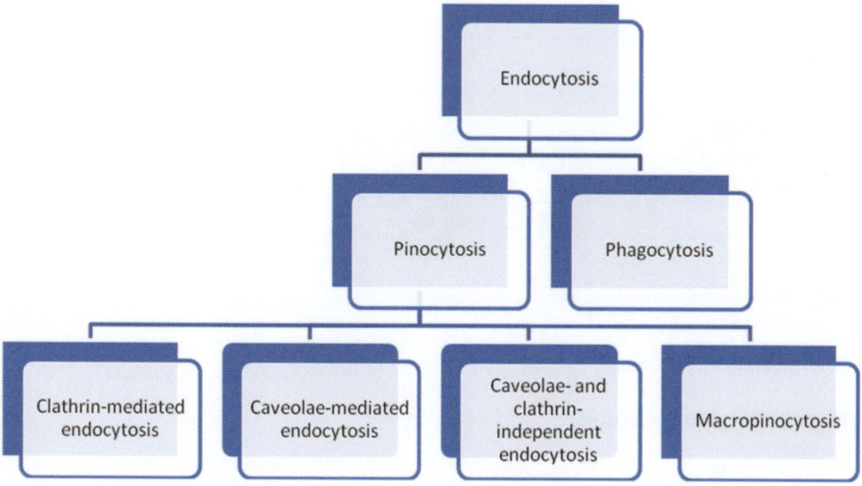

Fig. 2 General classification of endocytosis pathways (adapted from [6, 7])

1.1 Cell Internalization Process

The relevance of the physico-chemical properties of drug-loaded carriers for their efficiency as a DDS is directly related to the entry process into the cells, known as endocytosis. As shown in Fig. 2, endocytosis can be divided in two major pathways: phagocytosis (the uptake of large particles) and pinocytosis (the uptake of small particles, fluids and solutes) [6, 7].

Phagocytosis primarily occurs in specialized phagocytic cells, such as macrophages, monocytes, neutrophils and dendritic cells [6–8]. In contrast, pinocytosis occurs virtually in all types of cells and varies accordingly (Fig. 2). This pathway can be divided in four main distinct entry mechanisms: clathrin-mediated endocytosis (CME), caveolae-mediated endocytosis (CvME), caveolae- and clathrin-independent endocytosis, and macropinocytosis [6, 7, 9, 10]. Figure 3 presents a scheme describing the different endocytosis mechanisms.

The three main stages of the phagocytic pathway of entry into cells are: the recognition in the bloodstream by the reticuloendothelial system (RES) that tags the surface particles with opsonins (opsonization process) [10]; the adhesion of the opsonized particles to the macrophages through specific receptor-ligand interactions that leads to the formation of a cell-surface extension that surrounds the particle and induces internalization [6, 10]; and ingestion of the particle into the cell [6, 10]. This internalization process plays an important physiological role in the organism defense against non-self elements, infectious agents (mostly bacteria and some viruses) as well as exogenous inert particles, including the DDS [10], and represents the primary *in vivo* mode of particle clearance [11].

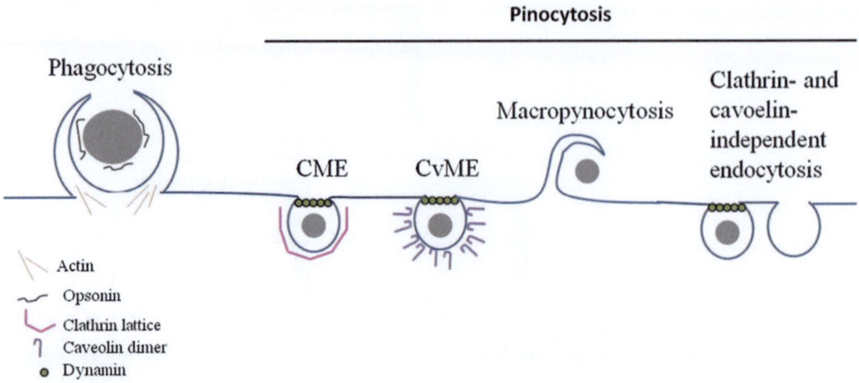

Fig. 3 Endocytosis mechanisms

Clathrin-mediated endocytosis (CME) acts as the main mechanism of internalization for plasma membrane constituents and macromolecules, with respect to pinocytosis [10]. In spite of CME being the predominant endocytic mechanism in most cells, other alternative pathways such as clathrin-independent endocytosis have recently been identified [10, 12]. The clathrin-independent endocytosis can be divided into three distinct categories: (a) caveolae-mediated endocytosis; (b) macropinocytosis; and (c) clathrin- and caveolae-independent endocytosis. Nanocarriers with surface ligands as folic acid, albumin and colesterols are internalized by caveolae-mediated endocytosis [13]. Macropinocytosis is a form of endocytosis involving cell ruffling, it is distinct in many ways from the better characterized micropinocytosis, which includes clathrin-coated vesicle endocytosis and small uncoated vesicles. Since macropinosomes are relatively large, they provide an efficient route for non-selective endocytosis of solute macromolecules and are classically observed in macrophages and tumor cells and in other cells after stimulation with growth factors or phorbol esters. Although it has not been reported to display any selectivity regarding the internalized molecules [14]. Clathrin- and caveolae-independent endocytosis differs [6, 7, 12] from the conventional endocytotic pathways described above, and are classically used by pathogens to invade healthy cells [12]. This clathrin independent is via dynamin dependent caveolae pathway or dynamin independent cell division control protein 42 (CDC42) [15] or ADP-ribosylation factor 6 (ARF6) [16] dependent pathway. Many other classical pathways are currently being explored and have not been fully elucidated [10, 17].

1.2 Key Properties of Nanocarrier DDS

As mentioned above, the internalization process of nanocarriers for drug delivery applications depends primarily on their size, shape, mechanical properties and surface properties of the agents. These are discussed in the following sections.

1.2.1 Size

The particle size is one of the most relevant and most extensively studied physical properties in designing nanocarriers for drug delivery. Particle size plays a key role in improving *in vivo* functional properties, such as enhanced circulation time, reduced immunogenicity, increased internalization, intracellular trafficking, clearance and cellular uptake [5, 18]. The selection of particle size influences the route of administration *in vivo*. For example, particles larger than 1.5 μm are generally avoided for intravenous administration due to the possibility of clogging the capillaries. Nevertheless, larger particles can be suitable for other routes of administration, for example by inhalation where 1–3 μm diameter particles have been proven to be suitable, for delivering DDS to lungs [19]. In cancerous tumors, vessels are structurally irregular, heterogeneous and leaky, and have a poorly developed lymphatic system when compared to vessels of healthy tissues. These features result in enhanced permeation and retention effect (EPR), which is described as increased permeability of the tumor blood vessels resulting in passive accumulation of large molecules, drugs and nanoparticles in the tumor tissue via extravasation of the blood vessels and capillaries [20]. Furthermore, some of the clearance mechanisms of drug-loaded carriers during transport in the body are also related to specific dimensions, such as renal (kidney) filtration (<5–10 nm) [5, 19], liver uptake (<100 nm) [9] or splenic filtration (>200 nm) [5, 9].

It is generally considered that particles smaller than 50 nm enter the cells by pinocytosis [5, 10, 19] and particles larger than 500 nm enter by phagocytosis [5, 19, 21]. Concerning the pinocytosis mechanism, each pathway is associated to a size range of vesicles formed to internalize nanocarriers: ~50–80 nm for caveolae-mediated endocytosis; ~90 nm for clathrin-independent and caveolin-independent endocytosis; ~100–120 nm for clathrin-mediated endocytosis; and larger than >1 μm for macropinocytosis [9, 10, 22]. In phagocytosis, particles with diameters between 2 and 3 μm exhibit a higher level of recognition by macrophages and were phagocytized more readily than smaller and or larger particles [11, 23].

It should be noted that even though nanocarriers display a certain size after synthesis, during *in vitro* and *in vivo* studies, they can aggregate into particles with vastly different sizes and shapes that may dictate the outcome and interpretation of the results [22]. The self-assembly process of nanoparticles will be discussed further in this chapter.

1.2.2 Shape

Another relevant factor in the design of drug-loaded nanocarriers is the shape of the particles. In the bloodstream, particles with different shapes are affected differently by the different hydrodynamic forces throughout the body. For instance, a spherical particle does not deviate from its streamlined motion unless it experiences an external force; whereas non-spherical particles undergo margination [24], which means that they are susceptible to tumbling and torque, resulting in a higher tendency to move towards the vessel wall. In addition, shape also influences particle flow and filtration through the spleen. Non-spherical particles larger than 200 nm are still likely to pass through the spleen as long as one of their dimensions is smaller than 200 nm [4].

Recent findings indicate that particle shape as well as geometry and orientation influence cellular uptake [5, 24]. In phagocytosis, despite the particle size, the geometry of the nanocarrier at the point of cell contact dictate macrophages to initiate cellular internalization. Macrophages that encountered an elliptical disc at the ends, fully internalized the particle in a few minutes, whereas macrophages that are attached to a flat region of the elliptical disc did not succeed in internalizing the particle until 12 h [19]. Doshi and Mitragotri [23] demonstrated that polymeric particles with shape, and size closer to that of bacteria exhibited highest attachment, showing a geometry-based target recognition by macrophages. Elliptical disk-shaped particles were found to have a lower rate of cellular internalization than spherical ones [25].

Recent research trends tend to target preparation of drug carriers that are biomimetic materials, designed and synthesized based on the physical properties of cells. For example, the discoidal shape and mechanical flexibility of erythrocytes avoid spleen filtration, while platelets use their disc-like shape to assist their function of adhesion and rolling on vascular endothelium [26].

1.2.3 Mechanical Properties

Mechanical properties, such as rigidity or hardness, are known to affect the cellular internalization of the nanocarrier [4]. A relevant characteristic of natural biological materials is their wide range of deformability, which can be seen by the red blood cells (RBC) that exhibit extreme reversible deformability under flow. In this respect, recent research studies demonstrated design and synthesis of biologically inspired polymeric particles that mimic RBCs [27, 28], having the capacity to flow through capillaries that are smaller than their own diameter [27, 28], and the capacity to carry oxygen [28].

Flexible particles have shown higher retention time in the bloodstream, [29] and it has been observed that macrophages recognize and incorporate rigid particles to an higher extent than soft and flexible ones, suggesting that the immune system response is greatly affected by both mechanical properties and physical properties of the drug carriers [5, 9, 21].

1.2.4 Surface Properties

The surface properties of drug-loaded nanocarriers such as solubility affect cell morphology, adhesion or motility [22]. Also, cell surface interactions with carriers are influenced mainly by their charge and inherent properties of the surface coating [4]. When administered in the body, the carrier surface adsorbs to the serum proteins and facilitate cellular internalization [22]. Non-specific adsorption of proteins on the carrier surface may lead to its agglomeration and clearance from the RES, negating any possibility of delivering to the target site [22].

Generally, drug carriers possessing a positively charged surface display better association and upregulate endocytosis. Due to the negative charge of the cell membrane, cationic polymers such as chitosan, polyethyleneimine or poly(β-amino ester) have been widely used [30, 31] to enhance cellular internalization. These polymers have the ability to disrupt the endosomal membrane, as a consequence of their buffering capacity [10]. Negatively charged nanocarriers will have more efficient targeted cellular uptake, if conjugated with targeting ligands [32, 33], whereas particles with slightly negative surface charge, which are swiftly cleared by macrophages of the RES.

In fact, an appropriate nanocarrier coating, property that is able to repel the opsonization and prevent rapid clearance via RES, should be considered during the design of a DDS [10]. For this purpose, nanocarriers coated with hydrophilic or neutral polymers, such as poly(ethylene glycol) (PEG), have been preferentially used [5, 10, 22]. PEG is a well-known synthetic polymer that possesses compatibility and stealth properties and is approved by Food and Drug Administration (FDA). A PEG coating on a DDS enhances the solubility, reduces the polymeric carrier cytotoxic effects, prevents protein complexation thereby decreasing RES uptake, and consequently prolongs the circulation half-life in the bloodstream, and also reduces immunogenicity and antigenicity [19]. It has been shown that a high concentration of PEG on the surface and low chain length are associated with a higher endocytosis rate, i.e., spatial conformation of PEG moieties on the surface can be used to modulate the stealthiness of these drug-loaded carriers [34, 35].

Another strategy in surface chemistry modification is the decoration of particles using targeting moieties to provide a more focused, and therefore more effective, delivery of the therapeutic medicines with less *off-target* effect. These molecules are able to recognize a specific biological target (surface receptor or protein) and promote drug delivery into a specific cell population and avoid healthy cells or tissues [5, 10]. Various biological targeting moieties, such as antibodies [35], peptides [36, 37], carbohydrates [38], aptamers [39] and vitamins [38], have been conjugated to particles to modulate their biodistribution and to increase local therapeutic concentrations [19].

2 Polymeric Segments in DDS

The first reported polymeric DDS dates back to early 1970s [40]. Since then, multiple new polymer based systems, with a myriad of architectures and physicochemical properties have been designed and successfully used as carriers for active compounds. DDS can be prepared from either natural or synthetic polymers, which can be biodegradable or non-biodegradable.

2.1 *Natural Polymers*

Some natural polymers, such as proteins and polysaccharides, have proven to be useful in designing synthetic DDS due to their excellent biocompatibility, biodegradability and low cost. Limitations on their broader application are related to strong immunogenic responses, possibility of disease transmission, batch-to-batch variation in properties and difficulties in their purification [41, 42].

2.1.1 Proteins

Proteins are the main component of human tissue and are composed of α-amino acid residues linked together by amide bonds. Their biodegradability, biocompatibility and low-immunogenicity make them natural candidates for DDS [43]. Additionally, due to the presence of free amine and carboxyl groups along the structure it is possible to prepare crosslinked protein based DDS [44]. Collagen, gelatin and albumin are germane proteins for use in the design of DDS.

Collagen is the main structural protein in the human body, being the major component of connective tissues [43, 45] and has been used as a therapeutic delivery system for a series of different active compounds, namely growth factors [46, 47], oligonucleotides [48–50], and proteins [43], in a variety of particle shapes and sizes [43, 44].

Gelatin is a hydrosoluble protein obtained by the thermal denaturation of collagen [43]. Gelatin based DDS have been used in the delivery of anti-cancer drugs [51–55], proteins [54–56] and growth factors [55, 57, 58].

Albumin is the most abundant protein in the human blood plasma. It transports the hydrophobic fatty acid molecules in the blood stream and maintains the blood pH [41]. Drug carriers based on albumin have proven to be useful in applications related to cancer treatment [59–61] and gene therapy [59, 62].

2.1.2 Polysaccharides

Polysaccharides are composed by monosaccharide repeating units, joined by glycosidic bonds. They are biodegradable, biocompatible, low toxic and non-immunogenic. Similarly to proteins, polysaccharides have a considerable amount of pendant functional groups in their structures, allowing further structural modifications [43, 63]. Chitosan, alginate and dextran (Fig. 4) are examples of polysaccharides that can be used in DDS.

Chitosan is a cationic polysaccharide obtained by the alkaline deacetylation of chitin. Its solubility, hydrophobicity and the ability of interacting electrostatically with polyanions are a function of the degree of deacetylation [41, 43, 64, 65].

Micro- and nanoparticles of chitosan, prepared by simple and mild methods, have been used in the encapsulation of different active compounds [54, 64, 66, 67].

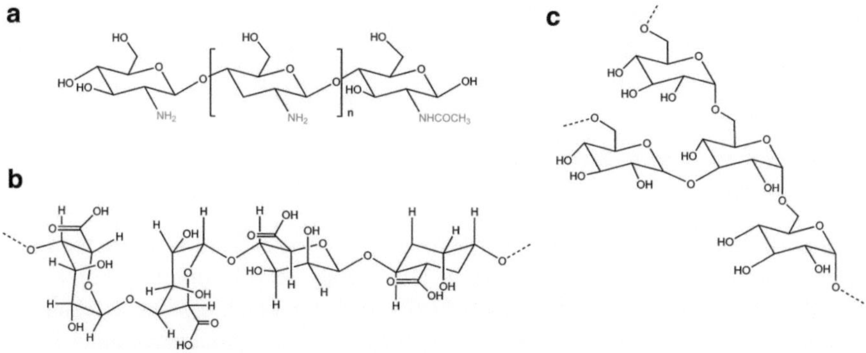

Fig. 4 Structures of: (a) chitosan; (b) alginate; (c) dextran

Ophthalmology [68], cancer [69, 70] and gene therapy [66, 71, 72] are fields were chitosan based carriers showed to be efficient in delivering active compounds.

Alginic acid is a polysaccharide with cationic nature that is extracted from the cell walls of brown algae, by an alkaline treatment, followed by an acidic treatment [41, 43, 73]. Due to the high content of acid groups along the polymeric backbone, alginic acid readily forms gels when placed in contact with divalent cations (e.g., Ca^{2+}), at ambient temperature. This property can be used for the encapsulation of active substances, without influencing their biological activity [41, 43, 74]. Alginate based carriers have been used in ophthalmology [75, 76], cancer treatment [77–79], tissue engineering [80–82] and gene therapy [83, 84].

Dextran is a water soluble homopolysaccharide of bacterial origin composed essentially of α-1,6-linked D-glucopyranose units, with branches extending mainly from the α-1,3- and occasionally from the α-1,2- and α-1,4- positions [63]. Dextran based carriers have been successfully used in the delivery of different drugs, in a wide range of treatment fields [85–88].

2.2 *Synthetic Polymers*

Synthetic polymers offer a great variety of compositions with adjustable chemical, mechanical, and biological properties. By changing the building blocks or the preparation techniques it is possible to develop a polymer with properties required for any given drug delivery application [41, 43].

2.2.1 Biodegradable Synthetic Polymers

Biodegradable synthetic polymers offer the advantage of being broken down into biologically compatible compounds or into products that can be easily excreted by the body. Thus, once they are introduced in the human body any further manipulation or removal steps are not necessary [41, 43].

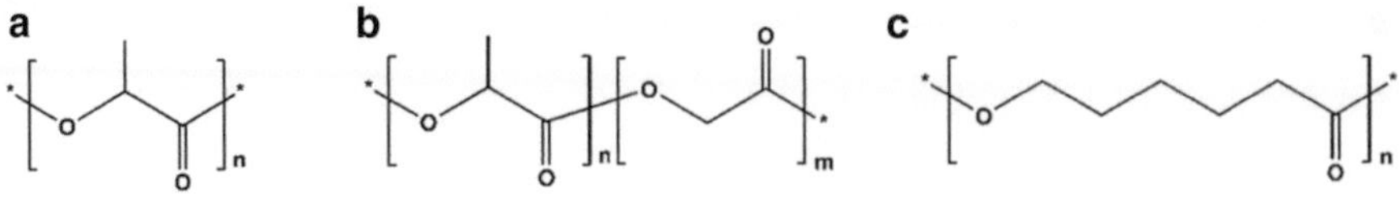

Fig. 5 Structures of: (**a**) PLA; (**b**) PLGA; (**c**) PCL

Poly(lactic acid) (PLA) and poly(lactic-co-glycolic acid) (PLGA) are the most widely used biodegradable polymers in DDS. They are biocompatible materials that degrade by hydrolysis of the ester bonds, through a bulk eroding mechanism. Upon degradation, they release lactic and glycolic acids, normal metabolites in the human body [41, 43, 54, 89]. PLA nanoparticles have been successfully used as carriers for many different drugs [54, 90]. However, due to its faster rate of *in vivo* degradation PLGA is sometimes preferred for DDS. Mundargi and co-authors [91] reviewed the application of PLGA particles for the controlled release of different active compounds and PLGA nanoparticles, being proven to be efficient as active compound carriers in the field of cancer treatment [54, 89].

Poly(ε-caprolactone) (PCL) is a semicrystalline polymer that provides a very low rate of *in vivo* degradation, which is why it has been mainly used for preparation of long-term drug delivery devices [41, 43]. Upon degradation, PCL does not release any acidic by-products, contrarily to what happens with PLA and PLGA [92]. The use of PCL nano- and micro-particles as carriers for several drugs was reviewed by Sinha and co-authors [92] and more recently by Kumari and co-authors [54].

Blending PCL with other polymers, or copolymerizing ε-caprolactone with other monomers, led to the preparation of PCL based materials with novel properties thereby enhancing its range of applications [43, 93, 94]. PCL based systems have been successfully used in the encapsulation of both genetic materials and anti-cancer drugs [95, 96].

Figure 5 shows the structures of the biodegradable polymers presented above.

Other families of biodegradable polymers, namely poly(ortho esters) [97, 98] and poly(alkyl cyanoacrylates) [99, 100], have been also successfully used in the design of DDS.

2.2.2 Other Synthetic Polymers

A considerable number of biocompatible synthetic polymers used in DDS do not degrade when placed in contact with human body. In such cases, once the drug is released, the device has to be removed by surgery or the polymer has to be cleared from the body through the kidneys [42, 65]. For instance, poly(methyl methacrylate) (PMMA) and poly(2-hydroxyethyl methacrylate) (PHEMA), Fig. 6, are biocompatible, biostable acrylic polymers, whose application for drug delivery is well recognized [42, 101–106].

Fig. 6 Chemical structure of PMMA (R: $-CH_3$) and PHEMA (R: $-CH_2CH_2OH$)

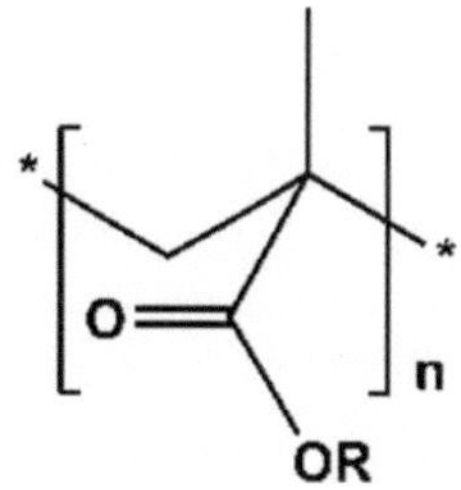

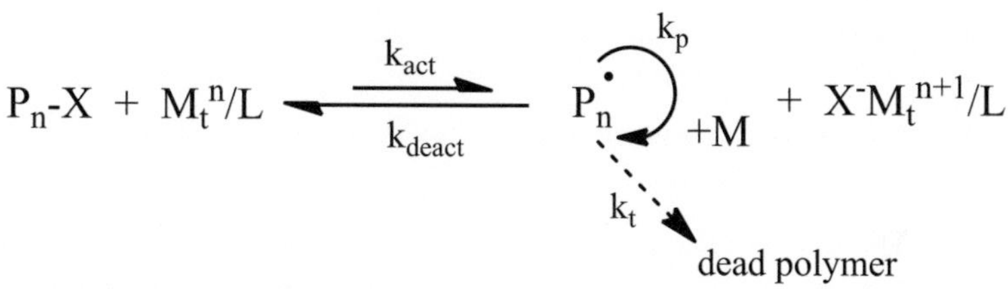

Fig. 7 General mechanism scheme for the ATRP method

Acrylic polymers with pendant acid groups, e.g. poly(acrylic acid) (PAA), and N-substituted acrylamide polymers, e.g. poly(N-isopropylacrylamide) (PNIPAAm), are polymers that can respond to external stimuli. This property has been found to have a great utility in several biomedical applications [107].

The development of CLRP methods have allowed extending radical polymerization to prepare homopolymers of well-defined structure, copolymers with controlled architectures or easily functionalizable polymeric materials, that are progressively gaining a greater level of interest from the scientific and industrial communities.

2.3 CLRP Methods

Over the last decade, significant progresses on CLRP methods have been achieved [108–110]. In order to obtain well controlled polymeric structures based on radical methods, it is required that the polymerization proceeds in a manner that allows all chains to grow simultaneously with the same probability of chain growth, which can generally be achieved if the initiation is much faster than propagation and if the contribution of transfer and termination reactions is negligible [111].

The most frequently used CLRP methods are stable free radical polymerization (SFRP), atom transfer radical polymerization (ATRP) and degenerative transfer (DT) based methods, such as the reversible addition fragmentation transfer (RAFT) or the iodine transfer (IT). Figure 7 provides a representative scheme for the ATRP

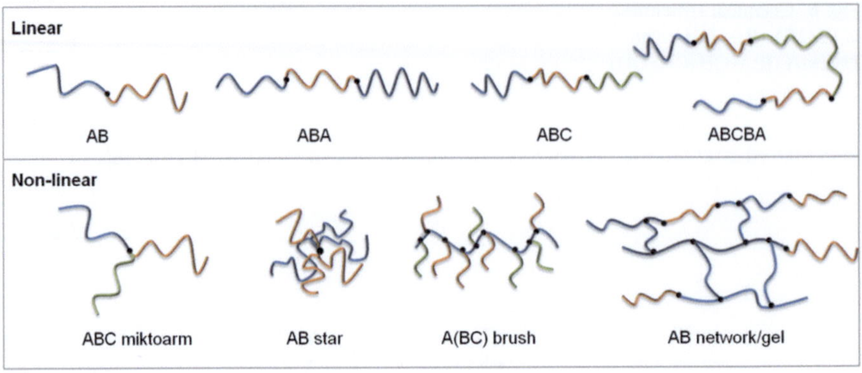

Fig. 8 Different block copolymer based molecular architectures (adapted from [109])

mechanism. The common feature of various CLRP strategies is the establishment of an equilibrium between the growing radicals and the dormant species, minimizing the fraction of the terminated chains [112–114].

In SFRP, the dormant covalent species are formed by the reversible reaction of stable radicals, initially 2,2,6,6-tetramethylpiperidinyl-1-oxyl (TEMPO), with the active radicals [110]. In DT, a chain transfer agent is used to reversibly react with the propagating macro-radicals [115]. Specifically for the RAFT mechanism, this is achieved by an addition-fragmentation process that enables the exchange of a moiety between two chains [116].

ATRP is the most commonly explored CLRP method due to its simplicity, efficiency, high tolerance to monomer functionality and to the commercial availability of required initiators and catalysts [117–121]. The polymerization mechanism is based in the formation of active radicals through a redox process that is catalyzed by a transition complex (M_t^n/L, see Fig. 7). This complex is responsible for controlling the activation and deactivation processes, keeping the concentration of radicals at a very low level.

The living character of most of CLRP methods has provided an excellent opportunity for the preparation of controlled block copolymers and materials with new molecular architectures based on functional polymeric segments.

2.4 New Polymeric Architectures Based on CLRP Methods

The possibility to re-initiate the chains formed in a CLRP opens the possibility to copolymerize monomers of a different nature and, therefore, to prepare block copolymers in a living manner. This means that accurate control of linear block copolymers' molecular weight, composition and segments sequence can be achieved using CLRP techniques, see Fig. 8. Other examples of copolymer topologies that can be prepared by CLRP methods include statistical, gradient and graft copolymers

[122, 123]. Such capacity to control the polymeric sequence distribution and spacing at a nanometric scale is a key aspect in the control of polymeric self-assembly structures, as it will be discussed later in this chapter.

The development of CLRP methods has also provided the possibility to prepare more complex molecular architectures. Star-shaped molecules are the most commonly synthesized CLRP materials displaying new architectures and are obtained when multifunctional initiators or multifunctional coupling agents are used. Other architectures are comb polymeric structures, in which copolymers with close packed branches are created from specific synthetic techniques; cyclic polymers, obtained from the reaction of two functional groups on the termini of the linear precursor, and dendrimers, in which all bonds converge to a single point and a large number of identical functional groups are present at peripheral chain-ends [124].

The ease of creation of segmented structures while employing CLRP methods has obviously extended such complex architectures to copolymers, giving rise to new possibilities of molecular arrangements at a nanometric scale. ATRP methods were found to be particularly useful in the preparation of a great variety of copolymeric architectures [109]. Figure 8 presents some examples of copolymeric structures that can be obtained from CLRP methods.

The capacity to introduce specific functional groups or branching points in polymeric structures obtained by CLRP is of great interest for further structure functionalization or simply to combine polymeric segments that are difficult to be copolymerized in a controlled manner.

2.5 CLRP as a Method to Prepare Functional Polymers

Polymer functionalization is a very useful tool to control the polymer topology or to provide specific functions to the polymeric based materials. The incorporation of site-specific branching points and distributed or site specific functionalities in the polymer backbone can contribute to the control and development of polymeric architectures that were otherwise inaccessible using ionic polymerization methods [125].

Polymers containing relevant chemical functionalities, also coined as functional polymeric materials, can be prepared by direct polymerization of functional monomers or by post-polymerization modification (see Fig. 9).

When the required chemical functionality is present in a given monomer, the polymer functionalization can be easily achieved by direct monomer polymerization. CLRP methods described previously, and particularly ATRP, have been shown to be efficient in the polymerization of several monomers containing specific functional groups to prepare functional polymeric materials. For instance, functional (meth)acrylic monomers have been polymerized using ATRP methods, including glycidyl acrylate, 2-hydroxyethyl acrylate, 2-hydroxyethyl methacrylate, 2-hydroxypropyl acrylate, allyl acrylate, 2-trimethylsilyloxyethyl acrylate, 2-(dimethylamino)ethyl methacrylate, protected acrylic acid and sugar-carrying (meth)acrylates [126].

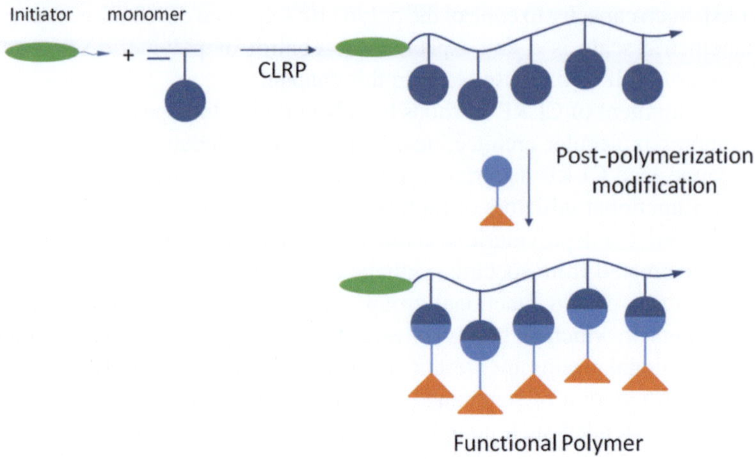

Fig. 9 Synthesis of functional polymers by post-polymerization modification

Additionally, the initiator that is used in the selected ATRP method establishes the end-groups on the polymer chain(s). Generally, with a mono-functional initiator one chain-end is an alkyl halide, whereas the other can be a predetermined functional group [126]. CLRP methods make possible to prepare end-functionalized polymers using an initiator that contains one or more functional groups. Nevertheless, it is not always possible to carry out the direct polymerization functionalization due to the nature of the species involved or the particular polymerization process. In those cases the functionalization is achieved by post-polymerization modification of the incorporated functional group [127].

In post-polymerization modification, functionalization can be achieved by polymerization of suitable protected monomers followed by a deprotection step or by modification of the final polymer, as shown in Fig. 9. In the latter case, the monomer functional groups can be quantitatively converted in successive reaction steps into a vast range of other functional groups according to the desired application. Many strategies can be applied when considering the modification of the synthesized polymer since it is possible to change the functionalities of the monomer units or even insert specific functionalities at the polymeric chain-ends. Some review articles present detailed information about the most important and effective reactions available for post-polymerization modification [127–129]. The most studied approaches for post-polymerization modification of polymers prepared by CLRP are the 'click' chemistry strategies [130], with the copper catalyzed Huisgen 1,3–dipolar cycloaddition of azides and terminal alkynes being the most representative [131–136]. Other interesting post-modification strategies include the modification of polymers by Michael-type addition reactions between thiols and activated alkenes [127, 137] and N-hydroxysuccinimide (NHS) side chain modifications [138]. In general these processes are highly selective and tolerant to a variety of functional groups and can be carried out under mild reaction conditions.

Therefore, the possibility to prepare these polymer-based structures having very well-defined structures at a nanometric scale and with incorporated key site selected functionalities and the capacity to self-assemble into larger and defined structures, constitute the basis for the preparation of new materials providing optimized synthetic DDS.

3 Self-Assembly of Block Copolymers

The synthesis of block copolymers using CLRP techniques yield products with controlled composition, molecular weight and architecture which have opened the possibility to prepare advanced DDS. The phase separation that occurs in these block copolymers at the nanometric scale provides nanostructures with controlled size, morphology, mechanical properties and interactions properties. Accurate control of nanostructured molecular assemblies is achieved by appropriately designing the copolymers in terms of their constituent polymeric segments, block length and architecture; by the preparation method that allows these structures to self-assemble in a given environment to a desired shape and size; or by responding to external stimulus, allowing the structure to re-assemble into a required structure for new conditions or environment.

The possibility of preparing nanostructured materials and their dynamic self-assembly is a critical matter for DDS, since it allows creating precisely controlled drug encapsulating structures and to control their release profile.

In this section, details of formation of block copolymers self-assembly structures, aspects for polymeric segments selection and some examples of responsive nanostructures based on block copolymers will be discussed.

3.1 Fundamentals of Self-Assembly

As stated by Whitesides and co-workers [139], "molecular self-assembly is the spontaneous association of molecules under equilibrium conditions into stable, structurally well-defined joined noncovalent bonds". This molecular self-assembly is the basis for the activity of all biological systems, but the recent development in the controlled synthesis of block copolymers has opened the possibility to prepare comparable systems based on synthetic materials.

Development of self-assembly systems based on block copolymers has always been mainly inspired by Nature. Although polypeptides are formed by only a few aminoacids, they can self-assemble to form a wide variety of proteins that can be combined in many different ways to build active biological cells. Figure 10 compares how complex and active structures are obtained from aminoacids in biological systems and from monomers in synthetic polymeric systems [140].

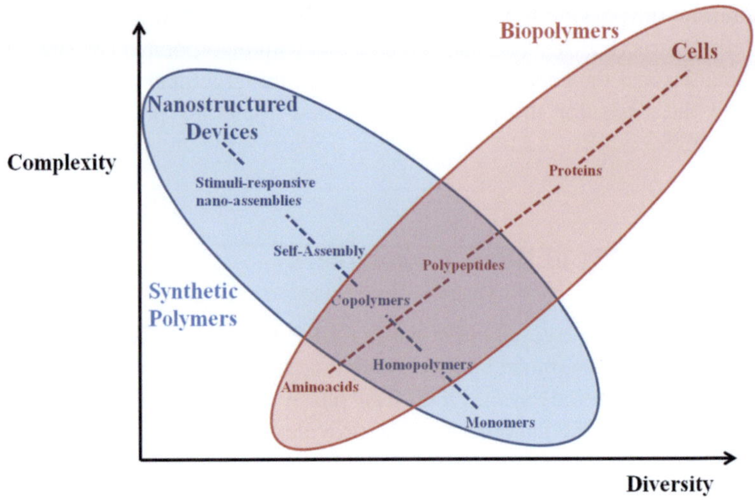

Fig. 10 Comparison of complexity and diversity of structures formed from synthetic and biological polymers (adapted from [140])

Similarly to aminoacids in biological systems, the vast range of existing synthetic monomers can be used to form different polymeric segments, which, when combined with each other, may form self-assembly block copolymers that may constitute the basis of responsive synthetic devices.

In block copolymers, polymeric segments that can have different levels of incompatibility are bonded covalently leading to phase separation. However, this covalent bonding between blocks restricts the phase separation on a molecular length scale, which is typically only a few nanometers [141, 142]. The possibility to control the phase separation is the key feature to prepare synthetic self-assembly systems having tailor-made properties. Phase separation at a nanometric scale can be tailored by effectively controlling the block copolymer architecture, which can be done using CLRP methods as discussed in the previous sections.

Generally, depending on the block copolymer composition, a coil-coil type diblock copolymer can self-assemble into two phases as the ratio of blocks increase progressively from sphere, cylinder, gyroid, and lamellar microdomains [143], as shown in Fig. 11.

One of the major benefits in the use of block copolymers systems for DDS is the possibility of self-assembly in solution. In this process, micelles of nanometric size are formed, making it possible to encapsulate drugs, to improve the apparent water solubility of hydrophobic drugs and to control the release profile of the drug [144]. In a solvent that selectively dissolves only one segment of a diblock copolymer, micelles having a core of the insoluble block and a shell of the soluble block are formed [145].

According to the Israelachvili's theory, which was initially developed for lipids with an optimal surface area, a_0, a hydrocarbon volume, v, and a critical chain

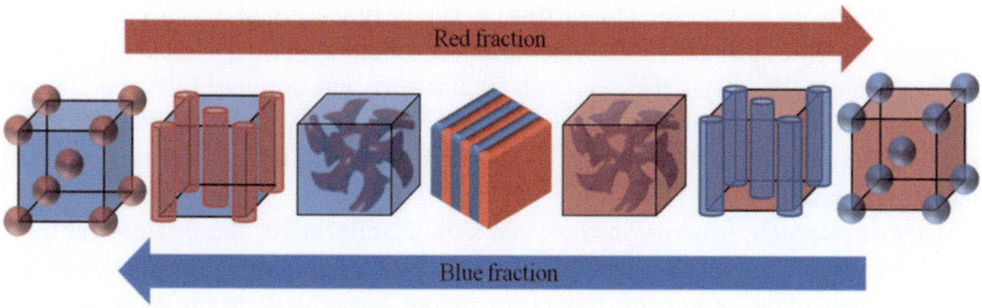

Fig. 11 Equilibrium ordered states obtained from diblock copolymers with different segments (*red* and *blue*) compositions (adapted from [141])

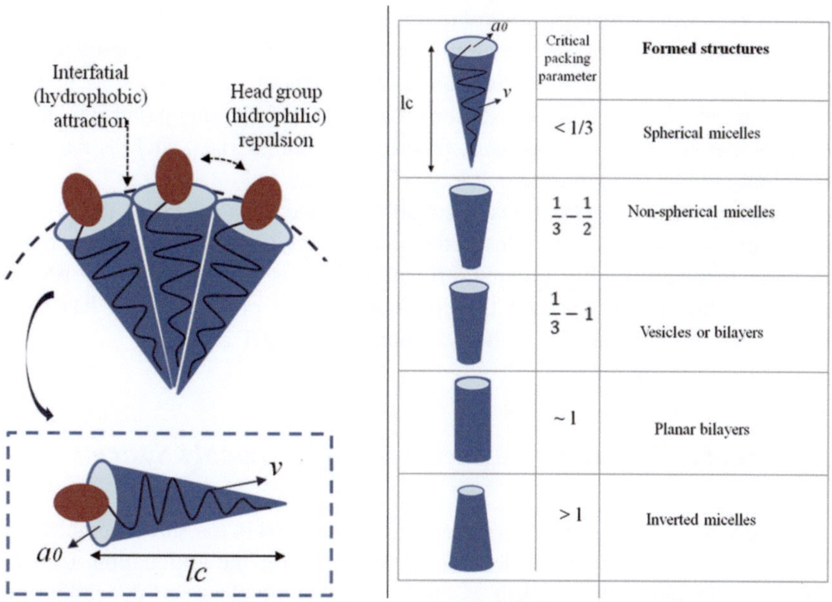

Fig. 12 Dependence of micellar structure with the Israelachvili's packing parameter (adapted from [146] and [140])

length, l_c, the value of the dimensionless packing parameter $p = v/(a_0.l_c)$ would determine the form of the self-assembled structure [146], as shown in Fig. 12. This theory has been generally expanded to amphiphilic block copolymers in aqueous solution, since the driving forces governing the assembly into low energy structures are the same [140].

These intrinsic molecular parameters may be altered, not only by the block copolymer molecular design but also during the course of the application due to the inclusion of polymeric segments that respond to external stimuli. Thus, by changing the surrounding environment conditions, it is possible to generate a great variety of morphologies and architectures from a single block copolymer system [147], to form, for example, polysome-like structures from spherical micelles [148], or to interchange the polymeric segments in the core and corona [149]. The use of stimuli-responsive segment in block copolymers is a very interesting tool for the preparation of improved and novel nanostructures [150], which will be discussed later in this section.

Other molecular structure factors can significantly affect the self-assembly of block copolymers, such as the polydispersity [141]. High polydispersity values are likely to negatively affect the properties of the assembled structures in targeted applications, since a wider dispersion of molecular arrangements, and thus of properties, is expected to occur. CLRP methods allows the preparation of block copolymers with very low polydispersity, extending the range of compositions of the polymeric segments that can be effectively utilized for self-assembly systems thereby providing enhanced control of application properties.

The development of self-assembly systems based on block copolymers may open new opportunities in a wide range of applications fields such as in coating formulations [151], chemical sensors, catalysts [152], nanopatterning systems such as nanolithography templates, functional nanomaterials, such as nanoporous membranes [143], and in drug delivery [153].

For any of these applications, a suitable balance between all of the polymeric segments that play different roles in the block copolymer self-assembly is required in order to obtain highly ordered structures. This is accomplished by providing critical, stringent control over the copolymerization process which has become gradually more efficient due to the development of CLRP methods.

3.2 *Role of Polymeric Segments in Self-Assembly Systems*

The selection of the polymeric segments to be included in the self-assembled block copolymer is driven by the roles that are required for the application. Generally, polymeric segments used in self-assembly systems are classified depending on its main function in the formed structure. Functions such as steric stabilizers, core forming (including adhesion promoters for loaded agents), crosslinkable and stimuli responsive segments can usually be found in block copolymer micellar systems (see scheme in Fig. 13).

3.2.1 Steric Stabilizer

For applications where the nanoparticles are dispersed in a liquid phase such in coatings formulations [151] or for DDS [144], the micellization process is of great importance and depends on the affinity of the steric stabilizer to the solvent medium and on the interaction with the other polymeric constituents of the self-assembly system [154]. A list of common solvating blocks from aqueous to fluorinated

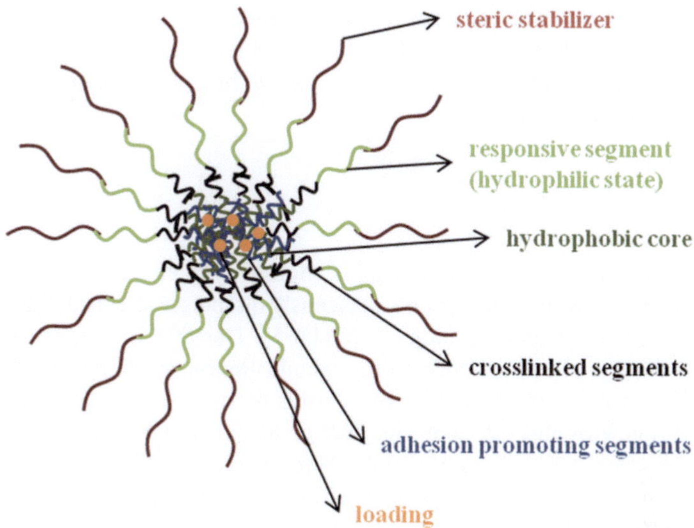

Fig. 13 Schematic representation of possible functions in segments forming a micellar system

media has been reviewed [145]. PEG is a widely used steric stabilizer allowing, for example, the preparation of controlled water soluble metallic nanoparticles [155] or stable pH-tunable vesicles [156].

3.2.2 Core

The segments that constitute the core may provide additional features to the self-assembled structure. Polymeric segments based on butyl methacrylate were shown to provide greater flexibility [147] and negatively charged pores capability to self-assembly structures [157]. Due to their higher glass transition temperature, polystyrene segments are typically used when enhanced mechanical stability is required for the core [158]. When biocompatible and biodegradable cores are required, PCL [159] and PLA [160] may be used.

The core of self-assembly systems typically has the function of encapsulating specific loaded molecules. Self-assembly systems have been widely used to prepare hybrid nanocomposites based on block copolymers, by incorporating inorganic species in the core [143, 161]. Due to the natural incompatibility of polymers and inorganic materials, polymer functional segments that bind to the inorganic phase need to be added to the self-assembly system, stabilizing the polymer/inorganic interface [145]. Poly(vinylpyridine) have been widely used to prepare hybrid nanocomposites, due to its great affinity to a wide range of inorganic components, such as noble metals, semiconductors or magnetic materials [162]. The inorganic phase may also be included in self-assembly system by bonding it covalently to the polymeric segments [163].

Similar concepts have been used to encapsulate organic compounds, such as drugs in self-assembly systems, either by binding it physically with the aid of polymeric compatibilizing segments [144] or by binding it covalently to form polymer-drug conjugates [164].

3.2.3 Crosslinking

One of the major issues limiting the application of synthetic block copolymers in some applications is that micelles spontaneously dissociate at concentration below their critical micellar concentration (CMC) [165]. This effect is expected to occur in several DDS applications during drug administration, resulting in a premature release of the drug [166]. When it is expected that the micelles may dissociate during the application, crosslinkable segments may be included in the block copolymer to provide enhanced stability to the micelle. Typical micelles stabilization techniques include chemical reaction between block terminals end-groups resulting in shell and core segments crosslinking [140, 167–170].

3.2.4 Responsive Segments

Very often, the same block copolymer is used to form self-assembled structures having different sizes and morphologies. This is generally obtained by incorporating responsive polymeric segments into the precursor and by applying different environmental conditions (such as pH, temperature or the solvent) during the self-assembly process. However, most of the time, responsive polymeric segments are used when the initial self-assembled structure is required to be re-assembled during the application to provide a different function in a given environment. This is the case with DDS, where the re-assembly of the micellar structure on application of external stimuli can induce the release of the drug at the target site. Different types of responsive polymeric systems are discussed below.

3.3 Responsive Polymeric Systems

Responsive polymers, also called “stimuli-responsive polymers” or “smart polymers”, are polymers that can undergo reversible physical/chemical changes when exposed to one or more external signals. Some of those stimuli, which act as response triggers, include temperature, pH, magnetic and/or electric field, light, enzymes and glucose, among others [171]. By varying these parameters in the polymers’ surrounding environment (e.g., solvent), it is possible to induce alterations in the polymers structure and, consequently, modify their solubility, molecular interactions with the solvent, volume (by swelling/collapsing) and degradation profile [171]. This type of behavior is very interesting since it resembles that of the

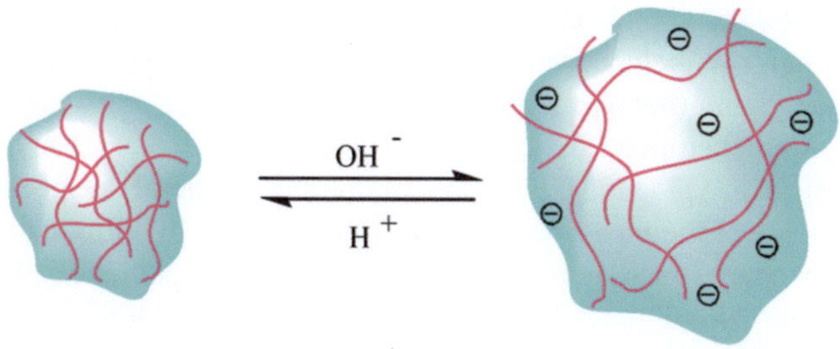

Fig. 14 pH-responsive behavior of an anionic hydrogel

macromolecules found in living systems, which are used to regulate the cells activity by responding to slight variations in their surrounding environment [171, 172]. In this sense, responsive polymers are excellent candidates to be used in biomedical applications such as tissue engineering, drug delivery, sensors, membranes and synthetic muscles, among others [172, 173].

Some examples of responsive polymers are provided in the following section with special focus in their mechanism of action.

3.3.1 pH-Responsive Polymers

Polymer structures which include basic or acidic moieties have the ability to undergo conformational alterations as a response of variations in the pH values of the surrounding environment [3]. In the case of anionic polymers this behavior is due to the protonation of basic functional groups at low pH, and, in the case of cationic polymers, this is due to the deprotonation of acidic functional groups at high pH values, inducing electrostatic repulsion between the polymer chains [172, 174]. As a consequence, pH-responsive polymer chains expand and contract, depending on the pH, leading to significant changes in the polymer volume (Fig. 14). Examples include pendant weak acid functional groups such as carboxylic acid and phosphoric acid, while weak basic functional groups are usually based on amine moieties [175].

Taking into account the properties exhibited by pH-responsive polymers, it is clear that the application of these materials in the biomedical field, namely on DDS, is of extreme utility. In the human body, some signals of unhealthy tissues are manifested by slight deviations of the pH values in comparison with the physiologic pH (~7.4) [175]. For instance, chronic wounds, sore tissue and cancer tissue exhibit an acidic character. Additionally, the gastrointestinal tract (GI) presents different pH environments along its length, the stomach shows pH ~2 and the intestine has pH ~6.2–7.5 [176]. These changes can be used as a trigger for oral drug delivery targeted treatments. For the development of such DDS, it is necessary to choose a suitable polymer taking into consideration the targeted local of release of the therapeutic agent. For instance, cationic hydrogels will swell and release drugs at low pH, such as in stomach environment, while anionic hydrogels will do that at high pH values, such as in intestine [3]. Common pH-responsive polymers used for this purpose include synthetic polymers such as poly(acrylic acid) (PAA), poly(2-(dimethylamino)ethyl methacrylate) (PDMAEMA), polymaleic anhydride (PMA), polyethyleneimine (PEI), as well as some biopolymers, such as chitosan and hyaluronic acid [174, 177]. In a view of optimizing the pH-responsive DDS, it is possible to adjust the phase transition of the polymers by introducing desirable acidic/basic pendant functional groups or including hydrophobic segments, which will influence the *pKa* of the polymers and, thus, their behavior when responding to pH variations [172].

3.3.2 Temperature-Responsive Polymers

The ability of certain polymers to respond to variations in temperature is related to alterations in the balance between intra- and inter-hydrogen bonding of polymer chains and in the solubilization capacity of the surrounding solvent [172]. As a consequence, around a temperature named critical solution temperature (CST), a value that is specific for each temperature-responsive polymer, the polymer undergoes an abrupt volume phase transition by chain expansion or contraction. The behavior exhibited by these temperature-responsive polymers differs according to the type of transition that the polymer undertakes at the CST: (i) a lower critical solution temperature (LCST); (ii) an upper critical solution temperature (UCST); (iii) a lower gel transition temperature (LGTT), or (iv) an upper gel transition temperature (UGTT) [175]. In water, a polymer that presents a LCST will adopt a globular conformation for temperatures above the LCST, due to the prevalence of hydrophobic interactions between the copolymer segments and the medium. For temperatures below the critical value, the polymer chains will be extended as the hydrogen bonding between them and water gain importance [3, 178]. Polymers that present a LCST are also named as negatively thermosensitive materials. Some of the most widely used LCST polymers are poly(N-isopropylacrylamide) (PNIPAAm) [178] and several poly(meth)acrylates [179–183]. An opposite behavior is that exhibited by positively thermosensitive polymers that present an UCST. Polymers exhibiting an UCST have their chains in an extended conformation at temperatures above the UCST that collapse upon cooling to temperatures below the UCST. Examples of polymers that have an UCST are PAA, poly(propylene oxide) (PPO), poly(ethylene oxide) (PEO) and polyacrylamide (PAAm) [172]. Finally, the last two types of CST, (iii) and (iv), are applied to non-covalent cross-linked hydrogels which present a viscosity phase transition [172], rather than the coil-to-globule transition that positive and negative thermosensitive polymers exhibit [3]. These hydrogels, called thermally reversible gels, change their viscosity from a low viscous aqueous solution into a rigid gel at a lower gel transition temperature (LGTT) or upper gel transition temperature (UGTT), depending on the nature of the polymers. Copolymers exhibiting such behavior are often composed of segments of PPO and PEO.

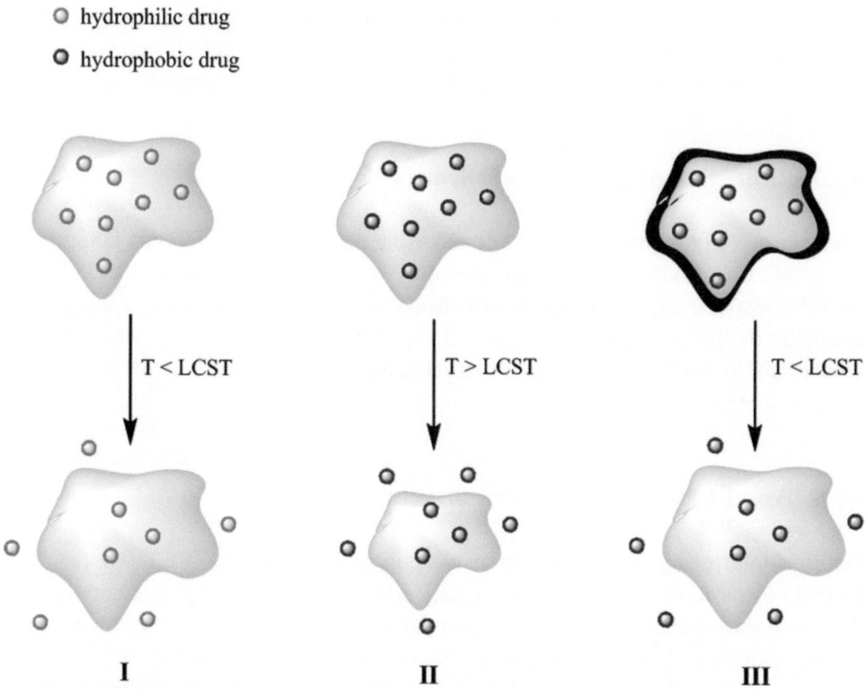

Fig. 15 Strategies used for targeted drug delivery using negatively thermosensitive hydrogels (adapted from [174])

As mentioned above, temperature-responsive polymers exhibit a wide range of behavior, depending on their nature. This, in combination with the fact that temperature could be changed by a human body response due to the presence of pathogens or pyrogens, [3] or by external devices, such as heat-triggered subdermal implants [172], makes temperature-responsive polymers good candidates for DDS. These deviations of temperature from the normal physiological value (37°C) are used to trigger the controlled release of pharmaceutical agents as the polymer undergoes swelling/deswelling. This is accomplished mainly by using hydrogels. Figure 15 presents three possible routes for delivery of therapeutic agents depending on the hydrophobic character of the DDS.

When using a negative thermosensitive hydrogel, it is possible to entrap a hydrophilic drug (Fig. 15), which will be released by diffusion when the hydrogel is in the swollen state, i.e. below its LCST. On the other hand, the same hydrogel can be loaded with a hydrophobic drug which is released upon collapse of polymer chains above the LCST, Fig. 15II. Finally, hydrogels containing a layer of a negative thermosensitive polymer segment can be used to encapsulate drug molecules at temperatures above the polymer LCST, Fig. 15III. In order to maximize the efficiency of temperature-responsive polymeric DDS, the polymers chosen for the purpose should have a CST near the temperature of the human body. An alternative is to adjust the CST of some polymers, with interesting properties for the application such as biocompatibility, by introducing hydrophilic or hydrophobic segments into their structure or even to change their molecular weight or concentration [175].

3.3.3 Other Stimuli-Responsive Polymers

The majority of the procedures developed using responsive polymers as DSS utilize temperature and/or pH as triggers. However it is possible to prepare DDS based on polymers that change their conformation when exposed, for instance to electric, magnetic or light triggers [184–189]. These stimuli can be controlled by external devices and provide an easy way for targeted delivery of the therapeutic agents that takes advantage of harmless non-invasive external energy sources [190].

3.3.4 Multi-Responsive Polymeric Structures

Another interesting group of responsive polymeric segments are those that can respond to more than one stimulus. CLRP have been used to prepare copolymers that form self-assembly structures containing polymeric segments that can respond to two stimuli. For instance, PDMAEMA is a polymer that displays a LCST and becomes hydrophilic at temperature below 35°C. It is a weak polybase with a pKa close to the physiological pH, making it a very interesting component for gene-delivery systems [191]. PAA is a water soluble polymer at physiological temperature but becomes insoluble at temperatures below 14°C, an UCST polymer, and since it is a weak polyacid in strongly acidic environments, below pH $\sim$2, it becomes uncharged and insoluble in water [147]. Poly(3-acrylamidophenylboronic acid) is a polymeric segment that is pH-responsive, becoming hydrophobic in acidic environments, and sugar-responsive due to its ability to reversibly form cyclic boronate esters with vicinal diols, meaning that it becomes hydrophilic in the presence of a given concentration of a diol, such as sugar [192].

Multi-responsive polymer systems can also be prepared by combining different stimuli-responsive polymers. This type of materials have gained significant scientific interest since they can combine different responsive properties which are of value in view of optimizing the efficiency of self-assembly systems, in a manner similar to what happens in biological systems.

CLRP methods have been used to prepare structures that can have, for instance, a triple-response to temperature (Fig. 16) by constructing block copolymers with three polymeric segments each with different LCST values [193].

Advances on CLRP methods have allowed the preparation of a vast range of well defined schizophrenic block copolymers, a concept that was first introduced by Armes' research group [194]. A schizophrenic self-assembly is based on the principle that copolymer chains can self-assemble in dilute solution in the absence of any organic co-solvent to form two distinct structures [195], and, therefore, can interchange the corona and core by changing the conditions of the surrounding

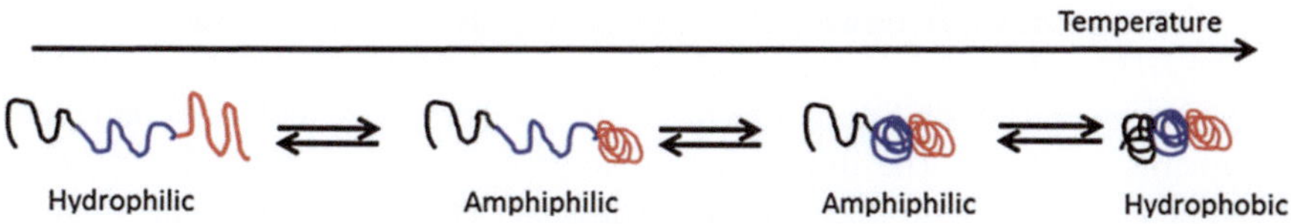

Fig. 16 Effect of temperature in a triple-responsive block copolymer (adapted from [193])

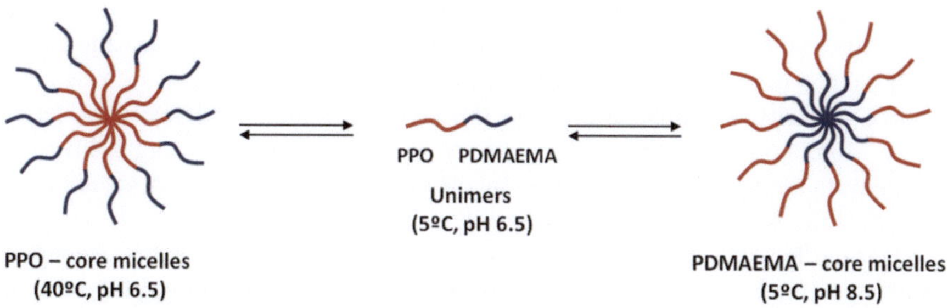

Fig. 17 Self-assembly of PPO-block-PDMAEMA into schizophrenic micelles (adapted from [196])

media. A schizophrenic self-assembly can be attained for example by copolymerizing a temperature-responsive segment with a pH-responsive segment [196]. While temperature and pH responsive structures are the mostly used schizophrenic structures [197–199], this type of self-assembly structures can also be obtained by copolymerization of segments that respond differently to pH and provide a zwitterionic character to the self-assembly [200, 201] (Fig. 17).

Systems that can respond simultaneously to temperature, pH and sugar have been shown to have a great potential for DDS [202] since major challenges in DDS involve the creation of structures that respond to several stimuli and provide precise and programmable drug release profiles [203]. CLRP methods have been widely used to prepare this multi-responsive structures for DDS [173], due to the ease of creation of complex and controlled polymeric structures and to the possibility for the initial copolymer to be further functionalized.

3.4 CLRP-Based Block Copolymers for DDS Applications

Nanostructures produced from the self-assembly of amphiphilic block copolymers stand out in the field of DDS [204, 205]. The different morphologies resulting from the solution assembly of the block copolymers allow the drug encapsulation resulting in the fact that polymeric micelles and vesicles are by far the most explored nanocarriers for DDS.

3.4.1 Drug Encapsulation and Release from CLRP-Based Systems

There are several possibilities for incorporation of therapeutic molecules into these polymeric structures. The different incorporation methods are mainly based on hydrophobic/hydrophilic interactions, electrostatic attractions, hydrogen bonding and/or covalent bonds. It is crucial to understand and explore the specific chemical and biological characteristics of the polymer and of the therapeutic agents in the design of a polymeric DDS [206]. It is well known that the fundamental properties of amphiphilic block copolymers, such as its architecture, the length and the nature of the different blocks, have a decisive influence on the efficiency of the nanocarrier while taking into consideration different desired DDS parameters, such as size, loading capacity, stability, circulation time, biodistribution and release kinetics [144]. Moreover, the drug loading capacity of the DDS and the drug's diffusion profile from the system are influenced by the drug's molecular weight, solubility in biological fluids, and miscibility in the polymer matrix [206].

Over the last decade, a significant amount of research work has been reported on the potential applications of self-assembly systems containing polymers in biomedical applications [205, 207]. The use of stimuli-responsive nanocarriers offers an interesting opportunity for drug and gene delivery in applications where the drug carrier becomes an active participant, rather than passive vehicle. This feature is particularly relevant when the stimuli are intrinsically correlated with the pathological environment. Generally, the main function of the stimuli-responsive segments in nanocarriers is to keep them stable under physiological conditions, but making the nanocarrier capable to respond to pathological triggers, such as pH, temperature, specific enzymes, or protein over-expression, provides a material for site specific targeted drug release and reduces undesired side-effects [208, 209].

For efficient targeted drug delivery in biological systems the release of the encapsulated species must occur in the presence of weak external signals. As discussed above, pH- and temperature-responsive polymers are the most studied among smart polymers for biomedical applications and DDS, since pH and temperature are important physiological indicators and are easily controlled. Considering the diverse pathophysiological conditions, this type of therapeutic response approach can be very promising and the controlled copolymerization of responsive segments via CLRP methods to prepare nanocarrier systems has greatly extended the number of design possibilities for new and more efficient DDS [210–212].

The most extensively studied nanocarriers are polymeric micelles and they have been proposed as nanocarriers for drug and gene delivery [153, 213]. In an aqueous environment, the hydrophilic segments of the block copolymers form the corona, whereas the hydrophobic blocks form the core of the micelles and provide a microenvironment for entrapment of hydrophobic drugs. The corona interacts with the external medium and should provide the required stabilization (Fig. 18). The controllability features of CLRP methods allow adjusting the physico-chemical properties of nanocarriers and ease further functionalization of block copolymeric systems with a view of optimizing their performance.

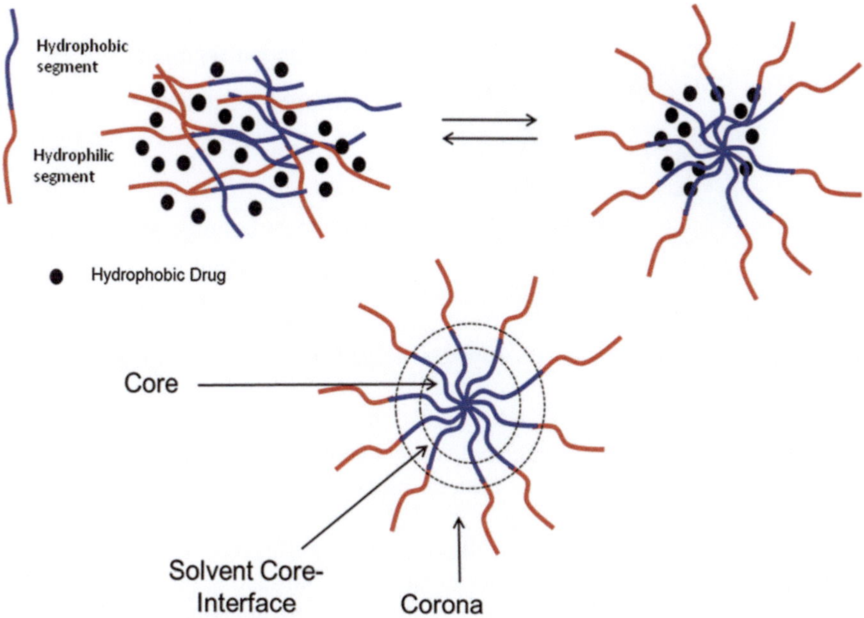

Fig. 18 Drug distribution and release from a spherical micelle formed from the self-assembly of an amphiphilic copolymer

In addition to the relevant nanocarrier properties described in the beginning of this chapter, there are other parameters that need to be considered that are specific of self-assembly systems based on block copolymers. Allen and co-authors [144] identified and critically discussed the influence of the corona's physical parameters in the performance of micelles as drug carriers, namely the surface density of the hydrophilic chains, the surface charge, the hydrophilicity and the block length and their derivatization. These parameters are known to affect very important properties of the micelles as DDS, such as: biodistribution, pharmokinetic parameters, biocompatibility, steric stability, specificity, surface adsorption to proteins and adhesion to biosurfaces. Other important parameters that influence the drug loading capacity and the release kinetics, in terms of the core properties, such as compatibility with drug, block length and core-corona block ratio, and the properties of the drug that has to be encapsulated such as compatibility with polymeric phase, size, shape, molecular weight, and physical state, were also discussed in that contribution [144].

Sant and co-authors [214] prepared diblock copolymers of PEG, *tert*-butyl methacrylate, ethyl acrylate or n-butyl acrylate by ATRP. After the hydrolysis of the *t*-butyl groups to form acrylic acid functionalities, the polymeric systems exhibited pH-dependent aggregation behavior and their critical aggregation concentration could be controlled with the block length of the hydrophobic segments. This stimuli-responsive system was designed for delivery of poorly water-soluble model drugs; indomethacin, fenofibrate and progesterone. The results indicated that such

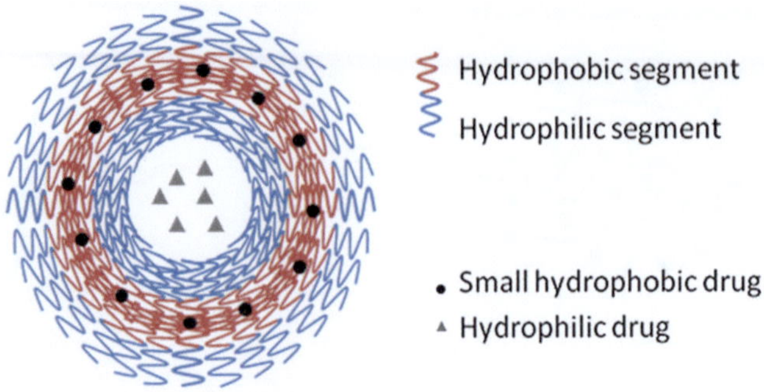

Fig. 19 Schematic representation of a polymeric vesicle from the self-assembly of amphiphilic block copolymer

supramolecular assemblies, with high drug loadings and pH-dependent release kinetics, can potentially enhance the oral bioavailability of poorly water-soluble drugs [214].

Studies on polymeric-based micelles that can be triggered by variation in pH include the controlled delivery of several bio-active substances, such as chemotherapeutic chlorambucil [215], anticancer taxomifen [216], antiparasitic miltephosine [217] and antithrombotic dipyridamole [218].

Polymer vesicles, also known as polymersomes, have significant interest for DDS applications due to their capacity to sequester small hydrophobic molecules in their lamellar bilayer structure or to encapsulate hydrophilic species in their aqueous interior cavity (Fig. 19). Due to their versatility, examples of the use of polymer vesicles for the delivery of both hydrophilic and hydrophobic anticancer drugs have been reported [219]. The release of the hydrophilic drug from the aqueous core of the polymersomes can occur by therapeutic diffusion through the membrane, or through vesicle degradation or vesicle disruption in the case of stimuli-responsive polymersomes [220].

Yang and co-workers reported the synthesis of poly(ethylene oxide)-block-poly(N-isopropylacrylamide) (PEO-b-PNIPAAm) vesicles using RAFT polymerization of a dithiobenzoyl end capped PEO macro chain-transfer agent in 1,4-dioxane [221]. The responsive block copolymer self assembled into stable vesicles and was shown to be able to encapsulate both hydrophilic drugs, such as *Doxorubicin*, and hydrophobic molecules into the polymeric membrane. The trigger release of the drugs occurs at temperatures above 32°C, due to LCST behavior of PNIPAAm segments.

More complex architectures for DDS applications such as comb-dendritic nanostructures can also be obtained using CLRP methods. One of the possible synthesis strategies to obtain such structures is the use of a dendritic molecule as a macroinitiator for CLRP. This type of structures can form different self-assembled structures and thus entrap drugs with different physico-chemical affinities. For example, Hammond and co-workers [222] developed an amphiphilic comb-dendritic block copolymer that self-assembles into polymeric micelles with very low CMC, 10^{-8} M.

The block copolymer was composed by one hydrophilic dendritic block with the corona modified with PEG chains and a hydrophobic comb-like structure made from poly(γ-n-dodecyl-L-glutamate). The cone-type morphology enhances the micelle stability and reduces the CMC [140, 222]. The authors also observed that a simple change in the hydrophilic part of the block copolymer, changes the critical packing parameter to a value close to one, and bilayer vesicles are formed by the self-assembly, instead of spherical micelles [223].

These self-assembly structures may be limited to the range of the molecular structures that can be obtained by the CLRP method. However, the combination of CLRP methods with organic chemistry based functionalization methods exponentially increases the range of possible DDS configurations. It is now possible to tailor the polymeric DDS behavior by post-polymerization adjusting polymer structural characteristics, such as the composition, molecular weight, topologies and also chemical functionalities. The consequent control over the polymer structures and molecular weight is known to allow the preparation of nanocarriers with similar size, resulting in DDS that have better reproducible pharmacokinetic behavior [206].

Another possibility is to use nanogels prepared in inverse mini emulsion [168–170]. They can entrap drugs and biomolecules, that can be delivered by degradation of cleavable crosslinkers (e.g., disulfide moieties) into well-defined primary chains.

3.4.2 Polymer Functionalization for CLRP-Based DDS

Functionalization of polymers emerges as a procedure to overcome some of the weaknesses in DDS performance inherent to the prepared polymer and allows the achievement of different polymer properties, otherwise inaccessible in the tailor made polymers [126]. Polymer modifications such as surface functionalization can significantly increase the efficiency of a DDS by promoting cellular targeting [208, 224] or by adapting the DDS surface to increase its circulation half-life [225]. The interface between inorganic carriers, as magnetic nano- and micro-particles [226–228] or nanotubes [229–231], and the biological molecules can also be enhanced by surface modifications, which greatly increase their *in vivo* biocompatibility. The possibility to attach biological/targeting residues to the outer surface of the nanostructures is of enormous interest due to the potential for targeted, receptor-mediated drug delivery.

In a simple concept, functionalization methods can be used to prepare copolymeric structures that would be difficult to prepare directly by conventional CLRP methods. For example, PDMAEMA grafted to PHEMA, via degradable linkages, copolymers were obtained by a combination of ATRP and "click chemistry" approaches and were promising DDS when evaluated as DNA carriers [232]. These polymeric segments were also used to prepare degradable brushes using a different combination of ATRP methods and "click chemistry". The brushes were composed of a PHEMA backbone onto which PDMAEMA was grafted via a hydrolysable linker. Both PDMAEMA and PHEMA were synthesized by ATRP with controlled molecular weight and narrow molecular weight distributions. A subsequent "click"

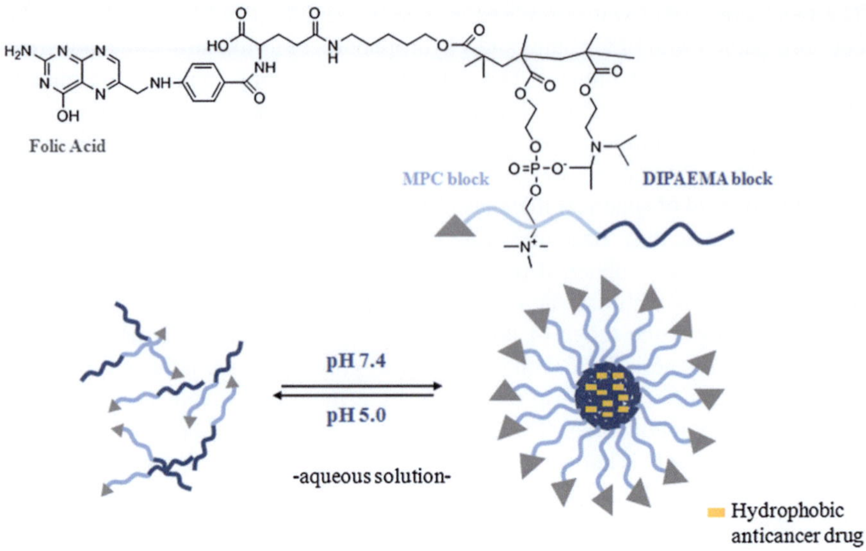

Fig. 20 Folic acid-functionalized PDIPAEMA-MPC block copolymer micelles and drug entrapment (adapted from [236])

reaction between PHEMA with alkyne side groups and the azide end-group on the PDMAEMA resulted in formation of a high molecular weight polymer composed of low molecular weight constituents via an easily degradable carbonate ester linkage. The degradable brushes were able to transfect cells efficiently in the presence of the endosomal-disrupting INF-7 peptide.

In another approach, PHEMA-*g*-(PCL-*b*-PEG) brush copolymers were prepared by the combination of ATRP and ring opening polymerization methods to form polymeric micelles that could be used as DDS [233]. These micelles showed lower CMC and exhibited higher hydrophobic drug loading capacity when compared to micelles formed by the self-assembly of the linear PCL-b-PEG. They encapsulated *Doxorubicin* as a hydrophobic anticancer drug and proved that the micelles can be effectively internalized by A549 human lung carcinoma cells and slowly released the drug.

One of the most used post-polymerization functionalization techniques is the inclusion of folic acid in the polymeric structure as targeting molecule for DDS in cancer therapy. Armes and co-workers [234] reported a biocompatible pH-responsive block copolymer composed of 2-methacryloyloxyethyl phosphorylcholine (MPC) and 2-(diisopropylamino)ethyl methacrylate (DIPAEMA) via ATRP. These copolymers were functionalized with folic acid to provide stimuli-responsive micellar vehicles for the delivery of highly hydrophobic anticancer drugs (among them the *Paclitaxel*) (Fig. 20) [235, 236].

CLRP methods have been used to combine polymeric segments with DDS-based liposomes. Liposomes, which are artificially prepared vesicles containing lipids, possess several features that make them very interesting for drug delivery

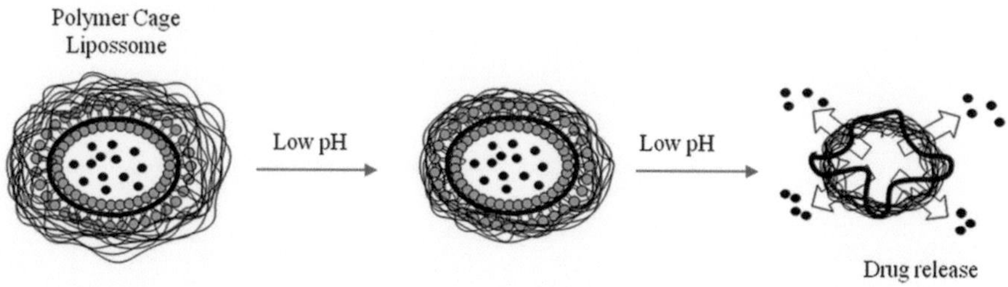

Fig. 21 Drug release from a pH-responsive liposome with a PAA crosslinked shell (adapted from [237])

purposes. Modified liposomes have been shown to have excellent pharmacokinetic profiles for the delivery of several biological agents such as nucleic acids, proteins and chemotherapeutic agents [237]. However, these nanocarriers have low stability and lack of tunable triggers for drug release. In order to overcome both problems, Lee and co-authors [237] proposed a very elegant strategy by describing a simple procedure using a cholesterol-functionalized PAA. The active carboxylate groups are crosslinked with telechelic 2,2′-(ethylenedioxy)bis(ethylamine) linkers, resulting in polymer-caged liposomes that are highly stable and have tunable pH-sensitive responses (Fig. 21). The cholesterol-terminated PAA was synthesized via nitroxide mediated polymerization (NMP) of *tert*-butyl acrylate followed by acidolysis.

3.4.3 Polymeric-Based Conjugates

An interesting approach for DDS is the preparation of nanoconjugate delivery vehicles, in which the drug is no longer encapsulated, but actively participates in the formation of the micelle. The polymer-drug conjugate is a chemically controlled DDS in which the drug molecules are attached to the polymer backbone, through a cleavable linkage, as a pendent moiety. It is also possible to attach other moieties of interest to the nanoconjugate delivery vehicles such as a solubilising agent or a targeting ligand. The polymer-drug linkage should be easily cleaved by hydrolysis or by enzymatic action, releasing the drug at the site of interest. Due to recent advances in macromolecular engineering, it is now possible to create polymer-drug conjugates with a variety of architectures. Monofunctional linear, polyfunctional linear, starlike- or dendritic-drug conjugates have been proposed [238]. Tong and Cheng [239] described the preparation of a polymer-drug conjugate containing a high loading of paclitaxel. This system was prepared by the ring opening polymerization of *d*, *l*-lactide initiated by the drug molecule [239]. In the final structure the paclitaxel was linked to PLA through a hydrolysable ester linkage. The polymer-drug conjugates formed nanoparticles having narrow particle size distribution, allowing an excellent control over drug loading and release.

Another very useful approach to prepare protein-based DDS is the combination of polymers and proteins to produce bioconjugates. These systems form stable

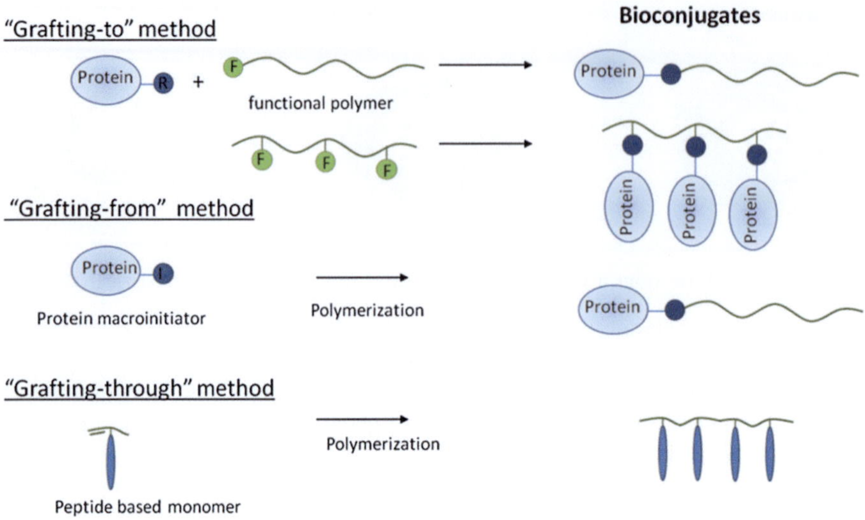

Fig. 22 Bioconjugate synthesis methods

protein complexes that reduce the protein immunogenicity, enhance its stability and increase its circulation half-life [238]. Retting and co-authors [240] reported a proof of concept, based on ATRP block copolymerization of n-butyl acrylate using an oligopeptide as macroinitiator, opening the possibility to integrate oligopeptides into a variety of polymeric materials, while maintaining controllability over molecular structure, low polydispersity and controllable polymers length, that is characteristic of CLRP methods.

The combination of CLRP methods and the functionalization strategies greatly increased the range of new tailor made bioconjugates [128]. Generally, bioconjugates can be prepared by three different strategies: "grafting-to", "grafting-from" or "grafting-through" methods, as shown in Fig. 22 [128, 241, 242]. Also they can involve genetically modified proteins [243–245] and be linked to DNA [246].

Moreover, polymer-protein conjugates based on stimuli-responsive polymers, which has been reviewed recently [247], allows improving the protein stability and their pharmacokinetic profiles by including further responsive functions.

The combination of these different strategies based on CLRP techniques to prepare controlled polymeric nanostructures, and the possibility to combine them with other relevant synthetic or even biological functionalities, have created a vast range of possibilities to achieve optimized and cutting edge DDS self-assembly structures.

Acknowledgements The authors gratefully acknowledge FP7-Health-2009-2.4.4-2-Project RdCVF for financial support.

References

1. Alonso MJ (2004) Nanomedicines for overcoming biological barriers. Biomed Pharmacother 58:168–172
2. Freichels H, Jerome R, Jerome C (2011) Sugar-labeled and PEGylated (bio)degradable polymers intended for targeted drug delivery systems. Carbohydr Polym 86:1093–1106
3. Bajpai AK, Shukla SK, Bhanu S, Kanjane S (2008) Responsive polymers in controlled drug delivery. Prog Polym Sci 33:1088–1118
4. Yoo JW, Doshi N, Mitragotri S (2011) Adaptive micro and nanoparticles: temporal control over carrier properties to facilitate drug delivery. Adv Drug Deliv Rev 63:1247–1256
5. Canelas DA, Herlihy KP, DeSimone JM (2009) Top-down particle fabrication: control of size and shape for diagnostic imaging and drug delivery. Wiley Interdiscip Rev Nanomed Nanobiotechnol 1:391–404
6. Sahay G, Alakhova DY, Kabanov AV (2010) Endocytosis of nanomedicines. J Control Release 145:182–195
7. Wang J, Byrne JD, Napier ME, DeSimone JM (2011) More effective nanomedicines through particle design. Small 7:1919–1931
8. Aderem A, Underhill DM (1999) Mechanisms of phagocytosis in macrophages. Annu Rev Immunol 17:593–623
9. Petros RA, DeSimone JM (2010) Strategies in the design of nanoparticles for therapeutic applications. Nat Rev Drug Discov 9:615–627
10. Hillaireau H, Couvreur P (2009) Nanocarriers' entry into the cell: relevance to drug delivery. Cell Mol Life Sci 66:2873–2896
11. Champion JA, Walker A, Mitragotri S (2008) Role of particle size in phagocytosis of polymeric microspheres. Pharm Res 25:1815–1821
12. Mayor S, Pagano RE (2007) Pathways of clathrin-independent endocytosis. Nat Rev Mol Cell Biol 8:603–612
13. Bareford LM, Swaan PW (2007) Endocytic mechanisms for targeted drug delivery. Adv Drug Deliv Rev 59:748–758
14. Swanson JA, Watts C (1995) Macropinocytosis. Trends Cell Biol 5:424–428
15. Sharma P, Varma R, Sarasij RC, Ira, Gousset K, Krishnamoorthy G, Rao M, Mayor S (2004) Nanoscale organization of multiple GPI-anchored proteins in living cell membranes. Cell 116:577–589
16. Kalia M, Kumari S, Chadda R, Hill MM, Parton RG, Mayor S (2006) Arf6-independent GPI-anchored protein-enriched early endosomal compartments fuse with sorting endosomes via a Rab5/phosphatidylinositol-3′-kinase-dependent machinery. Mol Biol Cell 17:3689–3704
17. Howes MT, Kirkham M, Riches J, Cortese K, Walser PJ, Simpson F, Hill MM, Jones A, Lundmark R, Lindsay MR, Hernandez-Deviez DJ, Hadzic G, McCluskey A, Bashir R, Liu L, Pilch P, McMahon H, Robinson PJ, Hancock JF, Mayor S, Parton RG (2010) Clathrin-independent carriers form a high capacity endocytic sorting system at the leading edge of migrating cells. J Cell Biol 190:675–691
18. Ferrari M (2008) Nanogeometry: beyond drug delivery. Nat Nanotechnol 3:131–132
19. Mitragotri S, Lahann J (2009) Physical approaches to biomaterial design. Nat Mater 8:15–23
20. Matsumura Y, Maeda H (1986) A new concept for macromolecular therapeutics in cancer chemotherapy: mechanism of tumoritropic accumulation of proteins and the antitumor agent smancs. Cancer Res 46:6387–6392
21. Beningo KA, Wang YL (2002) Fc-receptor-mediated phagocytosis is regulated by mechanical properties of the target. J Cell Sci 115:849–856
22. Verma A, Stellacci F (2010) Effect of surface properties on nanoparticle-cell interactions. Small 6:12–21
23. Doshi N, Mitragotri S (2010) Macrophages recognize size and shape of their targets. PLoS One 5:e10051

24. Decuzzi P, Pasqualini R, Arap W, Ferrari M (2009) Intravascular delivery of particulate systems: does geometry really matter? Pharm Res 26:235–243
25. Yoo JW, Doshi N, Mitragotri S (2010) Endocytosis and intracellular distribution of PLGA particles in endothelial cells: effect of particle geometry. Macromol Rapid Commun 31: 142–148
26. Frojmovic MM, Milton JG (1982) Human platelet size, shape, and related functions in health and disease. Physiol Rev 62:185–261
27. Haghgooie R, Toner M, Doyle PS (2010) Squishy non-spherical hydrogel microparticles. Macromol Rapid Commun 31:128–134
28. Doshi N, Zahr AS, Bhaskar S, Lahann J, Mitragotri S (2009) Red blood cell-mimicking synthetic biomaterial particles. Proc Natl Acad Sci U S A 106:21495–21499
29. Geng Y, Dalhaimer P, Cai S, Tsai R, Tewari M, Minko T, Discher DE (2007) Shape effects of filaments versus spherical particles in flow and drug delivery. Nat Nanotechnol 2:249–255
30. Merdan T, Kopecek J, Kissel T (2002) Prospects for cationic polymers in gene and oligonucleotide therapy against cancer. Adv Drug Deliv Rev 54:715–758
31. Pack DW, Hoffman AS, Pun S, Stayton PS (2005) Design and development of polymers for gene delivery. Nat Rev Drug Discov 4:581–593
32. He C, Hu Y, Yin L, Tang C, Yin C (2010) Effects of particle size and surface charge on cellular uptake and biodistribution of polymeric nanoparticles. Biomaterials 31:3657–3666
33. Xiao K, Li Y, Luo J, Lee JS, Xiao W, Gonik AM, Agarwal RG, Lam KS (2011) The effect of surface charge on in vivo biodistribution of PEG-oligocholic acid based micellar nanoparticles. Biomaterials 32:3435–3446
34. Venkataraman S, Ong WL, Ong ZY, Joachim Loo SC, Ee PL, Yang YY (2011) The role of PEG architecture and molecular weight in the gene transfection performance of PEGylated poly(dimethylaminoethyl methacrylate) based cationic polymers. Biomaterials 32:2369–2378
35. Cruz LJ, Tacken PJ, Fokkink R, Figdor CG (2011) The influence of PEG chain length and targeting moiety on antibody-mediated delivery of nanoparticle vaccines to human dendritic cells. Biomaterials 32:6791–6803
36. Blit PH, McClung WG, Brash JL, Woodhouse KA, Santerre JP (2011) Platelet inhibition and endothelial cell adhesion on elastin-like polypeptide surface modified materials. Biomaterials 32:5790–5800
37. Wang J, Tian S, Petros RA, Napier ME, Desimone JM (2010) The complex role of multivalency in nanoparticles targeting the transferrin receptor for cancer therapies. J Am Chem Soc 132:11306–11313
38. Spain SG, Cameron NR (2011) A spoonful of sugar: the application of glycopolymers in therapeutics. Polymer Chem 2:60–68
39. Chen CK, Shiang YC, Huang CC, Chang HT (2011) Using self-assembled aptamers and fibrinogen-conjugated gold nanoparticles to detect DNA based on controlled thrombin activity. Biosens Bioelectron 26:3464–3468
40. Vogelson CT (2001) Advances in drug delivery systems. Mod Drug Discov 4:49–50
41. Nair LS, Laurencin CT (2007) Biodegradable polymers as biomaterials. Prog Polym Sci 32:762–798
42. Grund S, Bauer M, Fischer D (2011) Polymers in drug delivery – state of the art and future trends. Adv Eng Mater 13:B61–B87
43. Ulery BD, Nair LS, Laurencin CT (2011) Biomedical applications of biodegradable polymers. J Polym Sci, Part B: Polym Phys 49:832–864
44. Maham A, Tang Z, Wu H, Wang J, Lin Y (2009) Protein-based nanomedicine platforms for drug delivery. Small 5:1706–1721
45. Pastorino L, Erokhina S, Soumetz FC, Bianchini P, Konovalov O, Diaspro A, Ruggiero C, Erokhin V (2011) Collagen containing microcapsules: smart containers for disease controlled therapy. J Colloid Interface Sci 357:56–62
46. Kanematsu A, Yamamoto S, Ozeki M, Noguchi T, Kanatani I, Ogawa O, Tabata Y (2004) Collagenous matrices as release carriers of exogenous growth factors. Biomaterials 25: 4513–4520

47. Nagai N, Kumasaka N, Kawashima T, Kaji H, Nishizawa M, Abe T (2010) Preparation and characterization of collagen microspheres for sustained release of VEGF. J Mater Sci Mater Med 21:1891–1898
48. Holladay C, Keeney M, Greiser U, Murphy M, O'Brien T, Pandit A (2009) A matrix reservoir for improved control of non-viral gene delivery. J Control Release 136:220–225
49. Viñas-Castells R, Holladay C, di Luca A, Díaz VM, Pandit A (2009) Snail1 down-regulation using small interfering RNA complexes delivered through collagen scaffolds. Bioconjug Chem 20:2262–2269
50. Krebs MD, Jeon O, Alsberg E (2009) Localized and sustained delivery of silencing RNA from macroscopic biopolymer hydrogels. J Am Chem Soc 131:9204–9206
51. Naidu BVK, Paulson AT (2011) A new method for the preparation of gelatin nanoparticles: encapsulation and drug release characteristics. J Appl Polym Sci 121:3495–3500
52. GuhaSarkar S, Banerjee R (2010) Intravesical drug delivery: challenges, current status, opportunities and novel strategies. J Control Release 148:147–159
53. Guo R, Cheng Y, Ding D, Li X, Zhang L, Jiang X, Liu B (2011) Synthesis and antitumoral activity of gelatin/polyoxometalate hybrid nanoparticles. Macromol Biosci 11:839–847
54. Kumari A, Yadav SK, Yadav SC (2010) Biodegradable polymeric nanoparticles based drug delivery systems. Colloids Surf B Biointerfaces 75:1–18
55. Yadav SC, Kumari A, Yadav R (2011) Development of peptide and protein nanotherapeutics by nanoencapsulation and nanobioconjugation. Peptides 32:173–187
56. Kuijpers AJ, van Wachem PB, van Luyn MJ, Engbers GH, Krijgsveld J, Zaat SA, Dankert J, Feijen J (2000) In vivo and in vitro release of lysozyme from cross-linked gelatin hydrogels: a model system for the delivery of antibacterial proteins from prosthetic heart valves. J Control Release 67:323–336
57. Huang S, Fu XB (2010) Naturally derived materials-based cell and drug delivery systems in skin regeneration. J Control Release 142:149–159
58. Kim MS, Shin YM, Lee JH, Kim SI, Nam YS, Shin CS, Shin H (2011) Release kinetics and in vitro bioactivity of basic fibroblast growth factor: effect of the thickness of fibrous matrices. Macromol Biosci 11:122–130
59. Kratz F (2008) Albumin as a drug carrier: design of prodrugs, drug conjugates and nanoparticles. J Control Release 132:171–183
60. Sebak S, Mirzaei M, Malhotra M, Kulamarva A, Prakash S (2011) Human serum albumin nanoparticles as an efficient noscapine drug delivery system for potential use in breast cancer: preparation and in vitro analysis. Int J Nanomedicine 5:525–532
61. Zhao D, Zhao X, Zu Y, Li J, Zhang Y, Jiang R, Zhang Z (2011) Preparation, characterization, and in vitro targeted delivery of folate-decorated paclitaxel-loaded bovine serum albumin nanoparticles. Int J Nanomedicine 5:669–677
62. Zöphel L, Eisele K, Gropeanu R, Rouhanipour A, Koynov K, Lieberwirth I, Müllen K, Weil T (2010) Preparation of defined albumin–polymer hybrids for efficient cell transfection. Macromol Chem Phys 211:146–153
63. Baldwin AD, Kiick KL (2010) Polysaccharide-modified synthetic polymeric biomaterials. Peptide Sci 94:128–140
64. Nagpal K, Singh SK, Mishra DN (2010) Chitosan nanoparticles: a promising system in novel drug delivery. Chem Pharm Bull(Tokyo) 58:1423–1430
65. Dash M, Chiellini F, Ottenbrite RM, Chiellini E (2011) Chitosan-a versatile semi-synthetic polymer in biomedical applications. Prog Polym Sci 36:981–1014
66. Peniche H, Peniche C (2011) Chitosan nanoparticles: a contribution to nanomedicine. Polym Int 60:883–889
67. Wang JJ, Zeng ZW, Xiao RZ, Xie T, Zhou GL, Zhan XR, Wang SL (2011) Recent advances of chitosan nanoparticles as drug carriers. Int J Nanomedicine 6:765–774
68. de la Fuente M, Raviña M, Paolicelli P, Sanchez A, Seijo B, Alonso MJ (2010) Chitosan-based nanostructures: a delivery platform for ocular therapeutics. Adv Drug Deliv Rev 62:100–117
69. Ta HT, Dass CR, Dunstan DE (2008) Injectable chitosan hydrogels for localised cancer therapy. J Control Release 126:205–216

70. Bhattarai N, Gunn J, Zhang MQ (2010) Chitosan-based hydrogels for controlled, localized drug delivery. Adv Drug Deliv Rev 62:83–99
71. Gaspar VM, Sousa F, Queiroz JA, Correia IJ (2011) Formulation of chitosan-TPP-pDNA nanocapsules for gene therapy applications. Nanotechnology 22:015101
72. Rudzinski WE, Aminabhavi TM (2010) Chitosan as a carrier for targeted delivery of small interfering RNA. Int J Pharm 399:1–11
73. Nair LS, Laurencin CT (2006) Polymers as biomaterials for tissue engineering and controlled drug delivery. Adv Biochem Eng Biotechnol 102:47–90
74. Augst AD, Kong HJ, Mooney DJ (2006) Alginate hydrogels as biomaterials. Macromol Biosci 6:623–633
75. Séchoy O, Tissié G, Sébastian C, Maurin F, Driot JY, Trinquand C (2000) A new long acting ophthalmic formulation of Carteolol containing alginic acid. Int J Pharm 207:109–116
76. Hassan MA (2007) A long acting ophthalmic gel formulations of atenolol. Drug Dev Ind Pharm 33:1192–1198
77. Azhdarinia A, Yang DJ, Yu DF, Mendez R, Oh C, Kohanim S, Bryant J, Kim EE (2005) Regional radiochemotherapy using in situ hydrogel. Pharm Res 22:776–783
78. Hori Y, Stern PJ, Hynes RO, Irvine DJ (2009) Engulfing tumors with synthetic extracellular matrices for cancer immunotherapy. Biomaterials 30:6757–6767
79. Hori Y, Winans AM, Huang CC, Horrigan EM, Irvine DJ (2008) Injectable dendritic cell-carrying alginate gels for immunization and immunotherapy. Biomaterials 29:3671–3682
80. Freeman I, Cohen S (2009) The influence of the sequential delivery of angiogenic factors from affinity-binding alginate scaffolds on vascularization. Biomaterials 30:2122–2131
81. Han YF, Han YQ, Pan YG, Chen YL, Chai JK (2010) Transplantation of microencapsulated cells expressing VEGF improves angiogenesis in implanted xenogeneic acellular dermis on wound. Transplant Proc 42:1935–1943
82. Silva EA, Mooney DJ (2010) Effects of VEGF temporal and spatial presentation on angiogenesis. Biomaterials 31:1235–1241
83. Tian JY, Sun XQ, Chen XG (2008) Formation and oral administration of alginate microspheres loaded with pDNA coding for lymphocystis disease virus (LCDV) to Japanese flounder. Fish Shellfish Immunol 24:592–599
84. Krebs MD, Salter E, Chen E, Sutter KA, Alsberg E (2010) Calcium phosphate-DNA nanoparticle gene delivery from alginate hydrogels induces in vivo osteogenesis. J Biomed Mater Res A 92:1131–1138
85. Hornig S, Bunjes H, Heinze T (2009) Preparation and characterization of nanoparticles based on dextran-drug conjugates. J Colloid Interface Sci 338:56–62
86. Coviello T, Matricardi P, Marianecci C, Alhaique F (2007) Polysaccharide hydrogels for modified release formulations. J Control Release 119:5–24
87. Shrivastava PK, Shrivastava SK (2009) Dextran polysaccharides: successful macromolecular carrier for drug delivery. Int J Pharm Sci 1:353–368
88. Shrivastava PK, Shrivastava SK (2010) Dextran carrier macromolecule for colon specific delivery of celecoxib. Curr Drug Deliv 7:144–151
89. Acharya S, Sahoo SK (2011) PLGA nanoparticles containing various anticancer agents and tumour delivery by EPR effect. Adv Drug Deliv Rev 63:170–183
90. Kumari A, Yadav SK, Pakade YB, Kumar V, Singh B, Chaudhary A, Yadav SC (2011) Nanoencapsulation and characterization of Albizia chinensis isolated antioxidant quercitrin on PLA nanoparticles. Colloids Surf B Biointerfaces 82:224–232
91. Mundargi RC, Babu VR, Rangaswamy V, Patel P, Aminabhavi TM (2008) Nano/micro technologies for delivering macromolecular therapeutics using poly(D, L-lactide-co-glycolide) and its derivatives. J Control Release 125:193–209
92. Sinha VR, Bansal K, Kaushik R, Kumria R, Trehan A (2004) Poly-epsilon-caprolactone microspheres and nanospheres: an overview. Int J Pharm 278:1–23
93. Natu MV, Gaspar MN, Ribeiro CA, Correia IJ, Silva D, de Sousa HC, Gil MH (2011) A poly(epsilon-caprolactone) device for sustained release of an anti-glaucoma drug. Biomed Mater 6:025003

94. Wei X, Gong C, Gou M, Fu S, Guo Q, Shi S, Luo F, Guo G, Qiu L, Qian Z (2009) Biodegradable poly(epsilon-caprolactone)-poly(ethylene glycol) copolymers as drug delivery system. Int J Pharm 381:1–18
95. Sun TM, Du JZ, Yao YD, Mao CQ, Dou S, Huang SY, Zhang PZ, Leong KW, Song EW, Wang J (2011) Simultaneous delivery of siRNA and Paclitaxel via a 'two-in-one' micelleplex promotes synergistic tumor suppression. ACS Nano 5:1483–1494
96. Xiong XB, Lavasanifar A (2011) Traceable multifunctional micellar nanocarriers for cancer-targeted co-delivery of MDR-1 siRNA and Doxorubicin. ACS Nano 5:5202–5213
97. Heller J, Barr J (2004) Poly(ortho esters)—from concept to reality. Biomacromolecules 5:1625–1632
98. Heller J, Barr J, Ng SY, Abdellauoi KS, Gurny R (2002) Poly(ortho esters): synthesis, characterization, properties and uses. Adv Drug Deliv Rev 54:1015–1039
99. Vauthier C, Dubernet C, Chauvierre C, Brigger I, Couvreur P (2003) Drug delivery to resistant tumors: the potential of poly(alkyl cyanoacrylate) nanoparticles. J Control Release 93: 151–160
100. Vauthier C, Labarre D, Ponchel G (2007) Design aspects of poly(alkylcyanoacrylate) nanoparticles for drug delivery. J Drug Target 15:641–663
101. Anguita-Alonso P, Giacometti A, Cirioni O, Ghiselli R, Orlando F, Saba V, Scalise G, Sevo M, Tuzova M, Patel R, Balaban N (2007) RNAIII-inhibiting-peptide-loaded in vivo Polymethyl-methacrylate prevents in vivo Staphylococcus aureus biofilm formation. Antimicrob Agents Chemother 51:2594–2596
102. Tao SL, Lubeley MW, Desai TA (2003) Bioadhesive poly(methyl methacrylate) microdevices for controlled drug delivery. J Control Release 88:215–228
103. Yuksel N, Baykara M, Shirinzade H, Suzen S (2011) Investigation of triacetin effect on indomethacin release from poly(methyl methacrylate) microspheres: evaluation of interactions using FT-IR and NMR spectroscopies. Int J Pharm 404:102–109
104. Dalmoro A, Lamberti G, Titomanlio G, Barba AA, d'Amore M (2010) Enteric micro-particles for targeted oral drug delivery. AAPS PharmSciTech 11:1500–1507
105. Anderson EM, Noble ML, Garty S, Ma H, Bryers JD, Shen TT, Ratner BD (2009) Sustained release of antibiotic from poly(2-hydroxyethyl methacrylate) to prevent blinding infections after cataract surgery. Biomaterials 30:5675–5681
106. Nyangoga H, Zecheru T, Filmon R, Baslé MF, Cincu C, Chappard D (2009) Synthesis and use of pHEMA microbeads with human EA.hy 926 endothelial cells. J Biomed Mater Res B Appl Biomater 89:501–507
107. Minoo-Rabeeh-Hobabi, Hassanzadeh D, Azarmi S, Entezami AA (2007) Effect of synthesis method and buffer composition on the LCST of a smart copolymer of N-isopropylacrylamide and acrylic acid. Polym Adv Technol 18:986–992
108. Braunecker WA, Matyjaszewski K (2007) Controlled/living radical polymerization: features, developments, and perspectives. Prog Polym Sci 32:93–146
109. Matyjaszewski K, Tsarevsky NV (2009) Nanostructured functional materials prepared by atom transfer radical polymerization. Nat Chem 1:276–288
110. Qiu J, Charleux B, Matyjaszewski K (2001) Controlled/living radical polymerization in aqueous media: homogeneous and heterogeneous systems. Prog Polym Sci 26:2083–2134
111. Matyjaszewski K (1996) Controlled radical polymerization. Curr Opin Solid State Mater Sci 1:769–776
112. Zetterlund PB, Kagawa Y, Okubo M (2008) Controlled/living radical polymerization in dispersed systems. Chem Rev 108:3747–3794
113. Matyjaszewski K (1995) Introduction to living polymerization. Living and/or controlled polymerization. J Phys Org Chem 8:197–207
114. Matyjaszewski K, Gaynor S, Greszta D, Mardare D, Shigemoto T (1995) 'Living' and controlled radical polymerization. J Phys Org Chem 8:306–315
115. Matyjaszewski K, Gaynor S, Wang JS (1995) Controlled radical polymerizations – the use of akyl iodides in degenerative transfer. Macromolecules 28:2093–2095

116. Moad G, Rizzardo E, Thang SH (2008) Radical addition-fragmentation chemistry in polymer synthesis. Polymer 49:1079–1131
117. Matyjaszewski K, Xia JH (2001) Atom transfer radical polymerization. Chem Rev 101: 2921–2990
118. Tsarevsky NV, Matyjaszewski K (2007) "Green" atom transfer radical polymerization: from process design to preparation of well-defined environmentally friendly polymeric materials. Chem Rev 107:2270–2299
119. Wang JS, Matyjaszewski K (1995) Controlled/"living" radical polymerization. atom transfer radical polymerization in the presence of transition-metal complexes. J Am Chem Soc 117:5614–5615
120. di Lena F, Matyjaszewski K (2010) Transition metal catalysts for controlled radical polymerization. Prog Polym Sci 35:959–1021
121. Kamigaito M, Ando T, Sawamoto M (2001) Metal-catalyzed living radical polymerization. Chem Rev 101:3689–3745
122. Davis KA, Matyjaszewski K (2002) Statistical, gradient, block, and graft copolymers by controlled/living radical polymerizations. Adv Polym Sci 159:1–13
123. Matyjaszewski K, Ziegler MJ, Arehart SV, Greszta D, Pakula T (2000) Gradient copolymers by atom transfer radical copolymerization. J Phys Org Chem 13:775–786
124. Hadjichristidis N, Iatrou H, Pitsikalis M, Mays J (2006) Macromolecular architectures by living and controlled/living polymerizations. Prog Polym Sci 31:1068–1132
125. Gao H, Matyjaszewski K (2009) Synthesis of functional polymers with controlled architecture by CRP of monomers in the presence of cross-linkers: from stars to gels. Prog Polym Sci 34:317–350
126. Coessens V, Pintauer T, Matyjaszewski K (2001) Functional polymers by atom transfer radical polymerization. Prog Polym Sci 26:337–377
127. Gauthier MA, Gibson MI, Klok HA (2009) Synthesis of functional polymers by post-polymerization modification. Angew Chem Int Ed Engl 48:48–58
128. Lutz JF, Boerner HG (2008) Modern trends in polymer bioconjugates design. Prog Polym Sci 33:1–39
129. Boyer C, Bulmus V, Davis TP, Ladmiral V, Liu J, Perrier S (2009) Bioapplications of RAFT polymerization. Chem Rev 109:5402–5436
130. Fournier D, Hoogenboom R, Schubert US (2007) Clicking polymers: a straightforward approach to novel macromolecular architectures. Chem Soc Rev 36:1369–1380
131. Huisgen R (1963) 1.3-Dipolare cycloadditionen – Ruckschau und ausblick. Angew Chem Int Ed Engl 75:604–637
132. Rostovtsev VV, Green LG, Fokin VV, Sharpless KB (2002) A stepwise Huisgen cycloaddition process: copper(I)-catalyzed regioselective "ligation" of azides and terminal alkynes. Angew Chem Int Ed Engl 41:2596–2599
133. Golas PL, Tsarevsky NV, Sumerlin BS, Walker LM, Matyjaszewski K (2007) Multisegmented block copolymers by 'click' coupling of polymers prepared by ATRP. Aust J Chem 60: 400–404
134. Golas PL, Matyjaszewski K (2010) Marrying click chemistry with polymerization: expanding the scope of polymeric materials. Chem Soc Rev 39:1338–1354
135. Sumerlin BS, Tsarevsky NV, Louche G, Lee RY, Matyjaszewski K (2005) Highly efficient \"click\" functionalization of Poly(3-azidopropyl methacrylate) prepared by ATRP. Macromolecules 38:7540–7545
136. Tsarevsky NV, Sumerlin BS, Matyjaszewski K (2005) Step-growth "click" coupling of telechelic polymers prepared by atom transfer radical polymerization. Macromolecules 38:3558–3561
137. Yan L, Ding J, Qi R, Yang L, Hu X, Huang Y, Jing X (2010) Versatile synthesis of functional biodegradable polymers by combining ring-opening polymerization and postpolymerization modification via Michael-type addition reaction. Macromolecules 43:201–207
138. Schuewer N, Klok HA (2011) Tuning the pH sensitivity of poly(methacrylic acid) brushes. Langmuir 27:4789–4796

139. Whitesides GM, Mathias JP, Seto CT (1991) Molecular self-assembly and nanochemistry – a chemical strategy for the synthesis of nanostructures. Science 254:1312–1319
140. Rodriguez-Hernandez J, Chécot F, Gnanou Y, Lecommandoux S (2005) Toward 'smart' nano-objects by self-assembly of block copolymers in solution. Prog Polym Sci 30:691–724
141. Lynd NA, Meuler AJ, Hillmyer MA (2008) Polydispersity and block copolymer self-assembly. Prog Polym Sci 33:875–893
142. Listak J, Jakubowski W, Mueller L, Plichta A, Matyjaszewski K, Bockstaller MR (2008) Effect of symmetry of molecular weight distribution in block copolymers on formation of "metastable" morphologies. Macromolecules 41:5919–5927
143. Kim JK, Yang SY, Lee Y, Kim Y (2010) Functional nanomaterials based on block copolymer self-assembly. Prog Polym Sci 35:1325–1349
144. Allen C, Maysinger D, Eisenberg A (1999) Nano-engineering block copolymer aggregates for drug delivery. Colloids Surf B Biointerfaces 16:3–27
145. Forster S, Antonietti M (1998) Amphiphilic block copolymers in structure-controlled nanomaterial hybrids. Adv Mater 10:195–217
146. Israelachvili JN (1992) Intermolecular and surface forces, 2nd edn. Academic, London
147. Tsitsilianis C, Roiter Y, Katsampas I, Minko S (2008) Diversity of nanostructured self-assemblies from a pH-responsive ABC terpolymer in aqueous media. Macromolecules 41:925–934
148. Xiong D, He Z, An Y, Li Z, Wang H, Chen X, Shi L (2008) Temperature-responsive multilayered micelles formed from the complexation of PNIPAM-b-P4VP block-copolymer and PS-b-PAA core-shell micelles. Polymer 49:2548–2552
149. Lee HI, Wu W, Oh JK, Mueller L, Sherwood G, Peteanu L, Kowalewski T, Matyjaszewski K (2007) Light-induced reversible formation of polymeric micelles. Angew Chem Int Ed Engl 46:2453–2457
150. De Clercq B, Laperre J, Ruys L (2005) The controlled radical polymerisation process as an instrument for tailor-made coating applications. Prog Org Coat 53:195–206
151. Zeng J, Shi K, Zhang Y, Sun X, Zhang B (2008) Construction and micellization of a noncovalent double hydrophilic block copolymer. Chem Commun (Camb) 32:3753–3755
152. Shenhar R, Norsten TB, Rotello VM (2005) Polymer-mediated nanoparticle assembly: structural control and applications. Adv Mater 17:657–669
153. Tyrrell ZL, Shen YQ, Radosz M (2010) Fabrication of micellar nanoparticles for drug delivery through the self-assembly of block copolymers. Prog Polym Sci 35:1128–1143
154. Forster S, Zisenis M, Wenz E, Antonietti M (1996) Micellization of strongly segregated block copolymers. J Chem Phys 104:9956–9970
155. Azzam T, Bronstein L, Eisenberg A (2008) Water-soluble surface-anchored gold and palladium nanoparticles stabilized by exchange of low molecular weight ligands with biamphiphilic triblock copolymers. Langmuir 24:6521–6529
156. Yu S, Azzam T, Rouiller I, Eisenberg A (2009) "Breathing" vesicles. J Am Chem Soc 131:10557–10566
157. du Sart GG, Rachmawati R, Voet V, Alberda van Ekenstein G, Polushkin E, ten Brinke G, Loos K (2008) Poly(tert-butyl methacrylate-b-styrene-b-4-vinylpyridine) triblock copolymers: synthesis, interactions, and self-assembly. Macromolecules 41:6393–6399
158. Bivigou-Koumba AM, Kristen J, Laschewsky A, Müller-Buschbaum P, Papadakis CM (2009) Synthesis of symmetrical triblock copolymers of styrene and n-isopropylacrylamide using bifunctional bis(trithiocarbonate)s as RAFT agents. Macromol Chem Phys 210:565–578
159. Nottelet B, Vert M, Coudane J (2008) Novel amphiphilic degradable poly(epsilon-caprolactone)-graft-poly (4-vinyl pyridine), poly (epsilon-caprolactone)-graft-poly (dimethylaminoethyl methacrylate) and water-soluble derivatives. Macromol Rapid Commun 29:743–750
160. You YZ, Hong C, Wang W, Lu W, Pan C (2004) Preparation and characterization of thermally responsive and biodegradable block copolymer comprised of PNIPAAM and PLA by combination of ROP and RAFT methods. Macromolecules 37:9761–9767

161. Bockstaller MR, Mickiewicz RA, Thomas EL (2005) Block copolymer nanocomposites: perspectives for tailored functional materials. Adv Mater 17:1331–1349
162. Fahmi A, Pietsch T, Mendoza C, Cheval N (2009) Functional hybrid materials. Mater Today 12:44–50
163. Oh JK, Park JM (2011) Iron oxide-based superparamagnetic polymeric nanomaterials: design, preparation, and biomedical application. Prog Polym Sci 36:168–189
164. Khandare J, Minko T (2006) Polymer-drug conjugates: progress in polymeric prodrugs. Prog Polym Sci 31:359–397
165. Read ES, Armes SP (2007) Recent advances in shell cross-linked micelles. Chem Commun 29:3021–3035
166. Smith AE, Xu X, McCormick CL (2010) Stimuli-responsive amphiphilic (co)polymers via RAFT polymerization. Prog Polym Sci 35:45–93
167. Siegwart DJ, Oh JK, Matyjaszewski K (2012) ATRP in the design of functional materials for biomedical applications. Prog Polym Sci 37:18–37
168. Oh JK, Bencherif SA, Matyjaszewski K (2009) Atom transfer radical polymerization in inverse miniemulsion: a versatile route toward preparation and functionalization of microgels/nanogels for targeted drug delivery applications. Polymer 50:4407–4423
169. Oh JK, Drumright R, Siegwart DJ, Matyjaszewski K (2008) The development of microgels/nanogels for drug delivery applications. Prog Polym Sci 33:448–477
170. Oh JK, Siegwart DJ, Lee HI, Sherwood G, Peteanu L, Hollinger JO, Kataoka K, Matyjaszewski K (2007) Biodegradable nanogels prepared by atom transfer radical polymerization as potential drug delivery carriers: synthesis, biodegradation, in vitro release, and bioconjugation. J Am Chem Soc 129:5939–5945
171. Roy D, Cambre JN, Sumerlin BS (2010) Future perspectives and recent advances in stimuli-responsive materials. Prog Polym Sci 35:278–301
172. Bajpai AK, Bajpai J, Saini R, Gupta R (2011) Responsive polymers in biology and technology. Polym Rev 51:53–97
173. Pasparakis G, Vamvakaki M (2011) Multiresponsive polymers: nano-sized assemblies, stimuli-sensitive gels and smart surfaces. Polym Chem 2:1234–1248
174. Mano JF (2008) Stimuli-responsive polymeric systems for biomedical applications. Adv Eng Mater 10:515–527
175. Schmaljohann D (2006) Thermo- and pH-responsive polymers in drug delivery. Adv Drug Deliv Rev 58:1655–1670
176. Liechty WB, Kryscio DR, Slaughter BV, Peppas NA (2010) Polymers for drug delivery systems. Ann Rev Chem Biomol Eng 1:149–173
177. Ma Y, Tang Y, Billingham NC, Armes SP, Lewis AL (2003) Synthesis of biocompatible, stimuli-responsive, physical gels based on ABA triblock copolymers. Biomacromolecules 4:864–868
178. Schild HG (1992) Poly(N-isopropylacrylamide): experiment, theory and application. Prog Polym Sci 17:63–249
179. Cho SH, Jhon MS, Yuk SH, Lee HB (1997) Temperature-induced phase transition of poly(N, N-dimethylaminoethyl methacrylate-co-acrylamide). J Polym Sci, Part B: Polym Phys 35:595–598
180. Yamamoto SI, Pietrasik J, Matyjaszewski K (2008) Temperature- and pH-responsive dense copolymer brushes prepared by ATRP. Macromolecules 41:7013–7020
181. Yamamoto SI, Pietrasik J, Matyjaszewski K (2007) ATRP synthesis of thermally responsive molecular brushes from Oligo(ethylene oxide) Methacrylates. Macromolecules 40: 9348–9353
182. Dong H, Matyjaszewski K (2010) Thermally responsive P(M(EO)2MA-co-OEOMA) copolymers via AGET ATRP in miniemulsion. Macromolecules 43:4623–4628
183. Dong H, Mantha V, Matyjaszewski K (2009) Thermally responsive PM(EO)2MA magnetic microgels via activators generated by electron transfer atom transfer radical polymerization in miniemulsion. Chem Mater 21:3965–3972

184. Filipcsei G, Fehér J, Zrínyi M (2000) Electric field sensitive neutral polymer gels. J Mol Struct 554:109–117
185. George PM, LaVan DA, Burdick JA, Chen CY, Liang E, Langer R (2006) Electrically controlled drug delivery from biotin-doped conductive polypyrrole. Adv Mater 18:577–581
186. Jeon G, Yang SY, Byun J, Kim JK (2011) Electrically actuatable smart nanoporous membrane for pulsatile drug release. Nano Lett 11:1284–1288
187. Katz JS, Burdick JA (2010) Light-responsive biomaterials: development and applications. Macromol Biosci 10:339–348
188. Narayana Reddy N, Murali Mohan Y, Varaprasad K, Ravindra S, Joy PA, Mohana Raju K (2011) Magnetic and electric responsive hydrogel–magnetic nanocomposites for drug-delivery application. J App Polym Sci 122:1364–1375
189. Filipcsei G, Csetneki I, Szilagyi A, Zrinyi M (2007) Magnetic field-responsive smart polymer composites. In: Oligomers—polymer composites—molecular imprinting. Springer, Berlin/Heidelberg
190. Kang H, Liu H, Zhang X, Yan J, Zhu Z, Peng L, Yang H, Kim Y, Tan W (2011) Photoresponsive DNA-cross-linked hydrogels for controllable release and cancer therapy. Langmuir 27:399–408
191. Bousguet A, Ibarboure E, Papon E, Labrugère C, Rodríguez-Hernández J (2010) Structured multistimuli-responsive functional polymer surfaces obtained by interfacial diffusion of Amphiphilic block copolymers. J Polym Sci A Polym Chem 48:1952–1961
192. Roy D, Cambre JN, Sumerlin BS (2009) Triply-responsive boronic acid block copolymers: solution self-assembly induced by changes in temperature, pH, or sugar concentration. Chem Commun 16:2106–2108
193. Weiss J, Laschewsky A (2011) Temperature-induced self-assembly of triple-responsive triblock copolymers in aqueous solutions. Langmuir 27:4465–4473
194. Butun V, Billingham NC, Armes SP (1998) Unusual aggregation behavior of a novel tertiary amine methacrylate-based diblock copolymer: formation of micelles and reverse micelles in aqueous solution. J Am Chem Soc 120:11818–11819
195. Butun V, Liu S, Weaver JVM, Bories-Azeau X, Cai Y, Armes SP (2006) A brief review of 'schizophrenic' block copolymers. React Funct Polym 66:157–165
196. Liu SY, Billingham NC, Armes SP (2001) A schizophrenic water-soluble diblock copolymer. Angew Chem Int Ed Engl 40:2328–2331
197. Naik SS, Ray JG, Savin DA (2011) Temperature- and pH-responsive self-assembly of poly(propylene oxide)-b-poly(lysine) block copolymers in aqueous solution. Langmuir 27:7231–7240
198. Jiang X, Zhao B (2008) Tuning micellization and dissociation transitions of thermo- and ph-sensitive poly(ethylene oxide)-b-poly(methoxydi(ethylene glycol) methacrylate-co-methacrylic acid) in aqueous solution by combining temperature and ph triggers. Macromolecules 41:9366–9375
199. Xiong DA, Shi L, Jiang X, An Y, Chen X, Lü J (2007) Composite worm-like aggregates formed from a pair of block-copolymers containing hydrogen-bonding donor and acceptor. Macromol Rapid Commun 28:194–199
200. Liu SY, Armes SP (2002) Polymeric surfactants for the new millennium: a pH-responsive, zwitterionic, schizophrenic diblock copolymer. Angew Chem Int Ed Eng 41:1413–1416
201. Du J, O'Reilly RK (2010) pH-Responsive vesicles from a schizophrenic diblock copolymer. Macromol Chem Phys 211:1530–1537
202. Wang L, Liu M, Gao C, Ma L, Cui D (2010) A pH-, thermo-, and glucose-, triple-responsive hydrogels: synthesis and controlled drug delivery. React Funct Polym 70:159–167
203. Timko BP, Dvir T, Kohane DS (2010) Remotely triggerable drug delivery systems. Adv Mater 22:4925–4943
204. Wiradharma N, Zhang Y, Venkataraman S, Hedrick JL, Yanga YY (2009) Self-assembled polymer nanostructures for delivery of anticancer therapeutics. Nano Today 4:302–317
205. Branco MC, Schneider JP (2009) Self-assembling materials for therapeutic delivery. Acta Biomater 5:817–831

206. Onaca O, Enea R, Hughes DW, Meier W (2009) Stimuli-responsive polymersomes as nanocarriers for drug and gene delivery. Macromol Biosci 9:129–139
207. Alarcon CDH, Pennadam S, Alexander C (2005) Stimuli responsive polymers for biomedical applications. Chem Soc Rev 34:276–285
208. Calderón M, Quadir MA, Strumia M, Haag R (2010) Functional dendritic polymer architectures as stimuli-responsive nanocarriers. Biochimie 92:1242–1251
209. Motornov M, Roiter Y, Tokarev I, Minko S (2010) Stimuli-responsive nanoparticles, nanogels and capsules for integrated multifunctional intelligent systems. Prog Polym Sci 35:174–211
210. York AW, Kirkland SE, McCormick CL (2008) Advances in the synthesis of amphiphilic block copolymers via RAFT polymerization: stimuli-responsive drug and gene delivery. Adv Drug Deliv Rev 60:1018–1036
211. Nishiyama N, Kataoka K (2006) Current state, achievements, and future prospects of polymeric micelles as nanocarriers for drug and gene delivery. Pharmacol Ther 112:630–648
212. Fu R, Fu GD (2011) Polymeric nanomaterials from combined click chemistry and controlled radical polymerization. Polym Chem 2:465–475
213. Kataoka K, Harada A, Nagasaki Y (2001) Block copolymer micelles for drug delivery: design, characterization and biological significance. Adv Drug Deliv Rev 47:113–131
214. Sant V, Smith D, Leroux J (2004) Novel pH-sensitive supramolecular assemblies for oral delivery of poorly water soluble drugs: preparation and characterization. J Control Release 97:301–312
215. Yuan W, Yuan J, Zheng S, Hong X (2007) Synthesis, characterization, and controllable drug release of dendritic star-block copolymer by ring-opening polymerization and atom transfer radical polymerization. Polymer 48:2585–2594
216. Licciardi M, Tang Y, Billingham NC, Armes SP, Lewis AL (2005) Synthesis of novel folic acid-functionalized biocompatible block copolymers by atom transfer radical polymerization for gene delivery and encapsulation of hydrophobic drugs. Biomacromolecules 6:1085–1096
217. Karanikolopoulos N, Pitsikalis M, Hadjichristidis N, Georgikopoulou K, Calogeropoulou T, Dunlap JR (2007) pH-responsive aggregates from double hydrophilic block copolymers carrying zwitterionic groups. Encapsulation of antiparasitic compounds for the treatment of Leishmaniasis. Langmuir 23:4214–4224
218. Tang Y, Liu SY, Armes SP, Billingham NC (2003) Solubilization and controlled release of a hydrophobic drug using novel micelle-forming ABC triblock copolymers. Biomacromolecules 4:1636–1645
219. Ahmed F, Pakunlu RI, Brannan A, Bates F, Minko T, Discher DE (2006) Biodegradable polymersomes loaded with both paclitaxel and doxorubicin permeate and shrink tumors, inducing apoptosis in proportion to accumulated drug. J Control Release 116:150–158
220. Du J, O'Reilly RK (2009) Advances and challenges in smart and functional polymer vesicles. Soft Matter 5:3544–3561
221. Qin S, Geng Y, Discher DE, Yang S (2006) Temperature-controlled assembly and release from polymer vesicles of poly(ethylene oxide)-block-poly(N-isopropylacrylamide). Adv Mater 18:2905–2909
222. Tian L, Hammond PT (2006) Comb-dendritic block copolymers as tree-shaped macromolecular amphiphiles for nanoparticle self-assembly. Chem Mater 18:3976–3984
223. Tian L, Nguyen P, Hammond PT (2006) Vesicular self-assembly of comb-dendritic block copolymers. Chem Commun 33:3489–3491
224. Yu WY, Zhang N (2009) Surface modification of nanocarriers for cancer therapy. Curr Nanosci 5:123–134
225. Veronese FM, Mero A (2008) The impact of PEGylation on biological therapies. BioDrugs 22:315–329
226. Reichelt S, Elsner C, Pender A, Buchmeiser MR (2011) Tailoring the surface of magnetic microparticles for protein immobilization. J Appl Polym Sci 121:3628–3634
227. Lu AH, Salabas EL, Schuth F (2007) Magnetic nanoparticles: synthesis, protection, functionalization, and application. Angew Chem Int Ed Engl 46:1222–1244

228. Lin PC, Ueng SH, Yu SC, Jan MD, Adak AK, Yu CC, Lin CC (2007) Surface modification of magnetic nanoparticle via Cu(I)-Catalyzed alkyne-azide 2+3 cycloaddition. Org Lett 9:2131–2134
229. Shim M, Wong Shi Kam N, Chen RJ, Li Y, Dai H (2002) Functionalization of carbon nanotubes for biocompatibility and biomolecular recognition. Nano Lett 2:285–288
230. Liu Z, Sun X, Nakayama-Ratchford N, Dai H (2007) Supramolecular chemistry on water-soluble carbon nanotubes for drug loading and delivery. ACS Nano 1:50–56
231. Prencipe G, Tabakman SM, Welsher K, Liu Z, Goodwin AP, Zhang L, Henry J, Dai H (2009) PEG branched polymer for functionalization of nanomaterials with ultralong blood circulation. J Am Chem Soc 131:4783–4787
232. Jiang X, Lok MC, Hennink WE (2007) Degradable-brushed pHEMA–pDMAEMA synthesized via ATRP and click chemistry for gene delivery. Bioconjug Chem 18:2077–2084
233. Du JZ, Tang LY, Song WJ, Shi Y, Wang J (2009) Evaluation of polymeric micelles from brush polymer with poly(epsilon-caprolactone)-b-poly(ethylene glycol) side chains as drug carrier. Biomacromolecules 10:2169–2174
234. Ma YH, Tang Y, Billingham NC, Armes SP, Lewis AL, Lloyd AW, Salvage JP (2003) Well-defined biocompatible block copolymers via atom transfer radical polymerization of 2-methacryloyloxyethyl phosphorylcholine in protic media. Macromolecules 36:3475–3484
235. Hu YQ, Kim MS, Kim BS, Lee DS (2007) Synthesis and pH-dependent micellization of 2-(diisopropylamino)ethyl methacrylate based amphiphilic diblock copolymers via RAFT polymerization. Polymer 48:3437–3443
236. Licciardi M, Giammona G, Du J, Armes SP, Tang Y, Lewis AL (2006) New folate-functionalized biocompatible block copolymer micelles as potential anti-cancer drug delivery systems. Polymer 47:2946–2955
237. Lee SM, Chen H, Dettmer CM, O'Halloran TV, Nguyen ST (2007) Polymer-caged lipsomes: a pH-Responsive delivery system with high stability. J Am Chem Soc 129:15096–15097
238. Haag R, Kratz F (2006) Polymer therapeutics: concepts and applications. Angew Chem Int Ed Engl 45:1198–1215
239. Tong R, Cheng J (2008) Paclitaxel-initiated, controlled polymerization of lactide for the formulation of polymeric nanoparticulate delivery vehicles. Angew Chem Int Ed Engl 47:4830–4834
240. Rettig H, Krause E, Borner HG (2004) Atom transfer radical polymerization with polypeptide initiators: a general approach to block copolymers of sequence-defined polypeptides and synthetic polymers. Macromol Rapid Commun 25:1251–1256
241. Nicolas J, Mantovani G, Haddleton DM (2007) Living radical polymerization as a tool for the synthesis of polymer-protein/peptide bioconjugates. Macromol Rapid Commun 28: 1083–1111
242. Lele BS, Murata H, Matyjaszewski K, Russell AJ (2005) Synthesis of uniform protein-polymer conjugates. Biomacromolecules 6:3380–3387
243. Averick S, Simakova A, Park S, Konkolewicz D, Magenau AJD, Mehl RA, Matyjaszewski K (2012) ATRP under biologically relevant conditions: grafting from a protein. ACS Macro Lett 1:6–10
244. Peeler JC, Woodman BF, Averick S, Miyake-Stoner SJ, Stokes AL, Hess KR, Matyjaszewski K, Mehl RA (2010) Genetically encoded initiator for polymer growth from proteins. J Am Chem Soc 132:13575–13577
245. Averick SE, Magenau AJD, Simakova A, Woodman BF, Seong A, Mehl RA, Matyjaszewski K (2011) Covalently incorporated protein-nanogels using AGET ATRP in an inverse miniemulsion. Polym Chem 2:1476–1478
246. Averick S, Paredes E, Li W, Matyjaszewski K, Das SR (2011) Direct DNA conjugation to star polymers for controlled reversible assemblies. Bioconjug Chem 22:2030–2037
247. Shakya AK, Sami H, Srivastava A, Kumar A (2010) Stability of responsive polymer-protein bioconjugates. Prog Polym Sci 35:459–486

Author Index

J. Coelho (ed.), *Drug Delivery Systems: Advanced Technologies Potentially Applicable in Personalised Treatment*, Advances in Predictive, Preventive and Personalised Medicine 4, DOI 10.1007/978-94-007-6010-3,

D

G

I

Q

T

U

Z

Subject Index

J. Coelho (ed.), *Drug Delivery Systems: Advanced Technologies Potentially Applicable in Personalised Treatment*, Advances in Predictive, Preventive and Personalised Medicine 4, DOI 10.1007/978-94-007-6010-3,

O

P

MIX
Papier aus verantwortungsvollen Quellen
Paper from responsible sources
FSC® C105338

If you have any concerns about our products,
you can contact us on
ProductSafety@springernature.com

In case Publisher is established outside the EU,
the EU authorized representative is:
Springer Nature Customer Service Center GmbH
Europaplatz 3, 69115 Heidelberg, Germany

Printed by Libri Plureos GmbH
in Hamburg, Germany